REVIEWS IN ECONOMIC GEOLOGY
Volume 2

GEOLOGY AND GEOCHEMISTRY OF EPITHERMAL SYSTEMS

B. R. Berger & P. M. Bethke, Editors

CONTENTS

THE GEOTHERMAL FRAMEWORK FOR EPITHERMAL DEPOSITS	R. W. Henley
A PRACTICAL GUIDE TO THE THERMODYNAMICS OF GEOTHERMAL FLUIDS AND HYDROTHERMAL ORE DEPOSITS	R. W. Henley & K. L. Brown
THE BEHAVIOR OF SILICA IN HYDROTHERMAL SOLUTIONS	R. O. Fournier
CARBONATE TRANSPORT AND DEPOSITION IN THE EPITHERMAL ENVIRONMENT	R. O. Fournier
FLUID INCLUSION SYSTEMATICS IN EPITHERMAL SYSTEMS	R. J. Bodnar, T. J. Reynolds, & C. A. Kuehn
LIGHT STABLE-ISOTOPE SYSTEMATICS IN THE EPITHERMAL ENVIRONMENT	C. W. Field & R. H. Fifarek
GEOLOGIC, MINERALOGIC, AND GEOCHEMICAL CHARACTERISTICS OF VOLCANIC-HOSTED EPITHERMAL PRECIOUS-METAL DEPOSITS	D. O. Hayba, P. M. Bethke, P. Heald, & N. K. Foley
GEOLOGIC CHARACTERISTICS OF SEDIMENT-HOSTED, DISSEMINATED PRECIOUS-METAL DEPOSITS IN THE WESTERN UNITED STATES	W. C. Bagby & B. R. Berger
RELATIONSHIP OF TRACE-ELEMENT PATTERNS TO ALTERATION AND MORPHOLOGY IN EPITHERMAL PRECIOUS-METAL DEPOSITS	M. L. Silberman & B. R. Berger
RELATIONSHIPS OF TRACE-ELEMENT PATTERNS TO GEOLOGY IN HOT-SPRING TYPE PRECIOUS-METAL DEPOSITS	B. R. Berger & M. L. Silberman
BOILING, COOLING, AND OXIDATION IN EPITHERMAL SYSTEMS: A NUMERICAL MODELING APPROACH	M. H. Reed & N. Spycher
USING GEOLOGICAL INFORMATION TO DEVELOP EXPLORATION STRATEGIES FOR EPITHERMAL DEPOSITS	S. S. Adams

Series Editor: James M. Robertson

SOCIETY OF ECONOMIC GEOLOGISTS

REVIEWS IN ECONOMIC GEOLOGY
(ISSN 0741-0123)

Published Annually by the
SOCIETY OF ECONOMIC GEOLOGISTS
Printed by BookCrafters, Inc., 140 Buchanan Street, Chelsea, MI 48118
Series Editor: James M. Robertson

Additional copies of this volume may be obtained from:
The Economic Geology Publishing Company
P.O. Box 637
University of Texas at El Paso
El Paso, TX 79968-0637 USA
(915) 533-1965

Vol. 1: FLUID-MINERAL EQUILIBRIA IN HYDROTHERMAL SYSTEMS (1984) ISBN 0-9613074-0-4
Vol. 2: GEOLOGY AND GEOCHEMISTRY OF EPITHERMAL SYSTEMS (1985) ISBN 0-9613074-1-2

Reviews in Economic Geology is a publication of the Society of Economic Geologists designed to accompany the Society's Short Course series. Like the Short Courses, each volume provides intensive updates on various applied and academic topics for practicing economic geologists and geochemists in exploration, development, research, and teaching. Volumes are produced annually in conjunction with each new Short Course, first serving as a textbook for that course, and subsequently made available to S.E.G. members and others at modest cost.

© Copyright 1985, Society of Economic Geologists

Permission is granted to individuals to make single copies of chapters for personal use in research, study, and teaching, and to use short quotations, illustrations, and tables from *Reviews in Economic Geology* for publication in scientific works. Such uses must be appropriately credited. Copying for general distribution, for promotion and advertising, for creating new collective works, or for other commercial purposes is not permitted without the specific written permission of the Series Editor.

Standing orders are accepted from libraries, institutions, and corporations who wish to automatically receive each new volume of *Reviews in Economic Geology* after it is published. An invoice is mailed with each volume. To place a standing order, notify the Economic Geology Publishing Company (PUBCO) business office at the address given above.

Address Change. Standing-order holders please note that the PUBCO business office must be notified of a change of address at least four weeks prior to mailing out a volume. It is essential to submit a copy of your mailing label for reference.

Replacement Policy. Missing volumes will be replaced without charge to standing-order holders who notify the PUBCO business office within six weeks (six months for India and Australia) of the date a new Short Course is given and new volume produced.

Remittances should be made payable to PUBCO, Reviews in Economic Geology, and should be mailed to the PUBCO business office at the address given above. Also all other business communications should be addressed to that office.

REVIEWS IN ECONOMIC GEOLOGY

(ISSN 0741–0123) Volume 2

GEOLOGY AND GEOCHEMISTRY OF EPITHERMAL SYSTEMS

ISBN 0-9613074-1-2

Volume Editors:

B. R. BERGER
Branch of Exploraton Geochemistry
U.S. Geological Survey
MS 973
Box 25046, Federal Center
Denver, CO 80225-0046

P. M. BETHKE
Branch of Resource Analysis
U.S. Geological Survey
MS 959, National Center
Reston, VA 22092

Series Editor: JAMES M. ROBERTSON
New Mexico Bureau of Mines & Mineral Resources
Campus Station
Socorro, NM 87801

SOCIETY OF ECONOMIC GEOLOGISTS

The Authors:

Samuel S. Adams
3030 Third Street
Boulder, CO 80302

William C. Bagby
Branch of Western Mineral Resources
U.S. Geological Survey
MS 901
345 Middlefield Road
Menlo Park, CA 94025

B. R. Berger
Branch of Exploration Geochemistry
U.S. Geological Survey
MS 973
Box 25046, Federal Center
Denver, CO 80225-0046

Philip M. Bethke
Branch of Resource Analysis
U.S. Geological Survey
MS 959, National Center
Reston, VA 22092

R. J. Bodnar
Department of Geological Sciences
Virginia Polytechnic Institute and State University
Blacksburg, VA 20461

K. L. Brown
Chemistry Division
D.S.I.R., Private Bag
Taupo
New Zealand

Cyrus W. Field
Department of Geology
Oregon State University
Corvallis, OR 97331-5506

Richard H. Fifarek
Department of Geology
Southern Illinois University
Carbondale, IL 62901

N. K. Foley
Branch of Resource Analysis
U.S. Geological Survey
MS 959, National Center
Reston, VA 22092

Robert O. Fournier
Branch of Igneous and Geothermal Processes
U.S Geological Survey
MS 910
345 Middlefield Road
Menlo Park, CA 94025

Daniel O. Hayba
Branch of Resource Analysis
U.S. Geological Survey
MS 959, National Center
Reston, VA 22092

Pamela Heald
Branch of Resource Analysis
U.S. Geological Survey
MS 959, National Center
Reston, VA 22092

R. W. Henley
Chemistry Divsion
D.S.I.R., Private Bag
Taupo
New Zealand

C. A. Kuehn
Department of Geosciences
The Pennsylvania State University
University Park, PA 16802

Mark H. Reed
Department of Geology
University of Oregon
Eugene, OR 97403

T. J. Reynolds
FLUID, Inc.
P.O. Box 6873
Denver, CO 80206

M. L. Silberman
Branch of Exploration Geochemistry
U.S. Geological Survey
MS 912
Box 25046, Federal Center
Denver, CO 80225-0046

N. Spycher
Department of Geology
University of Oregon
Eugene, OR 97403

GEOLOGY & GEOCHEMISTRY OF EPITHERMAL SYSTEMS

CONTENTS

FOREWORD x

PREFACE xi

BIOGRAPHIES xvi

CHAPTER 1

THE GEOTHERMAL FRAMEWORK OF EPITHERMAL DEPOSITS
R. W. Henley

INTRODUCTION 1

HYDROTHERMAL SYSTEMS IN GENERAL 1

    Collision-Related Amagmatic Hydrothermal Systems
    Terrestrial Magma-Related Hydrothermal Systems

TERRESTRIAL MAGMATIC-HYDROTHERMAL SYSTEMS 4

    Large Scale Structure
    Natural Discharges
    Hydrothermal Eruption Vents
    Heat and Mass Flow in Geothermal Systems

CHEMISTRY OF GEOTHERMAL DISCHARGES 11

EPITHERMAL ORE-FORMING SYSTEMS 12

    Requirememts for Ore Deposition
    Chemistry of Systems Responsible for Ore Formation
    Chemical and Physical Processes in Ore Formation
    Host-Rock Relations

SUMMARY 19

EPILOGUE 21

ACKNOWLEDGMENTS 21

REFERENCES 21

CHAPTER 2

A PRACTICAL GUIDE TO THE THERMODYNAMICS OF GEOTHERMAL FLUIDS AND HYDROTHERMAL ORE DEPOSITS
R. W. Henley and K. L. Brown

INTRODUCTION 25

GEOLOGICAL CHARACTERISTICS OF THE BROADLANDS GEOTHERMAL SYSTEM 25

FLUID CHEMISTRY 26

FLUID-MINERAL EQUILIBRIA: ALTERATION MINERALOGY 28

FLUID-MINERAL EQUILIBRIA: TRACE-METAL CONTENTS 32

    <u>Lead</u>
    <u>Gold</u>
    <u>Other Metals: Copper, Silver, and Arsenic</u>

MINERAL DEPOSITION 36

    <u>Silica</u>
    <u>Calcite</u>
    <u>Metal Sulfides and Gold</u>

ACKNOWLEDGMENTS 41

REVIEW QUESTIONS 41

REFERENCES 41

APPENDIX 43

CHAPTER 3

THE BEHAVIOR OF SILICA IN HYDROTHERMAL SOLUTIONS
R. O. Fournier

INTRODUCTION 45

SOLUBILITIES OF SILICA MINERALS 45

THE BEHAVIOR OF DISSOLVED SILICA IN HOT-SPRING SYSTEMS 46

ALKALINE WATERS 48

ACID WATERS 50

REACTION WITH GLASS 51

AMORPHOUS SILICA-CHALCEDONY RELATIONS 51

SPECULATIONS REGARDING SOME TEXTURES OF QUARTZ 51

    <u>Jasperoid and Massive Replacement of Limestone by Silica</u>
    <u>Quartz Solubility at High Temperatures</u>

CONCLUSIONS 55

ACKNOWLEDGMENTS 56

REFERENCES 56

APPENDIX 60

CHAPTER 4

CARBONATE TRANSPORT AND DEPOSITION IN THE EPITHERMAL ENVIRONMENT
R. O. Fournier

INTRODUCTION	63
CO_2 DISSOLVED IN AQUEOUS SOLUTIONS	63
THE SOLUBILITY OF CALCITE IN AQUEOUS SOLUTIONS	67
SUMMARY	71
REFERENCES	71

CHAPTER 5

FLUID-INCLUSION SYSTEMATICS IN EPITHERMAL SYSTEMS
R. J. Bodnar, T. J. Reynolds, and C. A. Kuehn

INTRODUCTION	73
INFORMATION AVAILABLE FROM FLUID-INCLUSION PETROGRAPHY	73
IDENTIFICATION OF FLUID INCLUSIONS TRAPPED FROM BOILING SOLUTIONS	79
IDENTIFICATION OF GASES IN FLUID INCLUSIONS FROM THE EPITHERMAL ENVIRONMENT	83
INTERPRETATION OF FLUID INCLUSIONS FROM THE EPITHERMAL ENVIRONMENT	93
APPLICATION OF FLUID INCLUSIONS IN EXPLORATION FOR EPITHERMAL PRECIOUS-METAL DEPOSITS	94
SUGGESTIONS FOR FUTURE FLUID-INCLUSION RESEARCH	95
REFERENCES	96

CHAPTER 6

LIGHT STABLE-ISOTOPE SYSTEMATICS IN THE EPITHERMAL ENVIRONMENT
C. W. Field and R. H. Fifarek

INTRODUCTION	99
CONVENTIONS, SYSTEMATICS, AND RATIONALE	99

    Fractionation
    Equilibrium Reaction
    Applications

GEOLOGIC DISTRIBUTIONS	110

    Hydrogen and Oxygen
    Carbon
    Sulfur

EPITHERMAL DEPOSITS	113

    Carbon
    Sulfur

    Hydrogen and Oxygen

SUMMARY 124

REFERENCES 125

CHAPTER 7

GEOLOGIC, MINERALOGIC, AND GEOCHEMICAL CHARACTERISTICS OF VOLCANIC-HOSTED EPITHERMAL PRECIOUS-METAL DEPOSITS
D. O. Hayba, P. M. Bethke, P. Heald, and N. K. Foley

INTRODUCTION 129

SUMMARY OF THE CHARACTERISTICS OF VOLCANIC-HOSTED EPITHERMAL ORE DEPOSITS 129

    Characteristics of Adularia-Sericite-Type Deposits
    Characteristics of Acid-Sulfate-Type Deposits
    Summary of Characteristics

THE ADULARIA-SERICITE ENVIRONMENT: CREEDE AS AN EXAMPLE 136

    Creede as an Exemplar
    Summary of Important Studies
    Geologic and Mineralogic Characteristics
    Geochemical Environment
    Hydrologic Environment
    Boiling and Mixing in the Ore Zone
    Summary of Creede Mineralization

THE ACID-SULFATE ENVIRONMENT: SUMMITVILLE AS AN EXAMPLE 151

    Geologic and Mineralogic Characteristics
    Geochemical Environment
    Summary of Summitville Mineralization

GEOTHERMAL INTERPRETATION OF VOLCANIC-HOSTED EPITHERMAL DEPOSITS 158

    Adularia-Sericite Deposits
    Acid-Sulfate Deposits

MECHANISMS OF ACID-SULFATE ALTERATION 159

ACKNOWLEDGMENTS 162

REFERENCES 162

CHAPTER 8

GEOLOGIC CHARACTERISTICS OF SEDIMENT-HOSTED, DISSEMINATED PRECIOUS-METAL DEPOSITS IN THE WESTERN UNITED STATES
W. C. Bagby and B. R. Berger

INTRODUCTION 169

CLASSIFICATION 169

REGIONAL GEOLOGIC CHARACTERISTICS OF DEPOSITS IN MINERAL TRENDS
AND ISOLATED DEPOSITS 172

    The Getchell Trend
    The Carlin Trend

    <u>The Cortez Trend</u>
    <u>Isolated Deposits</u>

GEOLOGIC CHARACTERISTICS OF THREE END-MEMBER, SEDIMENT-HOSTED,
DISSEMINATED PRECIOUS-METAL DEPOSITS 183

    <u>Carlin</u>
    <u>Taylor</u>
    <u>Preble</u>

GENERAL ASPECTS OF TRACE ELEMENT AND STABLE-ISOTOPE GEOCHEMISTRY 189

SUMMARY OF GEOLOGIC CHARACTERISTICS 192

    <u>Regional and District Scale</u>
    <u>Deposit Scale</u>

ENVIRONMENT OF FORMATION 195

EXPLORATION APPLICATION 195

INFLUENCE OF GEOLOGIC CHARACTERISTICS ON MINING 196

    <u>Grade and Tonnage</u>
    <u>Mineability</u>

REFERENCES 199

CHAPTER 9

RELATIONSHIP OF TRACE-ELEMENT PATTERNS TO ALTERATION AND MORPHOLOGY IN EPITHERMAL PRECIOUS-METAL DEPOSITS
M. L. Silberman and B. R. Berger

INTRODUCTION 203

GEOTHERMAL SYSTEMS 204

    <u>Morphology and Characteristics</u>
    <u>Alteration Patterns</u>
    <u>Geochemical Zones</u>

EPITHERMAL ORE DEPOSITS 208

    <u>Morphology and Characteristics</u>
    <u>Alteration Patterns</u>

NATURE OF FLUIDS INVOLVED IN GEOTHERMAL SYSTEMS AND EPITHERMAL
ORE DEPOSITS 213

TIMING 214

GEOCHEMICAL ZONING IN EPITHERMAL DEPOSITS 214

BODIE MINING DISTRICT 215

    <u>Large-scale Vertical Zoning at Bodie Bluff--The Big Picture</u>
    <u>Detailed Lateral Zoning</u>

PARAMOUNT MINING DISTRICT--VERTICAL ZONING 224

SUMMARY 227

ACKNOWLEDGMENTS 228

REFERENCES 230

CHAPTER 10

RELATIONSHIPS OF TRACE-ELEMENT PATTERNS TO GEOLOGY IN HOT-SPRING-TYPE PRECIOUS-METAL DEPOSITS
B. R. Berger and M. L. Silberman

INTRODUCTION 233

CONTROLS ON TRACE-ELEMENT PATTERNS 233

TRACE-ELEMENT PATTERNS IN STUDIED DEPOSITS 235

<u>Hasbrouck Mountain, Nevada</u>
<u>Round Mountain, Nevada</u>

DISCUSSION 245

REFERENCES 246

CHAPTER 11

BOILING, COOLING, AND OXIDATION IN EPITHERMAL SYSTEMS: A NUMERICAL MODELING APPROACH
M. H. Reed and N. F. Spycher

INTRODUCTION 249

BOILING 249

BOILING RESULTS 252

DISCUSSION OF BOILING AND COOLING 252

<u>Sulfide and Carbonate Mineral Precipitation</u>
<u>Precipitation of Silicates</u>
<u>Boiling Without Fractionation and Cooling Only</u>

SUPER- AND SUB-ISOENTHALPIC BOILING 258

BOILING AND GOLD PRECIPITATION 261

THE HOT-SPRING ENVIRONMENT 262

<u>Condensation of the Boiled Gas</u>
<u>Oxidation of Gases to Produce Acid-Sulfate Waters</u>
<u>Reaction of Gases with Meteoric Ground Water</u>
<u>Gold Precipitation from Mixing of Acid-Sulfate Water with Boiled Aqueous Phase</u>
<u>Gold Precipitation from Mixing of Oxygenated Ground Water with Boiled Aqueous Phase</u>

SUMMARY 269

ACKNOWLEDGMENTS 270

REFERENCES 270

CHAPTER 12

USING GEOLOGICAL INFORMATION TO DEVELOP EXPLORATION STRATEGIES FOR EPITHERMAL DEPOSITS
S. S. Adams

INTRODUCTION 273

SOME CONSIDERATIONS IN THE USE OF GEOLOGICAL INFORMATION
IN EXPLORATION 273

STRATEGIC FACTORS 274

    Organizational Objectives
    Commodity Prices
    Financial Resources
    Exploration Organization
    Regulations and Land Availability
    Competitor Activity
    Previous Exploration
    Geologic Information
    Exploration Methods
    Opportunities
    Risk

HUMAN FACTORS 279

    Personal Objectives
    Education and Training
    Problem Solving
    Intuition and Creativity
    Uncertainty
    Aversion to Loss

DEVELOPMENT OF MINERAL-DEPOSIT MODELS 282

    Organization of Geologic Information
    Model Terminology
    Level of Model Development

DATA-PROCESS-CRITERIA MODEL 286

    Definition of a Mineral-Deposit Type
    Compilation of Analog Deposits
    Selection of Geologic Data
    Data-Process Linking
    Identification of Formation Processes
    Evaluation of Data-Process Links
    Selection of Diagnostic Criteria
    Evaluation of Data-Process-Criteria Model
    Application of Data-Process-Criteria Model to Exploration
    Summary of Data-Process-Criteria Model

CONCLUSIONS 296

REFERENCES 297

TABLE OF CONVERSION FACTORS . . . . Inside Back Cover

FOREWORD

Geology and Geochemistry of Epithermal Systems--Volume 2 of <u>Reviews in Economic Geology</u>--was created to accompany a Society of Economic Geologists (SEG) short course of the same name that was given in October, 1985, prior to the annual meetings of the Geological Society of America and Associated Societies in Orlando, Florida. As was the case with Volume 1, the final published version of Volume 2 unfortunately postdates the short course by some months.

Geology and Geochemistry of Epithermal Systems presents a synthesis of the current understanding of the processes responsible for the concentration of metals (especially gold and silver) in near-surface environments, provides an overview of the systematics of the most important approaches to the study of epithermal ores and processes, and summarizes the geology of both sediment-hosted and volcanic-hosted epithermal precious-metal deposits.

After the volume editors, the most significant contributors to the production of this volume were the members of the Editorial Support Group, Branch of Exploration Geochemistry, U.S. Geological Survey, Denver, Colorado. These ladies, Marilyn A. Billone, Candace A. Vassalluzzo, and especially Pamela S. Detra and Dorothy B. Wesson, accomplished the long, arduous, and often frustrating job of assembling, editing, and formatting the book with a uniformly high level of professionalism and good cheer. Their efforts are gratefully acknowledged. Carol Hjellming of the New Mexico Bureau of Mines and Mineral Resources (NMBMMR) editing staff checked, balanced, and helped interpret the chemical equations; Lynne McNeil (NMBMMR) formatted the cutlines. Lastly, I wish to express my continuing appreciation to the New Mexico Bureau of Mines and Mineral Resources and its Director, Frank Kottlowski, who provide the Series Editor with time, space, and encouragement.

James M. Robertson
Series Editor
Socorro, NM
March, 1986

PREFACE

In a speech on May 10, 1911, before the Geological Society of Washington, Waldemar Lindgren described his systematic classification of all types of mineral deposits. One of his categories included deposits related to intrusive and eruptive igneous rocks that form veins at shallow depths that contain open-cavity filling textures and that have been a primary source of "bonanza" grades of gold and silver--the epithermal deposits. Historically, most of the ores in epithermal systems have been mined from quartz veins, breccias, or disseminations that are associated with non-marine volcanic rocks. Open-space filling textures and structures are common--comb structure, crustification, symmetrical banding, and crystal-lined vugs. Ore minerals include native gold, native silver, electrum, argentite, sulfosalts, tellurides, and selenides and often the common sulfides sphalerite, galena, and chalcopyrite. Common gangue minerals are quartz, adularia, calcite, barite, rhodochrosite, and fluorite. Alteration is commonly widespread in epithermal systems, particularly in the upper portions of the vein systems; among the alteration phases are quartz, adularia, illite, chlorite, alunite, and kaolinite.

Lindgren (1928) recognized the difficulty of developing a rigid subsidiary classification scheme for epithermal deposits; he separated them into six categories:

1. Gold deposits
2. Argentite-gold deposits
3. Argentite deposits
4. Gold selenide deposits
5. Gold telluride deposits
6. Gold telluride deposits with alunite

Nolan (1933) and Ferguson (1929) felt that few of these six characteristics were restricted enough to be diagnostic and proposed only two classes of epithermal systems based on the weight ratio of gold to silver, silver-gold, and gold-silver. Based on his experience with deposits in Nevada, Ferguson (1929) found that there is a bimodal distribution of gold-silver ratios, and Nolan (1933) felt that the bimodality was due to genetic processes.

For the silver-gold deposits, Nolan (1933) noticed that through-going fault fissures control the ore and felt that this implies a deep origin for the source of the metals. Nolan (1933) also noted that the precious-metal ores are very commonly sharply limited above and below by approximately parallel surfaces referred to as the ore "horizon." He suggested that these limits are related to temperature. Base metals tend to increase at and below the base of the lower surface of the precious-metal ore. Figure 1 is a longitudinal, vertical projection of the Last Chance-Confidence silver-gold vein in the Mogollon mining district, New

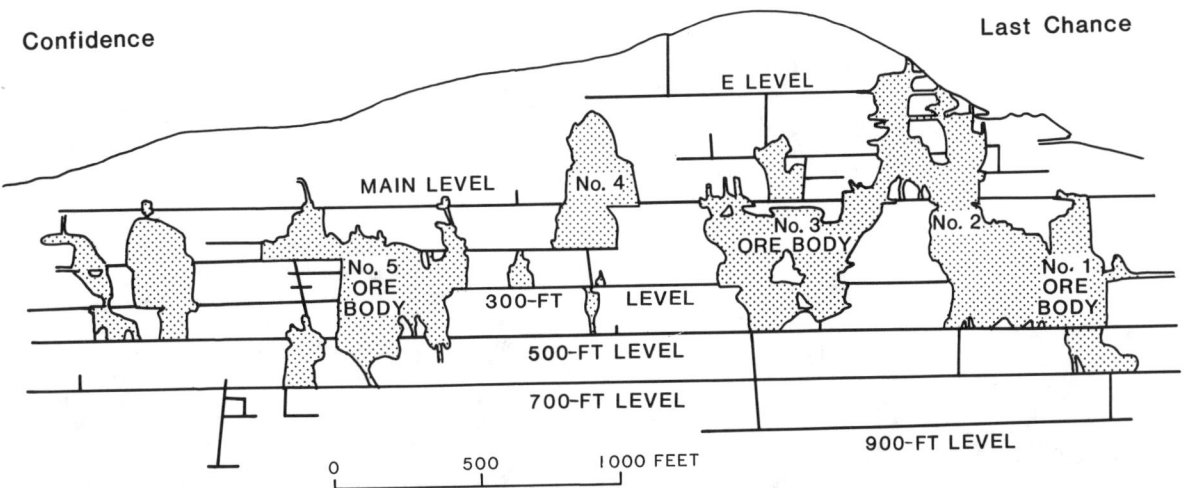

Figure P.1. Vertical, longitudinal projection of the Confidence-Last Chance vein in the Mogollon mining district, New Mexico (Ferguson, 1927). Banded quartz vein is continuous along strike with ore grade material occurring in specific masses (stippled areas) in the vein. The tops and bottoms of the silver-rich ore bodies describe near parallel surfaces referred to as the "ore horizon."

Mexico (Ferguson, 1927) illustrating the ore horizons, the shape of ore bodies, and the typical distribution of ore grades within a continuous banded quartz-adularia-sericite vein. Burbank (1933) reported that base metals appear to be more abundant in silver-gold deposits in regions of sedimentary rocks with overlying volcanic rocks and in thick, volcanic sequences with a long history of volcanic activity. In contrast to the silver-gold deposits, Nolan (1933) noted that gold-silver deposits are commonly within or close to small, shallow intrusive bodies and that the ore-controlling fracture systems are frequently more discontinuous than those associated with silver-gold deposits. The gold-silver ores are also more irregular in distribution than the silver-gold ores. Nolan felt that this irregularity may be related to the complex thermal regimes in these types of systems due to the shallow intrusive activity. Figure 2a shows a series of plan views of the January mine, Goldfield mining distrct, Nevada and a cross section through the January shaft (Ransome, 1909) showing the relationships of ore to quartz-alunite-kaolinite replaced wallrock ("ledge matter") and the host rocks. Figure 2b shows two cross sections from Ransome (1909, p. 154) of the Combination mine in Goldfield illustrating the irregular vertical distribution of bonanza-grade ore masses within the "ledge matter." Also, the ore bodies were not persistent along strike.

Although Waldemar Lindgren (1928) recognized the correlation between epithermal systems and active geothermal systems, it was Donald E. White (1955, 1981) who championed the detailed study of active systems and the application of the results and concepts derived from these studies to epithermal ore deposits. The impact of White's leadership in the study of hydrothermal systems, in general, and epithermal systems, in particular, was recognized by the Society of Economic Geologists when it held a symposium in

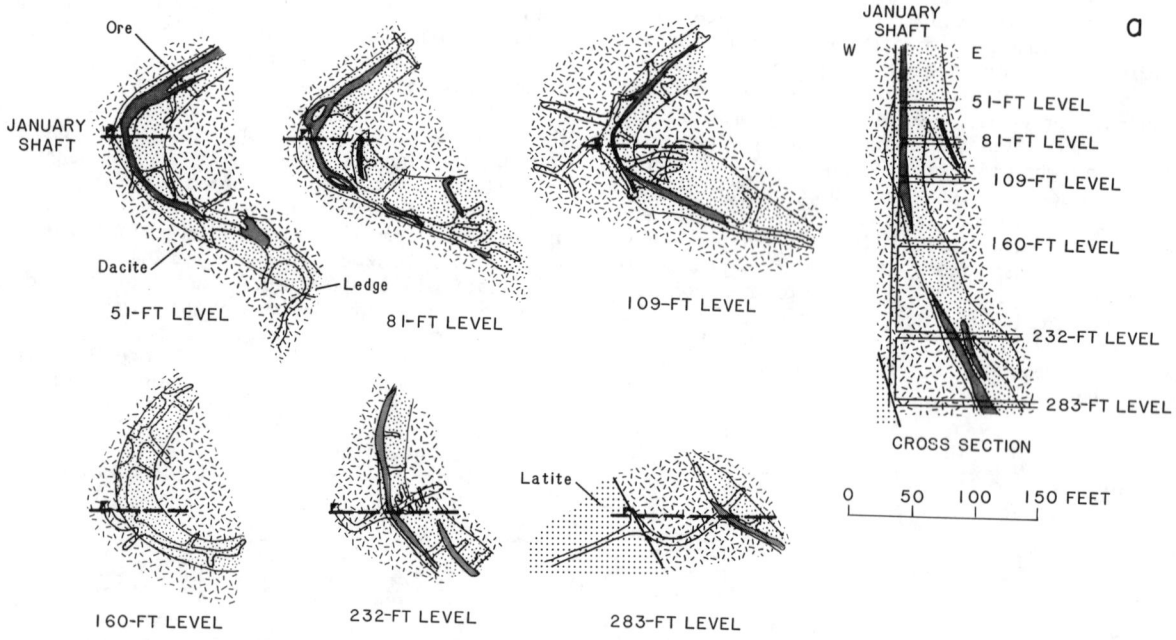

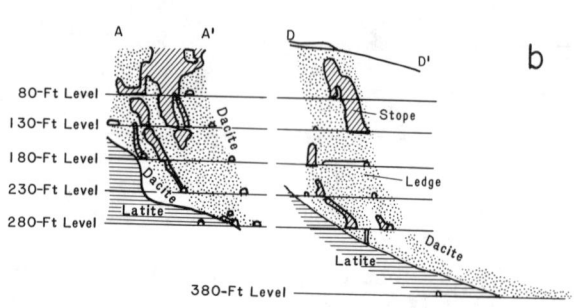

Figure P.2. a). Plan views of the January mine at selected mining levels and a cross section through the January shaft Goldfield mining district, Nevada (Ransome, 1909). Bonanza-grade gold ores occur in replaced dacite referred to as "ledge matter." The ore bodies are not persistent either down-dip or along strike, and occur both on the hanging wall side of the ledge and on the foot wall side. b). Two cross sections from Ransome (1909, p. 154) of the Combination mine in the Goldfield district. Bonanza-grade gold ores occur in irregular, discontinuous masses within the ledge. The ledge follows a lithologic contact and flattens with depth.

his honor in February, 1984 entitled: Geothermal Systems and Ore Deposits. It clearly emphasized the value of using active geothermal areas as models of fossil, ore-forming hydrothermal systems.

Thus, the evolution of understanding of the geology and genesis of epithermal precious-metal deposits has followed a pathway from the early, vividly descriptive studies of mining districts such as the Comstock Lode, Nevada (Becker, 1882), Cripple Creek, Colorado (Lindgren and Ransome, 1906), and Waihi, New Zealand (Bell and Fraser, 1912) to the later, topical studies on structure (Wisser, 1960), alteration (Hemley and Jones, 1964), stable isotopes (Taylor, 1973), and fluid chemistry (Barton et al., 1977). The most recent research on epithermal deposits has built on these past studies and has emphasized the thermal and compositional roles of volcanic rock terranes; the genesis, significance, and pattern of alteration mineralogies; the sources of the geothermal fluids and the paleohydrology of the systems; and, the chemical conditions surrounding the deposition of the ore minerals.

The present volume is an attempt to provide a synthesis of the current state of geological and geochemical knowledge of epithermal precious-metal systems. It follows on, and should be used in conjunction with, the first volume in this series: Mineral-Fluid Equilibria in Hydrothermal Systems by Henley et al. (1984). In the present volume we have attempted to provide a framework for understanding the systematics of controls on fluid compositions and of metal and gangue transport and deposition. The structure, dynamics, and transport properties of active geothermal systems are used as a starting point. With active systems as a reference, the evolution of fluid compositions and the constraints on metal and gangue transport and deposition in the epithermal environment are explored. The systematics of fluid inclusion and light stable-isotope applications is developed because these two approaches have been so useful in the development of our understanding of epithermal processes. The importance of boiling, cooling, and oxidation in transport properties of epithermal systems is evaluated through a numerical modelling approach. With the foregoing as background, the observational base and its interpretation for epithermal ore deposits in continental volcanic and sedimentary terranes is explored through summaries of the geologic, mineralogical, and geochemical characteristis of, and trace-element distributions in, some well-studied epithermal ore deposits. The final chapter is devoted to the use of our understanding of epithermal systems in the development of exploration strategies.

This volume does not attempt to be the final word on epithermal ore deposits, nor does it claim comprehensive treatment. The absence of a chapter on the hydrology of epithermal systems documents the fact that our current understanding of this aspect is woefully inadequate. It does not reflect a lack of recognition of the importance of hydrologic controls. Similarly, this volume focuses on volcanic- and sediment-hosted epithermal deposits in the cordillera of western North America, particularly the United States. It does not treat aspects of alkaline- or basaltic-rock related deposits such as Cripple Creek, Colorado, and Vatacoula, Fiji, nor does it treat the relationship of epithermal systems to deeper hydrothermal systems responsible for the formation of porphyry-type deposits. Again our reason is the lack of an adequate observational base. Our primary purpose in organizing this volume and the related Short Course has been to stimulate critical studies to improve our current understanding of epithermal deposits and processes rather than to document it. Perhaps our omissions will serve this purpose equally as well as our inclusions.

REFERENCES

Barton, P. B., Jr., Bethke, P. M., Roedder, E., 1977, Environment of ore deposition in the Creede mining district, San Juan Mountains, Colorado: III. Progress toward interpretation of the chemistry of the ore-forming fluid for the OH vein: Economic Geology, v. 72, p. 1-25.

Becker, G. F., 1882, Geology of the Comstock lode and the Washoe district: U.S. Geological Survey Monograph 3, 442 p.

Bell, J. M., and Fraser, C., 1912, The great Waihi gold mine: New Zealand Geological Survey, Bulletin 15.

Burbank, W. S., 1933, Epithermal base-metal deposits in Ore deposits of the Western States: American Institute of Mining Metallurgical Engineers, New York, Part VI, p. 641-652.

Ferguson, H. G., 1927, Geology and ore deposits of the Mogollon mining district, New Mexico: U.S. Geological Survey Bulletin 787, 100 p.

Ferguson, H. G., 1929, The mining districts of Nevada: Economic Geology, v. 24, p. 131-141.

Hemley, J. J., and Jones, W. R., 1964, Chemical aspects of hydrothermal alteration with emphasis on hydrogen metasomatism: Economic Geology, v. 59, p. 538-569.

Henley, R. W., Truesdell, A. H., and Barton, P. B., Jr., 1984, Fluid-mineral equilibria in hydrothermal systems: Society of Economic Geologists, Review in Economic Geology, v. 1, p. 267.

Lindgren, W., 1928, Mineral Deposits: Third Edition, McGraw Hill, New York, 1049 p.

Lindgren, W., and Ransome, F. L., 1906, Geology and gold deposits of the Cripple Creek district, Colorado: U.S. Geological Survey, Professional Paper 54, 516 p.

Nolan, T. B., 1933, Epithermal precious-metal deposits in Ore deposits of the Western States: American Institute of Mining Metallurgical Engineers, New York, Part VI, p. 623-640.

Ransome, F. L., 1909, The geology and ore deposits of Goldfield, Nevada: U.S. Geological Survey Professional Paper 66, 258 p.

Taylor, H. P., Jr., 1973, $^{18}O/^{16}O$ evidence for meteoric-hydrothermal alteration and ore deposition in the Tonopah, Comstock Lode, and Goldfield mining districts, Nevada: Economic Geology, v. 68, p. 747-764.

White, D. E., 1955, Thermal springs and epithermal ore deposits: Economic Geology, Fiftieth Anniversary Volume, p. 99-154.

White, D. E., 1981, Active geothermal systems and hydrothermal ore deposits: Economic Geology, Seventy-fifth Anniversary Volume, p. 392-423.

Wisser, E., 1960, Relation of ore deposition to doming in the North American Cordillera: Geological Society of America, Memoir 77.

ACKNOWLEDGMENTS

As is true for any effort of the scope of this volume, many people in addition to the editors played key roles along the road to final publication. The time and effort expended by each author is greatly appreciated as are the contributions of the large cadre of individual reviewers who have offered insights and alternative perspectives to the authors. Technical support to the editors including manuscript preparation and revision, final formatting for publication, and badgering of both editors and authors was provided by the Editorial Support Group, Branch of Exploration Geochemistry, U.S. Geological Survey. Within this group we would especially like to thank Pamela Detra, Dorothy Wesson, Marilyn Billone, and Candy Vassalluzzo. An earlier version of this text was assembled for use at the Society of Economic Geologists Short Course by the Branch of Exploration Geochemistry Clerical Support Group. Finally, we would like to express appreciation for the patience of Jamie Robertson, Series Editor, Reviews in Economic Geology, and the support of the Society of Economic Geologists.

Byron R. Berger
Philip M. Bethke

BIOGRAPHIES

BYRON R. BERGER received a B.A. degree in Economics/Geology from Occidental College in 1966 and a M.S. in Geology from the University of California, Los Angeles in 1975. He worked as a petroleum exploration geologist for Standard Oil Company of California from 1968-1970 and a minerals exploration geologist and research scientist for Continental Oil Company from 1971-1977. He joined the U.S. Geological Survey in 1977, and has been involved in research on epithermal precious-metal deposits and the relationship of magma genesis to ore genesis. He is currently the Chief of the Branch of Exploration Geochemistry. He is an adjunct assistant professor of geology in the Department of Geological Sciences at the University of Colorado, Boulder, where he has taught courses on the geology and geochemistry of epithermal ore deposits and exploration geochemistry. He is a member of several professional societies including the Geological Society of America and the American Geophysical Union.

PHILIP M. BETHKE received a B.A. degree in Geology from Amherst College in 1952 and a Ph.D. in Geology (specialization in Mineralogy and Ore Deposits) from Columbia University in 1957. He was Assistant Professor of Geology at the Missouri School of Mines and Metallurgy (now the University of Missouri-Rolla) from 1955 to 1959. He joined the U.S. Geological Survey as a WAE research geologist in 1957 and transferred to full time in 1959. His research has combined field and laboratory approaches to the study of hydrothermal ore deposits, particularly to epithermal vein systems. He has held several administrative positions with the U.S.G.S., most recently, Chief of the Branch of Experimental Geochemistry and Mineralogy. He is a member of several professional societies and is currently a Councillor of the Society of Economic Geologists. He has been active in the establishment of the SEG Short Course Series, and is currently Chairman of the Short Course Committee.

SAMUEL S. ADAMS received B.A. and M.A. degrees from Dartmouth College in 1959 and 1961, and a Ph.D. degree from Harvard University in 1967. From 1964 to 1977 he served as mine geologist, exploration geologist, exploration manager, and exploration vice president, employed by International Minerals and Chemical Corporation and then the Anaconda Company. During this period, his work emphasized sediment-hosted mineral deposits, particularly potash and uranium. Since 1977 he has served as a lecturer and consultant to industry, research organizations, and government agencies in the areas of mineral deposits, exploration, and resource assessment. His principal research interest is the representation of data and concepts for all types of mineral deposits in coherent and predictive models for exploration and resource studies. He is currently a Councillor of the Society of Economic Geologists and the Geological Society of America.

WILLIAM C. BABGY received a Ph.D. degree in Earth Science from the University of California, Santa Cruz, in 1979 based on petrogenetic research of Tertiary volcanic rocks in the Sierra Madre Occidental, Mexico. His industry experience includes geologic evaluation of volcanic-hosted uranium in the McDermitt caldera complex, Nevada, and the bulk mineability potential of the amythest silver vein system at Creede, Colorado. Industry research included development of an occurrence model for hot spring-related gold deposition based on the McLaughlin gold deposit in California. Present research interests are focused on the genetic aspects of sediment-hosted precious-metal deposits.

ROBERT J. BODNAR received an M.S. degree from the University of Arizona and a Ph.D. degree from The Pennsylvania State University and has been involved in various aspects of fluid-inclusion research for the past 10 years. He worked for 1 year as a research geochemist in the Ore Deposits Group of Chevron Oil Field Research Company and is currently an assistant professor in the Department of Geological Sciences at Virginia Polytechnic Institute and State University.

KEVIN BROWN received an M.S. degree in Chemistry in 1969 and a Ph.D. degree in Chemical Crystallography in 1972 from the University of Auckland, New Zealand. Except for a two-year sojourn at the E.T.H. in Zurich, he has worked at the Department of Scientific and Industrial Research, New Zealand. Initially in Wellington, his research interest centered around the crystal structures of organic reaction intermediates, but he gradually came down to earth with the crystal structures of some new epithermal minerals. In 1981, he shifted to the Geothermal Section at Wairakei, where his present research is concerned with experimental studies of mineral deposition from geothermal fluids.

CYRUS W. FIELD received a B.A. degree in Geology from Dartmouth College in 1956 and M.S. and Ph.D. degrees in Economic Geology, Geochemistry, and Petrology from Yale University in 1957 and 1961, respectively. He worked as an exploration geologist during the summers of 1955, 1956, and 1957 for the Oliver Iron Mining Company and Quebec Cartier Mining Company subsidiaries of the U.S. Steel Corporation, and served as a research geologist from 1960 to 1963 with the Bear Creek Mining Company division of Kennecott Copper Corporation. In 1963, he joined the faculty of Oregon State University where he is currently Professor of Geology. His research interests are largely concerned with the geology and geochemistry of hydrothermal mineral deposits; particularly the application of stable isotope and major-minor-trace element investigations to their genesis. He is a member of several professional societies and was Vice President of the Society of Economic Geologists in 1981.

RICHARD H. FIFAREK received a B.S. degree in Geology from the University of Washington in 1974, and M.S. and Ph.D. degrees in Geology (specialization in Economic Geology) from Oregon State University in 1982 and 1985, respectively. From 1974 to 1984, he worked periodically as an exploration geologist (4 yrs.) for several mining companies, as a research assistant/scientist (1 yr.) at the facilities of the Branch of Isotope Geology (Denver), U.S. Geological Survey, and as an instructor for Oregon State University. Presently, he is an assistant professor in the Department of Geology at Southern Illinois University where he teaches and conducts research in economic geology and isotope geochemistry. His research interests include integrated geologic (field) and geochemical investigation of massive sulfide and epithermal Au-Ag deposits, and modeling the isotopic evolution of fluids and rocks in hydrothermal systems.

NORA K. FOLEY received a B.S. degree in Geology and Mineralogy from the University of Michigan in 1978 and an M.S. degree in Geological Sciences from Virginia Polytechnic Institute and State University in 1980. She is currently working towards a doctoral degree in Geology through Virginia Polytechnic Institute and State University. Since 1980, she has been a research geologist at the U.S. Geological Survey in Reston, Virginia. Her research has included fluid-inclusion and isotopic studies of different types of ore deposits, including Ag- and base-metal-bearing, epithermal deposits, sediment-hosted, stratabound, Pb-Zn deposits, and Kuroko-type massive sulfides.

ROBERT O. FOURNIER received an A.B. degree in Geology in 1954 from Harvard College and a Ph.D. in Geology (specializing in Economic Geology, in general, and the Ely porphyry copper deposit, in particular) from the University of California at Berkeley in 1958. Since then, he has been a research geologist with the U.S. Geological Survey. His research interests have ranged from laboratory studies of mineral-water interactions at hydrothermal conditions appropriate for shallow levels in the crust, to field studies of presently active hydrothermal systems, including Yellowstone National Park, Coso and Long Valley, California, and Zunil, Guatemala. Experimental studies have emphasized solubilities of silica species in water and saline solutions. He has also been a leader in the development of several chemical geothermometers and mixing models that are now widely used in the exploration for geothermal resources. His present research focuses mainly on internally consistent chemical, isotopic, and hydrologic models of presently active hydrothermal systems. He has served on NATO committees to review geothermal energy development programs in Iceland, France, Greece, Portugal, and Turkey, and other committees to review geothermal exploration programs in Argentina and Thailand. He was Chairman of the Organizing Committee for the 1975 United Nations International Symposium on Geothermal Energy, and Chairman of the Technical Program Committee for the 1985 GRC International Symposium on Geothermal Energy. He now serves on panels to oversee geothermal developments in Costa Rica and Panama, and several U.S. Continental Scientific Drilling Committees. He is a member of several societies and has served on the Board of Directors of the Geochemical Society and the Geothermal Resources Council.

DANIEL O. HAYBA received a B.A. degree in Geology from the College of Wooster in 1976 and an M.S. degree in Geochemistry and Mineralogy from the Pennsylvania State University in 1979 following a study of the Salton Sea geothermal system. From 1978 to 1980, he worked for Exxon Production Research Company on computer modeling of ore deposits. Since that time, he has been a research geologist with the U.S. Geological Survey where his research has been directed towards understanding the ore-forming processes in epithermal systems.

PAMELA HEALD received a B.A. degree in Geology in 1971 from Vassar College and an M.S. degree in Geology from George Washington University in 1977. She has been a research geologist at the U.S. Geological Survey since 1972. Her research has included spectral reflectance and structural studies in Nevada, with a focus on ore deposits, and mineralogical and geochemical studies to evaluate ore-forming processes in epithermal precious- and base-metal deposits.

RICHARD W. HENLEY received a B.S. degree in Geology in 1968 from the University of London and a Ph.D. degree in Geochemistry from The University of Manchester in 1971 following experimental studies of gold transport in hydrothermal solutions and the genesis of some Precambrian gold deposits. He was Lecturer in Economic Geology Memorial University of Otago, New Zealand, from 1971 to 1975, and at Memorial University, Newfoundland, until 1977. Research interests have focused on the mode of origin of a number of different types of ore deposits including post-metamorphic gold-tungsten veins, porphyry copper, massive sulfide, and placer gold deposits. He is currently with the Geothermal Chemistry Section of the Department of Scientific and Industrial Research at Wairakei, New Zealand, and a visiting lecturer at the Auckland Geothermal

Institute. Through 1983-84, he was a Fulbright Fellow and Guest Investigator at the U.S. Geological Survey and during that time produced Volume 1 of this Review series. His present research includes a number of isotope and chemical studies relating to the exploration and development of geothermal systems and geothermal implications for the origin of ore deposits.

C. A. KUEHN received an M.S. degree from the Pennsylvania State University and has 7 years of experience in exploration for sediment-hosted gold deposits. He is currently an NSF Research Assistant and Ph.D. candidate at the Pennsylvania State University and part-time employee of the U.S. Geological Survey working on the Carlin gold deposit.

MARK H. REED received a B.A. degree in Chemistry and in Geology from Carleton College in 1971 and M.A. and Ph.D. degrees in Geology at the University of California, Berkeley, in 1977. His Ph.D. research was on the geology and geochemistry of the massive sulfide deposits of the West Shasta District, California. From 1977 through 1979, he worked for the Anaconda Minerals Company at Butte, Montana. Since that time, he has taught and conducted research at the University of Oregon, where he is currently Associate Professor of Geology. His research has focused on alteration and metal zoning in the porphyry copper and large vein deposits at Butte and the geochemistry of hydrothermal alteration, metal transport, and ore deposition in massive sulfide and epithermal systems.

T. J. REYNOLDS received an M.S. degree from the University of Arizona and has been an exploration geologist specializing in the application of fluid inclusions to mineral exploration for the past 5 years.

MILES L. SILBERMAN received a B.S. degree from the City University of New York and M.S. and Ph.D. degrees from the University of Rochester, New York. He is a member of the Branch of Exploration Geochemistry of the U.S. Geological Survey, with current assignments to the Redding, California (CUSMAP) project, and to the study of the geochemistry of volcanic and metamorphic-hosted gold deposits in the western U.S. and northern Mexico. Previous work for the U.S.G.S. included geochronological, geochemical, and regional geological studies of precious- and base-metal deposits in the Great Basin and Alaska, and tectonic syntheses with particular focus on the relationships of hydrothermal precious-metal deposits to magmatic and metamorphic evolution. Between tours at the U.S.G.S., he designed and supervised exploration programs for precious-metal deposits in the Great Basin for the Anaconda Minerals Company.

NICOLAS F. SPYCHER received a B.S. degree in Earth Sciences in 1979 and a Dipl. es Sc. in Exploration Geophysics in 1980 from the University of Geneva, Switzerland. He is now a Ph.D. candidate and research assistant at the University of Oregon. His present research includes studies of the transport of arsenic and antimony in hydrothermal solutions, the mixing properties of geothermal gases, and the geochemical modeling of hot spring systems.

Chapter 1
THE GEOTHERMAL FRAMEWORK OF EPITHERMAL DEPOSITS
R. W. Henley

INTRODUCTION

In the context of exploration for epithermal deposits, why study geothermal systems at all? After all, not one exploited system to date has been shown by drilling to harbor any economically significant metal resource--but then until recently not one had been drilled for other than geothermal energy exploration.* The latter involves drilling to depths of 500-3000 meters in search of high temperatures and zones of high permeability which may sustain fluid flow to production wells for steam separation and electricity generation. In many cases such exploration wells have discovered disseminated base-metal sulfides with some silver and argillic-propylitic alteration equivalent to that commonly associated with ore-bearing epithermal systems (Browne, 1978; Henley and Ellis, 1983; Hayba et al., 1985, this volume). In general, however, geothermal drilling ignores the upper few hundred meters of the active systems and drill sites are situated well away from natural features such as hot springs or geysers, the very features whose characteristics (silica sinter, hydrothermal breccias) are recognizable in a number of epithermal precious-metal deposits (see, for example, White, 1955; Henley and Ellis, 1983; White, 1981; Berger and Eimon, 1983; Hedenquist and Henley, 1985a; and earlier workers such as Lindgren, 1933). Knowledge of the upper few hundred meters of active geothermal systems is scant and largely based on interpretation of hot-spring chemistry. Tantalizingly, in a number of hot springs, transitory red-orange precipitates occur which are found to be ore grade in gold and silver and which carry a suite of elements (As, Sb, Hg, Tl) now recognized as characteristic of epithermal gold deposits (Weissberg, 1969).

---

*Kennecott has recently announced significant gold discoveries in still active geothermal fields on Lihir and Simberi Islands, Papua, New Guinea.

Today's active geothermal systems occupy the same tectono-volcanic niche as those hydrothermal systems, preserved from the past, which hosted the near-surface (0-1000 m) formation of epithermal ore deposits in the Tertiary volcanic terranes of the Circum-Pacific region and elsewhere--the relatively shallow origin of these deposits resulting in their loss by erosion from erstwhile similar, but older, terranes. Formed at deeper levels (2-5 km or so) beneath calc-alkaline volcanoes in these same volcanic terranes (Sillitoe, 1973; Henley and McNabb, 1978), porphyry-type copper and molybdenum deposits are preserved in both Tertiary and much older hydrothermal systems. The purpose of this chapter is to review some of the principal chemical and physical characteristics of the active geothermal systems which are essential to the understanding of the origin of epithermal ore deposits and therefore to their successful exploration. For more detailed information, the reader is referred to the publications cited in the text.

HYDROTHERMAL SYSTEMS IN GENERAL

The term "hydrothermal" encompasses all types of hot-water phenomena in the earth's crust although most commonly the term is used in reference to those associated with impressive geyser activity, aesthetically attractive hot pools, etc. These features are most common in volcanic areas such as Yellowstone National Park, U.S.A., Iceland, or in the Taupo Volcanic Zone of New Zealand, but other terranes also host hydrothermal activity even though subsurface temperatures may be relatively low and surface features less impressive. Warm springs in the Rocky Mountains, the European or New Zealand Alps, or in the sedimentary massifs of central Europe are examples, and it is clearly important for mineral exploration to discriminate these types of systems from those in more favorable geological environments.

Geothermal systems are extraordinarily abundant in the tectonically active zones of the earth's crust and may be broadly classified according to their plate tectonic setting and principal source of heat (Table 1.1). Chemical differences arise from the sources of recharge water and contribution of gases from magmatic or metamorphic sources. Warm springs also occur in the tectonically stable crust where the deep crustal penetration of groundwater occurs in favorable sedimentary formations such as limestones and the heat supply is the ambient continental heat flow.

Each of these classes of geothermal systems appears to have some correlative preserved in the geologic past and most commonly recognized as one or another of the various families of hydrothermal ore deposits. For magma-related hydrothermal systems, these range from ophiolite-hosted massive sulfides through the polymetallic massive sulfides of island arcs to the porphyry copper and epithermal precious-metal deposits of terrestrial continental terranes, while for amagmatic systems these range from the Mississippi Valley and related base-metal deposits in sedimentary basins to the post-metamorphic vein deposits associated with orogeny.

Table 1.1—Crustal setting of hydrothermal systems classified according to principal heat-source and crustal host.

CRUSTAL HOST/ HEAT SOURCE	MAGMATIC	AMAGMATIC
Oceanic	Ridge, hot spot, back-arc basin	-----
	Magmatic arc	
Continental	Crustal extension (Hot spot, rift)	Plate collision
		Plate-interior basins

Collision-Related Amagmatic Hydrothermal Systems

Only recently have data become available from geothermal investigations in mountain belts. In the Southern Alps of New Zealand, for example, hot springs occur in the central, relatively aseismic region with the highest uplift rate (10-20 mm/year) where the combination of uplift and erosion "exposes" a thermal anticline with near-surface gradients up to 150°C/km (Allis et al., 1979). A similar environment is proposed for hot springs in other collision-related mountain belts. Recent drilling at Yangbajing (Tibet) and in the Parbati Valley (N. India), for example, has located hot waters up to 170°C (Giggenbach et al., 1983) which are predominantly meteoric in origin, but contain low ^{3}He to ^{4}He ratios typical of helium of deep crustal origin.

The uplift setting of these hydrothermal systems is perhaps analogous to that of Late Mesozoic post-metamorphic gold and scheelite veins on the South Island (New Zealand) and, by inference, similar deposits in much older terranes. Examples are the gold veins of the Valdez Group (S. Alaska), Mother Lode (California), Yellowknife (Northwest Territories) and Kalgoorlie (W. Australia). In each of these, in contrast to the epithermal precious-metal deposits discussed below, vein quartz is enriched in ^{18}O relative to host rocks. This feature has led many workers (e.g., Henley et al., 1976; Fyfe and Kerrich, 1984) to suggest a metamorphic origin for the hydrothermal fluid; vein formation occurring from fluids of metamorphic dehydration origin in response to post-metamorphic uplift and/or overthrusting. It may also be possible, however, to generate these same isotope characteristics by interaction of meteoric water and rocks at a low water to rock ratio opening the possibility that such deposits may be much shallower* in origin than the 10 to 20 km generally considered.

---

*In this paper the term "shallow" is used rather irreverantly to refer to depths less than about 500 meters. In ore deposit research, depths (estimated perhaps from fluid-inclusion data) are generally also used irreverantly, taking no account of the importance of topographic relief; of the order of ±100 m in silicic volcanic terranes, ±1000 m in andesitic volcanic terranes and for the mountain belt systems, 2000 m.

Terrestrial Magma-Related Hydrothermal Systems

By contrast, systems in volcanic terranes have high ^{3}He to ^{4}He ratios and the δ^{18}O of alteration minerals are, with few exceptions, depleted relative to primary minerals. Temperatures encountered during drilling range up to 400°C (Batini et al., 1983a) and waters are predominantly meteoric in origin, and typified by the presence of chloride ion with m_{Cl^-} > > $m_{SO_4^=}$ —they are here, for convenience, designated chloride waters. Although some highly saline fluids are evolved in rift zones such as the Imperial Valley (California), salinities are typically low, clustering around 10,000 mg/kg Cl (1.6 wt.-% NaCl equivalent) in andesitic volcanic terranes, 1000 mg/kg in rhyolitic volcanic terranes and much lower in basaltic volcanic terranes. Dissolved gas, always preponderantly CO_2, affects a major contrast between systems and ranges from very low (0.01 wt.-% CO_2) at Wairakei (New Zealand) and Ahuachapan (El Salvador) to several wt.-% at Broadlands and Ngawha (New Zealand) (see Table 1.3). Other dissolved components are controlled by mineral-fluid and gas-gas reactions. Alteration assemblages in these types of geothermal systems correspond closely to those encountered in epithermal and porphyry-style mineral deposits.

The deep hydrologic structure of the terrestrial geothermal systems is controlled by the convective upflow of chloride waters (evolved by water-rock ± magma interaction at depths of 5 to 8 km) but above depths of around 1 km surface topography plays a major role in the dispersion of the chloride water by introducing a lateral flow component toward topographic lows. Boiling occurs as chloride water rises through the system, the resultant steam migrating to the surface independently where near-surface condensation and oxidation of co-transported H_2S produces sulfate-dominated steam-heated waters. These features are incorporated in the general model of the structure of a geothermal system reproduced in Figure 1.1a (from Henley and Ellis, 1983).

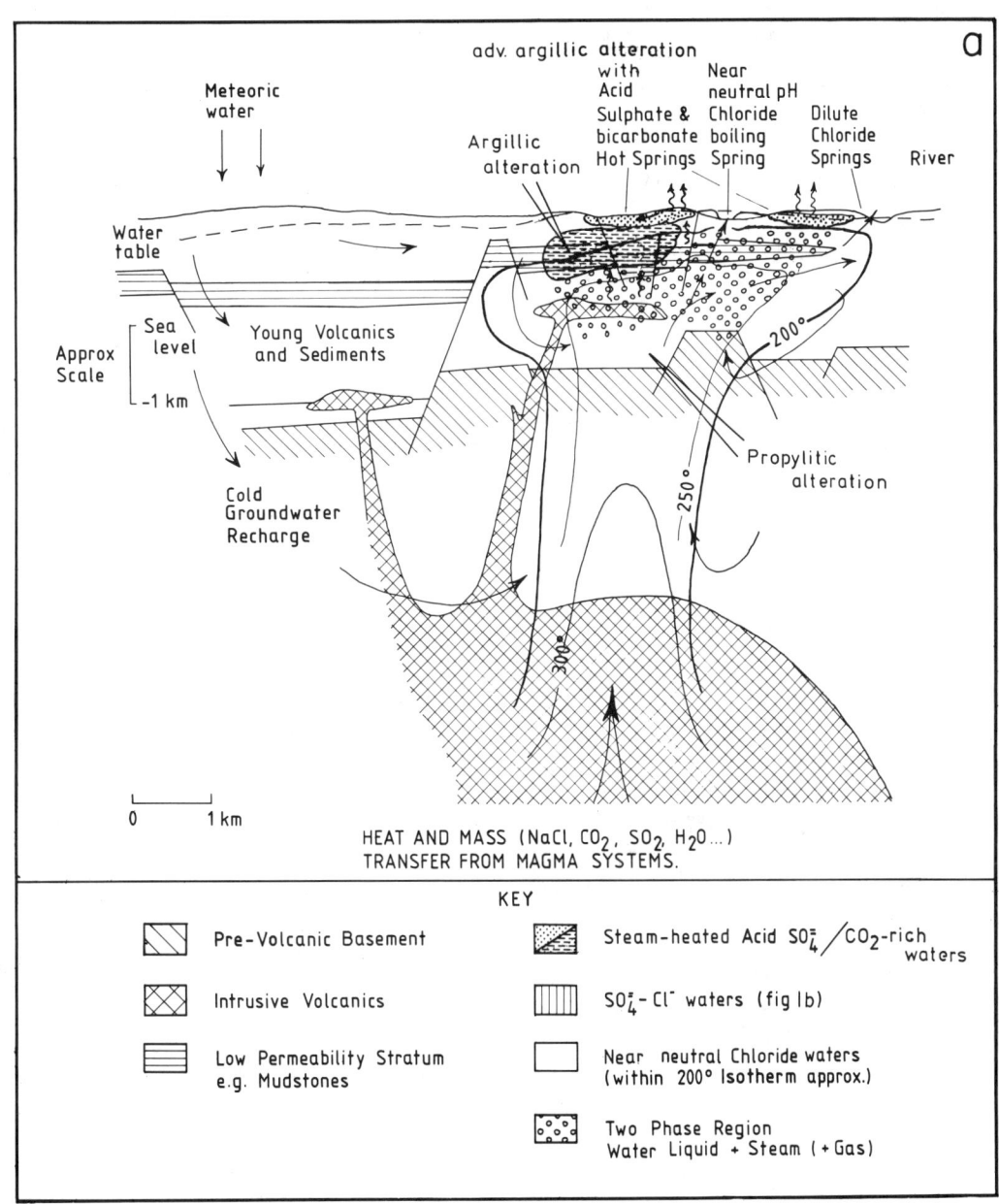

Figure 1.1a. Generalized structure of a typical geothermal system in silicic-volcanic terrane. Notice the overall size of the system relative to the size of the discharge features (i.e., hot springs, etc.). The temperature distribution shown is based on the Wairakei system where a west-to-east flow occurs in the upper portion of the system and boiling occurs above about 500 meters. In other systems such as those in Figure 1.2, more or less lateral flow may occur. Boiling may extend to much greater depths if CO_2 contents are high (see text), and higher temperatures may occur at shallower depths than shown in this figure, as at Mokai (Fig. 1.2d).

The relatively high relief of andesite volcanic terranes results in lateral flows of hot chloride water for up to 20 km while the occurrence of near-surface magmas exsolving gases (HCl, SO_2, etc.) often produces high temperature fumaroles and/or acid sulfate-chloride crater lakes, such as those on Mount Ruapehu, New Zealand (Plate 1.1) and El Chichon, Mexico (Giggenbach, 1974; Kyosu and Kurahashi, 1984; Casadevall et al., 1984). These latter features, with their associated intense advanced argillic alteration, are possible correlatives of the upper portions of the type of hydrothermal systems responsible for the formation of gold--(enargite) sulfide deposits of the "Goldfield type" (Ransome, 1909) such as Goldfield (Nevada), Summitville (Colorado), Bor (Yugoslavia), and elsewhere. Figure 1.1b provides a general structural model for this geothermal environment. They may also be related in some cases to the upper portions of developing porphyry copper deposits (Sillitoe, 1983).

The geochemistry and structure of magma-related hydrothermal systems have been reviewed in a number of recent texts to which the reader is referred for background reading and discussion of hydrothermal chemistry--see, for example, Ellis and Mahon, 1977; Henley and Ellis, 1983; Henley et al., 1984. A brief summary of hydrothermal chemistry is given in Henley and Brown (1985, this volume). In the remainder of this chapter attention is focused on those aspects of the chemistry and structure of geothermal systems relevant to the understanding of the formation of epithermal ore deposits.

TERRESTRIAL MAGMATIC-HYDROTHERMAL SYSTEMS

Large-Scale Structure

Early in the commercial development of the Wairakei geothermal field in New Zealand, the accumulating data from exploration wells showed (a) that the fluids present were not directly exsolved from shallow bodies of crystallizing magma and (b) that the hydrothermal activity seen at the surface was a minor phenomenon associated with the discharge of a very large, deeply convecting body of heated groundwater (Elder, 1966). Using analog and numerical modelling, Elder and other research scientists showed that convection, with a depth scale of at least 5 km, was

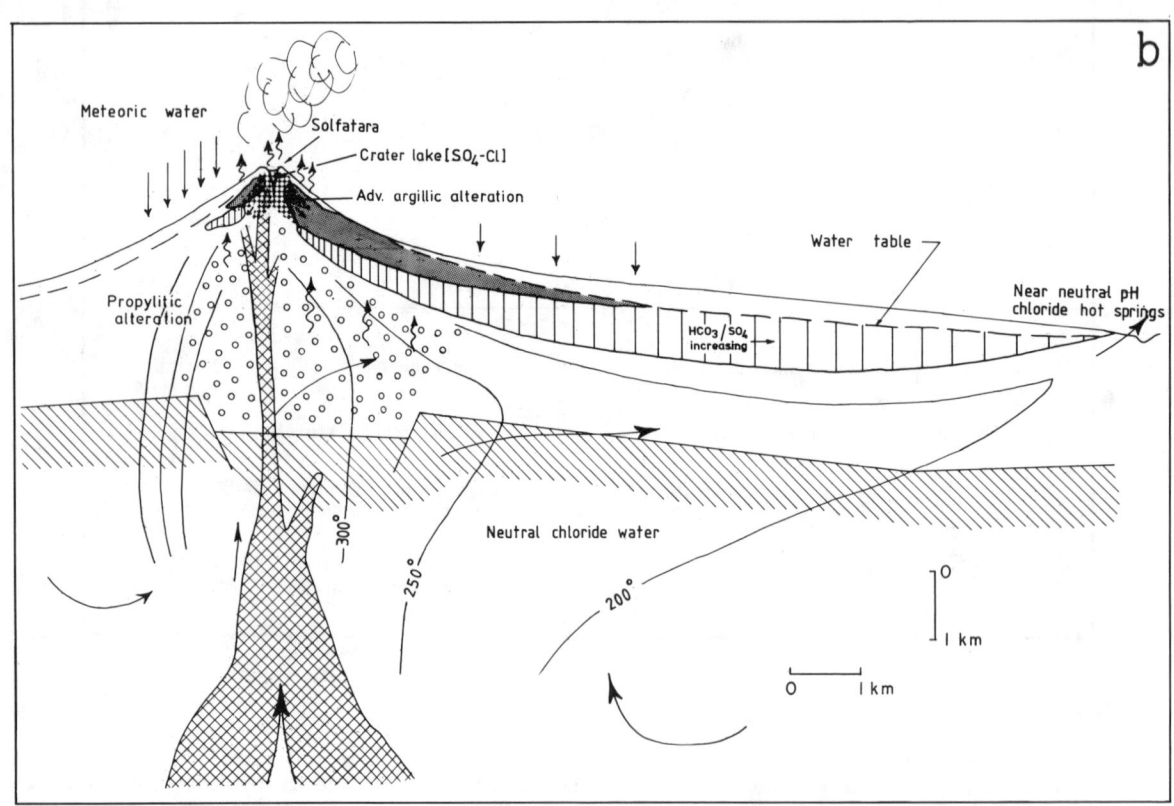

Figure 1.1b. Structure of a typical geothermal system in andesitic-volcanic terranes emphasizing (1) extensive lateral flow and (2) generation of a large advanced-argillic alteration zone in response to high-level volcanism.
(Modified and reproduced with permission from Henley and Ellis, 1983.)

Plate 1.1. Oblique, aerial view of the Waiotapu system, New Zealand from the southeast. Topographic features may be related to the system map (Fig. 1.2b). Mount Tarawera (on the horizon) is a composite rhyolite dome which, in 1886, violently erupted basalt through an axial rift. Associated phenomena were the destruction of the Pink and White Silica Terraces (Henley et al., 1984, Plate 1.2), and a number of hydrothermal eruptions in the Rotomahana-Waimangu geothermal system.

The natural discharge dominating the surface expression of the system is the Champagne Pool (middle right) which occupies a hydrothermal eruption vent formed 900 years ago and which may overlie some 0.1 million ounces of gold formed by boiling in the conduit of the pool. Surface antimony-arsenic precipitates occur which are ore-grade in silver and gold (photo D. L. Homer, N. Z. Geological Survey).

responsible for the extreme thermal gradients and temperature patterns observed in the exploration drilling. At the same time, at Wairakei and in other fields, the effects of near-surface (depths less than 1000 m) stratigraphy and structure and of relief-controlled groundwater flow became evident largely through geophysical techniques, especially resistivity surveying (Healy and Hochstein, 1973). Hanaoka (1980) has numerically modelled the effects of topographic relief on near-surface hot-water flow and its dispersion by cold groundwater. This effect is partly responsible for the mushrooming of isotherms shown in many convective models and field cross sections.

Figure 1.2 provides some examples of the lateral-flow characteristics and distribution of natural discharges in a number of geothermal fields explored by drilling, with perhaps the Mokai field in New Zealand (Fig. 1.2c) being a particularly good illustration of lateral flow as shown in cross-section in Figure 1.2d. In geothermal systems hosted by silicic volcanic rocks, surface topography is primarily controlled by block-faulting or caldera collapse providing relief of a few hundred meters and consequent lateral flow over distances of up to about 5 km. In the higher relief terrane typical of andesitic volcanism, more extreme lateral flow occurs up to about 20 km. An additional feature of active andesitic volcanic terranes is the occurrence of high-level volcanism which allows volcanic gas to vent to summit fumaroles or to summit crater lakes (Giggenbach, 1974) and to maintain high-level "perched" aquifers containing very acid sulfate-dominated waters. Exploration wells at high elevation in such terranes often encounter vapor-dominated geothermal environments.

In the majority of systems, liquid water provides the continuum for fluid flow but in other, far less common systems, water vapor dominates the discharges of deep exploration wells. The pre-exploitation states of these "vapor-dominated" systems are poorly known and various models have been produced based on production data from exploited fields. For example, for the Geysers (California) and for Larderello (Italy), White et al. (1971) suggest the presence of a very deep convecting brine overlain by an "alteration-sealed" cap of vapor. Of particular interest is the association of these systems with epithermal mercury and gold mineralization (e.g., McLaughlin, California), but both the Geysers and Larderello also contain base-metal sulfides and other "ore-related" mineral phases in drill core (Belkin et al., 1983; Sternfeld, 1981) which suggest that the present system has evolved from some previous liquid-dominated state. Others have suggested that elevated gas-content (dominantly CO_2) perhaps coupled with relatively low host-rock porosity, may account for the vapor-dominated character of well-discharges and post-exploitation pressure data. It is interesting to note that most of the explored "liquid-dominant" geothermal systems, in silicic volcanic terranes especially, are associated with tectonic subsidence (about -5 mm per year in the Taupo Volcanic Zone in New Zealand), but both the Geysers and Larderello occur in regions of high tectonic uplift associated with volcanism. Quantitative data from the Geysers region are not available, although regional topography and erosion are suggestive of high uplift rates. At Larderello uplift rates are of the order +5 to 10 mm per year (M. Puxeddu, personal communication) and are evidenced by the coastline migration of the Pisa area. The high heat flow and geothermal activity of the Larderello region appears to be related to the emplacement of a post-orogenic batholith into continental crust (Batini et al., 1983b; Puxeddu, 1984).

<u>Natural Discharges</u>

Hot water convecting into the near-surface part of a large hydrothermal system may be dispersed by mixing with laterally flowing cold groundwater or discharged directly to the surface. Only a minor amount of heat energy is lost by conduction, but most is dispersed as hot water and vapor flows at the surface. The processes affecting a deep fluid penetrating to the surface depend on a variety of factors. Direct discharge depends on the availability of a suitable fracture system (or hydrothermal eruption vent, see below) and gives rise to a boiling spring, high in chloride and mantled by silica sinter. Examples are the Champagne Pool, Waiotapu and the Pink and White Terraces of Rotomahana, New Zealand (Plate 1.1; and see Henley et al., 1984). Geysers are a special class of boiling discharge which have a periodic discharge due to the geometry of the conduit (Kieffer, 1984). Often dilution precedes boiling of the mixed fluid as it finally moves to the surface as in the Ohaaki Pool at Broadlands (Ohaaki) or the boiling springs of the Wairakei and Tauhara systems (Fig. 1.4a).

Fluids which are diluted with respect to the deep chloride water form where interaction with near-surface aquifers occurs either due to high surface relief and groundwater flow or to the proximity of the system margin. The natural discharges of the Wairakei-Tauhara and Mokai systems are examples (Fig. 1.2).

Figure 1.3a shows schematically the pressure distribution associated with various discharge phenomena. Drill-hole data suggest that pressure gradients in the deeper system are generally about 10% above hydrostatic pressure with the excess pressure due to the buoyancy of hot water relative to surrounding cold groundwater (Elder, 1966; Cathles, 1977; Grant et al., 1982), and in some cases (e.g., Mokai) a demonstrable component of hydrostatic head due to recharge from areas of relatively high relief. An excess pressure gradient is a requirement for flow through permeable media. Below a hot-spring vent, fluid expansion leads to two-phase flow in the high-permeability conduit. Phase separation may occur with the vapor discharging independently at the surface as a fumarole or interacting with groundwater to produce a steam-heated water. As suggested in Figure 1.3a, minor throttling may occur along the flow path, but pressure drops are unlikely to be greater than 1 bar. Where silicification isolates the conduit from the surrounding groundwater system, boiling, deep-system fluid exits the surface; but, where only partial isolation occurs through mineral deposition, the liquid may itself interact with surficial groundwater before reaching the surface as a hot or warm spring. In the

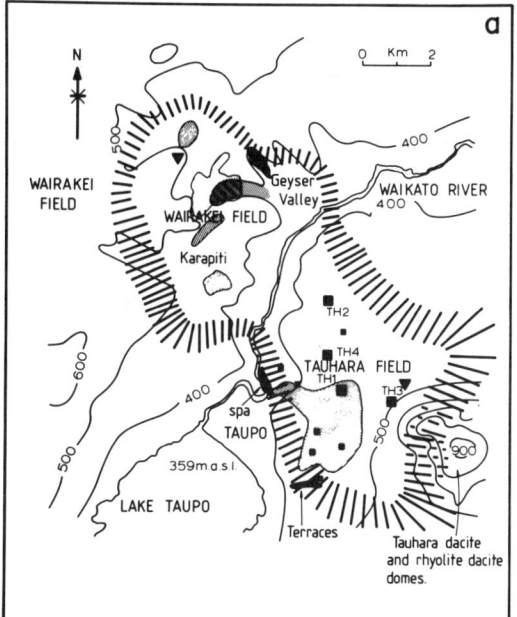

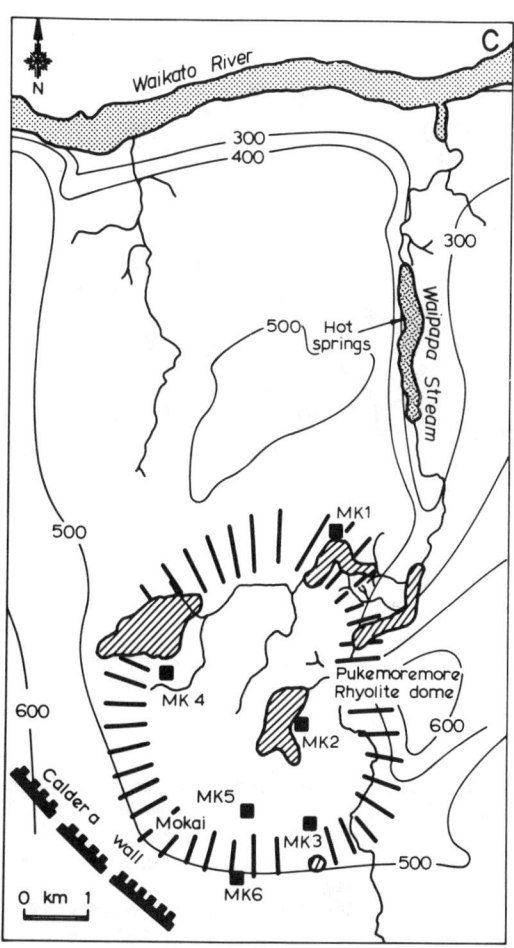

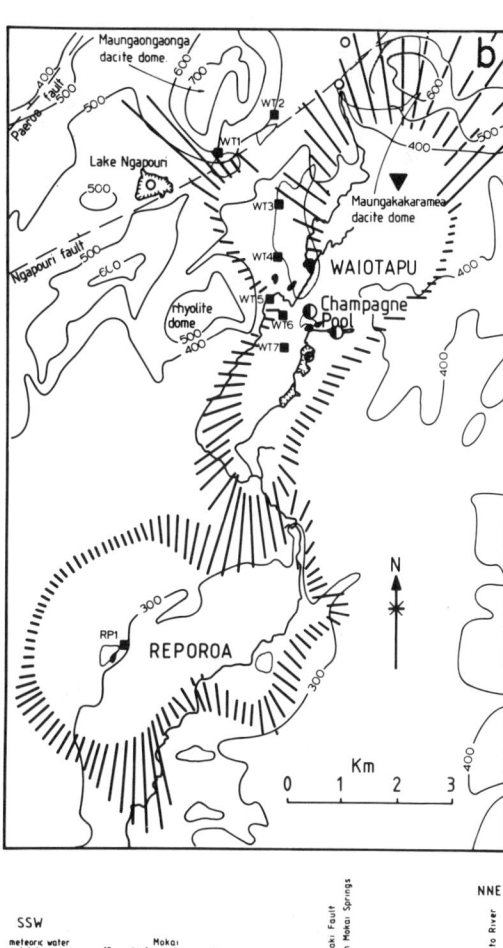

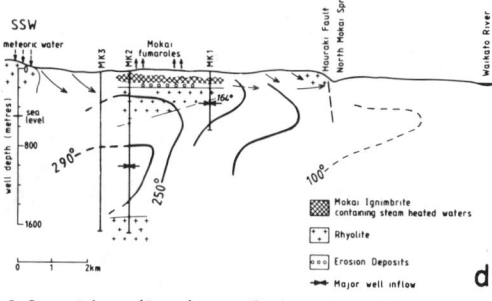

Figure 1.2. Distribution of natural discharges in some active geothermal systems. The field boundaries shown are based on the maximum resistivity gradient located by field surveys reflecting the contrast between unmineralized groundwater and the chloride water present in the upper 500 meters of the geothermal systems. Fumaroles, steaming ground, and outflows of steam-heated waters are indicated by the O symbol and hot-water discharges by the ◐ symbol. The location of the principal convective upflow for each field is indicated by the ▼. Notice that geothermal exploration and production wells are situated well away from natural features. Numerals designate features shown in the mixing diagrams of Figure 1.4.

Figure 1.2 (cont'd) a). Wairakei-Tauhara, New Zealand. These two fields are interconnected as shown by the resistivity boundary and both show the occurrence of vapor discharge in the central region and hot-water discharge on the margins following dilution. There is no evidence that water from Lake Taupo penetrates either field, recharge being derived from groundwaters to the east and west. b). Waiotapu, New Zealand. This field has an extensive north-to-south lateral flow originating in the vicinity of the 160,000 years b.p. dacite domes to the north. Thermal features are related to major faults and a number of hydrothermal eruption craters have been recognized (major centers shown by the circles)--for full discussion see Hedenquist and Henley, 1985a. c). Mokai, New Zealand. Extensive lateral flow occurs from the vicinity of the caldera wall in the south toward the Waikato River to the north. Dilute hot springs occur north of the "field boundary" in the gorge of a stream following a major fault. d). Cross-section of the Mokai geothermal field showing the effect of lateral flow and dispersion on the thermal structure of the system and distribution of natural features. (The cross section runs from the top right-hand corner of Figure 1.2c to the caldera wall south of well MK6).

Plate 1.2. Crater Lake, Ruapehu, New Zealand. Condensation of volcanic gas into the Crater Lake waters produces a fluid of pH 1.5 at about 55°C. The lake seldom overflows despite the presence of an incised channel (foreground) suggesting that much of the acid fluid drains through the core of the active andesite volcano producing an extensive high-level zone of advanced-argillic alteration. Interaction of this fluid with an underlying near-neutral pH hydrothermal system may generate a gold deposit of the Goldfield type (photo by permission, R. B. Glover, DSIR).

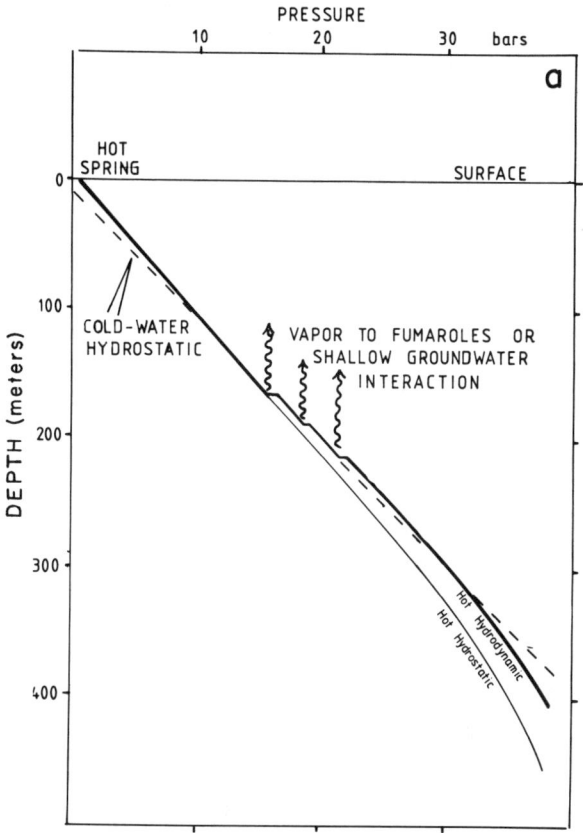

Figure 1.3a. Pressure-depth relations in the upper portion of a geothermal system. The diagram shows the transition between the deep system pressure and the pressure within the high permeability fracture network or conduit below a hot spring. Below the hot springs, the pressure at a specified depth is due to the weight of a <u>standing</u> column of hot water; the pressure-depth relation is here designated "hot hydrostatic". Deeper in a system pressures exceed hydrostatic so that flow is maintained through the permeable aquifer—this is shown as the "hot hydrodynamic" curve. Some minor pressure discontinuities are shown to indicate the possible occurrence of minor throttles which may occur due to fracture geometry or silicification, but these are probably rare. Phase separation may occur resulting in the presence of fumaroles or (acid) steam-heated waters in the vicinity of a boiling hot spring (e.g., Norris Geyser Basin, Yellowstone, Champagne Pool, Waiotapu). The effect of raising or lowering the ambient groundwater piezometric surface may be gauged by redrawing the curve for cold-water hydrostatic pressure. For example, if the cold-water piezometric surface is at +20 meters and the hot-spring conduit is not isolated by mineral deposi-

tion, dilution may occur near surface. Dilution occurs on the margin of a hydrothermal system due to the relative pressure of cold groundwater over that of the hot-water system.

example shown (Fig. 1.3a), deep mixing may occur where the pressure of cold water exceeds that of the hot upflow.

Exercise: The effect of relief, through a higher or lower piezometric surface, may be gauged by adding cold water pressure curves to Figure 1.3a corresponding to higher and lower piezometric surfaces. Try it by drawing curves parallel to the reference cold-water curve in the figure.

Haas (1971) has described the limiting hydrostatic conditions for temperature as a function of depth in hydrothermal systems. The limiting condition (Fig. 1.3b) is the phase change to vapor; liquid water rising within a system boils at the phase boundary with consequent formation of a low-density vapor fraction and a decrease in temperature (for a discussion of reversible and irreversible boiling in hydrothermal systems, see Barton and Toulmin, 1961). As discussed above, <u>hydrodynamic</u> pressures prevail at depth in geothermal systems so that at, for example, 250°C the boiling-point depth is at about 400 rather than 462 meters. The effect of salinity on the boiling point-depth relation is well known, but more recently the effect of gas pressure has been recognized (Sutton and McNabb, 1977) as shown in Figure 1.3b. The latter effect makes it particularly difficult to obtain reliable depth information from estimates of temperature (e.g., from fluid inclusions) in fossil hydrothermal systems (Hedenquist and Henley, 1985b; Bodnar et al., 1985, this volume).

The distribution of springs relative to the geothermal system as a whole is evident from the field maps shown in Figure 1.2. Areas occupied by hot-water discharge seldom represent more than about 5% of the area of the hydrothermal field itself. It is also evident from these field examples that the distribution of discharges is strongly controlled by topography, the presence of faults, etc.

In general, features associated with vapor-flow from the deep system occupy higher ground. They range from fumaroles to hot springs fed by steam-heated surficial groundwaters to steaming ground which results from the boiling of steam-heated waters. The latter originate above two-phase ('boiling') zones in the deep convective system from which CO_2 and H_2S-rich vapor escapes, but are adsorbed into surficial groundwater or condensate. Where H_2S oxidation occurs due to shallow interaction with the atmosphere, low pH steam-heated waters occur which are characterized by the presence of sulfate and absence of significant chloride in solution as well as the lack of significant silica sinter around the hot spring. CO_2-rich steam-heated waters, associated with illitic alteration, are also common marginal to many fields (Mahon et al., 1980; Hedenquist and Stewart, 1985) to depths of several hundred meters; pH's are around 5 due to dissolved CO_2 and often result in extreme corrosion of

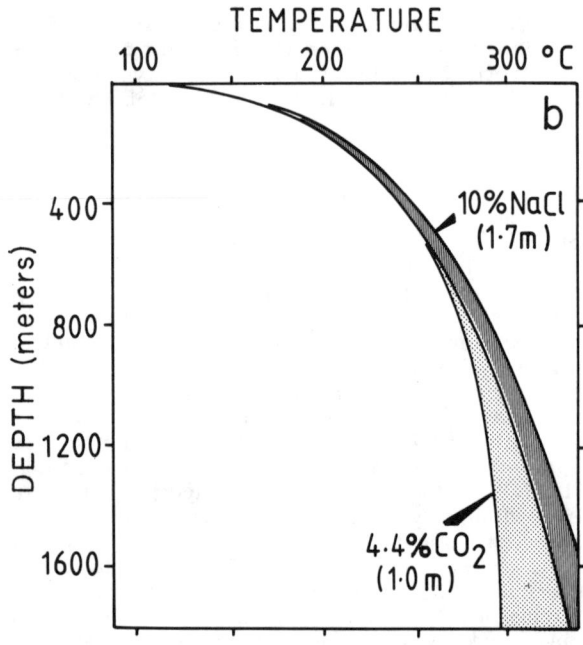

Figure 1.3b. Hydrostatic boiling-point versus depth relations of hydrothermal fluids, showing the contrasting effects of salinity and gas content. As discussed in the text, observations from active systems suggest that pressure gradients at depth are about 10% greater than hydrostatic allowing higher temperatures at shallower depth than shown here.

geothermal well casings. CO_2 exsolution following mixing of hydrothermal fluid with cool groundwater, and dissolution of the CO_2 into groundwater may be the dominant process in their formation rather than adiabatic boiling.

Note: The term "solfatara" encompasses steam discharges such as fumaroles, but is now most commonly used to refer to volcanic gas/steam discharges associated with sulfur deposition and advanced-argillic alteration. It is often used incorrectly in the discussion of metal transport in sub-sea floor systems!

As noted above, acid sulfate-chloride waters derived by condensation of volcanic gas occur at high levels in andesitic terrane and sometimes they may mix with meteoric water and accumulate to form crater lakes. Downward movement of such high-level waters is of special interest and is now well known in explored geothermal fields in the Philippines and Taiwan. At Ruapehu, New Zealand, for example, the summit crater lake (Plate 1.2) overflows discontinuously despite continuous input of volcanic vapor, meteoric water, and glacial melt. The crater lake itself has been shown to be over 300 meters deep and is occupied by an acid sulfate-chloride water with pH = 1.25 at 55°C. Presumably, the bulk of this acid fluid drains downward through the flanks of the volcano causing advanced-argillic alteration en route, and may encounter a normal hydrothermal system at depth. Silica- and iron-enriched springs occur on the flank of the volcano at lower elevations. These processes, as noted, may be responsible for the acid-sulfate type ore environments (see Hayba et al., 1985, this volume) and raises all sorts of problems with respect to terminology like hypogene and supergene!

Hydrothermal Eruption Vents

The hot springs described above are passive features of the topography, but in some cases the system itself may generate high-permeability flow paths to the surface. For example, at Waiotapu (Fig. 1.2, Plate 1.1) the largest single discharge of liquid from the deep system--the Champagne Pool--is independent of the stratigraphy and original topography and occupies the vent of a hydrothermal eruption crater formed some 900 years b.p. Such eruption vents are now known to have formed in almost all of the New Zealand geothermal systems, but are less well known elsewhere due to the frequent confusion of the eruptive products with volcaniclastic breccias which may also be common in the vicinity. Hydrothermal eruption breccias are characterized by an absence of primary volcanic material and are generally polylithic and matrix supported. Clasts have a range of alteration styles and, together with stratigraphic data, indicate an origin from depths up to about 300 meters. An origin by gas exsolution has been proposed by Henley et al. (1984) and by Hedenquist and Henley (1985a). Eruption breccias of shallower origin are also common in geothermal areas and result from the interaction of vapor with surficial groundwater or local removal of confining pressure, as appears to be the case for eruptions in Yellowstone (Muffler et al., 1971).

Heat and Mass Flow in Geothermal Systems

Table 1.2 shows heat and mass output data from some geothermal systems. These are obtained by integration of ground temperature data and physical measurements of the outflow rates and temperatures of discharging hot springs and fumaroles. In some cases an independent estimate of the upflow is obtainable using measurements of the chloride content of river water up and downstream of a geothermal field (Ellis and Wilson, 1955; Fournier et al., 1975). These data may be related to the convective upflow of high-temperature fluid in the system assuming some knowledge of the upflow temperature. For example, at Waiotapu the measured surface heat flow is 600MW(h) (600 MW(h) = 600 x 10^6 Joules/s). Exploration drilling and geochemical data suggest that the fluid feeding the field is at about 300°C. The enthalpy of steam-saturated water at 300°C is about 1350 Joules/gm (see Henley and Brown, 1985, this

Table 1.2--Summary of heat and mass flows in some New Zealand geothermal fields

Field	Total Heat Flow	Equivalent upflow
Wairakei	400 MW	350 kg/s (260°C)
Tauhara	100 MW	88 kg/s (260°C)
Waiotapu	600 MW	440 kg/s (300°C)
Ohaaki (Broadlands)	100 MW	75 kg/s (300°C)
Mokai	100 MW	75 kg/s (300°C)

volume, Fig. 2.3) so that the mass flux of 300°C fluid is obtained by

$$(600 \times 10^6)/1350 \times 10^3 = 440 \text{ gm/s}.$$

Exercise: The Champagne Pool at Waiotapu (Plate 1.1) has a discharge of about 10 kg/s of 70°C water and, we estimate, about 7 kg/s of steam. Calculate the proportion of the total convective upflow of heat and mass in the system which is discharged by this feature alone (the enthalpy of 70°C water is about 300 Joules/gm and of steam about 2600 Joules/gm).

Within the heat-flow budget the most difficult factor to assess is the proportion of the upflow which may be dissipated by subsurface groundwater flow. On the basis of the size of the field and its deep temperatures, at Broadlands (Ohaaki), the total heat flow is thought to be greater than 100 MW(h) and equivalent to >75 kg/s of chloride water, but the observed surface heat flow is less than a third of this estimate. The principal outflow of hot water from this field is the Ohaaki Pool but, at 10 kg/s, this accounts for only about 5 MW(h).

CHEMISTRY OF GEOTHERMAL DISCHARGES

Table 1.3 compares the chemistry of waters from natural features with the chemistry of deep waters encountered by deep drilling. The principal discriminating features with respect to origin have already been noted, i.e., deep fluid characterized by $Cl \gg SO_4$ and surficial steam-heated waters by $SO_4 \gg Cl$. In a given field, comparison of chloride contents in hot springs provides information about mixing processes and flow directions. Careful application of chemical geothermometer techniques may also provide some unique insights into temperature patterns in the underlying system and processes occurring during outflow (Fournier, 1981), and therefore provide an important guide for geothermal exploration.

Chemical relations between natural discharges and the deeper chloride-water system are most commonly illustrated by means of "mixing-diagrams." Figure 1.4 provides examples where two conservative quantities are compared; in this case chloride concentration and heat content (enthalpy). The latter is frequently assumed to be conservative during

Table 1.3--Summary of the chemistry of hot springs and geothermal fluids

				Concentrations in mg/kg			
Field	Feature	t°C	pH(t)	Cl	SO_4	H_2S	CO_2
Waiotapu	Champagne Pool	70	5.7	1934	103		
	Well 7	225	5.8	1026	68	65	1389
Waiotapu	Champagne Pool	98	8.0	1770	26		
	Well 80	200	6.4	1260	28	5	230
Tauhara	Crow's Nest	98	8.0	1250	55		
	Kathleen Spring	61		10	140		
	Well 1	257	6.0	1525	21	29	549
Mokai	Northern Springs	56	6.3	370	8		
	Well 3	315		1850	6	8	196
Tongonan	Banati Spring	98	8.25	3397	31		
	Well 405	260	5.4	8000	22	34	4060

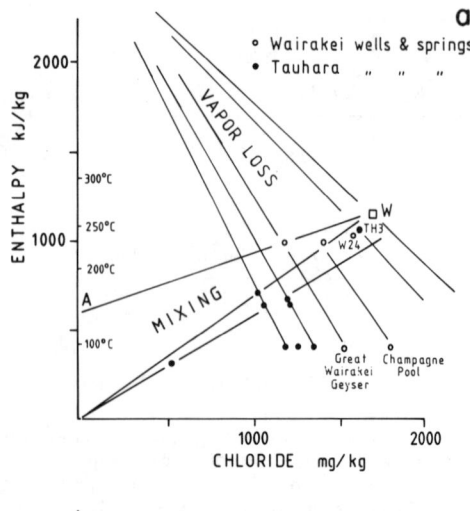

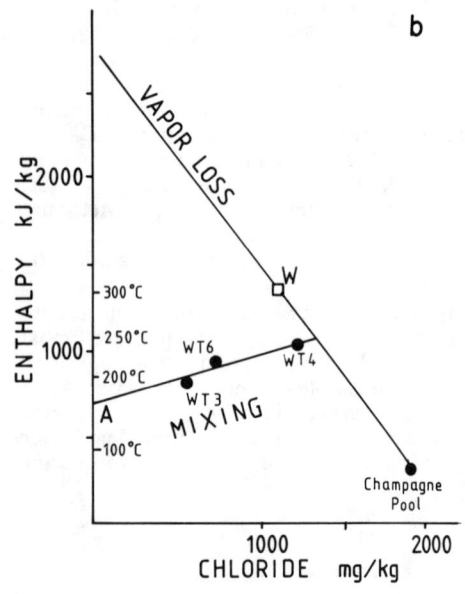

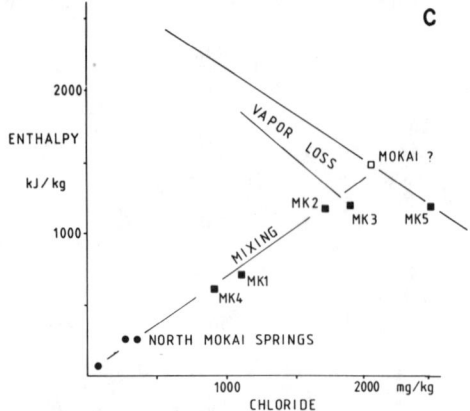

Figure 1.4a, b, and c. Fluid-mixing relations for the geothermal systems shown in Figure 1.2.

hydrothermal processes since conductive heat transfer is a minor component of the overall heat budget and systems are assumed to be in a steady-state with respect to heat and mass transfer (heat released by alteration reactions is also a minor component). The principal processes occurring are dilution (mixing) and adiabatic boiling during the irreversible expansion of the deeper system fluid as it rises and is subject to less confining pressure. As discussed above, chemical trends due to these processes are clearly shown in the diagrams which then allow the interpretation of the origin of individual thermal features. (In many cases, local mixing involves a steam- or conductively heated water with a temperature in the vicinity of 150°C.) Some of the Wairakei-Tauhara springs, for example, show evidence of dilution prior to boiling below a hot spring, while others, as at Mokai, are simply derived by dilution (Figs. 1.4a and c). The Champagne Pool at Waiotapu is an example of the direct discharge to the surface of deep fluid (Fig. 1.4b).

EPITHERMAL ORE-FORMING SYSTEMS

Requirements for Ore Deposition

The transport chemistry of the epithermal group of metals has been reviewed by Barnes (1979), Weissberg et al. (1979), and Henley and Brown (1985, this volume). In this chapter, these data are built into the geothermal system framework to provide an understanding of the origin of epithermal mineral deposits in general based on the solution chemistry of the metals and their response to the two principal processes operating in the upper levels of these systems--dilution and adiabatic boiling. It is also important to discuss two important interrelated criteria for ore deposition. These are (a) the availability of metals in solution and (b) the time required for the ore depositing system to operate.

Figures 1.5a and b show the solubilities of gold, silver, and lead (representing the base metals) as functions of temperature and ligand concentration. As shown elsewhere (Ellis, 1970; Giggenbach, 1981; Henley et al., 1984; Henley and Brown, 1985, this volume), the pH of hydrothermal fluids in active and fossil systems is buffered by fluid + alumino-silicate reactions such as the conversion of plagioclase to mica and/or clay minerals. For low salinity fluids (Cl = 1000 mg/kg), pH's are around 6.1 at 250°C (i.e., on the alkaline side of neutral pH) but about 1 pH unit more acid (5.1) for fluids an order of magnitude higher in salinity at the same temperature.

For lead the dominant dissolution reaction in chloride solutions is

$$PbS + 2H^+ + 2Cl^- = PbCl_2 + H_2S$$

so that pH, chloride concentration and H_2S content are the solubility controlling variables at a given temperature. Clearly, the higher the salinity and lower the pH, the more metal is dissolved, whereas high H_2S contents limit the solubility (Fig. 1.5a). (Note that in low chloride concentrations at near

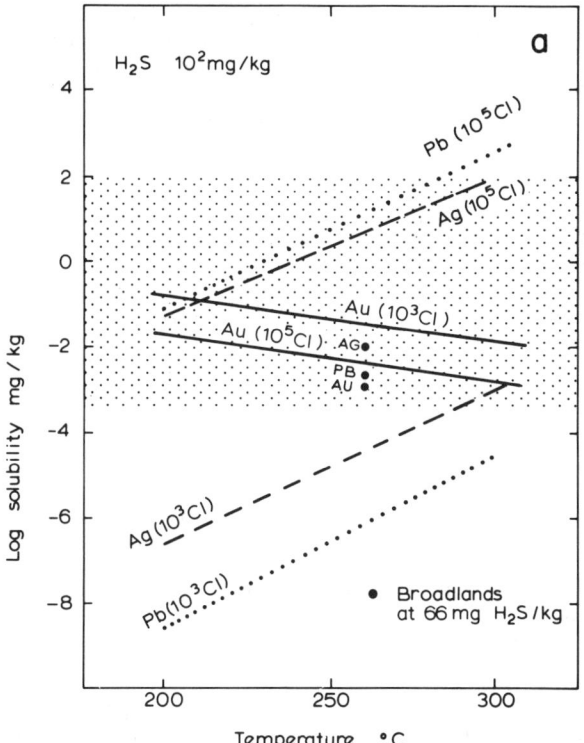

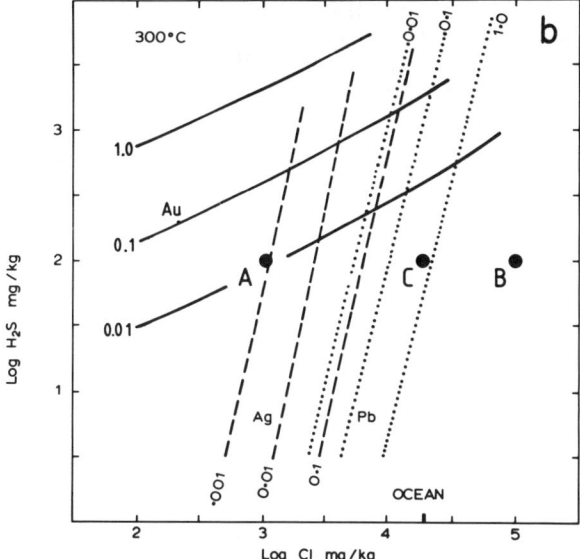

Figure 1.5. Gold, galena, and argentite solubility (mg/kg) versus (a) temperature and (b) ligand concentration for mineral-buffered hydrothermal fluids (calculated from the data summarized in Henley et al., 1984, Chapter 9). In Figures 1.5a, b, and c, the fluid is considered buffered with respect to pH by the assemblage Kmica-Kfeldspar-quartz and with respect to f_{H2} by the empirical relation for the assemblage pyrite-Fe-silicate-quartz derived by Giggenbach (1980). The pH of the fluid decreases to the right and f_{H2} increases upward. Figures 1.5a, b, and c refer to examples discussed in the text. Metal contents of the Broadlands system fluids (66 mg/kg H_2S) are shown for reference. As an exercise, convert these data to values representing solutions containing 100 mg/kg H_2S.

The slopes of the solubility curves shown relate to well-established thermodynamic data for the metal complexes considered, but relative solubilities may be in error due to the absence of reliable solubility constants for PbS and Ag_2S. In the low-salinity fluids (see text), bisulfide complexes of silver and hydroxy-carbonate complexes of lead may allow higher solubilities than calculated on the basis of chloride complexing alone.

The stippled region in Figure 1.5a emphasizes the temperature-salinity-metal concentration range of principal interest in epithermal studies.

Figure 1.5c, for 250°C, includes an estimate of the solubility of Ag_2S as $Ag(HS)_2^-$ at low salinities and as chloride complexes at higher salinity. The ordinate in the case shows total reduced sulfur rather than H_2S, and this introduces the curvature at low salinities (from Henley, 1986).

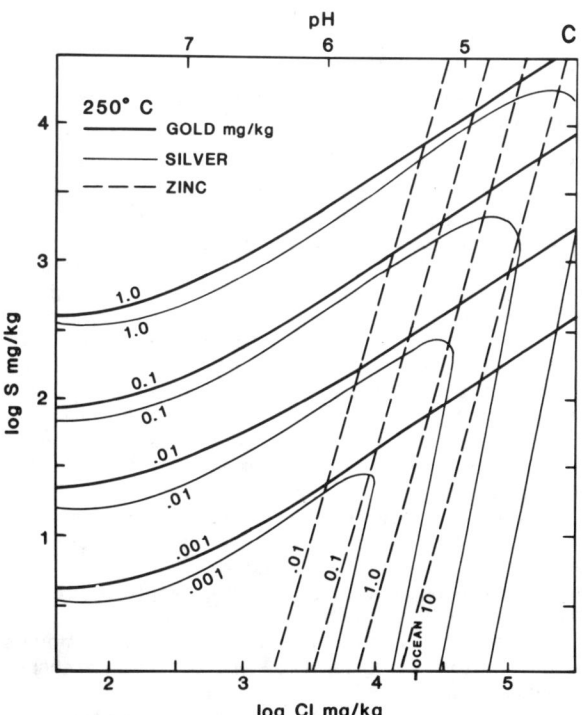

neutral pH, hydrolyzed hydroxy- or carbonate complexes may be more important than chloride complexes, but in these media the total dissolved base metal is only about 10 parts per billion, too low for the formation of base metal ores.)

For gold, transported as a bisulfide complex, the relative redox state--or f_{H_2}--of the fluid is also a controlling variable;

$$Au + 2H_2S = Au(HS)_2^- + H^+ + 0.5H_2$$

As shown by Giggenbach (1980), the ratio of H_2 to H_2S in natural hydrothermal systems appears to be constrained by iron sulfide + iron alumino-silicate reactions. The empirical relationship between H_2/H_2S and temperature may be used to obtain the redox state of the fluid for the calculation of the gold content of hydrothermal system fluids as a function of H_2S, Cl, and temperature. (Note that f_{H_2} may also be expressed as f_{O_2}--a vanishingly small number by comparison--see Henley et al., 1984, Chapter 8). For metals transported as bisulfide complexes, the relative effects of H_2S and pH are opposite to those for chloride-complexed metals. For these metals, therefore, increasing H_2S increases solubility while increasing salinity lowers the solubility because of the pH decrease due to the silicate buffer.

Using the same buffering constraints, Figure 1.5b shows the solubilities of these metals as a function of temperature at 100 mg/kg of H_2S--a typical value for gassy geothermal systems. For comparison the metal contents of fluids in the Broadlands system (66 mg H_2S/kg) are shown--see Henley and Brown, 1985, this volume. Note that solubilities range only up to 100 parts per billion but that because gold has such a high economic value compared to say lead, this range is quite sufficient to result in ore deposition.

There is strong field evidence (Brown, 1985; Henley and Brown, 1985, this volume) and some low-temperature experimental data (Barnes and Czamanske, 1968) that a bisulfide complex of silver may contribute to the solubility in low-salinity, high-gas geothermal fluids. It is clear from Figure 1.5a that its contribution must be negligible at salinities of more than about 10^4 mg/kg, where pH's are 1 to 2 units lower than in the relatively low-salinity Broadlands fluid. Its contribution is, however, important in considering silver transport in low salinity systems. The topology of the solubility of argentite at 250°C in mineral-buffered systems is explored in Figure 1.5c (Henley, 1986) based on the Broadlands field data and an assumption of $Ag(HS)_2^-$ as the dominant silver species. The solubility of sphalerite is shown on this diagram in place of galena. As a rule of thumb, its solubility is about ten times that of galena under the same conditions.

Figures 1.5a and b provide important information on the chemical nature of the hydrothermal systems responsible for ore deposition, but they also provide some basic constraints on the time required for an ore deposit to form. There is, as already indicated, a variable economic factor inherent in determining the amount and distribution of metal required for "ore", but the calculations below are well above this boundary condition.

Consider the two discharges A and B located in Figure 1.5b. A has the salinity and gas content of systems like Broadlands and Waiotapu, and B that of the Salton Sea system. For case A, with dissolved gold = 14×10^{-3} mg/kg, a single discharge to the surface at 10 kg/s would, with quantitative deposition of metals, deposit 0.13 million ounces of gold per 1000 years or, 1.3 million ounces in only 10,000 years. 10 kg/s represents only the flow of a single spring like the Waiotapu Champagne Pool, so it is evident that the time required for deposition of say 6 million ounces of gold--a respectable deposit--is only about 50,000 years. The time required to bring a typical geothermal system to a steady-state condition is also of the order 10^4 years (Henley and Ellis, 1983), so this figure may usefully be taken as the minimum time for ore deposition to occur (Fig. 1.7). White et al. (1964) have suggested that hydrothermal activity at Steamboat Springs, Nevada, was spread intermittantly over 10^6 years.

For case A (Fig. 1.5b), the relative amounts of metal which may be quantitatively deposited are as follows: gold, 40×10^6 gm (1.3 million ounces); silver, 9×10^6 gm (300,000 ounces); and lead 3000 gm. Since only chloride complexing of silver has been assumed, the ratio, Au/Ag, for the resulting deposit would be about 4.5. As noted above, field evidence specifically for this type of fluid indicates that silver bisulfide complexes dominate silver transport under most epithermal conditions, so that total gold and silver transport are of the same magnitude and the resulting Au/Ag ratio is close to 1. At these very low salinities, chloride transport of lead and zinc is insignificant and some hydroxy or bicarbonate complexing may allow transport of these metals to be two to three orders of magnitude higher--still insufficient to constitute a base-metal ore body. In this calculation a saturation of 1 is assumed for each metal, but it is clear that, based on typical (i.e., modest) discharge rates for active geothermal system features, a major gold deposit may be produced in a relatively short timespan.

For case B (Fig. 1.5b), with chloride = 10^5 mg/kg, a modest discharge for 10^4 years would produce the following deposit: 10^7 tonnes lead; 9×10^5 tonnes (3×10^4 million ounces) silver; and 1 tonne (0.03 million ounces) gold. Notice that for these high-salinity fluids, silver transport is dominated by chloride complexing because of the relative abundance of chloride over H_2S in solution and the lower pH of the more saline solution relative to case A. An important limitation appears in this case as the molal metal content approaches the molality of dissolved sulfur and there is, therefore, an interesting trade-off between the H_2S content and the mass of sulfide which may be deposited as a given salinity.

Case C in Figure 1.5 represents the Kuroko-type deposits which were formed through the seafloor venting of sea-water based geothermal systems--the Kuroko-type deposits are, afterall, simply telescoped sea-floor equivalents of the terrestrial epithermal systems! For seawater salinity, the relative amounts of metals formed during the same 10^4 year discharge history at say 10 kg/s are about 10^5 tonnes lead, 30 million ounces silver, and 0.11 million ounces gold.

Table 1.4—Summary of precious- and base-metal production or reserves from a number of geothermal deposits, natural features and the test cases discussed in the text

Deposit	Au	Ag	Pb	Au/Ag
	million	ounces	tonnes	
Carlin, Nevada	3.1	0.3	trace	344
Round Mountain, Nevada	8.7	16.2	0.01%	0.53
Goldfield, Nevada	4.3	1.5	1%	2.89
Creede, Colorado	0.15	79.3	1.5×10^5	1.9×10^{-3}
Pachuca, Mexico	6.2	1232	?	4.13×10^{-3}
Fresnillo, Mexico	0.096	290	3.2×10^{-4}	3.3×10^{-4}
Waihi, New Zealand	4.33	30	?	0.14
Kuroko	3	300	3×10^4	1×10^{-2}
Case A per 10^4 years	1.33	0.3*	3×10^{-3}	4.5*
Case B "	0.03	3×10^4	1×10^7	1.1×10^{-6}
Case C "	0.11	30	1×10^3	0.003
Ohaaki Pool	85 mg/kg	500 mg/kg	25 mg/kg	0.17
BR 22 scale	3%	30%	0.1%	0.1
Magmamax, Calif., scale		80 mg/kg	1.3%	

*As noted in the text, field data suggest that higher silver solubilities occur in low-salinity, high-H_2S solutions than indicated by chloride complexing alone; a more realistic ratio for case A is therefore about 1 as observed in the precipitate from well BR 22.

Again these figures are comparable to metal production figures for this type of deposit (Table 1.4) notwithstanding the inherent assumptions in the calculation and the single reference temperature, 300°C, employed.

It is evident from Figures 1.5a and b, again assuming quantitative deposition of metals, that the metal ratio of a given ore deposit is a function of temperature, salinity, H_2S content, chemical process, and proportion of remaining metal. This is reflected in the data of Ewers and Keays (1977) from the Broadlands field which shows a clear gradation of Au/Ag from 1 at about 200 m to 10^{-4} at 1200 m. Therefore, a precise correspondence is not to be expected in comparing the ratios calculated in cases A through C to those of exploited ore deposits (Fig. 1.6, Table 1.4). The range of Au/Ag ratios (and a suggestion of bimodality) relative to (Au + Ag) is consistent with the solution chemistry as well as the distribution of salinities for crustal fluids discussed above. These data do, however, suggest that the calculated solution compositions are broadly correct for the metal suite considered. Lead and silver production data for Creede, Colorado, and Pachuca, Mexico, are consistent with the solution model; fluid inclusion data for these deposits suggest salinities of the order 4×10^4 mg Cl$^-$/kg.

Figure 1.7 relates the assumptions of depositional efficiency and saturation to deposited ore metal and depositional time for case A. Order of magnitude changes in either of these parameters still allow ore deposition to be completed well within the lifespan estimated for the active geothermal systems, i.e., 0.3 to 1.0 million years.

The most significant discovery, which bears on the modelling of epithermal ore deposition, is that chalcopyrite scales containing 4.5 wt.-% gold and 25 wt.-% silver occur in surface pipework of discharging wells at Broadlands (Brown, 1985). These scales and other field experiments also imply that gold concentrations in the deep fluid at Broadlands are close to saturation (Brown, 1985). Prior to this, data from atmospheric pressure discharges had suggested that saturations for gold in the deep Broadlands fluids were less than 0.01 mg/kg, so that reconciliation of the chemistry of active systems and their discharge rates with the formation of significant gold deposits required either that the system host-rocks should be conveniently pre-enriched in gold--or some pretty nifty chemical footwork was necessary! This of course provided also the opportunity for a wide range of speculations concerning other complexes of gold and the role of host rocks. The new data not only confirm boiling as an extraordinarily efficient deposition process for gold and silver, but also infers that <u>all</u> geothermal system fluids are close to saturation with gold (as a bisulfide complex) and probably silver, the absolute concentrations reflecting H_2S content and salinity-pH control through deep fluid-rock interaction.

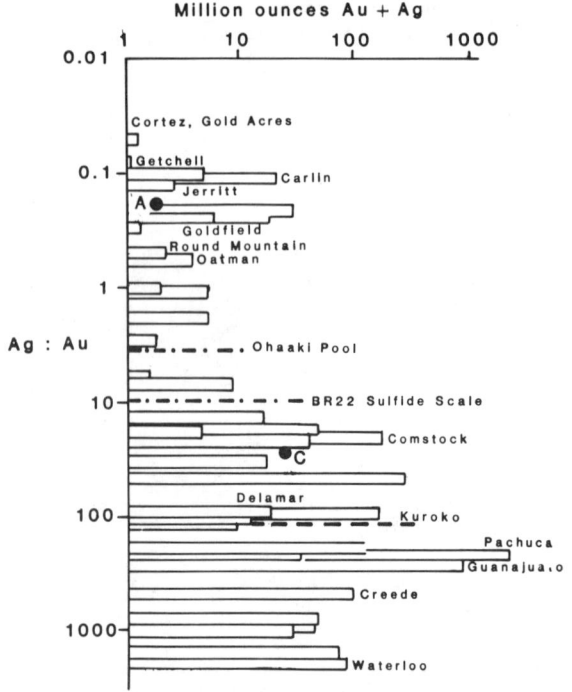

Figure 1.6. Au-Ag as a function of total gold and silver for a number of precious-metal deposits and the examples discussed in the text (modified from Graybeal, 1981). Note that the ratio shown for case A is based on chloride complexing of silver, so that a more appropriate estimate for low-salinity fluids may be gauged from the ratio observed in the Broadlands BR 22 precipitate.

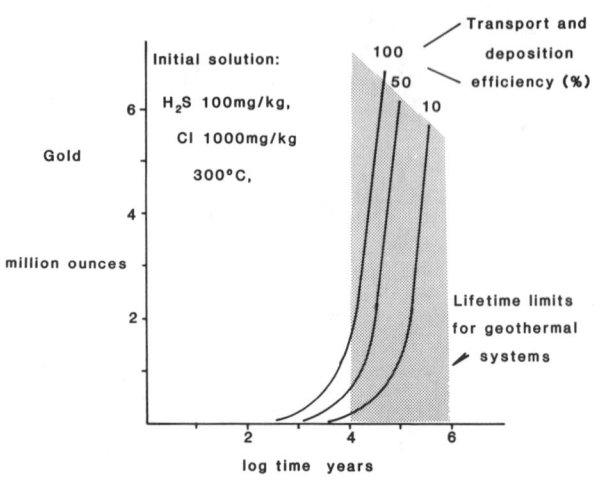

Figure 1.7. Gold deposition as a function of time, flow rate (10 kg/s) and process efficiency for a fluid initially at 300°C (see Fig. 1.5). The flow rate relates to that of a typical hot spring whereas the total flow of most geothermal systems is ten to forty times larger. How does this affect the position of the curves? Maximum and minimum lifetimes for hydrothermal systems are shown for reference (for discussion, see Henley and Ellis, 1983).

The host rocks of the Broadlands system are really quite unremarkable silicic volcanic rocks and, at depth, greywackes. Irrespective of the host rock composition, as shown above, significant gold transport and deposition can occur in relatively short time periods. The formation of an economic deposit is therefore more a function of the hydrology and chemistry of the system than it is of the availability of unusual host-rock gold contents. Availability of the metal(s) to solution is, however, a factor which may contribute to the transport and deposition efficiency of the system as a whole. In essence, for ore exploration the recognition of the source of ore metals may pale into insignificance relative to the recognition of a source for the metal-transporting ligand.

At Waiotapu, Hedenquist and Henley (1985a) showed, using appropriate estimates of the pH, f_{H_2}, and m_{H_2S} of the deep fluid, that the size of the system was such as to be able to supply all the required metal for at least 10^4 years. In that case in 10^4 years, with a constant heat and mass flux equivalent to the present, about 3.6×10^7 grams (1.2 million ounces) of gold could have been transported, but since the hydrology of the near-surface system appears to allow only one site with the focused depositional process, the overall efficiency of the hydrothermal system as a metal concentrator is quite low, around 10%, the remaining 90% being disseminated widely through the shallow boiling zones.

Chemistry of Systems Responsible for Ore Formation

It is immediately obvious from Figure 1.5a, that the prime requirement for the formation of a gold deposit is a fluid relatively high in H_2S but of low salinity (10^4 mg/kg Cl). The system temperature is of lower significance, as shown by Figure 1.5b. By contrast, silver-rich base-metal deposits require fluids of high salinity (seawater) and the effect of H_2S concentration is secondary. The temperature coefficients of solubility for these metals are large, but of less significance than salinity in determining metal transport capability.

In reviewing fluid-inclusion data from available studies on epithermal systems, Hedenquist and Henley (1985b) confirmed the validity of these salinity criteria. In many cases the low salinities of the fluids responsible for gold deposition were obscured by the presence in solution of dissolved gas, predominantly CO_2, which contributes to the freezing point of the inclusion fluid in the same way as other solutes. For example, 4 wt.-% CO_2 in solution depresses the freezing point by the same amount, $-1.7°K$, as 2.8 wt.-% NaCl. Since the molar CO_2/H_2S ratio of active geothermal-system fluids range from 10 to 100 (Giggenbach, 1980), the high CO_2 implies high H_2S as required by the solution model. Due to the scarcity of reliable gas analyses for fluid inclusions, the correlation of high-salinity, low-gas fluids with base metal-silver deposition has yet to be fully demonstrated.

Although no statistical analysis of fluid compositions is possible, it appears from the data available from well-explored geothermal systems that salinities reflect host rock and crustal setting. This is shown schematically in Figure 1.8, which makes the ad hoc assumption of normal frequency distributions for fluid compositions in different crustal environments. Salinities for basalt-hosted systems are lower than those for systems hosted by silicic volcanic and these in turn are lower than for andesite-hosted systems. Doesn't this broadly reflect the ore-host relationships seen in many districts? Fluid-inclusion data suggest that silver deposits of the Creede-type formed from fluids of even higher salinity. As discussed elsewhere (Hedenquist and Henley, 1985b), such high-salinity fluids are encountered in some terrestrial geothermal systems. Such fluids, some with extremely high salinity, occur in systems typified by those of the Imperial Valley, California, within which evaporites are present (Rex, 1983), reflecting both the tectonic setting, crustal rifts, and ambient climate. Using regional geologic data, therefore, it may be possible to discriminate hydrothermal systems, both ancient and modern, which could host silver-base metal mineralization from those potentially hosting gold, if their gas contents were high enough.

Chemical and Physical Processes in Ore Formation

If it is accepted that present-day active geothermal systems are the archetypes of those responsible for epithermal ore deposition in the past, the discussion of the physics of natural geothermal discharges (above) becomes immediately relevant to the discussion of ore-forming processes. The natural discharge of these large geothermal systems is focused on highly localized features such as hot springs whose locations reflect both topography and underlying geological structure. As shown above, these features are usually confined to an area of the order $10^5 m^2$, less than 5% of the surface area of the parent system (say $7.5 \times 10^7 m^2$). The feature common to all these discharge paths was the progressive pressure drop from that of the upflow system to the ambient pressure of near-surface aquifers or to atmospheric pressure. Boiling and dilution by near-surface waters are the accompanying processes which lead to mineral deposition along these flow paths.

Although the solubility and solution chemistry of a few metals have been outlined experimentally to an extent sufficient that their gross transport in these

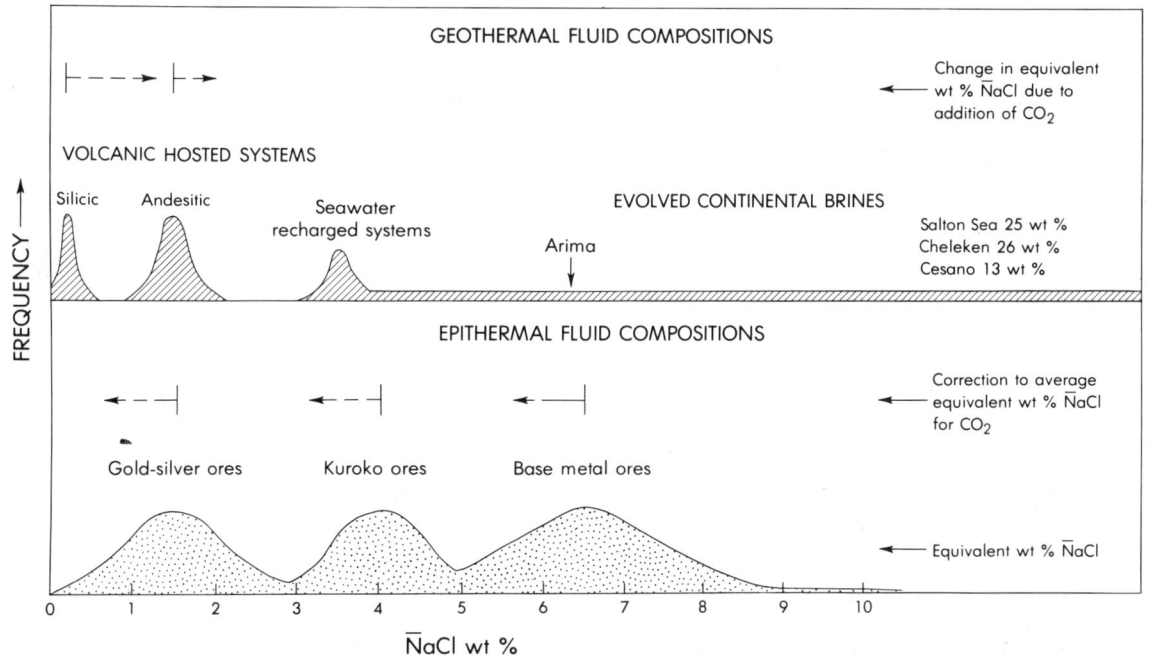

Figure 1.8. Distribution of fluid salinities in the earth's crust in relation to host-rock and crustal environment. A normal frequency distribution has been assumed for each fluid type in the absence of evidence for a continuity of compositions. For discussion, see text and Hedenquist and Henley, 1985b.

hydrothermal systems may be mapped, the understanding of ore-depositional mechanisms is far less advanced; indeed, no experimental studies have been attempted in view of inherent difficulties. Simple kinetics suggest, however, that optimum deposition rates are achieved from the most supersaturated solutions. This allows that discussion of ore deposition may concentrate on the major changes in solution chemistry consequent on boiling or dilution.

The effects of adiabatic boiling and dilution on solution chemistry have been discussed elsewhere (Ellis, 1970; Henley and Brown, 1985, this volume; Henley et al., 1984; Drummond and Ohmoto, 1985). Boiling is an especially important process because the formation of only a few percent of vapor allows the loss of more than 90% of dissolved CO_2, with a concomitant pH increase by more than one pH unit and the loss of H_2S.

Consider the effects of boiling and dilution on the case history solutions A and B discussed above. In both cases the temperature change due to these processes is taken for illustration as 50°C. Lead and silver supersaturations, attained by dilution with fresh water (with host-rock pH buffering), are 500 and 100, respectively, due to changes in temperature, pH, and chloride concentration.

At Creede, Colorado, for example, Hayba (1984) and Hayba et al. (1985, this volume) have shown that the mixing of high-salinity (7.2×10^4 mg Cl^-/kg) fluid with surficial steam-heated water was contemporaneous with ore deposition. The exsolution of CO_2 during such mixing cooling may also contribute to base-metal and silver deposition in some environments. The supersaturations attained due to adiabatic boiling ($\Delta pH=+1.5, \Delta H_2S=-80\%$) are 2000 and 300 for Pb and Ag (chloride species), respectively.

Figure 1.5a suggests that the mineral-buffered solubility of gold increases with dilution. This occurs in response to the increasing pH of the silicate-buffered solution as temperature falls and clearly gold deposition cannot be related to simple dilution. If acidic fluids are the dilutant, as may be the case in the high-level andesite-hosted systems, fluid mixing can cause gold deposition. In this case, a relative pH decrease lowers the solubility of gold allowing a paragenesis of gold-kaolinite-alunite to occur in the mixing zone between high-level acid-sulfate-chloride waters (with associated advanced-argillic alteration) and deeper near-neutral pH chloride waters (with associated propylitic alteration). The high arsenic content of the ore may also reflect this environment (see below).

In many epithermal deposits, gold is clearly associated with mineralogical indicators of boiling (open-space filling calcite, adularia, sulfides). The fluid data from Broadlands (Brown, 1985) also very clearly demonstrate the effectiveness of boiling as a process for the deposition of gold, silver, and copper, each in solution as a bisulfide complex (Henley and Brown, 1985, this volume). In this case, the initial pH increase due to CO_2 loss may undersaturate the fluid but sustained loss of the more soluble H_2S with the formation of only a few percent steam leads to supersaturation and deposition.

The loss of H_2 due to boiling gives an <u>apparent</u> increase in the redox state of the fluid (relative to the pyrite-pyrrhotite stability boundary) and also leads to an increase in solubility inside the H_2S stability field. Quantitative gold deposition may be expected if the final redox state corresponds to some point in the $SO_4^=$ stability field but this poses some severe headaches with respect to the stabilities of accessory minerals and the kinetics of $H_2S-H_2O-SO_4^=$ reactions. (For further discussion, see Henley and Brown, 1985, this volume, and Reed and Spycher, 1985, this volume). As discussed by Thorstenson (1984), modelling of the redox response of a fluid subject to such a non-equilibrium process as boiling is fraught with difficulty both in concept and in dealing with the redistribution of electrons over the large number of electroactive couples available in natural fluids.

In the discharge of well BR22 at Broadlands, gold- and silver-ore deposition is largely complete within a few seconds of boiling at the well-head throttling plate (Brown, 1985). This observation suggests that the loss of ligands (e.g., HS^- as H_2S) is most important. Reaction equations like

$$Au(HS)_2^- + 2H^+ = Au^+ + 2H_2S$$

illustrate this and emphasize that in discussing the deposition of gold under non-equilibrium conditions it is necessary to identify sources of electrons for the further reduction of aurous ions to the metal. With gold at concentration of the order of $\mu g/kg$ (3×10^{-9} molal), there are a number of possibilities because of the extremely low electronegativity of gold. The oxidation of other dissolved metals and of reduced sulfur are all possible sources. The deposition of silver in electrum proceeds in the same manner.

A role for arsenic in gold transport and deposition has often been suspected. Some experimental data suggest that arsenic complexing may increase the solubility of gold in reduced sulfur solutions, but the field data from Broadlands comparing observed gold content with the solubility calculated from bisulfide complexes suggest that this effect is minor. The high solubilities of arsenic sulfides in alkaline-sulfide solutions (Seward, 1984) suggest that thioarsenide complexes occur, but these may well be supplanted by arsenites in moderately acid-neutral pH deep, system waters. If so, thioarsenides may provide a sink for H_2S in residual, relatively high-pH waters derived by boiling. Arsenic concentrations in geothermal waters are usually in the range 1-10 mg/kg and could adequately lock up residual H_2S.

Field experiments (Brown et al., 1983) involving geothermal waters suggest that amorphous arsenic sulfide may be precipitated by acidification through reactions of the form

$$2AsS_3^{3-} + 6H^+ = As_2S_3 + 3H_2S$$

Natural examples of this process occur in the Tamagawa Hot Springs, Japan (Nakagawa, 1971) and in a number of springs in the New Zealand and Yellowstone geothermal areas. At Champagne Pool, Waiotapu, for example, the large surface area results in heat loss to the atmosphere and internal convection to a depth of about 75 meters. With a mass influx of about 20 kg/s, deep fluid entering the pool is quickly cooled and the pH buffered internally to about 5 by CO_2-HCO_3^- equilibration (Hedenquist and Henley, 1985a). A similar internally buffered process could be invoked for deposits like the sediment-hosted deposits at Getchell, Nevada.

Both in natural occurrences in New Zealand and in the geothermal field experiments, the amorphous arsenic sulfide precipitate is ore grade in gold and silver. Seward (D.S.I.R., personal communication, 1985) and Ellis (1969) have suggested that colloidal arsenic sulfide scavenges gold from solution. Recrystallization to realgar and orpiment with free gold and associated minerals such as stibnite and cinnabar may then account for the late-stage, low-grade ore assemblages occurring at Getchell and elsewhere. Together with the zonation of trace metals observed in the upper part of the Broadlands and other geothermal systems (Ewers and Keays, 1977), the association of gold with arsenic is commonly indicative of shallow depositional environments.

As noted above, the elemental association of gold and arsenic is also common in the Goldfield-type deposits, e.g., Goldfield, Nevada; Summitville, Colorado. In these environments mixing of descending high-level acid-sulfate waters with normal hydrothermal chloride-dominant waters may account for the deposition of gold and arsenic (as enargite) in association with an advanced-argillic alteration assemblage, the metals derived from the deep system.

The association of gold with organic matter has raised questions concerning gold transport and deposition in deposits such as Carlin and others in northern Nevada. Interaction of normal hydrothermal-system waters with organic-rich calcareous host rocks leads to thermal maturation of the hydrocarbon as suggested by Ilchik (1984) for the Alligator Ridge deposit and documented for the Cerro Prieto geothermal system (Barker and Elders, 1979). Little else distinguishes the possible mineralogical response of these rocks to hydrothermal alteration and it is most likely that the Nevada deposits result from sub-hot spring boiling as discussed above. Finely dispersed carbon may have a secondary role to play in the extraction of gold from solution, although even this interaction is not well supported by mineralogical data (Wells and Mullens, 1973).

Host-Rock Reactions

Whereas mixing and boiling are processes common to all hydrothermal systems, specific interactions with host rocks which may deposit ore are far less common. An exception is the interaction of relatively low-pH, high-salinity fluids with carbonate rocks which may, through loss of acidity, deposit massive replacement ores of silver and base metals (e.g., Taxco, Mexico). Interaction of a high-H_2S fluid with an iron-rich sediment may be a rare possibility for the formation of a gold-pyrite assemblage.

SUMMARY

The original question posed at the beginning of this chapter was "why study geothermal systems in the context of the origin of epithermal ore deposits?" It is clear from the foregoing paragraphs that such studies highlight a number of important factors necessary for the understanding of chemical and physical processes in epithermal systems. These may be summarized as follows:

1. The active high-temperature geothermal systems in volcanic-rock terranes are the archetypes of those systems responsible for epithermal precious- and base-metal ore deposits in analogous ancient terranes. Porphyry copper-molybdenum deposits represent crust-magma interactions at somewhat deeper levels within calc-alkalic volcanic-rock hosted systems and Kuroko-type massive sulfides are essentially sea-floor telescoped equivalents of the terrestrial epithermal deposits.

2. Studies of active geothermal systems provide insight into the physical processes governing flow to surface discharge features such as hot springs and near-surface interaction with ground and other hot waters. Geological structure provides a principal control, but in many geothermal fields hydrothermal eruptions focused at depths of about 300 meters provide focused flow paths to the surface. The characteristic hydrothermal eruption breccias formed by these events have been recognized in a number of epithermal precious-metal deposits such as Round Mountain, Nevada (Tingley and Berger, 1985), and McLaughlin, California. Figure 1.9 provides an interesting case history for the reconstruction of an epithermal system based on field observations, fluid inclusions, and isotope techniques. Such reconstructions are important to the exploration geologist for targeting drilling and for comparison with other districts under exploration where complementary but piece-meal data may be available.

3. Studies of active geothermal systems provide insight into the range of fluid compositions present in the crust. In conjunction with fluid and laboratory studies it may be shown that the formation of gold-silver (base-metal poor) epithermal deposits requires fluids of low salinity and high gas content (i.e., H_2S in association with CO_2), whereas silver-rich base-metal deposits (relatively low in gold) require high-salinity, low-gas fluids in their respective hydrothermal systems.

The salinities of hydrothermal fluids reflect their volcanic-tectonic environment. For example, andesite-hosted hydrothermal systems are, perhaps through high-level interaction with volcanic gases or the involvement of deep, connate waters, more saline than those of silicic volcanic-rock terranes, but less saline than those encountered where evaporite sequences occur (grabens, some caldera moats); the latter reflecting also a climatic control. Although acid-sulfate waters are locally encountered in silicic-rock hosted systems, they are quite abundant in high-level andesitic-rock terranes, and may be important components in the formation of certain types of gold deposits.

Such studies also highlight the important role of

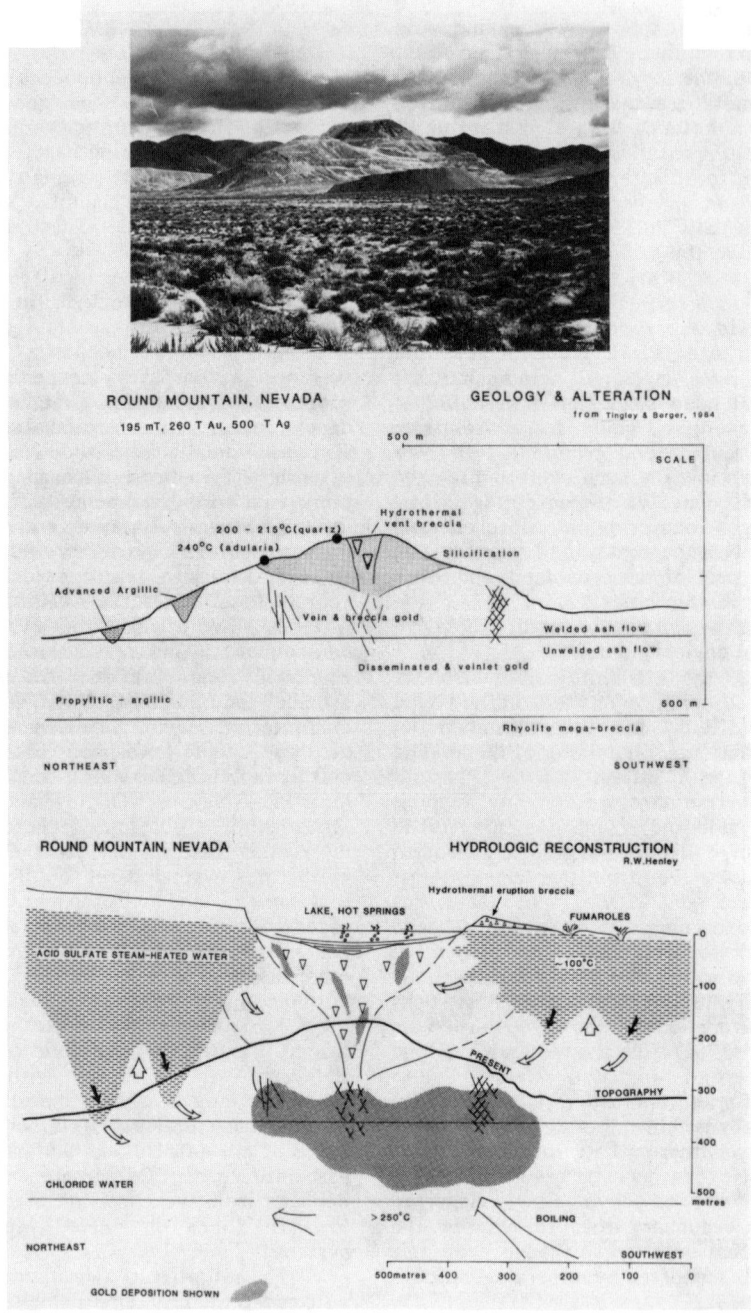

Figure 1.9a, b, and c. Distribution of alteration types and fluid-inclusion data for the Round Mountain gold deposit, Nevada. Data from Tingley and Berger, 1985. The topographic outline shown is drawn from the viewpoint shown in a).

Hypothetical reconstruction of the Round Mountain hydrothermal system (25 million years b.p.) emphasizes the dissection of the system by erosion. Much controversy has centered on the so-called hot-spring origin of this deposit; clearly the deposit is hot-spring related but is not itself a hot spring! There seems to be little advantage to the use of the term "hot-spring-type deposit" over the generally applicable "epithermal" unless very specific evidence is available for a given deposit relating it to a hot-spring environment (e.g., silica sinter, eruption breccia) as at McLaughlin, California. The hot-spring terminology may indeed be far too restrictive for the guidance of exploration programs even though the recognition of hot-spring features is a very important tool for the exploration geologist.

gas composition in controlling the physical processes within systems and their metal transport capability, as well as demonstrating that two processes (adiabatic boiling and mixing) provide the principal controls on fluid chemistry in the epithermal environment. Access to some gas source (CO_2 + H_2S) becomes, therefore, paramount in determining whether a given low-salinity system may or may not develop a gold deposit. As discussed elsewhere (Henley, 1986), there may be tectonic controls here (e.g., access to gas from subduction zone metamorphism?) which introduce the link between tectonic setting, fluid chemistry, and ore formation, which together provide guides to ore search.

4. Active geothermal systems provide the opportunity for studying the deposition of trace metals such as gold and the initial concentrations of ore-forming elements in the deep system. Geothermal systems such as Broadlands show that metal concentrations may be regarded as close to saturation and demonstrate that host rocks play only a passive role in determining whether systems are likely to deposit ore near the surface. The principal control on the latter is the provision of permeable features within which flow may be focused and adiabatic boiling and/or dilution may occur. The kinetics of metal deposition during irreversible boiling are not well understood and hard to model realistically, but pH and H_2S concentration appear to be more important than relatively slow redox reactions. More specialized environments involving gas buffering of pH or mixing with acid waters may be indicated by the association of gold with orpiment-realgar or enargite-alunite assemblages.

EPILOGUE

If physical and chemical processes alone are the control on whether or not a system may deposit an ore body, what chance does the exploration geologist have in targeting exploration effectively? The answer is in three parts:

(a) Since active geothermal systems are relatively similar in terms of the distribution of alteration minerals, these assemblages may be used to provide key data on the level of exposure within a system; clays and chlorites, for example, are quite sensitive indicators of temperature of formation (Browne, 1978; Henley and Ellis, 1983). Recognition of other depth indicators such as hydrothermal eruption or vent breccias is also a powerful targeting tool. Similarly the recognition of trace-element enrichments (As, Sb, Hg, Tl) provides a clue to structural level in the system.

(b) Good "old-fashioned" structural studies have been much neglected in recent decades, but they probably still provide the most effective means of targeting drilling since structure, particularly fracture analysis provides the main clue to flow structure in the fossil system.

Knowledge of the hydrodynamic characteristics common to active geothermal systems also provides a guide to the exploration of fossil systems. Additional factors to remember are the effects of contemporary or subsequent volcanism and tectonism which obscure and dismember the original structure. A useful exercise (or, in modern parlance, "thought experiment") is to imaginatively modify copies of Figure 1.1 by inclusion of layers of volcanic material and/or fault dissection to explore some of the problems which arise in reconstruction based on geological evidence. In tackling this exercise, remember the three-dimensional aspects of the systems.

(c) Discriminant analysis is a term used in statistics to describe a method for classifying multivariate observations into groups. In the above paragraphs, the variability of surface expression and chemistry has been stressed but as noted by Koch and Link (1970, p. 326) this is rather pointless if no ways are established to sort out these kinds of variability. This is the essence of the design of scientific experiments, but exploration for epithermal ore deposits is seldom effected in this manner, is it? Since the coincidence of a suitable flow structure, fracture pattern and system chemistry is responsible for ore-metal deposition rather than dispersion, the probability of exploration success is the factor requiring early determination in an exploration program. Factors such as a previous mining history are obvious high-score factors as is recognition of structural style. Modern techniques such as isotope analysis and fluid-inclusion studies should also be used in a discriminatory manner; e.g., fluid inclusions may very early in a project determine whether a system is relatively dilute and therefore not likely to host base metal-silver ore or relatively gassy and therefore quite likely to have transported and perhaps deposited gold. Understanding of the geochemistry of metal transport and deposition may be used in the same way.

ACKNOWLEDGMENTS

Discussions with K. L. Brown, R. G. Allis, P. B. Barton, and P. M. Bethke have been valuable in preparing this overview. The musical accompaniment of James Galway was especially soothing to the writer and may well provide solace to the reader!

REFERENCES

Allis, R. G., Henley, R. W., and Carman, A. F., 1979, The thermal regime beneath the Southern Alps; in Walcott, R., and Cresswell, M. (eds.), Origin of the Southern Alps: Royal Society of New Zealand Bulletin 18, p. 79-85.

Barker, C. E., and Elders, W. A., 1979, Vitrinite reflectance geothermometry in the Cerro Prieto geothermal field, Baja California, Mexico: Geothermal Resources Council Transactions, v. 3, p. 27-30.

Barnes, H. L., 1979, Solubilities of ore minerals; in Barnes, H. L. (ed.), Geochemistry of Hydrothermal Ore Deposits: 2nd ed., John Wiley and Sons, New York, p. 404-460.

Barnes, H. L., and Czamanske, G. K., 1968, Solubilities and transport of ore minerals; in Barnes, H. L.

(ed.), Geochemistry of Hydrothermal Ore Deposits: 1st ed., Holt, Rinehart, and Winston, New York, p. 334-381.

Barton, P. B., Jr., and Toulmin, P., 1961, Some mechanisms for cooling hydrothermal fluids: U.S. Geological Survey, Professional Paper, 424-D, p. 348-352.

Batini, F., Bertini, G., Gianelli, G., Pandeli, E., and Puxeddu, M., 1983a, San Pompeo 2 Deep Well--A high-temperature and high-pressure geothermal system: Proceedings 3rd International Seminar, Results of EC Research and Demonstration Projects in the Field of Geothermal Energy: European Geothermal Update, Extended Summaries, European Patent Office, Munich, p. 341-353.

Batini, F., Bertini, G., Gianelli, G., Pandeli, E., and Puxeddu, M., 1983b, Deep structure of the Larderello Field--contribution from recent geophysical and geological data: Memoir, Society of Geology, Italy, v. 27 (in press).

Belkin, H., Vivo, B. D., Gianelli, G., and Lattanzi, P., 1983, Fluid inclusion reconnaissance study of hydrothermal minerals from geothermal fields of Tuscany (Italy): Extended Abstracts, 4th International Symposium on Water Rock Interaction, Misasa, Japan, p. 43-47.

Berger, B. R., and Eimon, P., 1983, Conceptual models of epithermal precious-metals deposits; in Shanks, W. C., III (ed.), Cameron Volume on Unconventional Mineral Deposits: Society of Mining Engineer, A.I.M.E., p. 191-205.

Bodnar, R. J., Reynolds, T. J., and Kuehn, C. A., 1985, Fluid inclusion systematics in epithermal systems; in Berger, B. R., and Bethke, P. M. (eds.), Geology and Geochemistry of Epithermal Systems: Society of Economic Geologists, Reviews in Economic Geology, v. 2.

Brown, K. L., 1985, Gold deposition from New Zealand geothermal wells: Economic Geology (in press).

Brown, K. L., McDowell, G. D., Lichti, K. A., and Bijnen, H. J. C., 1983, pH control of silica scaling: Proceedings 5th New Zealand Geothermal Workshop, p. 157-162.

Browne, P. R. L., 1978, Hydrothermal alteration in active geothermal fields: Annual Review of Earth and Planetary Sciences, v. 6, p. 229-250.

Casadevall, T. J., dela Cruz-Reyna, S., Rose, W. I., Bagley, S., Finnegan, D. L., and Zoller, W. H., 1984, Crater Lake and post-eruption hydrothermal activity: Journal of Volcanology and Geothermal Research, v. 23, p. 169-191.

Cathles, L. M., 1977, An analysis of the cooling of intrusives by groundwater convection which includes boiling: Economic Geology, v. 72, p. 804-826.

Drummond, S. E., and Ohmoto, H., 1985, Chemical evolution and mineral deposition in boiling hydrothermal systems: Economic Geology, v. 80, p. 126-147.

Elder, J. W., 1966, Hydrothermal systems, heat and mass transfer in the earth: New Zealand D.S.I.R. Bulletin 169, 115 p.

Ellis, A. J., 1969, Present-day hydrothermal systems and mineral deposition: Ninth Common Mining and Metallurgy Congress, Institute of Mining and Metallurgy (London) 1-30.

Ellis, A. J., 1970, Quantitative interpretation of chemical characteristics of hydrothermal systems: Proceedings from the United Nations Symposium on the Development and Utilization of Geothermal Resources, v. 2, p. 516-528.

Ellis, A. J., and Mahon, W. A. J., 1977, Chemistry and Geothermal Systems: Academic Press, New York, 392 p.

Ellis, A. J., and Wilson, S. H., 1955, The heat from the Wairakei-Taupo thermal region calculated from the chloride output: New Zealand Journal of Science and Technology, Section B., v. 36, (6), p. 622-631.

Ewers, G. R., and Keays, R. R., 1977, Volatile and precious-metal zoning in the Broadlands geothermal field, New Zealand: Economic Geology, v. 72, p. 1337-1354.

Fournier, R. O., 1981, Application of water geochemistry to geothermal exploration and reservoir engineering; in Rybach, L., and Muffler, L. P. J. (eds.), Geothermal Systems: Principles and Case Histories: John Wiley and Sons, New York, p. 109-143.

Fournier, R. O., White, D. E., and Truesdell, A. H., 1975, Convective heat flow in Yellowstone National Park: Second United Nations Symposium on the Development and Use of Geothermal Resources, San Francisco, v. 1, p. 731-749.

Fyfe, W. S., and Kerrich, R., 1984, Gold: natural concentration processes; in Foster, R. P. (ed.), Gold '82, The Geology, Geochemistry, and Genesis of Gold Deposits, Geological Society of Zimbabwe Special Publication No. 1, p. 99-128.

Giggenbach, W. F., 1974, The chemistry of Crater Lake, Mt. Ruapehu (New Zealand) during and after the 1971 active period: New Zealand Journal of Science, v. 17, p. 33-45.

Giggenbach, W. F., 1980, Geothermal-gas equilibria: Geochimica et Cosmochimica Acta, v. 44, p. 2021-2032.

Giggenbach, W. F., 1981, Geothermal-mineral equilibria: Geochimica et Cosmochimica Acta, v. 45, p. 393-410.

Giggenbach, W. F., Geonfiantini R., Jangi, B. L., and Truesdell, A. H., 1983, Isotopic and chemical composition of Parbati Valley geothermal discharges, N.W. Himalaya, India: Geothermics, v. 12, p. 199-222.

Grant, M. A., Donaldson, I. A., and Bixley, P. F., 1982, Geothermal Reservoir Engineering: Academic Press, New York, 369 p.

Graybeal, F. T., 1981, Characteristics of disseminated silver deposits in the Western United States; in Dickinson, W. R., and Payne, W. D. (eds.), Relations of Tectonics to Ore Deposits in the Southern Cordillera: Arizona Geological Society Digest, v. 14, p. 271-281.

Haas, J. L., Jr., 1971, The effect of salinity on the maximum thermal gradients of a hydrothermal system at hydrostatic pressure: Economic Geology, v. 66, p. 940-946.

Hanaoka, N., 1980, Numerical model experiment of

hydrothermal system-topographic effects: Geological Survey Bulletin, Japan, v. 31 (7), p. 321-332.
Hayba, D. O., 1984, Documentation of thermal and salinity gradients and interpretation of the hydrologic conditions in the OH vein, Creede, Colorado: Geological Society of America, Abstracts with programs, v. 16, p. 534.
Kyosu, Y., and Kurahashi, M., 1984, Isotopic geochemistry of acid thermal waters and volcanic gases from Zao volcano, Japan: Journal of Volcanology and Geothermal Research, v. 21, p. 313-332.
Koch, G. S., and Link, R. F., 1970, Statistical Analysis of Geological Data: John Wiley and Sons, New York, 438 p.
Lindgren, W., 1933, Mineral Deposits: McGraw-Hill Book Company, New York, Fourth Edition, 930 p.
Mahon, W. A. J., Klyen, L. E., and Rhode, M., 1980, Natural sodium bicarbonate sulphate hot waters in geothermal systems: Chinetsu (Journal Japan Geothermal Energy Association), v. 17, p. 11-24.
Muffler, L. J. P., White, D. E., and Truesdell, A. H., 1971, Hydrothermal explosion craters in Yellowstone National Park: Geological Society of America Bulletin 82, p. 723-740.
Nakagawa, R., 1971, Solubility of orpiment (As_2S_3) in Tamagawa Hot Springs, Akita Prefecture: Nippon Kagaku Zasshi, v. 92, p. 154-159.
Ozima, M., Takayanagi, M., Zashu, S., and Amari, S., 1984, High $^3He/^4He$ ratio in ocean sediments, Nature, v. 311, p. 448-450.
Puxeddu, M., 1984, Structure and Late Cenozoic evolution of the upper lithosphere in southwest Tuscany (Italy): Tectonophysics, v. 101, p. 357-382.
Ransome, F. L., 1909, The geology and ore deposits of Goldfield, Nevada: U.S. Geological Survey, Professional Paper 66, 258 p.
Reed, M. H., and Spycher, N. F., 1985, Boiling, cooling, and oxidation in epithermal systems: Numerical modeling approach; in Berger, B. R., and Bethke, P. M. (eds.), Geology and Geochemistry of Epithermal Systems: Society of Economic Geologists, Reviews in Economic Geology, v. 2.
Rex, R. W., 1983, Origin of the brines of the Imperial Valley, California: Annual Meeting, Geological Society of America, Program with Abstracts, Indianapolis.
Roedder, E., 1984, Fluid inclusion evidence bearing on the environments of gold deposition; in Foster, R. P. (ed.), Gold '82, The Geology, Geochemistry and Genesis of Gold Deposits: Geological Society of Zimbabwe Special Publication, no 1, p. 129-164.
Seward, T. M., 1984, The transport and deposition of gold in hydrothermal systems; in Foster, R. P. (ed.), Gold '82, The Geology, Geochemistry and Genesis of Gold Deposits: Geological Society of Zimbabwe Special Publication, no. 1, p. 165-181.
Sillitoe, R. H., 1973, The tops and bottoms of porphyry copper deposits: Economic Geology, v. 68, p. 799-815.
Sillitoe, R. H., 1983, Enargite-bearing massive sulfide deposits high in porphyry copper systems: Economic Geology, v. 78, p. 348-352.
Sternfeld, J. N., 1981, Hydrothermal petrology and stable isotope geochemistry of two wells in the Geysers geothermal field, Sonoma County, California: Unpublished M.S. thesis, University of California (Riverside), 202 p.
Hayba, D. O., Bethke, P. M., Heald, P., and Foley, N. K., 1985, Geologic, mineralogic, and geochemical characteristics of volcanic-hosted epithermal precious-metal deposits; in Berger, B. R., and Bethke, P. M. (eds.), Geology and Geochemistry of Epithermal Systems: Society of Economic Geologists, Reviews in Economic Geology, v. 2.
Healy, J., and Hochstein, M. P., 1973, Horizontal flows in geothermal systems: Journal of Hydrology (New Zealand), v. 21, p. 71-82.
Hedenquist, J. W., and Henley, R. W., 1985a, Hydrothermal eruptions in the Waiotapu geothermal system, New Zealand. Origin, breccia deposits and effect on precious-metal mineralization: Economic Geology, v. 80, p. 1640-1668.
Hedenquist, J. W., and Henley, R. W., 1985b, Effect of CO_2 on freezing-point depression measurements of fluid inclusions--evidence from active systems and application to epithermal studies: Economic Geology, v. 80, p. 1379-1406.
Hedenquist, J. W., and Stewart, M. K., 1985, Natural CO_2-rich steam-heated waters in the Broadlands-Ohaaki geothermal system, New Zealand. Their chemistry, distribution and corrosive nature: Geothermal Resources Council, Transactions, v. 9 (2), p. 247-250.
Henley, R. W., 1986, Ore transport and deposition in epithermal environments; in Herbert, H. (ed.), Stable Isotopes and Fluid Processes in Mineralization: Geological Society of Australia Special Issue (in press).
Henley, R. W., and Brown, K. L., 1985, A practical guide to the chemistry of geothermal and epithermal systems; in Berger, B. R., and Bethke, P. M., Geology and Geochemistry of Epithermal Systems: Society of Economic Geology, Reviews in Economic Geology, v. 2.
Henley, R. W., and Ellis, A. J., 1983, Geothermal systems, ancient and modern: Earth Science Reviews, v. 19, p. 1-50.
Henley, R. W., and McNabb, A., 1978, Magmatic vapor plumes and ground water interaction in porphyry copper emplacement: Economic Geology, v. 73, p. 1-20.
Henley, R. W., Norris, R. J., and Paterson, C. J., 1976, Multistage ore genesis in the New Zealand geosyncline--a history of post-metamorphic lode emplacement: Mineralium Deposita, v. 11, p. 180-196.
Henley, R. W., Truesdell, A. H., and Barton, P. B., Jr., 1984, Fluid-Mineral Equilibria in Hydrothermal Systems: Society of Economic Geologists, Reviews in Economic Geology, v. 1, 267 p.
Ilchik, R. P., 1984, Hydrothermal maturation of indigenous organic matter at the Alligator Ridge gold deposits, Nevada: Unpublished M.S. thesis, University of California (Berkeley), 77 p.

Kieffer, S. W., 1984, Seismicity of Old Faithful Geyser--an isolated source of geothermal noise and possible analogue to volcanic seismicity: Journal of Volcanology and Geothermal Research, v. 22, p. 59-96.

Sutton, F. M., and McNabb, A., 1977, Boiling curves at Broadlands geothermal field, New Zealand: New Zealand Journal of Science, no. 20, p. 333-337.

Thorstenson, D. G., 1984, The concept of electron activity and its relation to redox potentials in aqueous geochemical systems: U.S. Geological Survey, Open-File Report 84-072, 67 p.

Tingley, J. V., and Berger, B. R., 1985, Lode gold deposits of Round Mountain, Nevada: Nevada Bureau of Mines and Geology Bulletin 100, 62 p.

Weissberg, B. G., 1969, Gold-silver ore-grade precipitates from New Zealand thermal waters: Economic Geology, v. 64, p. 95-108.

Weissberg, B. G., Browne, P. R. L., and Seward, T. M., 1979, Ore metals in active geothermal systems; in Barnes, H. L. (ed.), Geochemistry of Hydrothermal Ore Deposits, 2nd ed.: John Wiley and Sons, New York, p. 738-780.

Wells, J. D., and Mullens, T. E., 1973, Gold-bearing arsenian pyrite determined by microprobe analysis, Cortez and Carlin gold mines, Nevada: Economic Geology, v. 68, p. 187-201.

White, D. E., 1981, Active geothermal systems and hydrothermal ore deposits: Economic Geology, 75th anniversary volume, p. 392-423.

White, D. E., Muffler, L. J. P., and Trusdell, A. H., 1971, Vapor-dominated hydrothermal systems compared with hot-water systems: Economic Geology, v. 66, p. 75-97.

White, D. E., Thompson, G. A., and Sandberg, C. H., 1964, The rocks, structure, and geologic history of the Steamboat Springs thermal area, Washoe County, Nevada: U.S. Geological Survey, Professional Paper 458-B.

White, D. E., 1955, Thermal springs and epithermal ore deposits; in Bateman, A. M. (ed.): Economic Geology, 50th anniverary volume, p. 99-154.

Chapter 2
A PRACTICAL GUIDE TO THE THERMODYNAMICS OF GEOTHERMAL FLUIDS AND HYDROTHERMAL ORE DEPOSITS

R. W. Henley and K. L. Brown

INTRODUCTION

In trying to understand the depositional processes which led to ore deposition in fossil hydrothermal systems, we attempt to reconstruct the chemistry of the fluid phase from observation of its relics (e.g., alteration minerals, fluid inclusions). We may also attempt to thermodynamically model the chemical changes experienced by this fluid as it passes upward through a vein, vents to the seafloor, boils or mixes with other waters, etc. A number of important assumptions are made; one is the assumption of equilibrium and another is that the thermodynamic data base is sound.

Analyses of fluids discharged from geothermal wells, together with drill-core data, allow the opportunity to independently check the validity of the thermodynamic data base and to observe directly, chemical processes leading to the deposition of gold, base-metal sulfides and common gangue minerals like quartz and calcite. The calculations involved are not trivial, but are essential to the understanding of epithermal or any other type of hydrothermal ore deposit.

To illustrate these procedures, we shall examine the discharge of <u>one</u> production well in the Broadlands geothermal field in New Zealand. Through the use of thermodynamics, the amount of information we shall retrieve about the reservoir and depositional processes is quite astonishing. We shall then turn to some review questions to consider implications for the formation of some hydrothermal ore deposits.

In this chapter we have tried to follow a pragmatic course, avoiding the temptation to overindulge in the (essential) nuances of thermochemistry at the expense of the proscribed goal--a practical understanding of hydrothermal processes. A wider discussion of geothermal chemistry, a broader data base, and references to the current literature are presented in Henley et al. (1984).

GEOLOGICAL CHARACTERISTICS OF THE BROADLANDS GEOTHERMAL SYSTEM

The geology and alteration mineralogy of the Broadlands field have been recently reviewed by Weissberg et al. (1979). Figure 2.1 shows the location of exploration and production wells in the Broadlands field and Figure 2.2 shows in cross-section the structure and stratigraphy of the field in relation to measured subsurface temperatures. The host rocks for the hydrothermal system are mostly silicic ash-flow originally containing quartz and andesine with a glassy or fine grained groundmass and minor hornblende, biotite, magnetite and other common accessories. The geology and alteration mineralogy of cores from several wells are summarized in Table 2.1. As well as ubiquitous pyrite and lesser pyrrhotite, copper, lead and zinc sulfides have been recognized in veins and vugs in core from some wells, and are commonly associated with adularia and calcite. At the surface in the Ohaaki Pool, intermittent deposition of an amorphous sulfide precipitate has occurred on the silica sinter formed at the edge of the pool. This orange precipitate is ore grade with respect to gold and silver and rich in As, Sb, Tl, and Hg. A similar

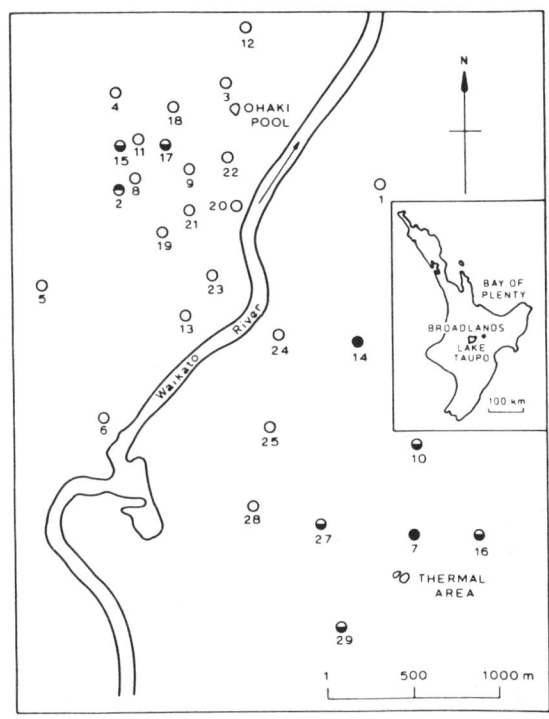

Figure 2.1. Location of geothermal wells and hot springs at Broadlands New Zealand. Wells marked, ◐ , intersected base-metal sulfide minerals at depth; wells marked, ◑ , deposited precious-metal-bearing, antimony-rich precipitates at the surface; and wells marked, ● , have both surface precipitates and base-metal sulfides at depth (reproduced with permission from Weissberg et al., 1977).

Table 2.1--Distribution of base-metal sulfides in Broadlands drill holes
(reproduced from Weissberg et al., 1979)

Well No.	Depth of well (m)	Base-metal sulfide zone				Rock types in base-metal sulfide zone	Associated hydrothermal minerals
		Depth range (m)		Temperature range (°C)			
		Minimum	Maximum	Minimum	Maximum		
7	1125	802	925	272	276	rhyolite tuff, greywacke	quartz, adularia, albite, calcite, pyrite, illite, chlorite
10	1092	954	1045	265	275	conglomerate, greywacke	quartz, adularia, chlorite, illite
14	1289	1234	1235	294	294	argillite, conglomerate	quartz, adularia, chlorite, illite, pyrite
15	2421	1620	2314	286	298	ignimbrite, tuff	calcite, quartz, chlorite, adularia, albite, pyrite
16	1406	280	1393	120	276	Broadlands Dacite, Rautawiri Breccia, greywacke	quartz, illite, chlorite, calcite, pyrite, pyrrhotite
17	1084	607	609	273	273	Upper Waiora pumiceous tuff breccia	calcite, illite, wairakite, quartz, pyrite
27	1158	1031	1033	280	280	tuff	quartz, pyrite, sphene, zoisite
29	1027	914	994	265	265	tuff, tuffaceous sandstone	illite, pyrite, chlorite, sphene

precipitate formed in the discharge disposal channels of a number of exploration wells. The depth zonation indicated by these observations is also reflected in chemical analyses of sulfide separates (Ewers and Keays, 1977) and recalls the general pattern of metal zonation encountered in many epithermal ore deposits.

FLUID CHEMISTRY

Table 2.2 contains analyses of the liquid water and steam separated from the discharge of BR22 at Broadlands. The discharge from this well is fairly typical, although higher temperature zones are encountered at depth elsewhere in the field.

Exercise 1. For later convenience, complete the conversion of the liquid sample analysis to molal units. (n.b. Σ indicates analytical totals for dissociated species; e.g.

$$\Sigma NH_3 = NH_3 + NH_4^+)$$

In order to examine chemical relationships between the reservoir mineral assemblage and the fluid phase, the data must first be recombined to the single high-temperature and pressure phase which fed the well. For this purpose a <u>heat balance</u> equation is used (with the assumption that negligible heat loss occurs) to calculate the mass of liquid water converted to vapor as the fluid rises, at several m/sec to the wellhead.

Table 2.2--Analytical data for fluids discharged from well BR22, Broadlands, New Zealand

Liquid sample

Collection Pressure	pH$_{20}$	Na	K	Ca	Mg	Cl	B	SO$_4$	SiO$_2$	ΣNH$_3$	ΣHCO$_3$	
11 b.a.	7.39	870	188	1.2	0.1	1432	42	2	705	6.1	216	mg/kg
									11.75			10^3 x moles kg^{-1}

Steam sample

Collection Pressure	CO$_2$	H$_2$S	NH$_3$	H$_2$	CH$_4$	N$_2$	
11 b.a.	534.3	10.3	2.6	0.57	6.6	5.0	10^3 x moles kg^{-1}

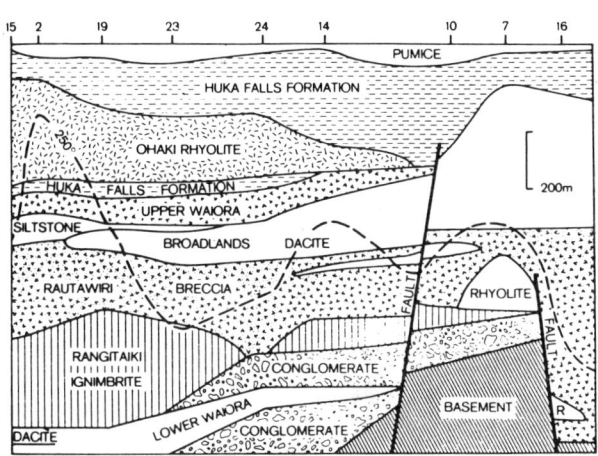

Figure 2.2. Geologic cross-section of the Broadlands, New Zealand, geothermal field (reproduced with permission from Weissberg et al., 1977).

$$H_{TD} = H_l(1 - y) + H_v y \quad (1)$$

where H is the enthalpy of the total discharge (T.D.), and of liquid water (l) and vapor (v) at the separation pressure; y is the steam fraction at the sampling pressure. [Note that 100 y is the percent of steam by mass in the two-phase mixture.]

H_{TD} is determined by physical measurement and for this well is 1160 (±20) J/g. Substituting appropriate values of H_l and H_v from tables of thermodynamic data for water (see Henley et al., 1984, Appendix III), equation (1) gives $y = 0.19$.

Digression No. 1

For convenience in this chapter, Figure 2.3 shows steam fractions for pure water as a function of enthalpy, pressure and temperature. To use this diagram locate the discharge enthalpy on the two-phase coexistence curve--this point represents conditions in the reservoir. Then project downward at constant enthalpy to the sampling pressure and read off the steam fraction present under the reduced pressure conditions. This procedure, or more precisely the heat balance equation, is the basis for the calculation of changes in chemistry and temperature during boiling (adiabatic, closed system) of hydrothermal fluids in an ore zone (for example, see Fournier (1985, this volume) and Reed and Spycher (1985, this volume)). Heat and mass balance equations are also the basis for similar calculations involving the mixing of fluids.

A <u>mass balance</u> equation is required to recombine the analytical data. For a component, X

$$X_{TD} = (1 - y)X_l + yX_v \quad (2)$$

so that, for example

$$SiO_{2,TD} = 0.81 \times 705 \text{ mg/kg}$$

and converting to molality by dividing by the molecular weight of SiO$_2$

$$SiO_{2,TD} = 9.52 \times 10^{-3} \text{ moles/kg}$$

where a volatile component such as CO$_2$ (534.3 x 10^{-3} moles/kg in the vapor in this example) is considered

$$CO_{2,TD} = 0.81 \times m_{HCO_3} + 0.19 \times 534.3 \times 10^{-3} \text{ moles/kg}$$

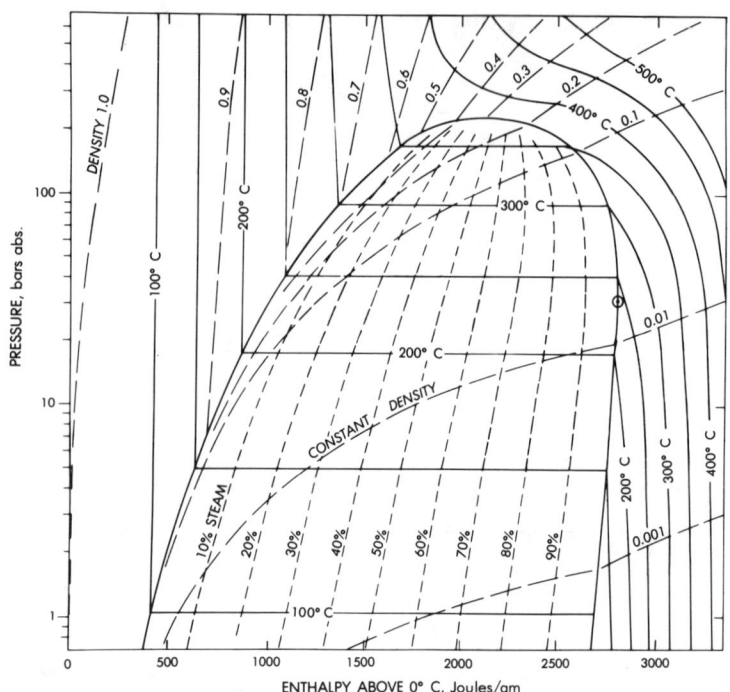

Figure 2.3. Enthalpy-temperature-pressure-density relations for water to 500°C and 600 bars (from Henley et al., 1984).

Notice that for CO_2 we combine all the analytical carbonate carbon (i.e., CO_2, HCO_3^-, $CO_3^=$, and $CO_{2,v}$), whereas for silica we may reasonably assume that the silica content of the steam is negligible.

Exercise 2. Complete Table 2.3 to obtain the total discharge composition of the well.

FLUID-MINERAL EQUILIBRIA: ALTERATION MINERALOGY

What controls the concentration of each component represented in the analysis?

First we examine silica. It is generally assumed that since quartz is ubiquitous in the reservoir, its solubility controls m_{SiO2} in the fluid. We can check this through available experimental data for the solubility of quartz in water (reaction 3)

$$SiO_{2,Qz} + 2H_2O = Si(OH)_4 \quad (3)$$

Figure 2.4 shows original solubility data for quartz (Kennedy, 1950; Morey et al., 1962). The observed concentration of silica in the aquifer fluid is 570 mg/kg (Table 2.3) so that inspection of Figure 2.4 yields an aquifer temperature of 265°C.

Exercise 3. Now check the measured enthalpy of the discharge against steam tables or Figure 2.3 to find at what temperature it corresponds to the enthalpy of vapor-saturated liquid. You have probably guessed the answer anyway; t = 265°C.

Can you now infer that m_{SiO2} is controlled by the solubility of quartz?

Equilibrium constants for this and other useful reactions are given as temperature dependent regression equations in Appendix Table 2.A1. These equations apply over limited temperature ranges. For example, for quartz, the equation given assumes a linear relationship between log K and 1/T from 200 to 280°C and conforms to the solubility data recommended by Fournier and Potter (1982). Various other equations have been published in the past and relate to the extent of experimental data available, the authors' bias, the regression technique used, and the assessment of errors. In this case, the equation yields a slightly higher aquifer temperature which may or may not be real. The enthalpy measurement and experience in a large number of wells confirm that quartz controls the silica content of the aquifer fluid above about 200°C (see Fournier, 1985, this volume).

Other components may be examined in a similar way. For example, for the hydrolysis reaction

$$3KAlSi_3O_{8} + 2H^+ \atop Kspar$$

$$= KAl_3Si_3O_{10}(OH)_2 + 6SiO_{2,Qz} + 2K^+ \quad (4) \atop Kmica$$

which is here written with conservation of aluminum in the solid phases.

Appendix Table 2.A1 gives log K_{265} = 7.897 at 265°C

Table 2.3—Total discharge composition for well BR22

moles kg^{-1} × 10^3

Na	K	Ca	Mg	Cl	B	SiO$_2$	ΣNH$_3$	ΣCO$_2$	H$_2$S	H$_2$	CH$_4$	N$_2$
						9.52		103.9				

and

$$\log K = 2 \log a_{K^+} + 2 \text{ pH} \quad (5)$$

where a is the thermodynamically effective concentration of, in this case, potassium ion.

Digression No. 2

We need for a moment to consider activities, a_i, and activity coefficients, γ_i, for solution species designated i. γ_i provides the link between the real messy world and that of ideal thermodynamics.

$$a_i = \gamma_i \, m_i \quad (6)$$

For the silica problem, we adopted Si(OH)$_4$ = 1; but this is allowable only for neutrally charged species in relatively dilute solutions up to say 1 molal. Individual ion activity coefficients are tricky to calculate, but are usually approximated in dilute solutions up to say 3 molal using the extended Debye-Hückel equation. For the high-temperature calculations discussed here, the following approximate values are quite satisfactory

$$\gamma_{H^+} = 0.8, \quad \gamma_{Ca^{++}} = 0.3$$

$$\gamma_{K^+} \simeq \gamma_{Cl^-} = \gamma_{HCO_3^-} = \gamma_{H_3SiO_4^-} = \gamma_{HS^-}$$

$$= \gamma_{BO_2^-} = \gamma_{Na^+} = 0.7 \simeq \gamma_{K^+}$$

Equation (5) becomes

$$\log K_{265} = 2 \log m_{K^+} + 2 \log \gamma_{K^+} + 2 \text{ pH} \quad (7)$$

Digression No. 3

Equation (5) contains a pH term (pH = −log a_{H^+}). Because the liquid sample taken at the surface is greatly depleted in CO$_2$ and H$_2$S relative to the reservoir fluid, the pH measured on the laboratory sample is really of no use to us and we need to calculate a new value, pH$_t$, for the reservoir fluid at 265°C. This is not a trivial task and it requires an iterative calculation best performed on a computer or programable calculator. The usual procedure involves using an ion or proton balance for all the pH-sensitive ions in solution. For more details, see Henley et al. (1984). Through these methods, pH$_{265}$ for the BR22 discharge is found to be 6.1.

A simplistic calculation often provides a useful approximation of pH. Consider the equilibrium

$$H_2CO_3^* = HCO_3^- + H^+ \quad (8)$$

where H$_2$CO$_3^*$ represents undissociated dissolved CO$_2$.

At 25°C log K = −6.36 so that in the liquid analysis we may calculate (with $\gamma_{HCO_3^-} \simeq 0.8$ at laboratory temperature) that

$$m_{HCO_3^-} = 3.29 \times 10^{-3} \text{ and } m_{H_2CO_3^*} = 0.25 \times 10^{-3}$$

Recalculating the discharge composition through (2), at 265°C, $m_{HCO_3^-}$ = 2.67 × 10^{-3} and $m_{H_2CO_3^*}$ = 101.2 × 10^{-3}. Log K for reaction (8) is −7.84 at 265°C and since

$$\log K = \log m_{HCO_3^-} + \log \gamma_{HCO_3^-} + \log a_{H^+} - \log m_{H_2CO_3^*}$$

we may substitute values to obtain pH. From these data our pH estimate is 6.1! (There are a number of hidden assumptions in this approach so that it may

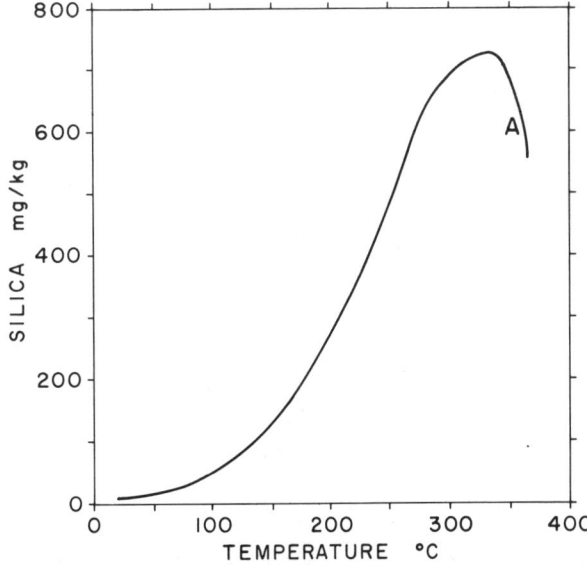

Figure 2.4. Experimental solubility data for quartz in water as a function of temperature.

be quite erroneous for some wells, particularly those discharging from a heavily exploited reservoir.)

Giggenbach (1980) has developed an approach to mineral-fluid equilibria which avoids the need to conserve Al as required in reaction (4) or use pH itself. In this approach the log ratio Al^{+3}/a_{H^+}, designated Al_H, is used as a principal variable. This makes for fun reading as you can imagine - but the method is extremely valuable!

The log analytical quotient a_{K^+}/a_{H^+} for the reservoir fluid is $-2.56 + 6.1 = 3.54$. This value, compared to the equilibrium quotient (3.94), suggests that the reservoir fluid lies in the 'mica' stability field, 0.4 log units away from the mica-Kspar boundary. This is compatible with the petrography since sericitization is widespread, but are there some data checks to be made?

1. Experimental determination of free energies for the reactants and products, particularly ions, is very difficult. Direct determination of the equilibrium constant is equally difficult due to unfavorable kinetics. The reaction coefficients required in hydrolysis reactions also require extreme precision in the thermodynamic data.
2. At 265°C, sericite is better described as illite and the activity of the $KAl_3Si_3O_{10}(OH)_2$ component should be determined, if possible, by x-ray and this value added into equation 7.
3. The pH_t calculation is dependent upon precise thermodynamic data for ion association reactions and reliable estimates of activity coefficients.

Exercise 4. Is the reservoir fluid close to equilibrium with respect to Kmica and Kspar at quartz saturation or is the pH_t too high or low?

Other hydrolysis reactions may be assessed similarly and the results are usually expressed on activity ratio diagrams such as those shown in Figures 2.5 and 2.6.

Exercise 5. Locate the BR22 fluid composition point in Figures 2.5 and 2.6 using the data of Table 2.3 and the estimated reservoir fluid pH_t.

The calcium aluminosilicates are a particular problem because they occur in solid solution with iron and because their thermodynamic data are poorly known. For this reason Figure 2.5 has been constructed using field data for the calcium aluminosilicate phase boundaries.

Exercise 6. Scan the introductory section and decide which is the dominant calcic phase in the alteration assemblage.

Digression No. 4.

The mineral is, of course, calcite, which, due to the high CO_2 content of the fluid, fixes most of the available Ca^{++}. To assess its solubility in terms of $a_{Ca^{++}}/a^2_{H^+}$ you will need an appropriate log K derived from Appendix Table 2.A1.

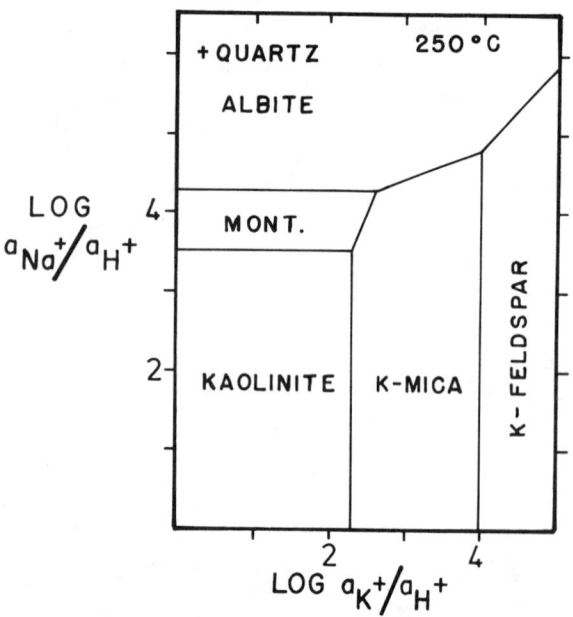

Figure 2.5. Activity-activity relations for the system $Na_2O-K_2O-Al_2O_3-SiO_2-H_2O$.

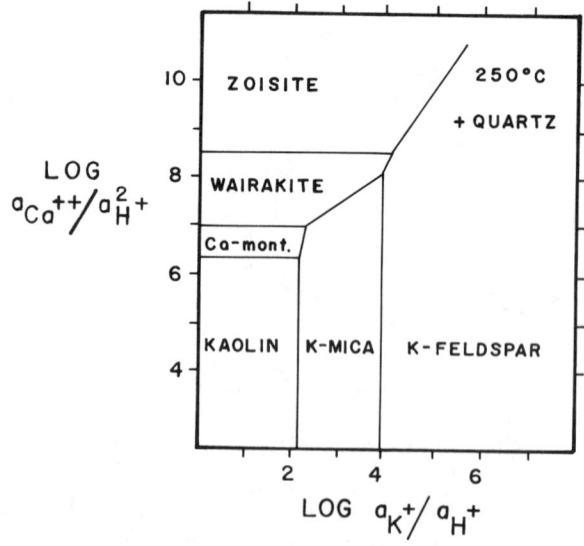

Figure 2.6. Activity-activity relations for the system $CaO-K_2O-Al_2O_3-SiO_2-H_2O$.

$$CaCO_3 + H_2CO_3^* = Ca^{++} + 2HCO_3^-$$
$$\underline{2(HCO_3^- + H^+ = H_2CO_3^*)}$$
$$CaCO_3 + 2H^+ = Ca^{++} + H_2CO_3^*$$

The coefficients of the regression equations (Appendix Table 2.A1) may be summed to give

log K =
$$-235.56 + 10676.81/T + 89.084 \log T - 0.0392 T$$

At 265°C this equation gives log K = 6.46. The analytical quotient, $a_{Ca^{++}} m_{H2CO3^*}/a^2_{H^+}$, for the BR22 fluid is 6.1, which suggests that the fluid is close to saturation with calcite. The difference between the analytical and equilibrium quotients is 0.36 log units, a small value compared with the sum of possible errors on pH, $\gamma_{Ca^{++}}$, and, if underground boiling has occurred, m_{H2CO3^*}.

Digression No. 5.

Control of fluid pH through silicate reactions has an important implication. Taking the Kspar-Kmica reaction, for example

$$K = (a_{K^+}/a_{H^+})^2$$

it is self-evident that m_{K^+} is a function of pH. Potassium and sodium are the predominant cations in these high-temperature solutions and their ratio is limited by the reaction of plagioclase to Kspar and Kmica, and, below about 270°C, the formation of secondary albite. The latter, the coprecipitation of albite and Kspar, affords a limiting condition for the ratio of sodium and potassium, which is used as a geothermometer. Combining the two reaction constants, a temperature dependent function of pH and $m_{(Na + K)}$ is found, and since $(m_{Na} + m_K) \simeq m_{Cl^-}$, pH is related to salinity. Figure 2.7 shows this function together with some data points for other active geothermal systems.

From an assessment of fluid compositions from a number of wells, Giggenbach (1981) showed that the Kspar-Kmica reaction was important in the upper two-phase (i.e., boiling) zone of geothermal systems, but that in general another equilibrium prevailed

$$\text{plagioclase} + H_2CO_3^* = \text{'clay'} + \text{calcite} \quad (9)$$

Data from a number of New Zealand fields are summarized in this context in Figure 2.8. Giggenbach (1980) has also shown that it may be more appropriate to refer to geothermal mineral-fluid interactions with reference to a <u>steady state</u> rather than to equilibrium, but this does not affect the validity of the general relations discussed here.

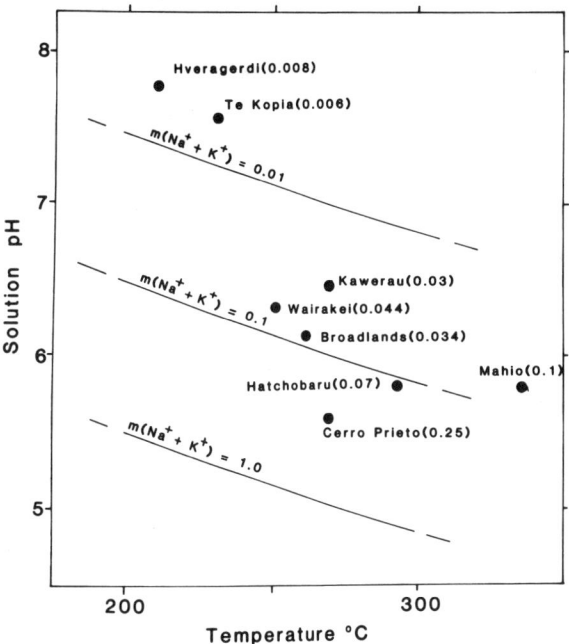

Figure 2.7. Relationship between fluid pH and salinity for fluids in equilibrium with Kspar-Kmica-quartz.

Another important group of constituents (the gases ammonia, methane, nitrogen, hydrogen, etc.) are included in Tables 2.2 and 2.3. These data provide a great deal of information about sources of various components, processes in the aquifer, etc.--see, for example, Giggenbach (1977, 1980). Although in the laboratory gas + gas reactions are normally slow, they appear to be as fast as silica precipitation in the fluid + mineral system of a geothermal aquifer. This provides a useful independent check on aquifer temperatures through reactions like

$$N_2 + 3H_2 = 2NH_3$$
$$CO_2 + 4H_2 = CH_4 + 2H_2O$$

although if the fluid boils en route to the well vapor loss or gain may occur in the discharge relative to the aquifer fluid. For this well, BR22, these effects are minor and the analysis quotients for these reactions are close to those expected for equilibrium at 265°C.

Reviewing data from Broadlands and other fields, Giggenbach (1980) showed that the ratio H_2/H_2S was a function of temperature (Fig. 2.9) reflecting equilibrium between pyrite and Fe-silicates represented by chlorite and/or epidote. Only near 300°C do the observed hydrogen contents approach those for the equilibrium between pyrite and pyrrhotite, or pyrrhotite + Fe-silicate.

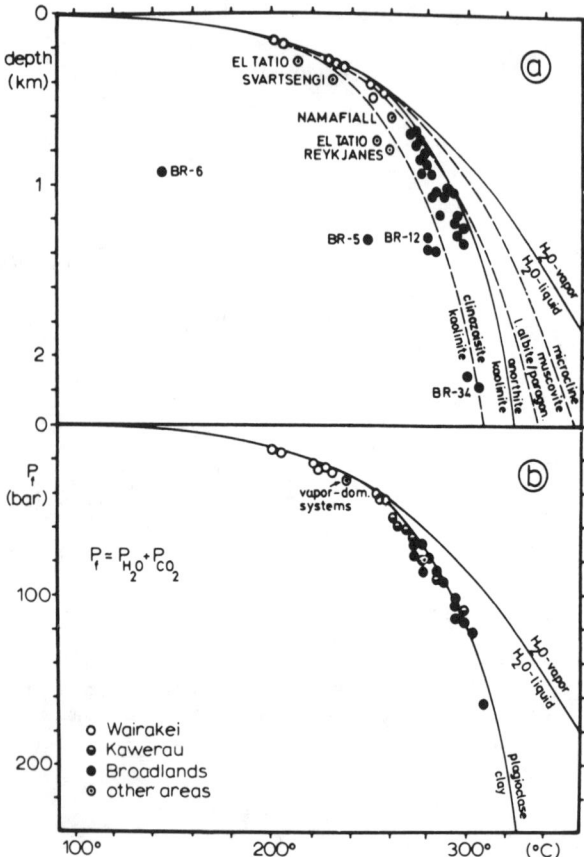
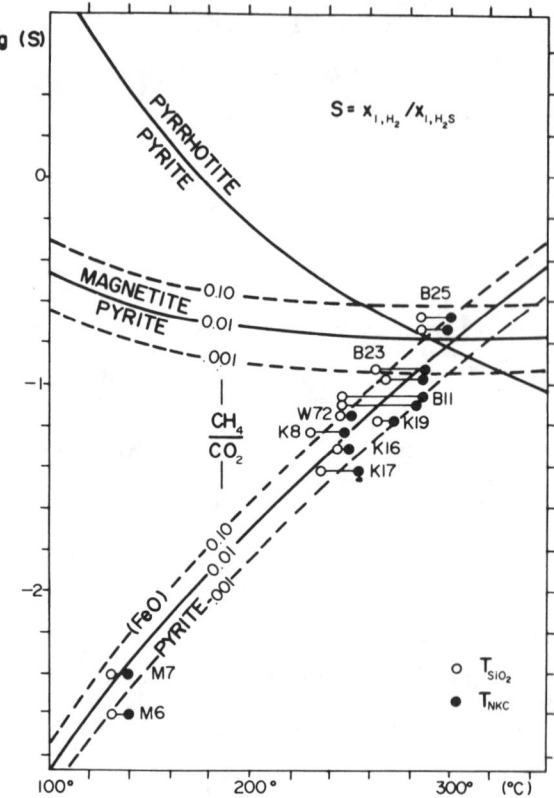

Figure 2.8. Boiling point vs depth (a) and fluid pressure P (b) curves for a series of mineral assemblages. The data points for Wairakei, Kawerau, Broadlands, El Tatio, and Iceland wells correspond to the point at which measured temperature/depth and temperature/pressure curves changed from (nearly) constant temperature with depth behaviour to that indicating boiling point vs depth conditions (reproduced with permission from Giggenbach, 1981).

Figure 2.9. The ratio H_2/H_2S as a function of temperature for selected mineral pairs in relation to data from active geothermal systems, plotted with respect to quartz and alkali geothermometer temperatures. (FeO) represents an iron-aluminium silicate (chlorite?, epidote?). CH_4/CO_2 contours are also shown. Note the differences between silica and alkali geothermometer temperatures (T_{NKC} is the NaKCa empirical geothermometer of Fournier and Truesdell, 1973). These differences reflect aquifer processes consequent on exploration and have high significance for monitoring geothermal field development (reproduced with permission from Giggenbach, 1980).

FLUID-MINERAL EQUILIBRIA: TRACE-METAL CONTENTS

Table 2.4a summarizes base- and precious-metal analysis from waters discharged at atmospheric pressure from two Broadlands wells (Weissberg et al., 1979).

As noted earlier, base-metal sulfides are to be found in drill core and cuttings from many of the wells at Broadlands, but gold has been reported only in a complex antimony-arsenic-mercury-thallium-sulfide precipitate in a hot spring. Free gold (as electrum) does occur in core from nearby systems, e.g, Kawerau (B. Christenson, personal communication) and, see below, is to be anticipated in Broadlands core--but no one has yet looked for it.

Recent work at BR22 has shown that the concentrations of trace metals discharged at atmospheric conditions do not always represent the concentrations of those metals in the deep geothermal fluid (Brown, 1985). Considerable deposition occurs in the piping so that corrected concentrations (Table 2.4b) are in fact much higher than would be judged from Table 2.4a.

Table 2.4a--Concentration of trace metals in water discharged at the surface from exploration well BR2 at Broadlands

	Fe	Cu	Pb	Zn	Ag	Au	As	Sb	
BR2	360	.9	1.3	1.0	.7	.04	5700	200	µg/kg

Table 2.4b--Minimum concentrations of trace and base metals in water discharged from BR22 at Broadlands (Brown, 1985). These data are based on empirical data for the metal contents of the total discharge, but for ease of comparison with Table 2.4a, these data are corrected to 1 b.a. separation pressure, although as discussed below, a large percentage of the copper, silver and gold is precipitated prior to discharge at atmospheric pressure.

Separation Pressure	Fe	Cu	Pb	Zn	Ag	Au	
1.0 b.a.	342	13.6	3.5	1.5	12.1	2.3	µg/kg

Exercise 7. Calculate the aquifer concentrations of these metals at BR22. What percentage of the gold and silver is precipitated by boiling during discharge to the surface? Water from BR22 flashed to 1 bar a. contains similar Au and Ag to those shown in Table 2.4a for BR2.

The concentration of these metals may be controlled by mineral deposition or dissolution in the aquifer, so first we must examine, as with calcite and the various silicates, whether the fluids are close to saturation with respect to common sulfides, or the native metal (for gold).

Lead

Figure 2.10 shows the distribution of the different lead chloride species as a function of m_{Cl^-} based on recent experimental data (Seward, 1984). From these data we see that $PbCl_2$ constitutes about 50% of the lead in solution as a chloride complex at 265°C in a low salinity fluid. The solubility of galena may then be written

$$PbS + 2H^+ + 2Cl^- = PbCl_2 + H_2S_{aq} \quad (10)$$

$$\log K_{265} = 2.64$$

Inserting analytical data for the reservoir fluid yields m_{Pb} = 0.015 µg/kg (as $PbCl_2$), so that the actual solubility of galena is, including all the chloride species, 0.03 µg/kg. This value is much lower than the 1-10 µg/kg actually found in these waters and indicates that in these low salinity fluids, some other complex species must be present; however, there are as yet no experimental data available for possible species like $Pb(OH)^+$, $PbHCO_3^+$, etc., to calculate which predominates under these conditions.

Digression No. 6

Figure 2.11 shows the calculated solubility of galena as a function of temperature for a 1.0 m chloride solution (~6 wt% NaCl). What is the increase in solubility to be expected at 265°C if species like $PbHCO_3^+$ are present? First write a simple reaction for the formation of $PbHCO_3^+$.

$$PbS + 2H^+ + HCO_3^- = PbHCO_3^+ + H_2S \quad (11)$$

Increasing chloride relative to the Broadlands fluid does not affect the reaction as written except through changes in activity coefficients, so that the observed contribution due to this or other species may be plotted directly onto Figure 2.10. It is clear that the chloride complexes dominate at these higher salinities. Notice also that in a hydrothermal system with fluid at this salinity, the pH_t of the fluid would be about 1 to 1.5 units less than for Broadlands (Fig. 2.7). The increased solubility due to this effect, reaction (10), is, however, partly offset by the decrease of $m_{HCO_3^-}$.

Incidentally, using data from Table 2.A1 and the observed lead content of the Broadlands fluid, an estimate of $\log K_{265}$ for this reaction is 7.58.

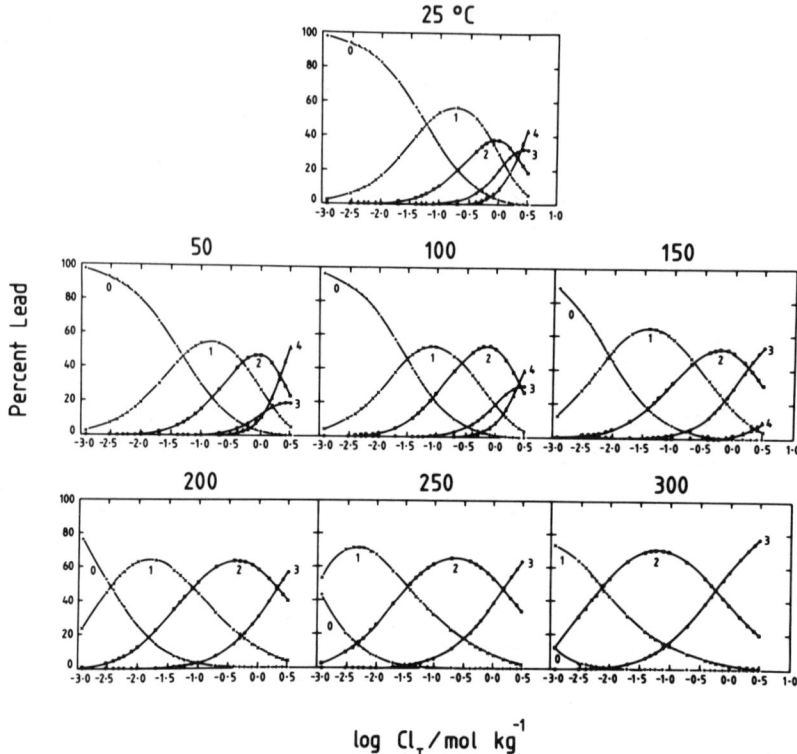

Figure 2.10. The percentage distribution of lead as chlorolead (II) complexes as a function of total chloride concentration, Cl_T, up to 300°C at the saturated vapor pressure of the system; the curve for each complex is labelled according to ligand number, e.g., curves labelled '2' refer to $PbCl_2^0$ (reproduced with permission from Seward, 1984).

Gold

In contrast to lead (and zinc), metals such as gold, arsenic and antimony have very stable bisulfide complexes. For gold the principle dissolution reaction (Seward, 1973) is

$$Au + 2H_2S = Au(HS)_2^- + 1/2H_2 + H^+ \quad (12)$$

and a regression equation for the equilibrium constants of this reaction is given in Appendix Table 2.A1.

Digression No. 7

Determination of the solubility of gold introduces a new solution parameter, the relative redox state, which is traditionally represented by the value of log f_{O_2} for the solution, where f is the fugacity. The choice of f_{O_2} is unfortunate since its fugacity is immeasurably small, whereas f_{H_2} can be more easily directly measured. Since gas concentrations are relatively low, it is permissible to substitute the partial pressure, P_i, for fugacity.

The concentration of H_2 in the reservoir fluid from Table 2.3 is 1.07×10^{-3} moles/kg. In order to calculate f_{H_2}, a knowledge of Henry's Law is required. This is given by

$P_i = X_i K_{H,i}$ where $K_{H,i}$ = Henry's Constant for i

X_i = mole fraction of i

P_i = Partial Pressure of i

The mole fraction of hydrogen in the reservoir fluid is

$$0.101 \times 10^{-3} \times \frac{18}{1000} = 0.182 \times 10^{-5}$$

Henry's constant for H_2 at 265°C is 26,000 bars/mole-fraction, and therefore

$f_{H_2} = 0.182 \times 10^{-5} \times 26000 = 0.0473$

At 265°C log K = -19.46 for the reaction

$$H_2O_{(l)} = H_{2(g)} + 1/2 O_{2(g)}$$

Therefore, log P_{O_2} = -36.27.

Locate the BR22 reservoir fluid on the f_{O_2}/pH diagram provided in Figure 2.12.

At the temperatures under discussion for Broadlands (≤300°C), magnetite is not present as an alteration mineral. Its stability field seems to be usurped by

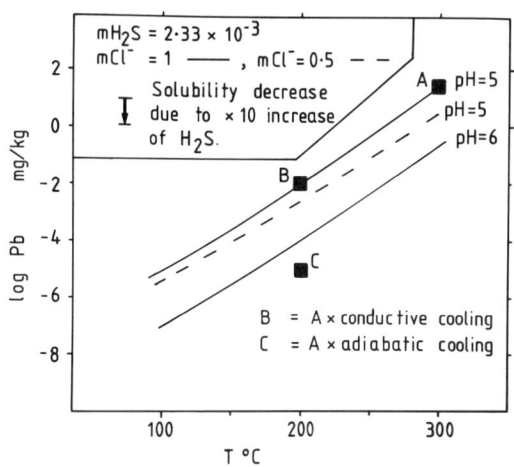

Figure 2.11. Solubility of galena in aqueous chloride solution.

that of chlorite, a feature noted in other systems and in many hydrothermal deposits (Barton et al., 1977).

Figure 2.12 contains solubility contours for $Au(HS)_2^-$.

Exercise 8. Compare the measured gold content of the Broadlands fluid with the solubility contours given in Figure 2.12. What conclusion may be drawn? Discount any suggestion of chloride complexes which are precluded (at this temperature, salinity, and f_{O_2}) on the basis of experimental data (Henley, 1984, unpublished data).
Sketch isosolubility curves (lead, mg/kg) in the H_2S field for galena as chloride complexes alone and note how differently they behave compared to those of gold.

Other Metals: Copper, Silver, and Arsenic

Other metals may be treated in the same manner although it may be necessary to consider a wider range of complexes. Available data for copper give the following species distribution in the Broadlands aquifer fluid at 265°C (Table 2.5), although there is some doubt as to the precision of the formation constants for the chloride species.

By far the predominant species is the $Cu(HS)_2^-$ complex and the concentration of this species is essentially equal to the measured aquifer concentration in Table 2.4b.

Exercise 9. By assuming coexistence of the same sulfides (Table 2.5), <u>estimate</u> the concentrations of $CuCl_2^-$ and $Cu(HS)_2^-$ for the Salton Sea geothermal fluid using the data below and that of Table 2.5.

Cl^- = 70,000 mg/kg

γ_{Cl^-} = 0.2 = γ_{HS^-}

pH = 5.4

t = 260°C

H_2S = 20 mg/kg

What are the predominant copper species here?

For silver, only data for the chloride complexes have been measured (Seward, 1976).

$Ag_2S_{(s)} + 2H^+ + 2Cl^-$
$= 2AgCl + H_2S_{(aq)}$; $\log K_{265°} = -5.5$ (13)

$Ag_2S_{(s)} + 2H^+ + 4Cl^-$
$= 2AgCl_2^- + H_2S_{(aq)}$; $\log K_{265°} = -2.1$ (14)

From these data, the solubility of argentite is

$10^{-9.10} + (10^{-9.09}/.694)$
 = 1.965×10^{-9} moles/kg
 = <u>0.21 ppb</u> .

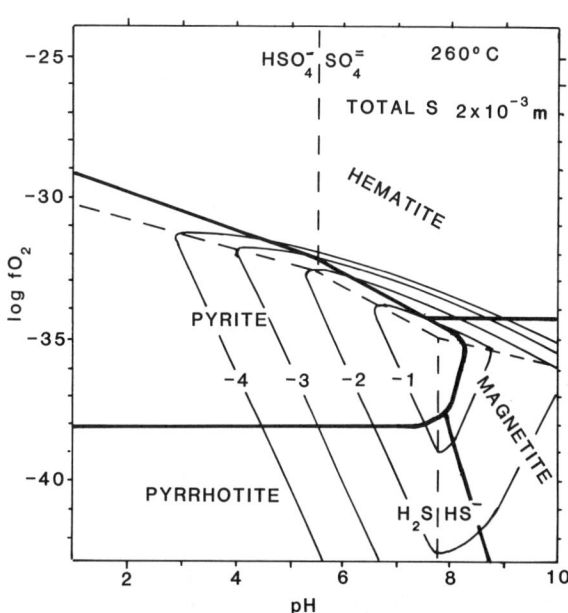

Figure 2.12. Log f_{O_2} versus pH diagram for fluids and alteration minerals at Broadlands at 260°C. Superimposed are solubility contours for gold in mg/kg.

Table 2.5--Distribution of copper bisulfide and chloride complexes in the BR22 aquifer fluid at 265°C. Thermodynamic data from Crerar and Barnes (1976) for the bornite-chalcopyrite-pyrite coexistence boundary. Units; g Cu/kg.

	$Cu(HS)_2^-$	$Cu(HS)_2H_2S^-$	Cu^+	$CuCl^0$	$CuCl_2^-$	$CuCl_2^{2-}$
g/kg	4.7	0.011	2.5×10^{-5}	6.2×10^{-3}	1.6×10^{-3}	6.6×10^{-5}

Exercise 10. Compare this value with the measured value in Table 2.4b. What conclusions may be drawn?

Assuming that silver forms an $Ag(HS)_2^-$ complex to account for the extra solubility, calculate a value for log K for the solubility of Ag_2S in the aquifer.

In all the New Zealand geothermal systems arsenic concentrations are high relative to other natural waters. The hydrothermal chemistry of arsenic is poorly known. Weissberg et al. (1966) found high solubilities for orpiment (As_2S_3) in alkaline solutions up to about 200°C. Arsenite (AsO_3^{3-}) or thioarsenite (AsS_3^{3-}) complexes are the most likely to account for arsenic transport in these fluids. Similarly the chemistry of antimony is poorly known. Elemental mercury has a relatively high volatility and may be significantly transported in any vapor fraction formed by boiling, however its speciation in solution is also very poorly understood at high temperatures.

MINERAL DEPOSITION

Silica

Whereas in the reservoir we showed that the silica concentration of the fluid was controlled by the dissolution of quartz, at lower temperatures (150°) silica solubility is controlled by the equilibrium with amorphous silica. The solubility of amorphous silica in the temperature range 0 to 250°C is given by (Fournier and Rowe, 1977).

$$\log S = -731/T + 4.52$$

where S = solubility in mg/kg, T = absolute temperature (°K).

When geothermal water arrives at the surface through a well or fissure, it becomes supersaturated with respect to amorphous silica due to both the reduction in temperature, and the concentrating effect of steam loss. At BR22 for example, at 98°C, the silica concentration in the separated water is about 800 mg/kg. The solubility of amorphous silica is only 355 mg/kg allowing possible deposition of almost 0.5 kg silica per ton of fluid. However, only a very small percentage of the excess silica is deposited. Most of the excess silica polymerizes to form a colloidal suspension (Plate 2.1A)--a process that may take several hours. Subsequent growth and aggregation of the colloidal particles forms silica sinter (Plates 2.1B and 2.1C). If the particles are ordered and all the same size, then gem quality opal is formed. Deposition processes for silica are considered in much more detail by Fournier (1985, this volume).

Calcite

Calcite sometimes forms inside the casing of geothermal wells during discharge and, as we discovered earlier, the reservoir fluid is saturated with respect to calcite. The equilibrium relation may be written

$$CaCO_3 + H_2O + CO_{2,g} = Ca^{++} + 2HCO_3^-$$

Exercise 11. Derive a regression equation for the temperature dependence of reaction (15), using data from Appendix Table 2.A1. Your result should be

$$\log K = -226.43 + 6552.81/T + 89.084 \log T - 0.0746 T$$

Since each mole of calcite that dissolves in reaction (15) produces one mole of Ca^{++} and two of HCO_3^-, the equilibrium constant can be modified to read

$$K = (4 m^3_{CaCO_3} \gamma_{Ca^{++}} \gamma^2_{HCO_3^-})/P_{CO_2} \quad (16)$$

With these two expressions we can draw solubility curves for calcite as a function of P_{CO_2} and t. Such a set of curves is shown in Figure 2.13 based on experimental data. Notice that calcite is more soluble at low temperature than high temperature.

Digression 8

The total fluid pressure in an aquifer is the sum of the water and gas pressures. P_{CO2} may be calculated from Henry's Law

$$H_2CO_3^* = CO_{2,g} + H_2O \quad (17)$$

At 265°C, $\log K = 1.917 = \log P_{CO_2}/m_{CO_2}$

$$\log P_{CO_2} = 1.917 + \log (0.104)$$

P_{CO_2} = 8.5 bars, and with P_{H_2O} = 50.8 bars,

$P_{reservoir}$ = 59.3 bars

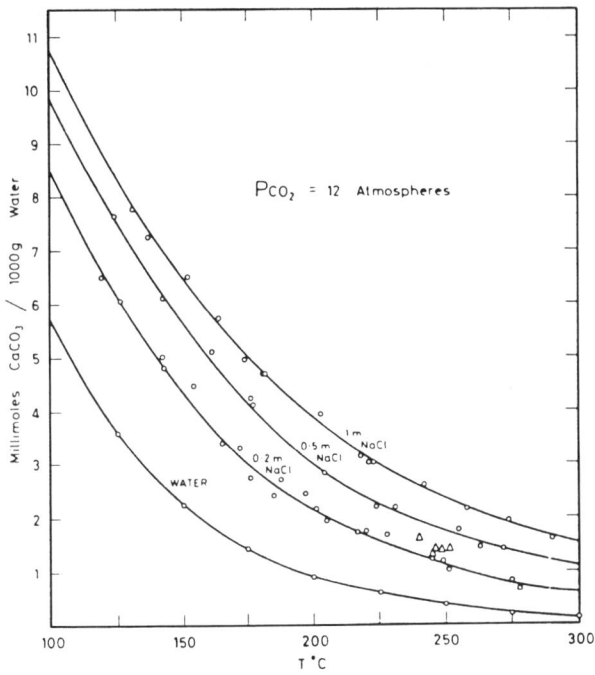

Figure 2.13. Experimentally determined solubility of calcite; notice the effect of dissolved salts (reproduced with permission from Ellis, 1963).

This, incidentally, means that the boiling-point depth for this solution is about 100 m deeper than for pure water, but that's a whole story in itself (Hedenquist and Henley, 1985). Figure 2.8b shows the effect of CO_2 on boiling depth where P_{CO_2} is buffered by the plagioclase, clay, calcite reaction. Consider a well discharging from this aquifer. If the wellhead pressure is held at 15 bars, the fluid experiences a pressure drop of nearly 45 bars during flow up the well. Since CO_2 is a relatively volatile species, it partitions strongly into any coexisting vapor phase as is clearly demonstrated by the original analytical data. The concentration of CO_2 in the liquid phase at equilibrium with vapor may be obtained from the following equation

$$m_{H_2CO_3}^* = m_{CO_2,initial}/[1 + y(B-1)] \quad (18)$$

where B is the partition coefficient for CO_2 between vapor and liquid. Regression equations for B are given in Appendix Table 2.A1.

Exercise 12. From (16) and (17), calculate the $m_{H_2CO_3}^*$ and $P_{CO_2,aq}$ of the liquid phase resulting from steam formation at 220°C as the well discharges, and then the state of saturation of calcite at this temperature.

You should find that (a) at 220°C the aqueous phase is about 4 times supersaturated with calcite and (b) each ton of fluid passing up the well could deposit 30 gm of calcite (15 cc). What would be the supersaturation at 200°C?

Digression No. 9.

The pH of the liquid separated at the wellhead at 184°C is 7.3 (this is a value recalculated from the analytical data as briefly outlined in Digression 3). Why is the flashed fluid more alkaline than the reservoir fluid?

Consider the reaction

$$CO_{2,g} + H_2O = HCO_3^- + H^+ \quad (19)$$

Clearly, even without considering the change of K for this reaction due to temperature drop, removal of $CO_{2,g}$ by boiling must result in a pH increase. Removal of 90% of the CO_2 increases pH_t by 1 unit.

Metal Sulfides and Gold

Galena, sphalerite and chalcopyrite sometimes occur within the calcite scale coating the insides of well casings. Metal sulfides have also been precipitated in the surface installations. Figure 2.14 is an x-ray diffraction pattern for a scale deposited inside wellhead equipment at BR27 and similar scales occur at BR22. Galena, pyrite, chalcopyrite, and sphalerite are all present. There are also peaks due to the presence of electrum. A scanning electron micrograph of the scale is shown in Plate 2.2. The scale was deposited in the wellhead plumbing where the pressure drops from ~40 bars to 11 bars with a concomitant temperature drop to 185°C. The boiling that accompanies the drop in pressure has a number of effects. As shown above, CO_2 partitions strongly into the vapor phase, so that the liquid phase experiences a rise in pH. H_2S is also lost from the solution, so that reactions such as:

$$Au + 2H_2S = Au(HS)_2^- + 1/2H_2 + H^+ \quad (20)$$

are driven to the left leading to gold deposition. The equilibrium constant for this reaction at 185°C is log K = -8.4 (Appendix Table 2.A1). Using the distribution coefficient, B = 174, from Appendix Table 2.A1 and the measured H_2S concentration of 0.103 moles/kg of steam, we can calculate

$$H_2S_{aq} = 0.059 \text{ mm/kg (2 mg/kg)}$$

Before we can calculate the solubility of gold from reaction (20), we need a value for P_{H_2} from the separated liquid. There are a number of problems involved in estimating P_{H_2} during irreversible processes such as flash boiling (Thorstensen, 1984), but one guess may be obtained from the reaction

$$HS^- + 4H_2O = SO_4^= + 4H_2 + H^+ \quad (21)$$

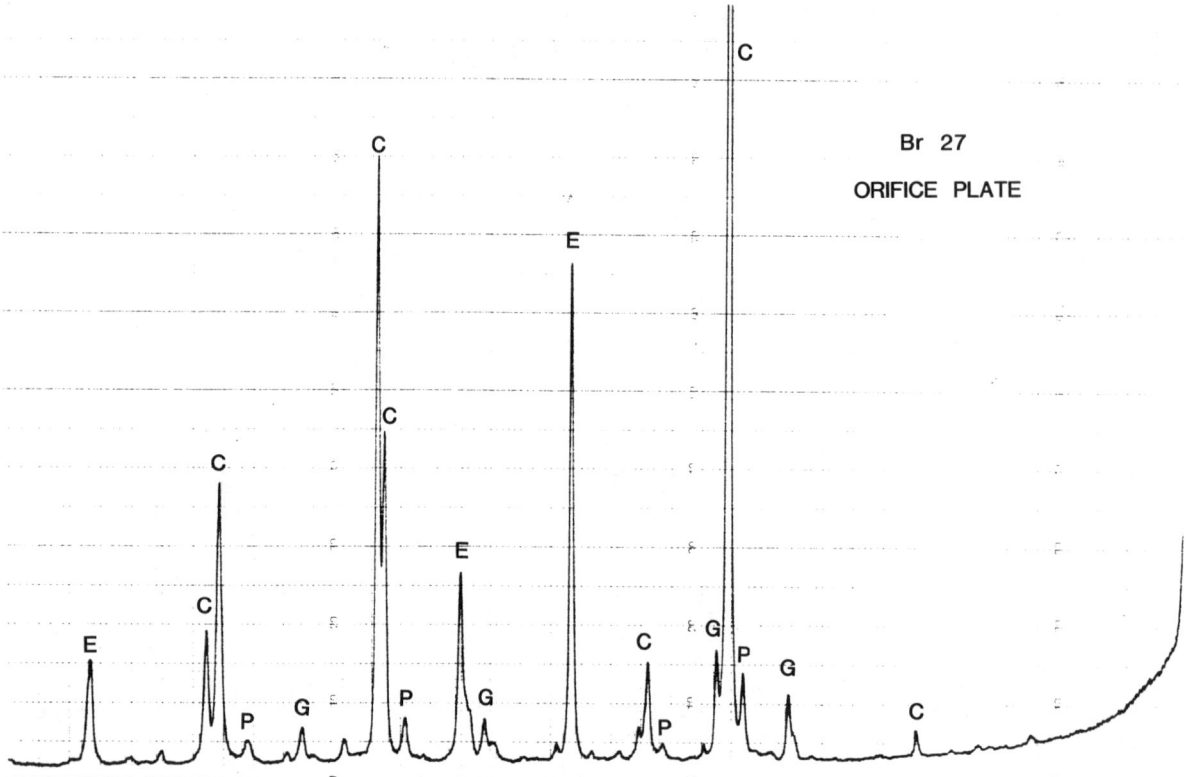

Figure 2.14. X-ray diffraction pattern of material deposited at BR22 well. C = chalcopyrite, G = galena, P = pyrite, M = magnetite, E = electrum.

Substituting values for m_{HS}, HS, $m_{SO_4^=}$, $SO_4^=$, and pH, assuming that the reaction is at equilibrium, and that no oxidation of H_2S to $SO_4^=$ had occurred in the sample, then using values from Appendix Table 2.A1.

$$\log P_{H2} = -2.72$$

An alternative method for estimating P_{H2} is by using the measured partial pressure of hydrogen. The amount of hydrogen separated at 11 bars into the gas phase is 0.57×10^{-3} moles/kg (from Table 2.2). The partial pressure of hydrogen (P_{H2}) is given by the product of the mole-fraction of H_2 and the total pressure

$$P_{H2} = 0.57 \times 10^{-3} \times \frac{18}{1000} \times 11$$
$$= 1.129 \times 10^{-4}$$

$$\log P_{H2} = -3.95$$

This value of P_{H2} places the solution outside the stability field for H_2S which clearly from the analysis is stable in the solution collected at 11 bars. It would, therefore, appear that this and other competing redox reactions in the solution have not come to equilibrium in the new pressure regime.

The measured gold concentration of 2.3 µg/kg (Table 2.4b) at 1 bar corresponds to a concentration of 1.9 µg/kg at 185°C. When this value is substituted in reaction (20) together with the H_2S activity, $\log P_{H2} = -2.85$. This is in remarkable agreement with the value of -2.72 derived from reaction (21) indicating that the "true" or effective $\log P_{H2}$ must lie around these values at least as far as gold is concerned.

Exercise 13. Compare the effects of boiling on dissolved metal content for lead, silver and copper.

The discussion of arsenic and antimony sulfide deposition is hindered by the absence of chemical data for the complexing of these metals. Natural occurrences (Weissberg, 1969; Hedenquist and Henley, 1985) and deposits formed during chemical processing of geothermal discharges suggest that pH is a controlling variable through a reaction of the form

$$2AsS_3^{3-} + 6H^+ = As_2S_3 + 3H_2S \qquad (22)$$

Observations on arsenic depositing in hot springs and well discharges (Hedenquist and Henley, 1985) suggest that local deposition may occur where normal near neutral pH fluids cool without loss of CO_2 so that pH is buffered by the bicarbonate-carbonic acid pair or where acidification occurs, perhaps by mixing with surficial acid waters as at Tamagawa Springs, Japan (Nakagawa, 1971).

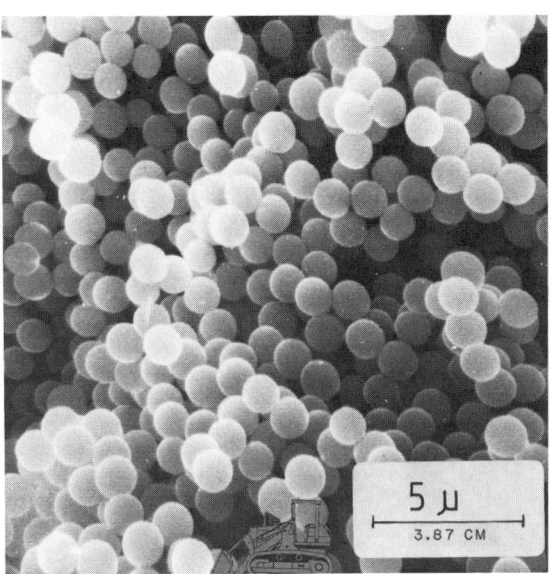
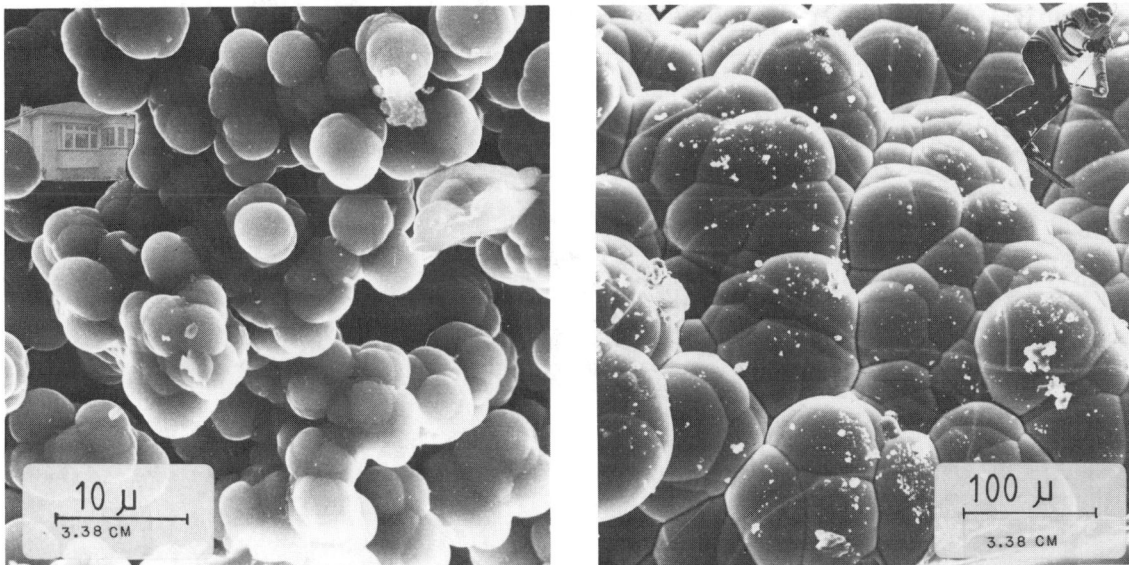

Plate 2.1. a). Scanning electron micrograph of colloidal silica from BR22 well after initial polymerization. b). After further growth and agglomeration – note the porous structure – and c). the deposited scale with the pores filled with silica to form a hard vitreous sinter.

Plate 2.2. Scanning electron micrograph of Broadland's sulfide scale. The x-ray diffraction pattern for this material is shown in Figure 2.14. Each segment of the bar corresponds to 50 μm and, for convenience, conversion to local Taupo units is provided.

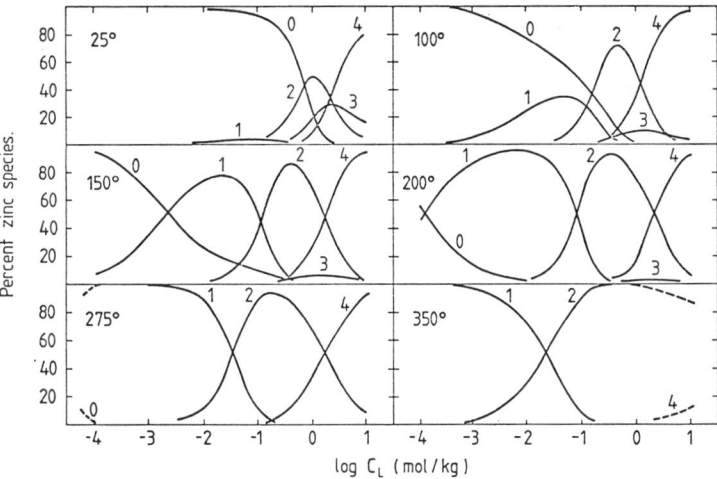

Figure 2.15. Distribution of zinc chloride species as a function of chloride concentration and temperature in acidic solutions. The numbers shown are the number of chloride ligands for each zinc atom (reproduced with permission from Ruaya and Seward, in press).

ACKNOWLEDGMENTS

We are grateful to Bruce Christenson for reviewing the manuscript, and to Sharon Thorne for cheerfully and painstakingly preparing the manuscript. T. M. Seward and Ruaya kindly allowed prepublication reproduction of the Zn complexing data. One of us (KLB) found no solace at all in the accompaniment of James Galway.

REVIEW QUESTIONS

1. Well BR22 discharges about 100 tonnes/hr; how long would it take to deposit a million ounces (2.5×10^6 gm) of gold at the wellhead? What does your answer imply about the chemistry of the fluids responsible for epithermal gold deposits? If all the potential silica and calcite coprecipitated with gold - and we continuously cleared the well! - what would be the gold concentration of the deposit?

2. From Figure 2.2 estimate the mass fraction of vapor formed <u>per 10°C</u> temperature drop as liquid (initially on the two-phase boundary) adiabatically boils from 300°C and 250°C to 200°C.

3. What differences do you recognize between the processes within a discharging well and in a fluid moving upward through a fissure to the surface?

4. Figure 2.15 shows the distribution of zinc chloride complexes as a function of chloride content and temperature. The equilibrium constant (as log K) for the dissolution of ZnS

$$ZnS + 2H^+ = Zn^{2+} + H_2S_{aq}$$

equals -2.7 (Helgeson, 1969) at 250°C and -2.23 at 300°C. Calculate the solubility of sphalerite in the fluid discharged from Broadlands, BR22, and compare it with the observed zinc content of the fluid. What conclusion may be drawn about zinc transport in the Broadlands aquifer fluid? Make a similar calculation for the Salton Sea brine using data provided earlier.

5. In Kuroko deposits (submarine polymetallic massive sulfides), metal sulfides are thought to be precipitated as the submarine hydrothermal fluid (300°C) mixes with cold seawater. Consider the chemistry of ore deposition as a result of cooling conductively or mixing with seawater. Are there any other possible depositional mechanisms?

6. Apart from boiling, what other precipitation mechanisms can you propose for silver and gold in epithermal deposits?

7. Consider the effect of salinity on the solubilities and depositional mechanisms of gold, silver, lead and zinc in hydrothermal fluids. Is dilution by cold, fresh water a process which could precipitate any of these metals?

8. In an epithermal vein adularia and calcite are found. Would this paragenesis provide information on physical conditions during ore deposition?

REFERENCES

Barton, P. B., Jr., Bethke, P. M., and Roedder, E., 1977, Environment of ore deposition in the Creede mining district, San Juan Mountains, Colorado: Part III. Progress toward interpretation of the chemistry of the ore-forming fluid for the OH vein: Economic Geology, v. 72, p. 1-24.

Brown, K. L., 1985, Gold deposition from New Zealand geothermal wells: Economic Geology (in press).

Crerar, D. A., and Barnes, H. L., 1976, Ore solution chemistry V. Solubilities of chalcopyrite and chalcocite assemblages in hydrothermal solution at 200° to 350°C: Economic Geology, v. 71, p. 772-794.

Ellis, A. J., 1963, Solubility of calcite in sodium chloride solutions at high temperatures: American Journal of Science, v. 261, p. 259-267.

Ewers, G. R., and Keays, R. R., 1977, Volatile and precious-metal zoning in the Broadlands geothermal field, New Zealand: Economic Geology, v. 72, p. 1337-1354.

Fisher, R. G., 1959, The natural heat flow from the upper Waiora Valley: Geophysics Division, DSIR, Geothermal Circular TSG 5.

Fournier, R. O., 1985, The behavior of silica in hydrothermal solutions; in Berger, B. R., and Bethke, P. M. (eds.), Geology and Geochemistry of Epithermal Systems: Society of Economic Geologists, Reviews in Economic Geology, v. 2.

Fournier, R. O., and Potter, R. W., 1982, A revised and expanded silica (quartz) geothermometer: Geothermal Resources Council Bulletin, v. 11, p. 3-9.

Fournier, R. O., and Rowe, J. J., 1977, The solubility of amorphous silica in water at high temperatures and high pressures: American Mineralogist, v. 62, p. 1052-1956.

Fournier, R. O., and Truesdell, A. H., 1973, An empirical Na-K-Ca geothermometer for natural waters: Geochimica et Cosmochimica Acta, v. 37, p. 1255-1275.

Giggenbach, W. F., 1977, The isotopic composition of sulphur in sedimentary rocks bordering the Taupo Volcanic Zone; in Ellis, A. J. (ed.), Geochemistry 77: New Zealand Department of Scientific and Industrial Research Bulletin 218, p. 57-64.

Giggenbach, W. F., 1980, Geothermal-gas equilibria: Geochimica et Cosmochimica Acta, v. 44, p. 2021-2032.

Giggenbach, W. F., 1981, Geothermal-mineral equilibria: Geochimica et Cosmochimica Acta, v. 45, p. 393-410.

Hedenquist, J. W., and Henley, R. W., 1985, Hydrothermal eruptions in the Waiotapu geothermal system, New Zealand. Origin, breccia deposits and effect on precious-metal mineralization: Economic Geology, v. 80, p. 1640-1668.

Helgeson, H. C., 1969, Thermodynamics of hydrothermal systems at elevated temperatures and pressures: American Journal of Science, v. 267, p. 729-804.

Henley, R. W., Truesdell, A. H., and Barton, P. B., Jr., 1984, Fluid-Mineral Equilibria in Hydrothermal Systems: Society of Economic Geologists, Reviews in Economic Geology, v. 1, , 267 p.

Kennedy, G. C., 1950, A portion of the system silica-water: Economic Geology, v. 45, p. 629-653.

Morey, G. W., Fournier, R. O., and Rowe, J. J., 1962, The solubility of quartz in water in the temperature interval from 25 to 300°C: Geochimica et Cosmochimica Acta, v. 26, p. 1029-1043.

Nakagawa, R., 1971, Solubility of orpiment (As_2S_3) in Tamagawa Hot Springs, Akita Prefecture: Nippon Kagaku Zasshi, v. 92, p. 154-159.

Reed, M. H., and Spycher, N. F., 1985, Boiling, cooling, and oxidation in epithermal systems: Numerical modeling approach; in Berger, B. R., and Bethke, P. M. (eds.), Geology and Geochemistry of Epithermal Systems: Society of Economic Geologists, Reviews in Economic Geology, v. 2.

Seward, T. M., 1973, Thiocomplexes of gold and the transport of gold in hydrothermal ore solutions: Geochimica et Cosmochimica Acta, v. 48, p. 121-134.

Seward, T. M., 1976, The stability of chloride complexes of silver in hydrothermal solutions up to 350°C: Geochimica et Cosmochimica Acta, v. 40, p. 1329-1341.

Seward, T. M., 1984, The formation of lead (II) chloride complexes to 300°C. A spectrophotometric study: Geochimica et Cosmochimica Acta, v. 48, p. 121-134.

Thorstenson, D. G., 1984, The concept of electron activity and its relation to redox potentials in aqueous geochemical systems: U.S. Geological Survey, Open-File Report 84-072, 67 p.

Weissberg, B. G., Dickson, F. W., and Tunell, G., 1966, Solubility of orpiment ($As_2 S_3$) in Na_2S-H_2O at 50-200°C and 100-1500 bars, with geologic appreciations: Geochimica et Cosmochimica Acta, v. 30, p. 815-827.

Weissberg, B. G., 1969, Gold-silver ore-grade precipitates from New Zealand thermal waters: Economic Geology, v. 64, p. 95-108.

Weissberg, B. G., Browne, P. R. L., and Seward, T. M., 1979, Ore metals in active geothermal systems; in Barnes, H. L. (ed.), Geochemistry of Hydrothermal Ore Deposits, Second edition: John Wiley and Sons, New York, p. 738-780.

Appendix Table 2.A1--Summary of thermodynamic data

t in °C
T in °K

Reaction	Equation	Temp. Range	Data Reference
1. $SiO_{2,qz} + 2H_2O = Si(OH)_4$	$\log K = 0.104 - 1162.87/T$	200-280°C	(1)
2. $3KAlSi_3O_8 + 2H^+ = KAl_3Si_3O_{10}(OH)_2 + 6SiO_{2,qz} + 2K^+$	$\log K = 5.1062 + 6.31 \times 10^{-4}T + 1302.5/T$	0-250°C	(2)
3. $H_2CO_3^* = HCO_3^- + H^+$	$\log K = 6.2 - 1.897 \times 10^{-2} T - 2.062 \times 10^3/T$	0-300°C	(3)
4. $CaCO_3 + H_2CO_3^* = Ca^{++} + 2HCO_3^-$	$\log K = -223.16 + 6552.81/T + 89.084 \log T - 0.0771 T$	100-300°C	(4)
5. $H_2CO_3^* = CO_{2,g} + H_2O$	$\log K_H = 3.2702 - 0.002515 T$	200-350°C	(5)
6. $PbS + 2Cl^- + 2H^+ = PbCl_2 + H_2S_{aq}$	$\log K = 19.417 - 9030.1/T$	200-300°C	(6,2,8)
7. $Au + 2H_2S = Au(HS)_2^- + H^+ + 1/2 H_2$	$\log K = 834.443 - 36945.17/T - 305.449 \log T + 0.11046 T$	200-300°C	(7,8)
8. $H_2S_{aq} = HS^- + H^+$	$\log K = -12.18 + 2377.5/T$	200-300°C	(8)
9. $H_2O_l = H_{2,g} + 1/2 O_{2,g}$	$\log K = 7.6 - 14564.13/T$	200-300°C	(9)
10. $HS^- + 2O_2 = SO_4^= + H^+$	$\log K = -31.75 + 49975.5/T$	200-300°C	(9)
Partition coefficients (for pure water solvent)			
11. $CO_{2,v}/CO_{2,l}$	$\log B = 4.7593 - 0.01092 t$	100-340°C	(10)
12. $H_2S_{,v}/H_2S_{,l}$	$\log B = 4.0547 - 0.00981 t$	100-340°C	(10)

Reference list for Appendix Table 2.A1

(1) Fournier, R. O., and Potter, R. W., 1982, A revised and expanded silica (quartz) geothermometer: Geothermal Resources Council Bulletin, v. 11, p. 3-9.

(2) Helgeson, H. C., 1969, Thermodynamics of hydrothermal systems at elevated temperatures and pressures: American Journal of Science, v. 267, p. 729-804.

(3) Read, A. J., 1975, The first ionization constant of carbonic acid from 25 to 250°C and to 2000 bars: Journal of Solution Chemistry, v. 4, p. 53-70.

(4) Ellis, A. J., 1963, Solubility of calcite in sodium chloride solutions at high temperatures: American Journal of Science, v. 261, p. 259-267.

(5) Ellis, A. J., and Golding, R. M., 1963, The solubility of carbon dioxide above 100°C in water and in sodium chloride solutions: American Journal of Science, v. 261, p. 47-60.

(6) Seward, T. M., 1984, The formation of lead (II) chloride complexes to 300°C: Geochimica et Cosmochimica Acta, v. 48, p. 121-134.

(7) Seward, T. M., 1973, Thiocomplexes of gold and the transport of gold in hydrothermal ore solutions: Geochimica et Cosmochimica Acta, v. 37, p. 379-399.

(8) Ellis, A. J., and Giggenbach, W. F., 1971, Hydrogen sulfide ionization and sulfur hydrolysis in high temperature solutions: Geochimica et Cosmochimica Acta, v. 35, p. 247-260.

(9) Henley, R. W., Truesdell, A. K., and Barton, P. B., 1984, Fluid mineral equilibria in hydrothermal systems, Chapter 8: Reviews in Economic Geology, Society of Economic Geologists, v. 1.

(10) Giggenbach, W. F., 1980, Geothermal gas equilibria: Geochimica et Cosmochimica Acta, v. 44, p. 2021-2032.

Chapter 3
THE BEHAVIOR OF SILICA IN HYDROTHERMAL SOLUTIONS
Robert O. Fournier

INTRODUCTION

Quartz and chalcedony are the silica minerals commonly found in hydrothermal ore deposits. However, in many places there is textural evidence that chalcedony formed after amorphous silica, probably with poorly crystalline cristobalite or opal-CT as an intermediate phase. Fournier (1973) and apparently White (1965) used the term β-cristobalite both for poorly crystalline cristobalite and for opal-CT. Poorly crystalline cristobalite shows broad X-ray diffraction peaks centered at about 4.1 and 2.5 Å. Opal-CT also shows these same X-ray peaks plus an additional low-tridymite peak at about 4.3 Å (Jones and Segnit, 1971).

Quartz is the stable form of silica at pressure-temperature conditions found in convecting hydrothermal systems. Faceted quartz crystals generally grow in solutions that are not greatly supersaturated with silica, indicating relatively slowly changing conditions. In contrast, the deposition of amorphous silica requires high degrees of silica supersaturation with respect to quartz, and generally indicates large and rapid changes in the physical or chemical nature of the solution. These large and rapid changes may also affect the capacity of a solution to transport and deposit ore. There are various ways to bring about this supersaturation such as rapid cooling (generally with decompressional boiling), mixing of different waters, pH changes, and reaction of the solution with volcanic glass. Each of these processes will be discussed in the subsequent sections. Much of what follows is taken from Fournier (1985).

SOLUBILITIES OF SILICA MINERALS

Experimentally determined solubilities of the common silica minerals in pure water at the vapor pressure of the solution up to 300°C are shown in Figure 3.1. Equations for those portions of these curves in the temperature range 0-250°C are given in the appendix. Equations expressing the molal solubility of quartz and amorphous silica in water at the vapor pressure of the solution over the temperature ranges 20°-330°C and 90°-340°C respectively are also given in the appendix, along with more general equations that can be used to calculate quartz solubilities in water and saline solutions at most temperatures and pressures of geologic interest. Increased pressure has little effect at 25°C, but causes significantly increased solubilities as the temperature is increased (Fournier and Potter, 1982a). Going from the vapor pressure of the solution to 1000 bars increases the solubility of quartz about 19 percent at 200°C, and about 36 percent at 300°C.

Cristobalite exhibits a wide range of crystallinities (Murata and Nakata, 1974), and different samples may exhibit a complete spectrum of solubilities ranging from α-cristobalite to that of glass. Well-crystallized tridymite should exhibit a solubility between those of α-cristobalite and chalcedony. When two or more silica minerals are in contact with a given solution, the most soluble silica phase will generally control aqueous silica until that phase completely dissolves, is converted to another more stable phase, or is taken out of contact with the circulating water by formation of intervening alteration products or precipitates (Fournier, 1973).

In the evaluation of hydrothermal processes, it is of particular interest to determine conditions leading to the precipitation of amorphous silica. Experimental determinations of the solubility of amorphous silica in water at high temperatures have been carried out by many investigators, including Fournier and Rowe (1977), and Marshall (1980). According to Weres et al.

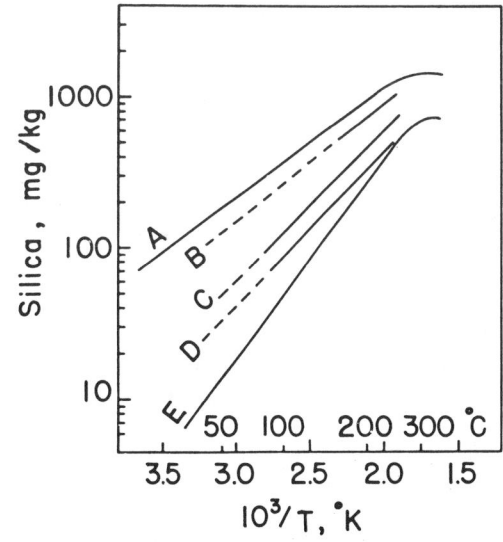

Figure 3.1. Solubilities of various silica phases in water at the vapor pressure of the solution. A - amorphous silica, B = opal-CT (incorrectly identified as α-cristobalite by Fournier, 1973), C = α-cristobalite, D = chalcedony, and E = quartz (from Fournier, 1973).

(1982), where degrees of silica supersaturation in respect to amorphous silica reach a factor of about 2.5, homogeneous nucleation is likely to occur throughout a solution, silica polymers grow past critical nucleus size, and colloidal particles grow by further deposition of silica on their surfaces. The colloidal particles finally coagulate or flocculate, producing gelatinous material or friable aggregates of weakly cemented particles. Eventually, additional deposition of silica in the interstices between these particles may transform a friable aggregate to relatively hard, dense material. If degrees of supersaturation are not great enough for homogeneous nucleation to occur, deposition of dissolved silica takes place through heterogeneous nucleation, and growth of amorphous silica takes place directly on solid surfaces. The resulting material is dense, vitreous silica that initially should contain much less water than does gelatinous material. High salinity and near-neutral pH favor more rapid rates of amorphous silica deposition (Morey et al., 1961; Makrides et al., 1980; Crerar et al., 1981; Weres et al., 1982).

Natural hydrothermal waters range from very dilute to concentrated salt solutions. The effects of dissolved salts on the solubility of amorphous silica have been investigated by Marshall (1980), Marshall and Warakomski (1980), Marshall and Chen (1982a, 1982b), Chen and Marshall (1982), and Fournier and Marshall (1983). Below about 300°C most dissolved salts, with the exception of Na_2SO_4, cause a slight decrease in the solubility of amorphous silica. Solubilities of quartz and other silica minerals should be similarly affected by dissolved salts.

Addition of Na_2SO_4 greatly increases the solubility of amorphous silica at all measured temperatures (Chen and Marshall, 1982; Marshall and Chen, 1982a; Fournier and Marshall, 1983), apparently by the formation of a silica-sulfate complex (Marshall and Chen, 1982b; Fournier and Marshall, 1983). It is not known if sodium is involved in that complex, or if other dissolved sulfates also form complexes with silica. The formation of a silica-sulfate complex also should increase the solubility of quartz in Na_2SO_4-rich solutions. These silica-sulfate complexes may be very important in some hydrothermal systems; at Cesano, Italy, a hot, Na-K-SO_4-rich brine with over 356,000 ppm total dissolved solids at 200° to 300°C was encountered in a 1435-m deep geothermal well (Calamai et al., 1976). Unfortunately, the concentration of dissolved silica in the brine at Cesano is not well known.

THE BEHAVIOR OF DISSOLVED SILICA IN HOT-SPRING SYSTEMS

Much is known about the behavior of dissolved silica in active hydrothermal systems as a result of studies of hot-spring waters (White et al., 1956; Morey et al., 1961; Fournier, 1973) and extensive drilling for geothermal resources (Mahon, 1966; Arnorsson and Sigurdsson, 1974; Arnorsson, 1975; Truesdell, 1976; Ellis and Mahon, 1977; Ellis, 1979; Henley and Ellis, 1983). In long-lived, presently active systems the solubility of quartz has been found to control dissolved silica in all geothermal reservoir waters at temperatures greater than about 180°C, most reservoir waters above about 140°C, and many above 90°C. Chalcedony has a slightly higher solubility than quartz (Fig. 3.1) and commonly controls silica at temperatures below 90° to 140°C, and sometimes as high as 180°C. However, under special conditions for short periods of time, cristobalite or volcanic glass may control dissolved silica, even at very high temperatures. Special conditions include active faulting, hydrothermal explosions, and "short-circuiting" by drilling that allows hot water to come into contact with glass- or cristobalite-rich rock that has not previously been in contact with circulating hot water.

In active hydrothermal systems very little silica appears to precipitate underground when water rises relatively quickly from reservoirs where temperatures are less than 230° to 250°C. This behavior allows the silica concentration in a hot-spring or well water to be used as a chemical geothermometer to estimate underground reservoir temperature (Fournier and Rowe, 1966; Mahon, 1966). Where reservoir temperatures are in excess of 230° to 250°C, enough quartz, chalcedony, or amorphous silica generally precipitates from an ascending solution to cause calculated silica geothermometer temperatures applied to hot-spring waters to be significantly low. These empirically derived generalizations, based on field observations, are corroborated by experimental studies (Rimstidt and Barnes, 1980).

Self-sealing by deposition of amorphous silica at the tops of active geothermal systems has been discussed by Bodvarsson (1964), Facca and Tonani (1967), White et al. (1971), Coplen et al. (1973), and Grindley and Browne (1976). Where quartz controls dissolved silica at depth, considerable cooling is generally required before amorphous silica can precipitate. Amorphous silica at 100°C has the same solubility in pure water as quartz at about 225°C. However, if an ascending solution rises fast enough to cool adiabatically (through boiling in response to lower pressure and without transfer of heat to the wallrock) and silica does not precipitate from solution during that ascent, the silica concentration in the residual liquid will increase as steam separates and a water in equilibrium with quartz at a reservoir temperature of 210°C will yield a solution just saturated with respect to amorphous silica at 100°C (Fig. 3.2).

In hot-spring systems at Yellowstone National Park, Fournier and Truesdell (1970) found that when silica does not precipitate from a thermal water during ascent from an underground reservoir, there generally is an exchange of cations between the solution and the wallrock during upward flow. The reverse is true where significant amounts of silica precipitate from an ascending thermal water, little or no cation exchange occurs. It appears that where a fracture is completely coated by precipitated silica minerals, the wallrock may be effectively removed from contact with the hydrothermal solution flowing through that fracture. In that event, ions in solution will not be able to react with minerals in the wallrock in response to changing temperature. Thus, the precipitation of silica may

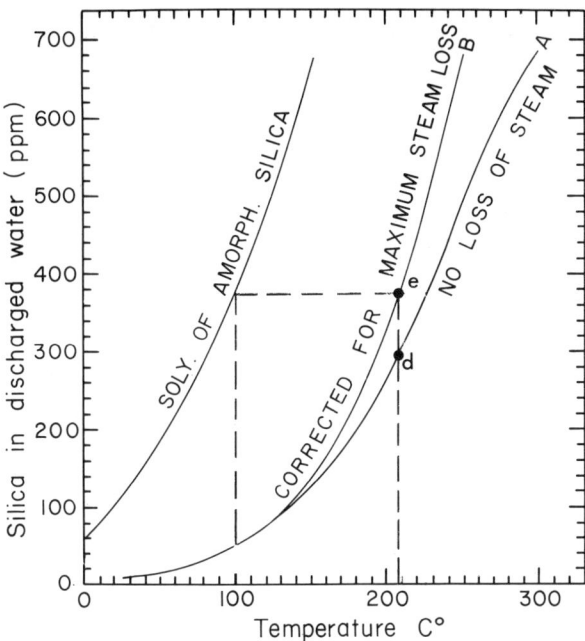

Figure 3.2. Solubilities of amorphous silica and quartz (curve A) at the vapor pressures of the solutions. Curve B shows the amount of dissolved silica that would be present after adiabatic cooling (boiling) to 100°C, plotted as a function of the initial temperature of the solution before boiling. Point d shows the initial concentration of dissolved silica in equilibrium with quartz at 210°C, and point e shows the concentration of dissolved silica after adiabatic cooling of that water from 210° to 100°C.

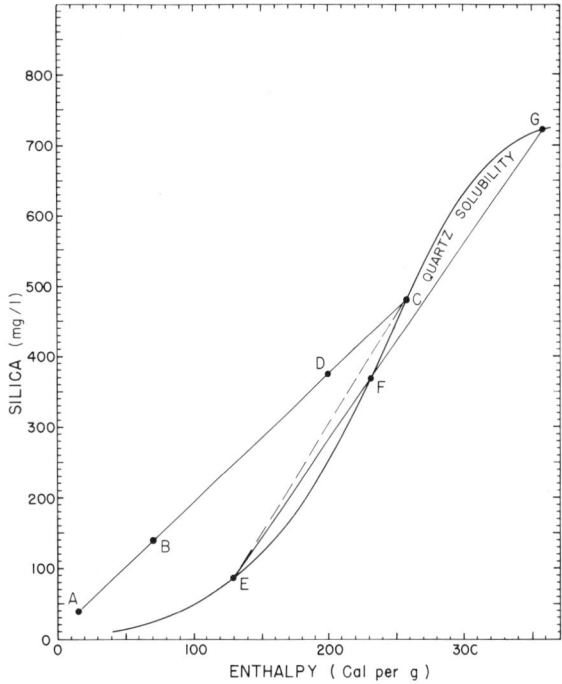

Figure 3.3. Solubility of quartz as a function of enthalpy at the vapor pressure of the solution. The straight lines A-C, E-C, and E-G show enthalpy-silica relations in solutions that result from mixing various proportions of hot and cold waters. See text for additional discussion.

adversely affect geochemical models applied to fluid flow that assume continuous water-rock chemical equilibrium, or buffering of the pH of the solution by reactions among silicate minerals, unless those minerals precipitate along with quartz, chalcedony, or amorphous silica. Where gelatinous silica coats the walls of a vein, some reaction of the solution with the wallrock might continue at a very slow rate by diffusion of ions through what amounts to a chromatographic column.

In some presently active hydrothermal systems, such as Broadlands, New Zealand, quartz appears to precipitate at the interface where relatively saline, hot water from deep in the system mixes with overlying, less saline, colder water (W. A. J. Mahon, oral communication, 1982). Quartz precipitation also may occur at the cool margins of hydrothermal systems where hot and cold waters mix (Mahon et al., 1980). The amount of silica supersaturation that will occur as a result of mixing waters having different initial temperatures can be determined graphically using an enthalpy-silica diagram (Fig. 3.3). In enthalpy-concentration diagrams, if heat of mixing effects are negligibly small, solutions that result from mixing cold and hot components lie along approximately straight lines joining the end members (Fournier, 1979), such as line A-C in Figure 3.3. Cold groundwaters commonly have dissolved silica concentrations ranging from 10 to 50 mg/kg. If the solution after mixing has a relatively high enthalpy (high temperature), such as point D in Figure 3.3, quartz is likely to precipitate. Note that even when two hot waters mix, each in equilibrium with quartz at different temperatures, the resulting solutions may be supersaturated with respect to quartz (line E-C in Fig. 3.3). However, mixing of some thermal waters, such as E and G may give either supersaturated (segment E-F) or undersaturated (segment F-G) silica solutions, depending on the proportion of the two waters in the mixture.

The cold-water component before mixing may be of relatively recent meteoric origin and still rich in oxygen from prior contact with the atmosphere. Mixing of this oxygen-rich, meteoric water with hot water could result in the formation of amethystine quartz because oxidizing conditions are required to produce ferric iron that gives amethyst its purple color (Frondel, 1962).

Chalcedony might precipitate directly from some mixed waters at appropriate low temperatures, but if the enthalpy of the solution after mixing is 100 Cal/g, such as point B in Figure 3.3, it is likely that little, if any, silica will deposit as a result of the mixing. This conclusion is based on the observation that the silica mixing model of Truesdell and Fournier (1977) works very well when applied to many hot-spring waters.

The thick, siliceous sinters that are being deposited from presently active hot-spring systems are composed predominantly of amorphous material that precipitates from neutral to slightly alkaline waters. These waters are neutral to slightly acid in the reservoirs at depth and commonly become alkaline during and after upward movement, owing to loss of CO_2 during boiling and evaporation. In contrast, highly acidic waters (pH below about 3) tend not to form thick siliceous deposits because hydrogen ions appear to inhibit the polymerization of dissolved silica (Rothbaum et al., 1979; Weres et al., 1982). Hydrogen ions have been added to some geothermal waters to prevent the formation of silica scale (Rothbaum et al., 1979). From the above information, it can be concluded that thick siliceous sinter deposits generally imply neutral to slightly alkaline waters (alkaline by loss of CO_2) that have risen relatively quickly from underground reservoirs where the waters were in equilibrium with quartz at temperatures greater than about 210°C. Waters coming from reservoirs with temperatures as low as 160°C may become supersaturated in respect to amorphous silica as a result of cooling below 100°C and atmospheric evaporation. However, silica that precipitates below 100°C is likely to be soft, easily eroded diatomaceous mud rather than hard sinter deposits that are relatively resistant to erosion.

It is not known what effect hydrogen ions might have upon precipitation of silica from Na-K-SO_4-rich brines. Because silica forms a complex with sulfate, where sulfate is removed from solution as by formation of alunite, exceedingly high-silica supersaturations may result, and silica may precipitate in an acid environment.

Although boiling and cooling of slightly acid to slightly alkaline waters appears to be the usual method of generating silica solutions supersaturated with respect to amorphous silica, other mechanisms are possible. These mechanisms involve very alkaline or very acid waters, reactions with volcanic glass, or sudden decreases in fluid pressure.

ALKALINE WATERS

High pH is a possible, though unlikely, cause of high concentrations of dissolved silica in most natural hydrothermal waters. Dissolved silica hydrolyzes to form silicic acid, H_4SiO_4, and some of this silicic acid dissociates to form $H_3SiO_4^-$ (equation (o) in the Appendix). The amount of dissociation of silicic acid can be calculated using equations (q), (r), and (s) that are given in the Appendix. High pH favors increased dissociation and, therefore, increased amounts of dissolved silica in solution. Figures 3.1 and 3.2 show the solubilities of silica minerals in water at near-neutral to acidic conditions. The effect of increased pH on the solubility of quartz at the vapor pressure of the solution is shown in Figure 3.4. For comparison, the solubility of amorphous silica at neutral pH also is shown in Figure 3.4. At temperatures above 130°C, quartz in alkaline solutions (pH 9.2 to 9.5) is more soluble than is amorphous silica in neutral solutions at the same temperature. Therefore, neutralization at constant temperature of an alkaline solution initially in equilibrium with quartz could lead to precipitation of amorphous silica. This mechanism requires (1) that a highly alkaline solution exist at high temperature, and (2) that little or no silica be consumed in the neutralization process.

Where water-rock chemical equilibrium is reached and quartz plus mica or clay are present in the rock, hydrogen ions are buffered by reactions such as

$3KAlSi_3O_8 + 2H^+$
K-feldspar

$= KAl_2(AlSi_3O_{10})(OH)_2 + 6SiO_2 + 2K^+$ (1)
  Kmica   Quartz

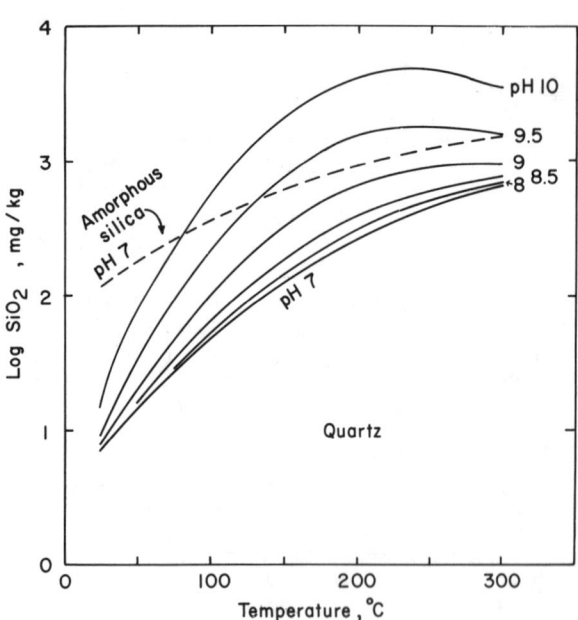

Figure 3.4. Solubility of quartz from 25° to 300°C at pH values ranging from 7 to 10 (solid lines). The dashed line shows the solubility of amorphous silica at neutral pH. Quartz solubilities at neutral pH are from Fournier and Potter (1982a), and ionization constants from Fleming and Crerar (1982) and Sweeton et al. (1974). Amorphous silica solubilities at neutral pH are from Fournier and Rowe (1977).

$$2.33 \text{NaAlSi}_2\text{O}_6 \cdot \text{H}_2\text{O} + 2\text{H}^+$$
$$\text{Analcime}$$

$$= \text{Na}_{.33}\text{Al}_2(\text{Al}_{.33}\text{Si}_{3.67}\text{O}_{10})(\text{OH})_2$$
$$\text{Na-montmorillonite}$$

$$+ \text{SiO}_2 + 2.33\text{H}_2\text{O} + 2\text{Na}^+ \quad (2)$$
$$\text{Quartz}$$

$$1.17\text{CaAl}_2\text{Si}_4\text{O}_{12} \cdot 2\text{H}_2\text{O} + 2\text{H}^+$$
$$\text{Wairakite}$$

$$= \text{Ca}_{.17}\text{Al}_2(\text{Al}_{.33}\text{Si}_{3.67}\text{O}_{10})(\text{OH})_2$$
$$\text{Ca-montmorillonite}$$

$$+ \text{SiO}_2 + 2.33\text{H}_2\text{O} + \text{Ca}^{+2} \quad (3)$$
$$\text{Quartz}$$

and

$$2\text{KAl}_2(\text{AlSi}_3\text{O}_{10})(\text{OH})_2 + 3\text{H}_2\text{O} + 2\text{H}^+$$
$$\text{Kmica}$$

$$= 3\text{Al}_2\text{Si}_2\text{O}_5(\text{OH})_4 + 2\text{K}^+ \quad (4)$$
$$\text{Kaolinite}$$

The equilibrium constants for the silicate pH-buffering reactions are such that pH values are generally less than 7, with higher salinity waters having lower pH values at any given temperature, and waters with given cation activities having lower pH values at higher temperatures. At water-rock equilibrium conditions, pH values greater than 7 are likely to be found in hydrothermal solutions only where no phyllosilicates are present to buffer the pH, or where there is no normative or modal quartz in the rock. When the precipitation of quartz controls aqueous silica concentrations, the activity of silica can be defined equal to unity so that the equilibrium constant, K_{eq}, for the reaction given by equation (1) is

$$K_{eq} = [\text{K}^+]^2 / [\text{H}^+]^2 \quad (5)$$

with square brackets indicating activities. When dissolved silica is not controlled by quartz or a more soluble silica mineral, such as in ultramafic and nepheline-bearing rocks that are devoid of silica minerals, the equilibrium constant is

$$K_{eq} = [\text{K}^+]^2 [\text{SiO}_2]^6 / [\text{H}^+]^2 \quad (6)$$

and hydrogen ions decrease (pH increases) drastically in response to the decreased silica activity. In near neutral and acid waters, dissolved silica appears to be predominately H_4SiO_4. With increasing pH, the solution is capable of dissolving increasing amounts of silica, but the activity of H_4SiO_4 remains relatively low because much of that dissolved silica is present as H_3SiO_4^-.

If the recharge water for a hydrothermal system comes from an alkaline, playa lake, continued water-rock reaction at high temperature (as recharge water flows through the system) could eventually convert all the pH-buffering clay and mica in the rock to other minerals that would allow sustained high pH values to be reached deep in the system. In that event, an alkali feldspar-carbonate assemblage would likely evolve and the isotopic composition of the oxygen in the minerals would be relatively rich in ^{18}O, reflecting the evaporated nature of the recharge water.

Subsurface boiling with removal of the evolved gas phase is another mechanism that can generate high pH in a hydrothermal system. Chemically, this is a non-equilibrium situation in which continuing water-mineral reactions (such as equations 1-4, proceeding from right to left) consume alkalies and liberate hydrogen ions at a slower rate than that at which hydroxyl is generated by reactions such as

$$\text{HCO}_3^- = \text{CO}_2 + \text{OH}^- \quad (7)$$

Down-hole water samples collected from research wells drilled into active hot-spring systems in Yellowstone National Park (White et al., 1975) generally have pH values of 6 to 7, measured at air temperature immediately after collection. However, one sample from the Y13 drill hole was found to have a pH of 9.2 (R. Fournier, unpublished data). That drill hole "short-circuited" a shallow permeable zone beneath the bottom of the casing and a deeper permeable zone, separated from the shallow zone by impermeable rock. The deeper zone had an initial temperature of about 205°C, and was slightly overpressured relative to the weight of an overlying column of water everywhere at boiling temperature. Because the shallow zone could accept water faster than the deep zone could supply it, a steady state developed in which boiling occurred in the deep aquifer and steam flowed up the drill hole and out into the shallow aquifer, even when the wellhead valve was closed. As steam flowed from the system, an alkaline residual solution accumulated in the bottom of the well, just below the deep permeable zone. A similar situation might occur as a result of earthquake activity fracturing the impermeable barrier. However, high pH generated by evolution of acid-forming gases from a boiling water is likely to be short-lived and limited in extent.

Where a high pH and a high dissolved silica concentration appropriate for that pH occur in a hydrothermal fluid, that fluid must be neutralized to cause the precipitation of amorphous silica. Furthermore, that neutralization process must be relatively fast to induce supersaturation with respect to quartz, and it must not consume a large amount of silica relative to the amount of hydrogen ion produced. Most water-mineral reactions are likely to proceed too slowly to produce high degrees of silica supersaturation. Also, reactions forming alkali feldspars, such as equation (1), consume three times as much silica as the amount of hydrogen ion that is liberated, and reactions involving montmorillonite and analcime or wairakite (equations 2 and 3) liberate two hydrogen ions for each silica consumed. Anorthite altering to epidote liberates hydrogen ion without consuming silica.

$$3\text{CaAl}_2\text{Si}_2\text{O}_8 + \text{Ca}^{+2} + 2\text{H}_2\text{O}$$
$$\text{Anorthite}$$

$$= 2\text{Ca}_2\text{Al}_3\text{Si}_3\text{O}_{12}(\text{OH}) + 2\text{H}^+ \quad (8)$$
$$\text{Epidote}$$

But, experimentally, epidote reacts very slowly, so quartz would be likely to precipitate if a solution were neutralized by this reaction.

Other mechanisms for decreasing the alkalinity of a hydrothermal solution include mixing with another neutral or acidic water, influx of acid gases evolved from another part of the system, oxidation of sulfides, and reactions involving native sulfur.

ACID WATERS

Acid alteration is found in and around many ore deposits. Quartz dissolves and precipitates very slowly in acid (pH <3) hydrothermal solutions at 200° to 350°C using aqueous HCl as the solvent (R. Fournier, unpublished data). Amorphous silica behaves similarly in HCl solutions (Kamiya et al., 1974; Rothbaum et al., 1979; Weres et al., 1982). Therefore, where large amounts of acid are added to a rock from an outside source, or quickly generated within a rock by oxidation reactions, acid attack upon feldspars and other silicates may release silica to solution, causing extreme supersaturation with respect to quartz and even with respect to amorphous silica. Neutralizing of such a solution could cause amorphous silica to precipitate. The possibility that silica might precipitate from an acid, sulfate-rich solution as a consequence of formation of alunite was discussed previously.

Acidity within hydrothermal systems can be generated in several ways. Acid-sulfate alteration is commonly observed above the water table in presently active hot-spring systems. The acidity results mainly from oxidation of H_2S that moves upward with steam that has separated from underground boiling water, and partly from oxidation of sulfides in the rock. The silicification that accompanies this acid attack is not due to influx of silica, but to strong leaching of alkalies that are flushed from the rock by condensed steam and meteoric water that percolate down to the water table.

Until very recently, drilling for production of geothermal energy in active hydrothermal systems had encountered extremely acid conditions at depth at only a few localities in zones of active volcanism, such as Matsao in Taiwan (Chen, 1970, 1975) and Onikobe (Yamada, 1976) and Matsukawa (Nakamura et al., 1970) in Japan. It now appears that there is deep, chloride-rich, acid thermal water and/or deep acid alteration (pyrophyllite plus quartz) in many active hydrothermal systems associated with active or relatively young andesitic volcanism. These include Biliran (Lawless and Gonzales, 1982), Nasuji-Sogonon (Seastres, 1982), Palimpinon (Leach and Bogie, 1982), and Baslay-Dauin (Harper and Arevalo, 1982) in the Philippines, Suretimeat (Heming et al., 1982) in the New Hebrides, and Miravalles in Costa Rica (R. Fournier, unpublished data). In some of these systems much of the acidity may come from reactions involving magmatic gases, including SO_2, H_2S, and HCl.

At Matsao and Onikobe acid-chloride waters with pH values less than 2 were found at depths greater than 1000 meters at temperatures exceeding 275°C. The reservoir at Matsao is in quartzite, and at Onikobe it is in andesite altered to pyrophyllite and quartz. Waters collected at intermediate depths, and alteration products found in cuttings from wells, show that these deep, acidic waters are neutralized by mixing with shallow ground water and by reaction with overlying volcanic rocks as the waters rise toward the surface. Therefore, the acidity does not appear to be the result of downward movement of directly overlying waters that had become acidic by surface oxidation. Ellis (1977) attributed the deep acidity at Matsao to the reaction of water with deeply buried native sulfur deposits, producing sulfuric acid and hydrogen sulfide,

$$4S + 4H_2O = 3H_2S + H_2SO_4 \qquad (9)$$

However, as mentioned above, some or all of the acidity could result from interaction of water with gases evolved from a crystallizing magma at depth, or from hydrolysis reactions between salt and water that occur at high temperatures and low pressures.

Iwasaki and Ozawa (1960) and Saki and Matsubaya (1977) present evidence for the generation of acidity by the reaction

$$4SO_2 + 4H_2O = 3H_2SO_4 + H_2S \qquad (10)$$

It is likely that sulfuric acid also can be generated by other mechanisms. The 1982 eruption of El Chichon volcano in Mexico contributed far more sulfuric acid to the atmosphere and stratosphere than is usual for comparably sized eruptions of other volcanoes, such as Mount St. Helens (B. Toon, oral communication, 1982). It is noteworthy that gypsum beds occur in the sedimentary section beneath El Chichon, but are not present beneath Mount St. Helens. Apparently relatively oxidized, sulfur-rich gases may be evolved where gypsum or anhydrite are involved in hydrolysis reactions at high temperatures

$$CaSO_4 + H_2O = H_2SO_4 + CaO \qquad (11)$$

The CaO that forms by reaction (11) will react with quartz and/or other silicates to form a variety of calcium silicates. The isotopic composition of sulfur in these volcanic systems should be useful for distinguishing between sulfate derived from gypsum and sulfate derived from volcanic SO_2.

The importance of HCl as a cause of acidity in hydrothermal systems should not be overlooked. Over 7000 mg/kg Cl as HCl were found in dry steam coming from a shallow well drilled at Hakone volcano in Japan (Kimio Noguchi, oral communication, 1970). Some or all of that HCl may have been generated by hydrolysis of NaCl at moderate to high temperatures and low pressures

$$NaCl + H_2O = NaOH + HCl \qquad (12)$$

Many investigators have found HCl in condensate after circulating dry steam over solid NaCl (Briner and Roth, 1948; Martynova and Samoilov, 1959; Galobardes et al., 1981). In experiments at 600°C I found that significant amounts of HCl are generated by reaction (12) at pressures below about 350 bars, with more HCl produced at lower pressures. Addition of quartz to the system greatly increased the yield of HCl. This occurs because NaOH is removed from the solution by reaction with quartz, with precipitated sodium silicates as products. Solubilities of sodium silicates decrease with increasing temperature (Rowe et al., 1967). In natural systems, where aluminum is available in clays and other minerals, albitization is likely to result from the hydrolysis of NaCl.

In some places, deep acidity may result from downward movement of water that has become acid at and near the water table (Oki and Hirano, 1970; Henley and Ellis, 1983). In the Norris Geyser Basin region of Yellowstone National Park, acid waters are generated high on a hillside where H_2S is oxidized to sulfate. Some of this water appears to percolate hundreds of meters underground where it mixes with high-temperature ($\sim 270°C$), neutral water rich in chloride. The resulting "acid-chloride-sulfate" waters, that issue as hot springs and geysers, have been extensively analyzed (Gooch and Whitfield, 1888; Allen and Day, 1935; Rowe et al., 1973).

REACTION WITH GLASS

It was noted above that the solubility of glass can control dissolved silica concentrations where new fractures bring hot water into contact with previously unaltered rock. Laboratory experiments at hydrothermal conditions show that volcanic glass contributes silica to solution to about the same extent as pure amorphous silica (Dickson and Potter, 1982). In the absence of rapid precipitation of quartz, chalcedony, or cristobalite, any cooling of a solution that has reacted extensively with volcanic glass could result in the precipitation of amorphous silica. The first stage in the precipitation process is likely to be polymerization of the dissolved silica, with formation of colloidal and gelatinous particles that can be swept along in a moving fluid.

Reaction of hot water with previously unaltered glass-rich volcanic rock may explain exceedingly high silica concentrations found in one vigorously discharging, boiling hot spring in Yellowstone National Park that is presently depositing gold-bearing sinter. That spring first appeared shortly after the magnitude 7.1 Hebgen Lake earthquake in 1959. When I first observed and sampled this hot spring, it was discharging more than 300 l/min., the water was fountaining as much as a meter above the vent, and it had an opalescent appearance owing to dispersed colloidal silica. The deposition of siliceous sinter around the edge of the pool and along the channels where water overflowed was especially rapid where water came in contact with organic material. One sample of this sinter was analyzed for gold by neutron activation; it contained 0.8 mg/kg. In this instance the gold could have been co-precipitated with the colloidal silica at depth and then transported to the surface along with that silica.

AMORPHOUS SILICA-CHALCEDONY RELATIONS

Recently deposited hot-spring siliceous sinters have been found to be composed predominantly of amorphous material. Sinters a few thousand years old tend to have an opaline appearance, especially if buried by younger deposits, and either a poorly crystalline cristobalite or opal-CT X-ray pattern. According to T. E. C. Keith (written communication, 1982), many X-ray diffraction patterns of older sinters from Yellowstone National Park indicate poorly crystalline cristobalite without the small tridymite peak that is characteristic of opal-CT. In contrast, the silica phase left as a residue in acid-altered and leached environments in Yellowstone National Park generally does exhibit a broad 4.3 Å tridymite X-ray peak in addition to the broad cristobalite peaks.

Many sinters tens of thousands of years old are chalcedonic, especially if buried and exposed to higher temperatures than those generally attained at the earth's surface. The transformation of amorphous silica to poorly crystalline cristobalite, opal CT, chalcedony, and quartz has been studied in the field and experimentally by many investigators (White et al., 1956; Carr and Fyfe, 1958; Fyfe and McKay, 1962; Heydemann, 1964; Ernst and Calvert, 1969; Mizutani, 1970; Murata and Nakata, 1974; Murata and Larson, 1975; Murata and Randall, 1975; Kastner et al., 1977; Hein and Scholl, 1978; Keith and Muffler, 1978; Kano and Taguchi, 1982). Time, high temperature, high pH, high salinity, and the presence of dissolved Mg all have been found to favor the transformation of amorphous silica to quartz (or chalcedony), with either poorly crystalline cristobalite or opal CT as an intermediate phase. It is possible that other dissolved constituents also may catalyze the transformation.

Apparently, chalcedony may form either by direct precipitation from solution without going through an amorphous silica stage, or by transformation of amorphous silica to crystalline material. It is important to know if chalcedony formed as a primary precipitate or by transformation of amorphous silica because of the implications about conditions required to precipitate one or the other. Morphologic features, such as dehydration cracks, slump structures, and thicker silica deposits on the bottoms of cavities than on the sides and tops (gravitational settling of amorphous silica particles) indicate that amorphous silica was present initially. In the absence of such features, the mechanism of formation of a given chalcedony is in doubt. The trace-element contents of primary chalcedonies and those formed by transformation of amorphous silica should be compared to determine if there are diagnostic differences. Elements that form small, highly charged, cations, such as Ge, Ga, B, and Al, might substitute for Si more readily in amorphous material than in crystalline and, therefore, might be useful in distinguishing among types of chalcedony.

SPECULATIONS REGARDING SOME TEXTURES OF QUARTZ

Where quartz precipitates in open spaces directly from hydrothermal solutions it exhibits crystal faces, and all the c crystallographic directions generally are roughly perpendicular to a surface upon which growth initially occurred. However, many hydrothermal veins exhibit bands of randomly oriented equant grains of, or uniform filling by, anhedral quartz that show no indications of earlier euhedral stages of growth. In deep deposits there tend to be no cavities in these veins. In shallower deposits some cavities may be present, and portions of quartz grains that extend into cavities are faceted. These textures suggest that the quartz may have formed by growth within a band that was initially amorphous silica, or chalcedony (less likely). If so, fluid inclusions within that quartz either were introduced after crystallization, or represent fluid that was trapped at the time of formation of the amorphous or chalcedonic silica and later remobilized during the transformation to quartz. Therefore, fluid-inclusion filling temperatures and salinities may give little or no information about the condition of the hydrothermal fluid that initially deposited the siliceous vein material.

The size and general appearance of quartz grains after amorphous silica probably depends upon many factors, including the nucleation mechanism and initial water content of the amorphous silica, the pH and chemical composition of the surrounding fluid, the temperature at which the amorphous silica deposited, and the rate at which the system cooled. In general, higher temperatures of transition of amorphous silica to quartz will result in coarser grained quartz. Equant, anhedral grains of milky-looking quartz with many tiny fluid inclusions might indicate crystallization from gelatinous amorphous silica that formed by homogeneous nucleation from a solution very supersaturated with silica. A likely mechanism for generating highly supersaturated silica at high temperatures is by a sudden decrease in fluid pressure (Fig. 3.5); possibly from nearly lithostatic to hydrostatic or lower. Grindley and Browne (1976) attributed the formation of strongly silicified breccias adjacent to open fissures at Wairakei and Broadlands, New Zealand to sudden decreases in fluid pressure resulting from hydraulic fractures propagating through self-sealed rock. The silicifying minerals are quartz plus adularia, usually accompanied by pyrite. Keith and Muffler (1978) suggested that simultaneous brecciation and deposition of amorphous silica occurred as a result of a sudden decrease in pressure, caused either by fracturing accompanying resurgent doming, or draining of a glacial lake that decreased the local hydrostatic pressure. Self-sealing and effects of sudden decreases in fluid pressure will be discussed further in the section on quartz solubility at high temperatures.

Jasperoid and Massive Replacement of Limestone by Silica

The term jasperoid is often used to categorize bodies of massive silica (commonly containing iron sulfides or oxides), irrespective of the size, shape, internal structure, or geologic setting. Lovering (1972) has described their characteristics and distribution in the United States. It is likely that they form by many mechanisms. Some may be recrystallized hot-spring sinter deposits. Others may have formed where rising, hot water mixed with shallow, cold water. Still others may have formed as a result of vigorous decompressional boiling, particularly where overpressured thermal fluid expanded into open, hydrostatically pressured cavities. Such throttling processes are discussed by Barton and Toulmin (1961) and Toulmin and Clark (1967). Most jasperoids, however, appear to be massive silica replacements of limestone or, much less commonly, dolomite (Lovering, 1972). They are particularly prevalent in carbonate-hosted gold deposits, such as Carlin (Radtke et al., 1980; Rye, 1985) and those in the Jerritt Canyon District (Hawkins, 1982), where massive, fine-grained silica (now quartz) replaces limestone with little or no associated calc-silicate. In those deposits the silicification of limestone is most pronounced near faults and where the limestone was initially most permeable.

Silica replacement of limestone requires the simultaneous dissolution of calcite and precipitation of silica. Below 300°C, at a constant partial pressure of CO_2, calcite becomes more soluble with decreasing temperature (Ellis, 1959, 1963; Holland and Malinin, 1979), while the solubilities of quartz, chalcedony, and amorphous silica decrease (Fig. 3.1). Also, at constant temperature, calcite becomes less soluble as the partial pressure of CO_2 decreases (Ellis, 1959). Therefore, slow cooling (without boiling) of a solution with a near neutral pH should promote replacement of limestone by silica. The silica phase that precipitates is likely to be quartz or chalcedony, because of the slow cooling. It is less likely that limestone will be replaced by silica where a near-neutral solution boils. Decompressional boiling results in loss of dissolved CO_2, as well as a rapid decrease in temperature, and calcite may dissolve or precipitate, depending on the composition of the hydrothermal solution, the pressure at which boiling is initiated, the drop in pressure, and whether the system is open or closed to loss of volatiles during that boiling. Escape of volatiles from the system during boiling is particularly likely to cause deposition of calcite, which would not favor replacement of limestone by silica.

It may be possible to accomplish massive replacement of limestone by silica with little or no cooling if the hydrothermal solution is acidic, even when that solution is just saturated with silica in respect to quartz. Reaction of a silica-saturated, acidic solution with limestone will generate CO_2, and some or all of it will dissolve in the solution, depending on several factors, including the temperature, partial pressure of CO_2, pH, and salinity. Dissolved CO_2 lowers the activity of water and decreases the solubility of quartz (Shettel, 1974). Therefore, quartz should precipitate at the interface of solution and limestone where CO_2 is generated. If the acidic solution were initially supersaturated with silica with respect to quartz, increasing the pH of the solution by reaction with calcite would probably cause precipitation of silica, even without the CO_2 effect. The deposited silica might be quartz, chalcedony, or amorphous, depending on the initial silica concentration in the acid solution and temperature at which the solution reacts with the limestone. By the above CO_2 mechanism, silica replacement of limestone at constant temperature would be more likely deeper in the system where fluid pressures are high enough for significant amounts of CO_2 to dissolve in the solution at the point where that gas is generated. At shallow levels, where the total fluid pressure is too low to allow much CO_2 to dissolve in the solution, silica replacement of limestone is more likely to occur as a result of reaction of that limestone with an acid solution that has become supersaturated with silica as a result of cooling, either by boiling or conductive heat loss.

Quartz Solubility at High Temperatures

It was previously noted that below about 300°C, pressure has a moderate affect and added salt has little affect upon the solubility of quartz. Above 300°C, both pressure and added salt are very important. Calculated solubilities of quartz in pure water over a wide range of temperatures and pressures, using the equation of Fournier and Potter (1982a), are shown in Figure 3.5. In that figure there is a solubility maximum (reported by Kennedy, 1950) that extends from about 340°C at the vapor pressure of solution to 520°C close to 900 bars. The shaded area in Figure 3.5 shows a region of retrograde solubility in pressure-temperature space. Where cold water is heated at constant pressure less than about 900 bars, it will dissolve more and more silica until either the solution starts to boil (at pressures below about 165 bars) or the solubility maximum is reached. With further heating that water will precipitate quartz. The precipitation of quartz in deep parts of a hydrothermal system may decrease the permeability to such an extent that little convecting meteoric water can attain temperatures much greater than those shown by the quartz solubility maximum in Figure 3.5 (Fournier, 1977, 1983a, 1983b). However, computer modeling by L. A. Keith and P. T. Delaney (written communication, 1984) shows that a completely impermeable seal is not likely to result in realistic times solely by deposition of quartz (or a mixture of quartz and other minerals) from a solution that is heated as it flows toward a heat source. This is because the process is self-limiting: as permeability is decreased by quartz deposition, the rate of flow decreases, which, in turn, decreases the rate of transport of silica to the place where deposition can occur. However, other factors also may contribute to the attainment of a completely impermaeble seal, such as quasi-plastic flow of rock that takes place at increasingly rapid rates as temperature increases. Temperature profiles calculated from heat-flow data for several localities in the western United States, and earthquake focal depths at those same locations, show that the temperature at which deformation changes from frictional (brittle fracture) to quasi-plastic flow ranges from about 300° to 450°C (Smith and Bruhn, 1984). This overlaps the 350° to 500°C temperature range in which self sealing by precipitation of quartz and other minerals is likely to occur when solutions are heated at constant pressure (Fournier, 1977, 1983a, 1983b; Sleep, 1983). Therefore, because of the above mentioned permeability reduction processes, the time interval over which meteoric water at hydrostatic pressure may interact directly with a shallow intruded

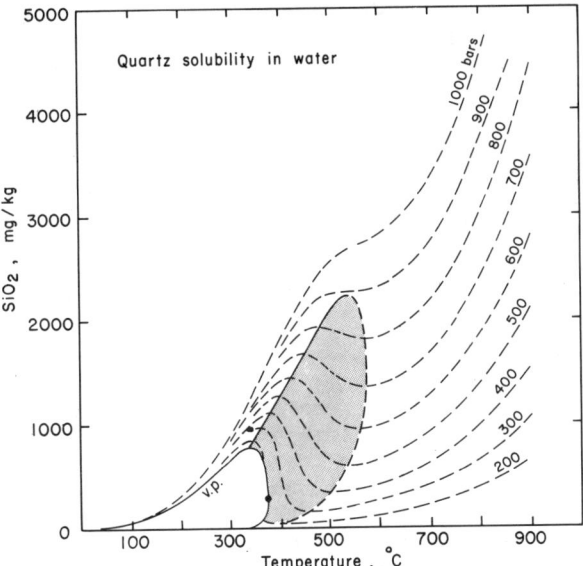

Figure 3.5. Calculated solubilities of quartz in water up to 900°C at the indicated pressures. The shaded area emphasizes a region of retrograde solubility.

body of magma (or still very hot rock) may be limited to the early stage of development of the hydrothermal system, or episodically thereafter with creation of new fractures by tectonic activity or thermal or hydraulic cracking (Secor, 1965; Phillips, 1973; Henley and McNabb, 1978).

This model has some interesting implications from the point of view of ore genesis. With the circulation of meteoric water through shallow, intrusive rocks cut off at an early stage in the cooling process, those intrusive rocks will cool at a slower rate, and hydrothermal activity will continue for a longer time. This is because heat must be transferred by conduction from the remaining very hot rock to the cooler convecting hot water. Conductive transfer of thermal energy is much less efficient than convective transfer.

Hydrothermal explosion activity is another possible consequence of the deposition of an impermeable quartz seal (Henley and McNabb, 1978). Large and steep temperature and pore-pressure gradients are likely to evolve where an impermeable zone becomes established about a heat source. Even though convective flow of meteoric water is cut off from the outside, the pore spaces within the zone between the quartz-sealed barrier and the remaining very hot rock are likely to contain water or brine. This fluid may be all or part meteoric or connate water, left over from before the silica sealing became complete. However, some or all of that fluid could be volatiles evolved from a crystallizing magma (Burnham, 1967, 1979). If volatiles continue to be evolved from a crystallizing magma, it is easy to envision a situation in which the fluid pressure on the high-temperature side of the quartz seal becomes very large (Phillips, 1973); sufficiently large to cause formation of a breccia pipe or even a conduit for a volcanic eruption (Morey, 1922).

Hydraulic fracturing will occur when the pore-fluid pressure exceeds the confining pressure (the least principle stress) by an amount equal to the tensile strength of the rock. The confining pressure may range from less than normal hydrostatic to lithostatic, depending on whether open fissures are present, the nature of the fluid in those fissures, and permeability relations. Propogation of either a hydraulic or tectonic fracture through impermeable rock from a region of high fluid pressure into a region of lower fluid pressure may cause a significant decompression of the high-pressure fluid. If the thermal energy in the decompressing liquid and surrounding rock is large, massive flashing of water to steam may result. The expanding steam may explosively propel rock fragments into the air where flashing occurs at relatively shallow levels, and into cavities and open fissures at deeper levels. Even without a magmatic contribution to the trapped fluids, pore pressures of those fluids could increase sufficiently to rupture the enclosing rock as a result of conductive heating.

Hydrothermal explosive activity may be an important contributing factor to ore deposition for various reasons, both physical and chemical. Brecciation greatly increases the permeability, providing easy access for later hydrothermal fluids that may deposit ores. When a hydrothermal explosion occurs, a lot of water is converted to steam. Other volatile constituents, such as H_2S and CO_2, initially dissolved in the liquid phase, are preferentially partitioned into that steam. This partitioning of volatiles, in turn, may increase the pH of the residual liquid. At the same time the concentrations of the non-volatile constituents remaining in the liquid phase increase as a result of the separation of steam, while the solubilities of minerals generally decrease as a result of the decrease in pressure. In Figure 3.5, note the large decrease in quartz solubility with decreasing pressure at temperatures above about 340°C. In addition, where boiling occurs, the temperature of the system should decrease because thermal energy is required to convert liquid water to steam. The above factors generally favor deposition of silica, sulfides, and noble metals. Whether or not ore is deposited will depend in part on the metal and sulfur content of the initial fluid. However, when initial temperatures are above about 340°C, quartz and other minerals should precipitate when and where there is a sudden drop in pore pressure. Also, any K-feldspar that precipitates along with the quartz as a result of a sudden drop in pressure is likely to be more potassium-rich than that which was in equilibrium with the fluid prior to the drop in pressure (Fournier, 1976). At Wairakei and Broadlands, quartz, adularia, and generally pyrite are the phases observed in the hydrothermal breccias that formed at about 200°-300°C (Grindley and Browne, 1976). Where exceptionally high degrees of silica supersaturation occur, particularly at lower temperatures, amorphous silica may precipitate and then alter to quartz or chalcedonic silica.

Many of the conclusions in the above discussion were based on the solubility behavior of quartz in pure water. The effects of added salts can be modeled using NaCl solutions. Calculated solubilities of quartz in aqueous NaCl, using the method of Fournier (1983b), show that adding dissolved salts should change the positions of the quartz solubility maxima and the extent of the field of retrograde quartz solubility that are shown in Figure 3.5. However, the conclusion that quartz deposition should contribute significantly to the formation of an impermeable barrier that prevents fluids at hydrostatic pressure from interacting directly with very hot rock or magma, is not changed by adding salt to the system. Figure 3.6 shows the effect of 5 and 18 weight percent NaCl upon quartz solubility at 500 bars pressure and high temperatures. Added NaCl greatly increases the solubility of quartz at temperatures above about 300°C, and shifts the solubility maximum toward higher temperatures. At very high concentrations of salt, the vapor-pressure curve may be intersected before a solubility maximum is attained, such as at point A in Figure 3.6. In that event, the solution will boil if the temperature is increased further without increasing the pressure. Wherever a solution boils, the concentration of dissolved silica in the residual solution should increase at a more rapid rate than can be accommodated by the increasing salinity. The amount of silica that can dissolve in the newly formed gas or steam phase is generally insufficient to offset the supersaturation generated in the residual brine, and quartz should precipitate because the temperature is high. Thus, quartz veining should occur either when a dilute solution is heated to a temperature above the quartz

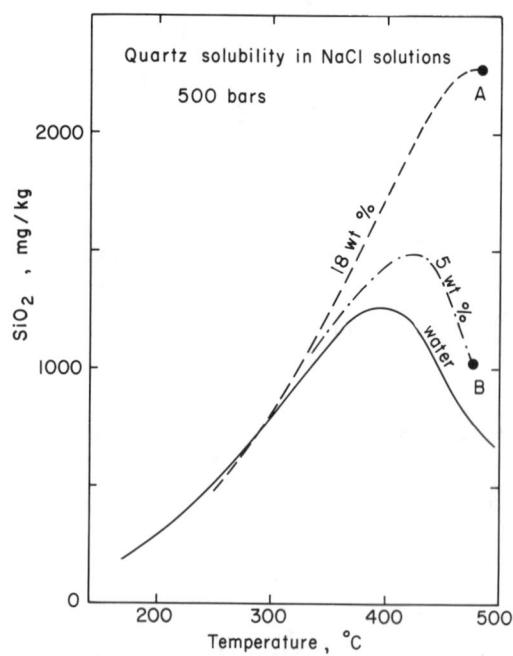

Figure 3.6. Comparison of calculated quartz solubilities (Fournier, 1983b) in water and 5 and 18 weight percent aqueous NaCl at 500 bars and the indicated temperatures.

solubility maximum or when a saline solution exceeds the temperature of the vapor-pressure curve at a given pressure.

Temperatures and approximate depths at which boiling will occur in water and 5 and 18 weight percent NaCl solutions are shown in Figure 3.7. Hydrostatic conditions are assumed in Figure 3.7, with pressure fixed by an overlying liquid with an average density of 1 throughout a vertical column up to the ground surface. Relative to the depth scale, lower assumed average densities will move the boiling point curves down, and higher assumed densities will move the curves up. The approximate positions of the quartz solubility maxima for water and 5 weight percent NaCl also are shown in Figure 3.7.

Because the initial permeability several kilometers deep in a hydrothermal system is likely to be limited to a few widely spaced fractures or fractured zones of rock, an impermeable zone resulting in great part from quartz deposition in those few fractures may go unrecognized as a significant feature. Also, in fossil hydrothermal systems where estimated temperatures at the time of vein formation are greater than 340°C, it may be difficult to determine whether a given quartz vein deposited as a result of increasing or decreasing temperature. If there is other hydrothermal alteration associated with the quartz deposition, that alteration may give an indication of the thermal history: albite is likely to form in veins and after K-feldspar where a solution is heating and K-feldspar or muscovite would be deposited in veins and after plagioclase where a solution is cooling (Hemley et al., 1971).

CONCLUSIONS

In well-established hydrothermal systems, where water remains in contact with the surrounding rock at a given high temperature for more than a few days or weeks, quartz controls aqueous silica (Rimstidt and Barnes, 1980). Slow cooling of a hydrothermal solution generally will result in the deposition of quartz if initial temperatures are between about 200° and 340°C. Rapid cooling allows supersaturated silica solutions to form, particularly when the cooling is predominantly the result of decompressional boiling. Supersaturated silica solutions also may evolve where hot water dissolves glass-bearing rocks, and where rocks are altered by very acid solutions. High alkalinity (high pH) is generally not important in most natural hydrothermal systems, but might be a factor in a few places for short periods of time.

Slight silica supersaturation in respect to quartz is required for chalcedony to precipitate directly from solution. Chalcedony appears to form and persist only at temperatures below about 180°C. Large degrees of silica supersaturation are required for amorphous silica to precipitate. Voluminous deposits of siliceous sinter generally indicate deposition from neutral to slightly alkaline (by loss of CO_2), chloride-rich waters that flowed quickly to the surface from a reservoir with a temperature in the range 200° to 270°C. Waters flowing from reservoirs with lower temperatures contain too little silica to form thick, hard, sinter deposits. Waters flowing from reservoirs with temperatures above 270°C contain so much dissolved silica that significant amounts precipitate in the channelways leading to the surface, stopping hot-spring activity before large sinter deposits can form. Very little silica precipitates from waters with pH values below about 3 to 4.

Amorphous silica is relatively unstable and transforms to poorly crystalline cristobalite, opal CT and chalcedony or quartz. The time required for these transformations depends upon temperature and the composition of fluid in contact with the amorphous material. Morphologic features show that some chalcedonies have formed after amorphous silica. Where such morphologic features are absent, the origin of a given chalcedony is in doubt. It is important to know if chalcedony formed as a primary precipitate or by transformation of amorphous silica because of the implications about conditions required to precipitate one or the other.

At temperatures above about 300°C, increased pressure and added salt greatly increase the solubility of quartz. However, when a solution is heated at constant pressure, eventually it will either boil or enter a field of retrograde quartz solubility. In either event, quartz should precipitate, decreasing the permeability in the hottest part of a convecting hydrothermal system. This precipitation of quartz is likely to occur at 300° to 550°C, depending on the depth of circulation and the salinity of the system. An impermeable barrier may form by a combination of

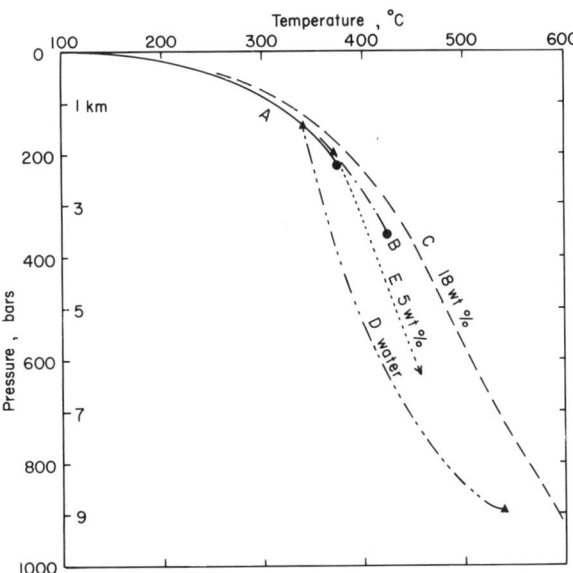

Figure 3.7. Temperature vs. depth (pressure) diagrams showing boiling point curves for pure water (curve A), 5 weight percent aqueous NaCl (curve B), and 18 weight percent aqueous NaCl (curve C). Also shown are the positions of quartz solubility maxima in water (curve D) and 5 weight percent aqueous NaCl (curve E).

quartz deposition and quasi-plastic flow that prevents subsequent direct contact of circulating water at hydrostatic pressure with very hot rock, or magma. This, in turn, will increase the time required to cool a shallow, intrusive magma, and might lead to very steep temperature and pore-pressure gradients across the impermeable barrier. If the impermeable barrier is then ruptured by seismic activity or increasing pore pressure in the confined, high-pressure side of the system, a hydrothermal explosion may occur. This may be a mechanism by which some breccia pipes form. The sudden decrease in density of the pore fluids, formation of steam, separation of gases, and decrease in temperature accompanying the hydrothermal explosion should cause silica and K-rich feldspar to precipitate along with other minerals.

ACKNOWLEDGMENTS

Portions of this manuscript were reviewed by B. R. Berger, J. W. Hedenquist, and D. E. White. It has benefited greatly from their comments and suggestions.

REFERENCES

Allen, E. T., and Day, A. L., 1935, Hot springs of the Yellowstone National Park: Carnegie Institute of Washington Publication 466, 525 p.

Arnorsson, S., 1975, Application of the silica geothermometer in low-temperature hydrothermal areas in Iceland: American Journal of Science, v. 275, p. 763-784.

Arnorsson, S., and Sigurdsson, S., 1974, The utility of water from the high-temperature area in Iceland for space heating as determined by their chemical composition: Iceland National Energy Authority, Department of National Heat, Report OSJHD 7426, 34 p.

Barton, P. B., and Toulmin, P., III, 1961, Some mechanisms for cooling hydrothermal fluids: U.S. Geological Survey, Professional Paper 424-B, p. 348-352.

Bodvarsson, G., 1964, Utilization of geothermal energy for heating purposes and combined schemes involving power generation, heating and/or by-products: Geothermal Energy II: United Nations Conference on New Energy, 1961, Proceedings, v. 3, p. 429-436.

Briner, E., and Roth, P., 1948, Recherches sur l'hydrolyse par la vapeur d'eau de chlorures alcalins seuls ou additions de divers adjuvants: Helvetica Chimica Acta, v. 31, p. 1352-1360.

Burnham, C. W., 1967, Hydrothermal fluids at the magmatic stage; in Barnes, H. L. (ed.), Geochemistry of Hydrothermal Ore Deposits: Holt, Rinehart and Winston, New York, p. 34-76.

Burnham, C. W., 1979, Magmas and hydrothermal fluids; in Barnes, H. L. (ed.), Geochemistry of Hydrothermal Ore Deposits, 2nd Edition: John Wiley and Sons, New York, p. 71-136.

Calamai, A., Cataldi, R., Dall'Aglio, M., and Ferrara, G. A., 1976, Preliminary report on the Cesano hot brine deposit (northern Latium, Italy): Proceedings, 2nd United Nations Symposium of Development and Use of Geothermal Resources, v. 1, p. 305-313.

Carr, R. M., and Fyfe, W. S., 1958, Some observations on the crystallization of amorphous silica: American Mineralogist, v. 43, p. 908-916.

Chen, C. H., 1970, Geology and geothermal power potential of the Tatun volcanic region: Proceedings from the United Nations Symposium on the Development and Utilization of Geothermal Resources: Geothermics, Special Issue 2, v. 2, pt. 2, p. 1134-1143.

Chen, C. H., 1975, Thermal waters in Taiwan, a preliminary study: Proceedings of the International Association of Hydrological Sciences, Grenoble, Publication 199, p. 79-88.

Chen, C. T. A., and Marshall, W. L., 1982, Amorphous silica solubilities--IV. Behavior in pure water and aqueous sodium chloride, sodium sulfate, magnesium chloride, and magnesium sulfate up to 350°C: Geochimica et Cosmochimica Acta, v. 46, p. 279-287.

Coplen, T. B., Combs, J., Elders, W. A., Rex, R. W., Burckhatter, G. C., and Laird, R., 1973, Preliminary findings of an investigation of the Dunes thermal anomaly, Imperial Valley, California: California Department of Water Resources, 48 p.

Crerar, D. A., Axtmann, E. V., and Axtmann, R. C., 1981, Growth and ripening of silica polymers in aqueous solutions: Geochimica et Cosmochimica Acta, v. 45, p. 1259-1266.

Dickson, F. W., and Potter, J. M., 1982, Rock-brine chemical interactions: Final Report, Electric Power Research Institute Project 653-2, AP-2258, 89 p.

Ellis, A. J., 1959, The solubility of calcite in carbon dioxide solutions: American Journal of Science, v. 257, p. 354-365.

Ellis, A. J., 1963, The solubility of calcite in sodium chloride solutions at high temperatures: American Journal of Science, v. 261, p. 259-267.

Ellis, A. J., 1977, Chemical and isotopic techniques in geothermal investigations: Geothermics, v. 5, p. 3-12.

Ellis, A. J., 1979, Explored geothermal systems; in Barnes, H. L. (ed.), Geochemistry of Hydrothermal Ore Deposits, 2nd Edition: John Wiley and Sons, New York, p. 632-683.

Ellis, A. J., and Mahon, W. A. J., 1977, Chemistry and Geothermal Systems: Academic Press, New York, 392 p.

Ernst, W. G., and Calvert, S. E., 1969, An experimental study of the recrystallization of porcelanite and its bearing on the origin of some bedded cherts: American Journal of Science, v. 267, p. 114-133.

Facca, G., and Tonani F., 1967, The self-sealing geothermal field: Bulletin of Volcanology, v. 30, p. 271-273.

Fleming, B. A., and Crerar, D. A., 1982, Silicic and ionization and calculation of silica solubility at elevated temperature and pH--Application to geothermal fluid processing and reinjection: Geothermics, v. 11, p. 15-29.

Fournier, R. O., 1973, Silica in thermal waters: Laboratory and Field investigations; in Proceedings of International Symposium on Hydrogeochemistry and Biogeochemistry, Japan 1970, Volume 1, Hydrogeochemistry: J. W. Clark (publisher), Washington, D.C., p. 122-139.

Fournier, R. O., 1976, Exchange of Na^+ and K^+ between water vapor and feldspar phases at high temperature and low-vapor pressure: Geochimica et Cosmochimica Acta, v. 40, p. 1553-1561.

Fournier, R. O., 1977, Constraints on the circulation of meteoric water in hydrothermal systems imposed by the solubility of quartz (abs.): Geological Society of America, Abstracts with Programs, v. 9, p. 979.

Fournier, R. O., 1979, Geochemical and hydrological considerations and the use of enthalpy-chloride diagrams in the prediction of underground conditions in hot-spring systems: Journal Volcanology Geothermal Research, v. 5, p. 1-16.

Fournier, R. O., 1981, Application of water geochemistry to geothermal exploration and reservoir engineering; in Rybach, L., and Muffler, L. J. P. (eds.), Geothermal Systems: Principles and Case Histories: John Wiley and Sons, New York, p. 109-143.

Fournier, R. O., 1983a, Self-sealing and brecciation resulting from quartz deposition within hydrothermal systems: Extended abstracts, Fourth International Symposium on Water-Rock Interaction, Misasa, Japan, p. 137-140.

Fournier, R. O., 1983b, A method of calculating quartz solubilities in aqueous sodium chloride solutions: Geochimica et Cosmochimica Acta, v. 47, p. 579-586.

Fournier, R. O., 1985, Silica minerals as indicators of conditions during gold deposition; in Tooker, E. W. (ed.), Geologic Characteristics of the Sediment- and Volcanic-hosted Types of Gold Deposits--Search for an Occurrence Model: U.S. Geological Survey, Bulletin 1646, p. 15-26.

Fournier, R. O., and Marshall, W. L., 1983, Calculation of amorphous silica solubilities at 25° to 300°C and apparent hydration numbers in aqueous salt solutions using the concept of effective density of water: Geochimica et Cosmochimica Acta, v. 47, p. 587-596.

Fournier, R. O., and Potter, R. W. II, 1982a, An equation correlating the solubility of quartz in water from 25° to 900°C at pressures up to 10,000 bars: Geochimica et Cosmochimica Acta, v. 46, p. 1969-1973.

Fournier, R. O., and Potter, R. W. II, 1982b, A revised and expanded silica quartz geothermometer: Bulletin Geothermal Resources Council, v. 11, no. 10, p. 3-12.

Fournier, R. O., and Rowe, J. J., 1966, Estimation of underground temperatures from the silica content of water from hot springs and wet-steam wells: American Journal of Science, v. 264, p. 685-697.

Fournier, R. O., and Rowe, J. J., 1977, The solubility of amorphous silica in water at high temperatures and high pressures: American Mineralogist, v. 62, p. 1052-1056.

Fournier, R. O., and Truesdell, A. H., 1970, Chemical indicators of subsurface temperature applied to hot-spring waters of Yellowstone National Park, Wyoming; in Proceedings from the United Nations Symposium on the Development and Utilization of Geothermal Resources: Geothermics (Special Issue 2), v. 2, Part 1, p. 529-535.

Frondel, C., 1962, The System of Mineralogy. Volume III, Silica Minerals: John Wiley and Sons, New York, 334 p.

Fyfe, W. S., and McKay, D. S., 1962, Hydroxyl ion catalysis of the crystallization of amorphous silica at 330°C and some observations on the hydrolysis of albite solutions: American Mineralogist, v. 47, p. 83-89.

Galobardes, D. R., Van Hare, D. R., and Rogers, L. B., 1981, Solubility of sodium chloride in dry steam: Journal of Chemical Engineering Data, v. 26, p. 363-366.

Gooch, F. A., and Whitfield, J. E., 1888, Analyses of waters of the Yellowstone National Park, with an account of the methods of analysis employed: U.S. Geological Survey, Bulletin 47, 84 p.

Grindley, G. W., and Browne, P. R. L., 1976, Structural and hydrologic factors controlling the permeabilities of some hot-water geothermal fields; in Proceedings of the Second United Nations Symposium on the Development and Use of Geothermal Resources, San Francisco, 1975: v. 1, p. 377-386, U.S. Government Printing Office.

Harper, R. T., and Arevalo, E. M., 1982, A geoscientific evaluation of the Basley-Danin prospect, Negros Oriental, Philippines: Proceedings of the Pacific Geothermal Conference, 1982, 4th New Zealand Geothermal Workshop, pt. 1, p. 235-240.

Hawkins, R. B., 1982, Discovery of the Bell Mine-Jerritt Canyon District, Elko County, Nevada: Mining Congress Journal, v. 68, p. 28-32.

Hein, J. R., and Scholl, D. W., 1978, Diagenesis and distribution of late Cenozoic volcanic sediments in the southern Bering Sea: Geological Society of America Bulletin, v. 89, p. 197-210.

Heming, R. F., Hochstein, M. P., and McKenzie, W. F., 1982, Suretimeat geothermal system: An example of a volcanic geothermal system: Proceedings of the Pacific Geothermal Conference, 1982, 4th New Zealand Geothermal Workshop, pt. 1, p. 247-250.

Hemley, J. J., Montoya, J. W., Nigrini, A., and Vincent, H. A., 1971, Some alteration reactions in the system $CaO-Al_2O_3-SiO_2-H_2O$: Society of Mining Geology Japan, Specical Issue 2, p. 58-63.

Henley, R. W., and Ellis, A. J., 1983, Geothermal systems ancient and modern: A geochemical review: Earth-Science Reviews, v. 19, p. 1-50.

Henley, R. W., and McNabb, A., 1978, Magmatic vapor plumes and ground-water interaction in porphyry copper emplacement: Economic Geology, v. 73, p. 1-19.

Heydemann, A., 1964, Untersuchungen uber die bildungsbeidugungen von quartz im temperaturbereich zwischen 100°C and 250°C: Beitrage zur Mineralogie und Petrographie, v. 10, p. 242-259.

Holland, H. D., and Malinin, S. D. 1979, On the solubility and occurrence of non-ore minerals; in Barnes, H. L. (ed.), Geochemistry of Hydrothermal Ore Deposits, 2nd Edition: John Wiley and Sons, New York, p. 461-508.

Iwasaki, I., and Ozawa, T., 1960, Genesis of sulfate in acid hot springs: Bulletin Chemical Society of Japan, v. 33, p. 1018-1019.

Jones, J. B., and Segnit, E. R., 1971, The nature of opal. I. Nomenclature and constituent phases: Journal of Geological Society of Australia, v. 18, p. 57-68.

Kamiya, H., Ozaki, A., and Imahashi, M., 1974, Dissolution rate of powdered quartz in acid solution: Geochemical Journal, v. 8, p. 21-26.

Kano, K., and Taguchi, K., 1982, Experimental study on the ordering of opal-CT: Geochemical Journal, v. 16, p. 33-41.

Kastner, M., Keene, J. B., and Gieskes, J. M., 1977, Diagenesis of siliceous oozes. I. Chemical controls on the rate of opal-A to opal-CT transformation--an experimental study: Geochimica et Cosmochimica Acta, v. 41, p. 1041-1059.

Keenan, J. H., Keyes, F. G., Hill, P. G., and Moore, J. G., 1969, Steam Tables (international edition-metric units): John Wiley and Sons, New York, 162 p.

Keith, T. E. C., and Muffler, L. J. P., 1978, Minerals produced during cooling and hydrothermal alteration of ash flow tuff from Yellowstone drill hole Y-5: Journal of Volcanology Geothermal Research, v. 3, p. 373-402.

Kennedy, G. C., 1950, A portion of the system silica-water: Economic Geology, v. 45, p. 629-653.

Lawless, J. V., and Gonzales, R. C., 1982, Geothermal geology and review of exploration, Biliran Island: Proceedings of the Pacific Geothermal Conference, 1982, 4th New Zealand Geothermal Workshop, pt. 1, p. 161-166.

Leach, T. M., and Bogie, I., 1982, Overprinting of hydrothermal regimes in the Palimpinon Geothermal Field, Southern Negros, Philippines: Proceedings of the Pacific Geothermal Conference, 1982, 4th New Zealand Geothermal Workshop, pt. 1, p. 179-184.

Lovering, T. G., 1972, Jasperoid in the United States. Its characteristics, origin, and economic significance: U.S. Geological Survey, Professional Paper 710, 164 p.

Mahon, W. A. J., 1966, Silica in hot water discharged from drill holes at Wairakei, New Zealand: New Zealand Journal of Science, v. 9, p. 135-144.

Mahon, W. A. J., Klyen, L. E., and Rhode, M., 1980, Neutral sodium/bicarbonate/sulphate hot waters in geothermal systems: Chinetsu (Journal Japanese Geothermal Energy Association), v. 17, p. 11-24.

Makrides, A. C., Turner, M., and Slaughter, J., 1980, Study of silica scaling from geothermal brines: Journal of Colloid and Interface Science, v. 73, p. 345-367.

Marshall, W. L., 1980, Amorphous silica solubilities. II. Activity coefficient relations and predictions of solubility behavior in salt solutions, 0-300°C: Geochimica et Cosmochimica Acta, v. 44, p. 925-931.

Marshall, W. L., and Chen, C. T. A., 1982a, Amorphous silica solubilities. V. Predictions of solubility behavior in aqueous mixed electrolyte solutions to 300°C: Geochimica et Cosmochimica Acta, v. 46, p. 289-291.

Marshall, W. L., and Chen, C. T. A., 1982b, Amorphous silica solubilities. VI. Postulated sulfate-silicic acid solution complex: Geochimica et Cosmochimica Acta, v. 46, p. 367-370.

Marshall W. L., and Warakomski, J. M., 1980, Amorphous silica solubilities. II. Effect of aqueous salt solutions at 25°C: Geochimica et Cosmochimica Acta, v. 44, p. 915-924.

Martynova, O. I., and Samoilov, Yu. F., 1959, Dissolution of sodium chloride in an atmosphere of water vapor of high parameters (Transactions): Zhurnal Neorganisheskoi Khimii, v. II, no. 12, p. 2829-2833.

Mizutani, S., 1970, Silica minerals in the early stage of diagenesis: Sedimentology, v. 15, p. 419-436.

Morey, G. W., 1922, The development of pressure in magmas as a result of crystallization: Washington Academy of Science Journal, v. 12, p. 219-230.

Morey, G. W., Fournier, R. O., Hemley, J. J., and Rowe, J. J., 1961, Field measurements of silica in water from hot springs and geysers in Yellowstone National Park; in Short Papers in the Geologic and Hydrologic Sciences: U.S. Geological Survey, Professional Paper 424-C, p. C333-336.

Murata, K. J., and Larson, R. R., 1975, Diagenesis of Miocene siliceous shales, Temblor Range, California: U.S. Geological Survey, Journal of Research, v. 3, p. 553-566.

Murata, K. J., and Nakata, J. K., 1974, Cristobalite stage in the diagenesis of diatomaceous shales, Temblor Range, California: Science, v. 184, p. 567-568.

Murata, K. J., and Randall, R. G., 1975, Silica mineralogy and structure of the Monterey Shale, Temblor Range, California: U.S. Geological Survey, Journal of Research, v. 30, p. 567-572.

Nakamura, H., Sumi, K., Katagiri, K., and Iwata, T., 1970, The geological environment of Matsukawa geothermal area; in Proceedings from the United Nations Symposium on the Development and Utilization of Geothermal Resources: Geothermics (Special Issue 2), v. 2, pt. 1, p. 221-231.

Oki, Y., and Hirano, T., 1970, The geothermal system at the Hakone Volcano; in Proceedings from the United Nations Symposium on the Development and Utilization of Geothermal Resources: Geothermics (Special Issue 2), v. 2, pt. 2, p. 1157-1156.

Phillips, W. J., 1973, Mechanical effects of retrograde boiling and its probable importance in the formation of some porphyry ore deposits: Institute Mining Metallurgy Transcripts, sec. B,

v. 82, p. B90-98.

Radtke, A. S., Rye, R. O., and Dickson, F. W., 1980, Geology and stable isotope studies of the Carlin gold deposit, Nevada: Economic Geology, v. 75, p. 641-672.

Rimstidt, J. D., and Barnes, H. L., 1980, The kinetics of silica-water reactions: Geochimica et Cosmochimica Acta, v. 44, p. 1683-1699.

Rothbaum, H. P., Anderton, B. H., Harrison, R. F., Rohde, A. G., and Slatter, A., 1979, Effect of silica polymerization and pH on geothermal scaling: Geothermics, v. 8, p. 1-20.

Rowe, J. J., Fournier, R. O., and Morey, G. W., 1967, The system water-sodium oxide-silicon dioxide at 200, 250, and 300°: Inorganic Chemistry, v. 6, p. 1183-1188.

Rowe, J. J., Fournier, R. O., and Morey, G. W., 1973, Chemical analysis of thermal waters in Yellowstone National Park, Wyoming, 1960-65: U.S. Geological Survey, Bulletin 1303, 31 p.

Rye, R. O., 1985, A model for the formation of carbonate hosted disseminated gold deposits as indicated by geologic, fluid inclusion, geochemical, and stable isotope studies of the Carlin and Cortez deposits, Nevada; in Tooker, E. W. (ed.), Geologic Characteristics of the Sediment- and Volcanic-hosted Types of Gold Deposits--Search for an Occurrence Model: U.S. Geological Survey, Bulletin 1646, p. 35-42.

Saki, H., and Matsubaya, O., 1977, Stable isotopic studies of Japanese geothermal systems: Geothermics, v. 5, p. 97-124.

Seastres, J. S., Jr., 1982, Subsurface geology of the Nasuji-Sogongon sector, Southern Negros Geothermal Field, Philippines: Proceedings of the Pacific Geothermal Conference, 1982, 4th New Zealand Geothermal Workshop, pt. 1, p. 173-178.

Secor, D. T., Jr., 1965, Role of fluid pressure in jointing: American Journal of Science, v. 263, p. 633-646.

Shettel, D. L., 1974, The solubility of quartz in supercritical H_2O-CO_2 fluids: Unpublished M.S. thesis, The Pennsylvania State University, 52 p.

Sleep, N. H., 1983, Hydrothermal convection at ridge axes; in Rona, P. R., and Lowell, R. L. (eds.), Hydrothermal Processes at Seafloor-Spreading Centers: Plenum Press, New York, p. 71-82.

Smith, R. B., and Bruhn, R. L., 1984, Interplate extensional tectonics of the eastern Basin-Range: Inferences on structural style from seismic reflection data, regional tectonics, and thermal mechanical models of brittle-ductile deformation: Journal of Geophysical Research, v. 89, p. 5733-5762.

Sweeton, F. H., Mesmer, R. E., and Baes, C. F., Jr., 1974, Acidity measurements at elevated temperatures. VII. Dissociation of water: Journal of Solution Chemistry, v. 3, p. 191-214.

Toulmin, P., III, and Clark, S. P., 1967, Thermal aspects of ore formation; in Barnes, H. L. (ed.), Geochemistry of Hydrothermal Ore Deposits: Holt, Rinehart and Winston, New York, p. 437-464.

Truesdell, A. H., 1976, Summary of section III geochemical techniques in exploration; in Proceedings of the Second United Nations Symposium on the Development and Use of Geothermal Resources, San Francisco, 1975, v. 1: Washington, D.C., U.S. Government Printing Office, p. liii-lxxxix.

Truesdell, A. H., and Fournier, R. O., 1977, Procedure for estimating the temperature of a hot-water component in a mixed water by using a plot of dissolved silica versus enthalpy: U.S. Geological Survey, Journal of Research, v. 5, p. 49-52.

Weres, O., Yee, A., and Tsao, L., 1982, Equations and type curves for predicting the polymerization of amorphous silica in geothermal brines: Society of Petrological Engineering Journal (Feb. 1982), p. 9-16.

White, D. E., 1965, Saline waters of sedimentary rock: American Association of Petroleum Geologists, Memoir 4, p. 342-366.

White, D. E., Brannock, W. W., and Murata, K. J., 1956, Silica in hot-spring waters: Geochimica et Cosmochimica Acta, v. 10, p 27-59.

White, D. E., Fournier, R. O., Muffler, L. J. P., and Truesdell, A. H., 1975, Physical results of research drilling in thermal areas of Yellowstone National Park, Wyoming: U.S. Geological Survey, Professional Paper 892.

White, D. E., Muffler, L. J. P., and Truesdell, A. H., 1971, Vapor-dominated hydrothermal systems compared to hot-water systems: Economic Geology, v. 66, p. 75-97.

Yamada, E., 1976, Geological development of the Onikobe caldera and its hydrothermal system; in Proceedings of the Second United Nations Symposium on the Development and Use of Geothermal Resources, San Francisco, 1975: v. 1, p. 665-672, U.S. Government Printing Office.

APPENDIX

Information For Use In Calculating Silica Solubilities

Approximate solubilities of selected silica species in liquid water at the vapor pressure of the solution can be calculated using equations (a)-(f) below (after Fournier, 1981; Fournier and Potter, 1982b). Concentrations of dissolved silica (S) are in mg/kg, t is temperature in degrees Celsius, and the temperature range of application is 0° to 250°C, except as noted.

Quartz $\log S = [-1309/(t+273.15)] + 5.19$ (a)

Chalcedony $\log S = [-1032/(t+273.15)] + 4.69$ (b)

α-Cristobalite

$$\log S = [-1000/(t+273.15)] + 4.78 \quad (c)$$

β-Cristobalite

$$\log S = [-781/(t+273.15)] + 4.51 \quad (d)$$

Amorphous silica

$$\log S = [-731/(t+273.15)] + 4.52 \quad (e)$$

Quartz (20°-330°C)

$$t = -42.196 + 0.28831\,S - 3.6685 \times 10^{-4}\,S^2 + 3.1665 \times 10^{-7}\,S^3 + 77.034 \log S \quad (f)$$

More precise solubilities of amorphous silica in the temperature range 90° to 340°C at the vapor pressure of the solution and at 1000 bars can be calculated using equations (g) and (h) (from Fournier and Marshall, 1983). T is temperature in Kelvin and m is the molality of dissolved silica.

Vapor Pressure of solution

$$\log m = -6.116 + (0.01625\,T) - (1.758 \times 10^{-5}\,T^2) + (5.257 \times 10^{-9}\,T^3) \quad (g)$$

1000 bars

Appendix Table 3.A1--Temperatures, enthalpies (Keenan et al., 1969), and quartz solubilities (Fournier and Potter, 1982b) in liquid and gaseous water (steam) at the vapor pressure of the solution

No.	T °C	H J/g	SiO$_2$ mg/kg	No.	T °C	H J/g	SiO$_2$ mg/kg	No.	T °C	H J/g	SiO$_2$ mg/kg
1.	20	84.0	6.1	18.	190	807.6	230.1	35.	360	1760.5	707.7
2.	30	125.8	8.1	19.	200	852.5	263.0	36.	370	1890.5	573.0
3.	40	167.6	10.7	20.	210	897.8	298.5	37.	374	2099.3	299.6
4.	50	209.3	14.0	21.	220	943.6	336.4	38.	370	2332.1	125.0
5.	60	251.1	18.2	22.	230	990.1	376.4	39.	360	2481.0	60.4
6.	70	293.0	23.5	23.	240	1037.3	418.3	40.	350	2536.9	35.4
7.	80	334.9	30.0	24.	250	1085.4	461.8	41.	340	2622.0	22.1
8.	90	376.9	37.9	25.	260	1134.4	506.2	42.	330	2665.9	14.1
9.	100	419.0	47.4	26.	270	1184.5	551.0	43.	320	2700.1	9.2
10.	110	461.3	58.6	27.	280	1236.0	595.3	44.	310	2727.3	6.0
11.	120	503.7	71.8	28.	290	1289.1	638.2	45.	300	2749.0	3.9
12.	130	546.3	87.1	29.	300	1344.0	678.4	46.	290	2766.2	2.5
13.	140	589.1	104.7	30.	310	1404.3	714.2	47.	280	2779.6	1.6
14.	150	632.2	124.6	31.	320	1461.5	743.6	48.	270	2789.7	1.0
15.	160	675.6	147.1	32.	330	1525.3	763.6	49.	260	2796.9	0.6
16.	170	719.2	172.1	33.	340	1594.2	770.1	50.	250	2801.5	0.4
17.	180	763.2	199.8	34.	350	1670.6	756.2	51.	240	2803.8	0.2

$$\log m = -7.010 + (0.02285\ T) - (3.262 \times 10^{-5}\ T^2)$$
$$+ (1.730 \times 10^{-8}\ T^3) \qquad (h)$$

The solubility of quartz in water in the temperature range 25° to 900°C at specific volume (V) of the solvent ranging from about 1 to 10 and from 300° to 600°C at specific volume of the solvent ranging from about 10 to 100 can be calculated using equations (i)-(l) (from Fournier and Potter, 1982a).

$$\log m = A + B\ (\log V) + C\ (\log V)^2 \qquad (i)$$

where

$$A = -4.66206 + 0.0034063\ T + 2179.7\ T^{-3}$$
$$- 1.1292 \times 10^6\ T^{-2} + 1.3543 \times 10^8\ T^{-3} \qquad (j)$$

$$B = -0.0014180\ T - 806.9\ T^{-1} \qquad (k)$$

$$C = 3.9465 \times 10^{-4}\ T \qquad (l)$$

The solubility of quartz in saline solutions can be calculated using equation (i) when $(-\log \rho F)$ is substituted for $(\log V)$, where ρ is the density of the solution and F is the weight fraction of water in that solution (Fournier, 1983b).

The molal solubility of amorphous silica in saline solutions (m_s) at temperatures ranging from 100° to 340°C, and pressures ranging from the vapor pressure of the solution to about 1000 bars can be calculated using the following equations where ρ_s is the density of the saline solution, ρ^o is the density of pure water at any given temperature and the indicated pressure (obtained from steam tables) and m^o is the molal solubility in pure water obtained either from equation (g) for the vapor pressure of the solution, or equation (h) for 1000 bars (Fournier and Marshall, 1983).

$$\log m_s = -n \log \rho_s F$$
$$+ n \log \rho^o(v.p.) + \log m^o\ (v.p.) \qquad (m)$$

$$n = \frac{[\log m^o(1000\ \mathrm{bar}) - \log m^o(v.p.)]}{[\log \rho^o(1000\ \mathrm{bar}) - \log \rho^o(v.p.)]} \qquad (n)$$

Appendix Table 3.A gives temperature, enthalpy, and quartz solubility data that are useful for calculating silica concentrations after mixing of waters with different initial temperatures and after boiling, as discussed by Fournier and Potter (1982b).

The dissociation of silicic acid as a function of pH can be determined as follows, where square brackets denote activities of the indicated species, K_1 is the first dissociation constant, m is molality, γ is the activity coefficient, and T is temperature in Kelvin

$$H_4SiO_4 = H_3SiO_4^- + H^+ \qquad (o)$$

$$K_1 = \frac{[H_3SiO_4^-][H^+]}{[H_4SiO_4]} \qquad (p)$$

$$K_1 = \left(\frac{m_{H_3SiO_4^-} 10^{-pH}}{m_{H_4SiO_4}}\right)\left(\frac{\gamma_{H_3SiO_4^-}}{\gamma_{H_4SiO_4}}\right) \qquad (q)$$

Values of $\gamma_{H_3SiO_4^-}$ can be calculated using the Debye Hückel equation. When $m_{H_4SiO_4}$ is determined using equation (i) and substituting $-\log \rho F$ for $\log V$, the value of $\gamma_{H_4SiO_4}$ is unity.

Values of K_1 in the temperature range 1 to 350°C can be calculated using the following equation

$$-\log K_1 = -631.8744 - .2967\ T$$
$$+ .000133266\ T^2 + 267.6478 \log T \qquad (r)$$

For most natural thermal waters values of pH are too low for the second dissociation of silicic acid to be important, and

$$m_{H_3SiO_4^-} = \frac{m_{SiO_2(total)}}{\dfrac{10^{-pH} \gamma_{H_3SiO_4^-}}{K_1} + 1} \qquad (s)$$

Chapter 4
CARBONATE TRANSPORT AND DEPOSITION IN THE EPITHERMAL ENVIRONMENT
Robert O. Fournier

INTRODUCTION

The factors affecting the transport and deposition of carbonate in hydrothermal systems have been discussed in detail by Holland and Malinin (1979). Solubilities of carbonates are strongly influenced by pH, P_{CO_2}, temperature, and the presence of other dissolved salts. The alkali carbonates, Na, K, and Li, are relatively soluble at all temperatures and generally precipitate only where there is extreme evaporation. In contrast, the alkaline earth carbonates, Ca, Mg, Sr, and Ba, are moderately to sparingly soluble and commonly precipitate in hydrothermal systems. Calcite is by far the most abundant and important carbonate found in the epithermal environment, and more solubility data at hydrothermal conditions are available for it than for any of the other carbonates. Therefore, after briefly reviewing the system CO_2-water, the discussion will focus on the transport and deposition of calcite in hydrothermal solutions. The behaviors of other moderately to sparingly soluble carbonates in hydrothermal solutions are similar to that of calcite.

CO_2 DISSOLVED IN AQUEOUS SOLUTIONS

There is extensive literature on pressure-volume-temperature measurements for the system CO_2-water, with and without additional dissolved salts (Bowers and Helgeson, 1983; and references therein). The experimental work that is most applicable to conditions appropriate for the formation of epithermal ore deposits was carried out by Ellis and Golding (1963), who used solubility data to calculate Henry's Law coefficients, K_H, for carbon dioxide in water and NaCl solutions (Fig. 4.1). According to Henry's Law

$$f_{CO_2} = K_H X \qquad (1)$$

where f_{CO_2} is the fugacity of carbon dioxide and X is the mole fraction of carbon dioxide dissolved in the liquid phase. Because fugacity coefficients for carbon dioxide in dilute aqueous solutions at temperatures below about 330°C are near unity (Ellis and Golding, 1963), f_{CO_2} in equation (1) can be replaced by the partial pressure of carbon dioxide, P_{CO_2}, with little error. An equation expressing Henry's Law coefficient for the system carbon dioxide-water as a function of temperature is given in Table 4.1, equation (a). The constants used in the equations of Table 4.1 are given in Table 4.2. Values of the Henry's Law coefficient for the system CO_2-H_2O at selected temperatures are given in Table 4.3. Adding salt to the system CO_2-H_2O increases the Henry's Law coefficient and decreases the solubility of carbon dioxide in the solution (Fig. 4.1). Salting-out coefficients for carbon dioxide in sodium chloride solutions generally are of the Steschenow type

$$k = \frac{1}{m} \log (K_H/K^o_H) \qquad (2)$$

where k is the salting-out coefficient, m is the molality of NaCl, and K^o_H and K_H are respectively the Henry's Law coefficients for pure water as solvent and for the saline solution (Ellis and Golding, 1963). Approximate salting-out coefficients can be obtained from equation (b) in Table 4.1, and at selected temperatures from Table 4.3.

Some dissolved carbon dioxide reacts with water (hydrates) to form carbonic acid

$$CO_{2(dissolved)} + H_2O = H_2CO_3 \qquad (3)$$

and some of the carbonic acid, in turn, dissociates according to the reactions

$$H_2CO_3 = HCO_3^- + H^+ \qquad (4)$$

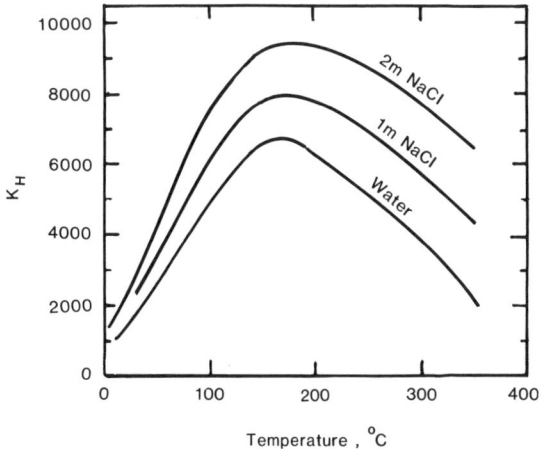

Figure 4.1. Values of Henry's Law constant, K_H, for the solution of carbon dioxide in water and sodium chloride solutions. (Redrawn from Ellis and Golding, 1963).

Table 4.1--Equations expressing the temperature dependence of various equilibrium constants and other coefficients that are described in the text. T is temperature in kelvin or °K, and t is temperature in Celsius or °C. The respective constants for equations (a) through (k) are given in Table 4.2. The data used in the derivation of these equations are shown in Table 4.3.

$$K_H = a + bT + cT^2 + dT^{-1} + e \log T \quad (a)$$

$$k = a + bT + cT^2 + dT^3 + e \log T \quad (b)$$

$$-\log K_1 = a + bT + cT^{-1} + d \log T \quad (c)$$

$$-\log K_2 = a + bT + cT^2 + dT^{-1} + e \log T \quad (d)$$

$$-\log K_0 = a + bT + cT^2 + dT^{-1} + e \log T \quad (e)$$

$$H_L = a + bt + ct^2 + dt^3 + et^4 + ft^5 + gt^{-1} + ht^{-2} + i \log t \quad (f)$$

$$H_G = a + bt + ct^2 + dt^3 + et^4 + ft^5 + gt^{-1} + ht^{-2} + i \log t \quad (g)$$

$$-\log K_c = a + bT + cT^2 + dT^3 + eT^{-1} \quad (h)$$

$$A = a + bT + cT^2 + dT^3 + eT^4 + fT^{-1} + g \log T \quad (i)$$

$$B = a + bT + cT^2 + dT^3 + eT^4 + fT^{-1} \quad (j)$$

$$-\log A_0 = a + bT + cT^2 + dT^3 + eT^{-1} + fT^{-2} \quad (k)$$

$$K_1 = \frac{[HCO_3^-][H^+]}{[H_2CO_3]} \quad (5)$$

$$HCO_3^- = CO_3^= + H^+ \quad (6)$$

$$K_2 = \frac{[CO_3^=][H^+]}{[HCO_3^-]} \quad (7)$$

where square brackets indicate activities of the enclosed species, and K_1 and K_2 are respectively the first and second dissociation constants of carbonic acid (Table 4.3). Equations (c) and (d) in Table 4.1 express the temperature dependence of K_1 and K_2. The reaction shown by equation (3) takes place relatively slowly, while the reaction shown by equation (4) is almost instantaneous (Kern, 1960). This information will be of use later when the consequences of boiling are discussed.

By tradition a distinction is not made between aqueous CO_2 and H_2CO_3, and total dissolved CO_2 is reported as H_2CO_3. A net reaction is generally written

$$CO_{2(gas)} + H_2O = H_2CO_3 \quad (8)$$

$$K_O = \frac{[H_2CO_3]}{f_{CO_2}[H_2O]} \quad (9)$$

where K_O is the equilibrium constant for the reaction shown by equation (8). Values of K_O at temperatures ranging from 100° to 300°C (Table 4.3) can be calculated using equation (e) in Table 4.1. The $[H_2CO_3]$ term in equations (5) and (9) includes the activity of dissolved, nonhydrated CO_2.

H_2CO_3 is less ionized at high temperatures compared to low temperatures (values of K_1 range from about $10^{-6.57}$ at 0°C to about $10^{-8.29}$ at 300°C). Therefore, as a hydrothermal solution cools, bicarbonate dissociates (equation (4)), liberating hydrogen ions that attack the minerals in the wall rock. Hydrolysis reactions involving feldspars generally buffer the pH at near neutral to slightly acidic conditions and cooling solutions become richer in cations (Na at higher temperatures and Ca at lower temperatures) as hydrogen ions are consumed by the formation of micas or clays.

Where boiling occurs, generally as a result of decreasing hydrostatic pressure exerted upon an ascending hydrothermal solution, CO_2 is strongly partitioned into the gas (steam-rich) phase. The total gas pressure is equal to the sum of the partial pressures of all the constituent gases, and these partial pressures are proportional to the respective mole fractions. By Raoult's Law, for a mixture of two components that exhibit ideal behavior (may be closely approximated when the components are similar in molecular structure, or when one of the components is present in great excess)

$$P_{Total} = P_a + P_b = P^*_a X_a + P^*_b X_b$$
$$= P^*_a X_a + P^*_b (1-X_a) \quad (10)$$

Table 4.2--Coefficients for use with equations listed in Table 4.1

	a	b	c	d
K_H	−7656970	−3122.11449	1.092229	1.880778×10^8
k	108.875	0.174114604	$-1.9845113 \times 10^{-4}$	1.0131668×10^{-7}
$-\log K_1$	−124.4478	−0.0056623	5.86972601×10^3	45.589821
$-\log K_2$	−143.4475	−0.0345539	1.9732326×10^{-5}	6.16187137×10^3
$-\log K^o$	130.5993	0.0517898	$-2.6766438 \times 10^{-5}$	-4.4176248×10^3
H_L	418.84	10.2859	−0.05092	2.63085×10^{-4}
H_G	2034.63	−5.04986	0.057399	-3.04263×10^{-4}
$-\log K_c$	−30.8131	0.1728295	-3.271501×10^{-4}	2.5529153×10^{-7}
A	−11092.1143	11.562984	0.01454757	$-1.2064489 \times 10^{-5}$
B	3.8096	0.01846467	4.8075571×10^{-5}	$-6.0838461 \times 10^{-8}$
$-\log A^o$	1823.663	−3.9000818	0.004142503	-1.75503×10^{-6}

	e	f	g	h	i
K_H	3.1771246×10^6				
k	−58.867703				
$-\log K_1$					
$-\log K_2$	57.248899				
$-\log K_o$	−51.47548				
H_L	-6.93025×10^{-7}	7.4566×10^{-10}	−1209.757	11.98996	−353.764
H_G	7.909545×10^{-7}	-8.69676×10^{-10}	1342.406	−13.2981	396.288
$-\log K_c$	2.95989303×10^3				
A	4.4598806×10^{-9}	1.76312033×10^5	5229.515		
B	$3.0473184 \times 10^{-11}$	-2.617130669×10^2			
$-\log A^o$	-4.2179661×10^5	3.9002725×10^7			

where P^*_a and P^*_b are the vapor pressures of pure components A and B respectively at the given temperature, and X_a and X_b are the respective mole fractions of A and B in the mixture.

When boiling is first initiated, the ratio of CO_2 to water in the gas phase tends to be relatively large because most of the carbon dioxide initially dissolved in the liquid exsolves quickly into the gas phase, while only a small amount of water changes to steam. With continued boiling, the mole fraction of CO_2 in the gas phase steadily decreases because little additional CO_2 is available to partition into the gas phase while the fraction of water that is converted to steam increases at a relatively constant rate. As the temperature of

Table 4.3--Values of dissociation constants, enthalpies of liquid water H_L and steam H_G, and Debye-Hückel Coefficients, A and B, for the indicated temperatures. Enthalpy units are J/g. Units for A are $kg^{1/2}mole^{-1/2}$ and for B are $kg^{1/2}mole^{-1}cm^{-1} \times 10^8$ (the product $\overset{o}{a} \times B$ cancels the 10^8 factor)

t (°C)	100	125	150	175	200	225	250	275	300	Reference
K_H	5245	6150	6670	6860	6620		5340		3980	(1)
k	.078		.078		.089		.120		.178	(2)
$-\log K_1$	6.42	6.57	6.77	6.99	7.23	7.49	7.75	8.02	8.29	(1)
$-\log K_2$	10.16	10.25	10.39	10.57	10.78	11.02	11.29	11.58	11.89	(1)
$-\log K_0$	1.97		2.08		2.08		1.98		1.8*	(1)
H_L	419	525	632	741	852	967	1085	1210	1314	(3)
H_G	2676	2714	2746	2774	2793	2803	2802	2785	2749	(3)
$-\log K_c$	9.33		10.07		11.02		12.27		14.0*	(1)
A	.5998		.6898		.8099		.9785		1.2555	(1)
B	.3422		.3533		.3655		.3792		.3965	(1)
$-\log A_0$	3.71	3.41	3.13	2.86	2.58	2.31	2.05	1.78	1.51	(2)

* Extrapolated.
(1) Henley et al. (1984).
(2) Ellis and Golding (1963).
(3) Keenan et al. (1969).

the ascending gas-water mixture decreases, the volume of the gas phase increases due to the decrease in hydrostatic load. The net effect is a drastic decrease in the partial pressure of CO_2 as a boiling fluid ascends toward the earth's surface.

Procedures for calculating the partitioning of relatively volatile constituents between coexisting liquids and gases, using hand-held, programmable calculators, are described in Henley et al. (1984). For relatively dilute systems and low initial dissolved gas concentrations, a distribution coefficient, B, is defined as the concentration of gas in the vapor divided by the concentration of gas in the liquid. Giggenbach (1980) derived the following equation that expresses the temperature dependence of B for carbon dioxide in dilute aqueous solutions

$$\log B = 4.7593 - .01092t \qquad (11)$$

where t is temperature in degrees Celsius. Equation (11) is valid from 100° to 340°C. Henley et al. (1984) give the following equation for calculating the concentration of CO_2 remaining in the liquid phase (C_l) for the situation in which all the evolved gas remains in contact with a boiling fluid during adiabatic decompression (single-step steam separation)

$$C_l = \frac{C_o}{1 + y(B - 1)} \qquad (12)$$

where C_o is the initial concentration of dissolved CO_2 before boiling, B is the distribution coefficient, and y is the fraction of separated steam. The corresponding equation that gives the concentration of CO_2 in the coexisting steam (C_v) is

$$C_v = \frac{B\,C_o}{1 + y(B-1)} \qquad (13)$$

Values of y are generally calculated using enthalpy data for pure boiling water and the relationship

$$y = \frac{H_o - H_L}{H_G - H_L} \qquad (14)$$

where H_o is the enthalpy of the initial liquid prior to boiling, and H_L and H_G are the enthalpies of coexisting liquid water and steam after boiling (Table 4.3). Enthalpies of liquid water and steam are generally obtained from steam tables (Keenan et al., 1969) or they can be calculated using equations (f) and (g) in Table 4.1. Equation (13), however, yields values of y that are slightly in error because the enthalpy of steam containing CO_2 is different from the enthalpy of pure steam. Other factors also may cause the calculated concentration of CO_2 in the liquid and steam fractions of a boiling solution to be in error. Assumptions implicit in the use of equations (12) and (13) are that dissolved CO_2 does not become supersaturated in the liquid phase as pressure is released, and that little or no H_2CO_3 converts to HCO_3^- as the boiling solution cools. The rapid transfer of most of the dissolved CO_2 into the steam fraction at an early stage of boiling and the relatively slow conversion of dissolved CO_2 to H_2CO_3 (previously discussed) will tend to limit the amount of HCO_3^- that can form, but some non-equilibrium partitioning of CO_2 between the liquid and gas phase is likely, particularly when the first boiling is initiated at a temperature below about 200°C. Another factor that must be considered is physical removal of the steam fraction from contact with the residual liquid as the boiling process proceeds. Compared to single-step steam separation, multistep and continuous steam separation result in much lower concentrations of CO_2 in the last liquid and steam fractions that are in contact. Henley et al. (1984) present methods and equations for dealing with multistep and continuous steam separation. Figure 4.2 shows values of C_l/C_o for single-step steam separation for a variety of initial and final temperatures, calculated using equations (12) and (14).

THE SOLUBILITY OF CALCITE IN AQUEOUS SOLUTIONS

Ellis (1959, 1963) determined experimentally the solubility of calcite in aqueous solutions at conditions appropriate for the formation of epithermal ore deposits. At a given partial pressure of CO_2, the solubility of calcite decreases with increasing temperature (Fig. 4.3). Adding NaCl to the solution increases the solubility of calcite (Fig. 4.4). At any given temperature the solubility of calcite in solutions in equilibrium with a vapor phase increases with increasing CO_2 pressure until $m_{CO2} \simeq 1$ mole/kg (Miller, 1952; Segnit et al., 1962). In solutions held at a constant total pressure, the solubility increases with increasing CO_2 concentration until $m_{CO2} \simeq 1$ mole/kg and then decreases at higher CO_2 concentrations (Sharp and Kennedy, 1965; Malinin and Kanukov, 1971).

The simplest equation representing the reaction by which calcite dissolves in aqueous solutions can be written

$$CaCO_3 = Ca^{++} + CO_3^= \quad (15)$$

and the equilibrium constant (K_c) for reaction (15) is

$$K_c = \frac{[Ca^{++}][CO_3^=]}{[CaCO_3]} \quad (16)$$

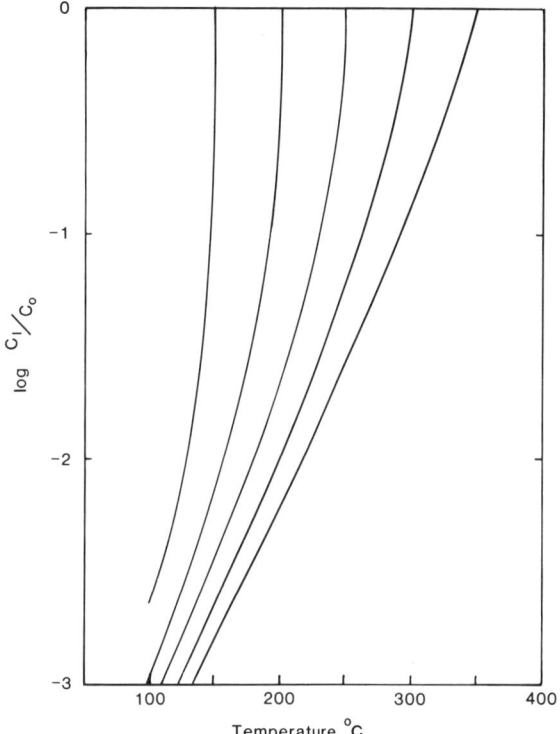

Figure 4.2. The ratio of CO_2 remaining in the residual liquid (C_l) after single-step steam separation at various temperatures to CO_2 in the initial liquid (C_o) before boiling.

Equation (16) is useful mainly for testing whether a solution of given composition is unsaturated, saturated, or supersaturated in respect to calcite. Values of K_c in the temperature range 100° to 300°C (Table 4.3) can be calculated using equation (h), Table 4.1. Because very little $CO_3^=$ is present in most natural hydrothermal solutions, the solubility of calcite is commonly expressed in terms of reactions involving H^+, HCO_3^-, and f_{CO_2} using equations (5), (7), (9), and (16).

$$CaCO_3 + H^+ = Ca^{++} + HCO_3^- \quad (17)$$

$$\frac{K_c}{K_2} = \frac{[Ca^{++}][HCO_3^-]}{[H^+][CaCO_3]} \quad (18)$$

$$CaCO_3 + 2H^+ = Ca^{++} + H_2O + CO_2 \quad (19)$$

$$\frac{K_c}{K_1 K_2 K_0} = \frac{[Ca^{++}][H_2O]f_{CO2}}{[H^+]^2[CaCO_3]} \quad (20)$$

$$CaCO_3 + CO_2 + H_2O = Ca^{++} + 2HCO_3^- \quad (21)$$

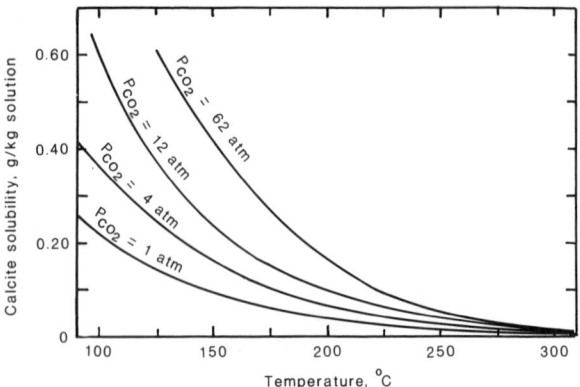

Figure 4.3. The solubility of calcite in water up to 300°C at various partial pressures of carbon dioxide. (Redrawn from Ellis, 1959).

$$\frac{K_c K_1 K_0}{K_2} = \frac{[Ca^{++}][HCO_3^-]^2}{[H_2O] \, f_{CO_2} \, [CaCO_3]} \quad (22)$$

In equations (16), (18), (20), and (22), the activity of calcite is unity if there is no significant substitution of other cations for calcium in solid solution, such as Mg, Fe, or Mn.

In order to evaluate equations (16)-(22), activities of the indicated aqueous species must be used. In dilute solutions, activities of dissolved constituents are about equal to the corresponding molalities. In saline solutions, however, the molality of each species i (m_i) must be multiplied by its activity coefficient (γ_i) to obtain the activity ($a_i = \gamma_i m_i$). Activity coefficients for solutions with ionic strengths less than about 2 molal can be calculated using an extended form of the Debye-Hückel equation

$$-\log \gamma_i = \frac{A z_i^2 \, I^{1/2}}{1 + \mathring{a}_i B I^{1/2}} + bI \quad (23)$$

where z_i is the ionic charge, I the ionic strength, and A, B, $\mathring{a}_i$ and b are constants (Henley et al., 1984). However, A and B vary with temperature. Their values from 100° to 350°C, in 25°C increments, are given in Table 4.3. The coefficients A and B also can be calculated to three decimal places using equations (i) and (j), Table 4.1. The ionic strength is defined as

$$I = 1/2 \, \Sigma \, m_i z_i^2 \quad (24)$$

For most hydrothermal waters I is approximately equal to the sum of m_{Na^+} and m_{K^+}. Values of $\mathring{a}_i$ and z_i are listed in Table 4.4. Up to 250°C, b has values in the range 0.03 to 0.05 when concentrations are up to 3 molal (Helgeson, 1969).

In natural hydrothermal solutions many dissolved constituents and a variety of chemical reactions involving solids, liquids and gases influence the

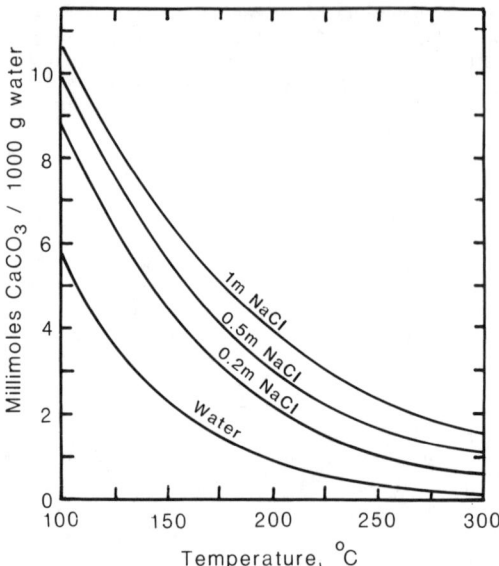

Figure 4.4. The solubility of calcite in water and sodium chloride solutions at a carbon dioxide pressure of 12 atmospheres (12.2 bars). (Redrawn from Ellis, 1963).

dissolution and deposition of calcite. The situation is particularly complex when boiling occurs. Computer programs can be used to evaluate these complex reactions (Truesdell and Singers, 1971, 1974; Morel and Morgan, 1972; Kharaka and Barnes, 1973; Truesdell and Jones, 1974; Plummer et al., 1975; Wolery, 1979; Reed, 1982; Reed and Spycher, 1984). These programs, however, generally require main-frame computers for their execution.

Arnorsson (1978) specifically calculated the amount and location of calcite deposition in geothermal wells in Iceland where natural thermal waters flash to steam during production of the resource. The results of his calculations, showing the degree of supersaturation with respect to calcite that occurs during single-step adiabatic flashing (boiling) in the wells, are shown in Figure 4.5 (supersaturated solutions that plot below the thick solid line).

Programs that can be used with programmable hand-held calculators, and that are applicable to calcite transport and deposition in natural waters, are given in Henley et al. (1984). These programs are very useful even though a few simplifying assumptions are required, and only the most important dissolved species are included. The ensuing discussion follows the general procedures presented in Henley et al. (1984).

From a consideration of the cation-anion charge balance that must be maintained in all solutions (and neglecting effects of oxidation-reduction reactions and pH-dependent ions that are likely to be present only in very small amounts in most natural hydrothermal solutions, such as $CO_3^=$, $H_2SiO_4^=$, and OH^-) a constant, Δ,

Table 4.4--Values of ionic charge, z, and ion size parameter, $\overset{o}{a}$, for the common ionic species in geothermal fluids (from Henley et al., 1984)

	H^+	Na^+	HCO_3^-	HS^-	$H_3SiO_4^-$	$H_2BO_3^-$	F^-	$SO_4^=$	NH_4^+	HSO_4^-
z	1	1	1	1	1	1	1	1	1	1
$\overset{o}{a}$	9.0	4.0	4.5	4.0	4.0	4.0	3.5	4.0	2.5	4.0

	OH^-	$CO_3^=$	Cl^-	Li^+	K^+	Ca^{++}	Mg^{++}
z	1	2	1	1	1	2	2
$\overset{o}{a}$	4.0	4.5	3.5	6.0	3.0	6.0	8.0

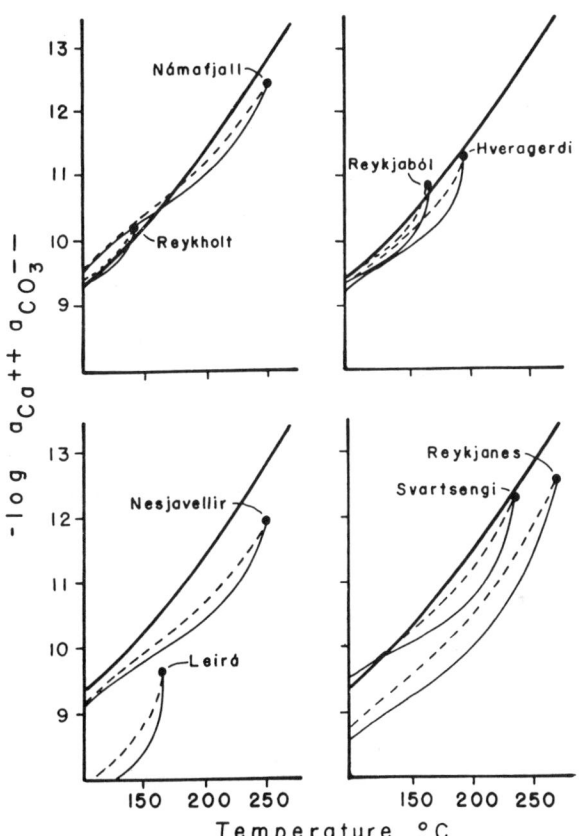

Figure 4.5. The computed activity product of Ca^{++} and $CO_3^=$ in geothermal waters during single-step adiabatic flashing in relation to the calcite solubility curve (thick solid line). The solid lines assume maximum degassing and the dashed lines 1/5 of maximum degassing. (From Arnorsson, 1978).

can be defined that is independent of temperature and equal to the sum of the principal pH-dependent ions

$$\Delta = [m_{HCO_3^-} + m_{H_3SiO_4^-} + m_{H_2BO_3^-}] - [m_{NH_4^+} + m_{H^+}] \quad (25)$$

The concentrations of the ionic species indicated in equation (25) can be calculated using the following relationships, where K_1 is the first dissociation constant for each of the respective weak acids

$$m_{HCO_3^-} = \frac{m_{CO_2(total)}}{\frac{[H^+] \gamma_{HCO_3^-}}{K_1} + 1} \quad (26)$$

$$m_{H_3SiO_4^-} = \frac{m_{SiO_2(total)}}{\frac{[H^+] \gamma_{H_3SiO_4^-}}{K_1} + 1} \quad (27)$$

$$m_{H_2BO_3^-} = \frac{m_{B(total)}}{\frac{[H^+] \gamma_{H_2BO_3^-}}{K_1} + 1} \quad (28)$$

$$m_{NH_4^+} = \frac{m_{NH_3(total)}}{\frac{K_1 \gamma_{NH_4^+}}{[H^+]} + 1} \quad (29)$$

The reader is referred to Henley et al. (1984) for derivation of equations (26)-(29).

If the pH, ionic strength, and chemical composition (particularly total dissolved carbon, silica, boron, and ammonia) of a solution are known at a given temperature, equations (25)-(29) can be used to estimate the indicated species concentrations at any other temperature up to the limit of the available thermodynamic data (now about 300° to 350°C). Note, however, that equations (25)-(29) do not take account of changing concentrations and partitioning of constituents between the liquid and gas phase during boiling.

For adiabatic boiling resulting from decompression, the value of Δ changes as a function of the fraction of steam (y) that forms.

$$\Delta_{(before\ boiling)} = (1-y)\ \Delta_{(after\ boiling)} \quad (30)$$

The effect of partitioning of CO_2 between liquid and gas can be accounted for by using the relationship

$$A_o = \frac{n_{CO_2,l}\ n_{H_2O,v}}{n_{CO_2,v}\ n_{H_2O,l}} \quad (31)$$

where n_l and n_v are the number of moles of the indicated species in the liquid and gas phases respectively, and values of A_o at various temperatures with boiling water and boiling NaCl solutions as solvents are given in Ellis and Golding (1963). A_o for dilute solutions also can be calculated using equation (k) in Table 4.1. Rearranging equation (31) and substituting

$n_{H_2CO_3}$ for $n_{CO_2,l}$, and $y/(1-y)$ for $n_{H_2O,v}/n_{H_2O,l}$

$$n_{CO_2,v} = \left(\frac{n_{H_2CO_3}}{A_o}\ \frac{y}{(1-y)}\right) \quad (32)$$

For one-step steam separation without carbonate precipitation (supersaturation allowed in the calculation) the total number of moles of CO_2-bearing species remains constant during the boiling process, even though CO_2 partitions between the gas and liquid phases

$$n_{CO_2(total)} = n_{H_2CO_3} + n_{CO_2(gas)} + n_{HCO_3^-} + n_{CO_3^=} \quad (33)$$

The concentration of $CO_3^=$ in equation (33) cannot be neglected when solutions boil because the pH may increase significantly (Fig. 4.6). Combining equations (32) and (33)

$$n_{CO_2(total)} = n_{H_2CO_3} + \left(\frac{n_{H_2CO_3}}{A_o}\ \frac{y}{(1-y)}\right) + n_{HCO_3^-} + n_{CO_3^=} \quad (34)$$

and dividing equation (34) by $n_{HCO_3^-}$

$$\frac{n_{CO_2(total)}}{n_{HCO_3^-}} = \frac{n_{H_2CO_3}}{n_{HCO_3^-}}\left(1 + \frac{y}{A_o(1-y)}\right) + 1 + \left(\frac{n_{CO_3^=}}{n_{HCO_3^-}}\right) \quad (35)$$

Equation (35) is expressed in terms of mole ratios, so molal concentration units can be substituted for the number of moles of the given species in the liquid fraction. Combining equations (5), (7), and (35), multiplying molalities by activity coefficients to obtain activities where required, and rearranging gives

$$m_{HCO_3^-} = \frac{m_{CO_2(total)}}{\left(\frac{[H^+]\gamma_{HCO_3^-}}{K_1}\right)\left(1 + \frac{y}{A_o(1-y)}\right) + 1 + \left(\frac{K_2\ \gamma_{HCO_3^-}}{[H^+]\ \gamma_{CO_3^=}}\right)} \quad (36)$$

Account can be taken of the partitioning of other volatile constituents in a similar manner. The resulting equations can be used to estimate by

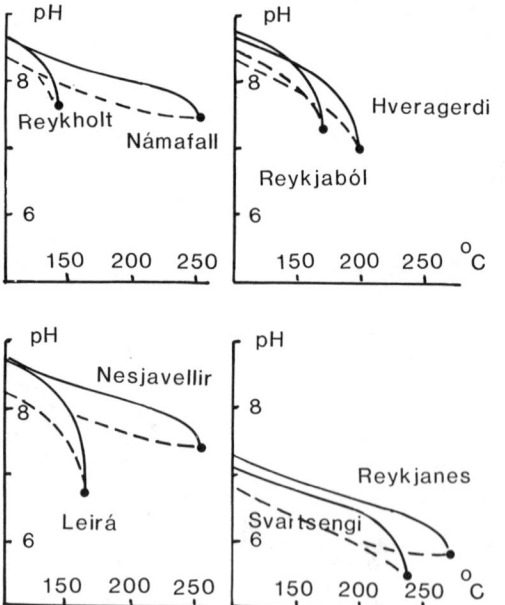

Figure 4.6. The variation in pH in geothermal waters during one-step adiabatic flashing in relation to the calcite solubility curve (thick solid line). The solid lines assume maximum degassing and the dashed lines 1/5 of maximum degassing. (Redrawn from Arnorsson, 1978).

iteration the pH and distribution of pH-dependent species at given temperatures for boiling conditions. The equations and procedures are cumbersome using a hand-held calculator, but are easily dealt with using table-top micro or personal computers.

Where cooling occurs adiabatically (generally too quickly for solution-mineral reactions involving silicates to buffer pH) the pH of the residual liquid usually rises as a result of partitioning of CO_2 and other acid-forming gases into the steam phase. Figure 4.6 shows calculated changes in pH that accompany single-step adiabatic flashing of natural thermal waters in geothermal wells in Iceland (Arnorsson, 1978). These are the same well waters used to illustrate the degree of supersaturation with respect to calcite that occurs during single-step flashing (Fig. 4.5).

SUMMARY

In most natural waters heating will cause calcite and other carbonates to precipitate, whereas cooling without boiling will cause them to dissolve (Fig. 4.4). However, where an ascending solution boils as a result of decompression (cooling adiabatically), carbonate is likely to precipitate as a result of the boiling (Fig. 4.5). The cooling that tends to move a solution toward a condition of undersaturation with respect to the various carbonate minerals is generally more than offset by the strong partitioning of CO_2 into the vapor phase (and concomitant decrease in partial pressure of CO_2) that decreases the solubility of carbonates. At present, calculations that take account of all the physical processes (mainly boiling) and chemical reactions that influence the transport and deposition of carbonate minerals can be carried out only with the aid of large computers. However, if only the most abundant dissolved species in natural waters are considered, and simplifying assumptions are made about enthalpies of coexisting liquids and gases (mainly steam), small table-top computers and hand-held, programmable calculators can be used effectively to calculate the approximate conditions for transport and deposition of calcite in hydrothermal solutions.

REFERENCES

Arnorsson, S., 1978, Precipitation of calcite from flashed geothermal waters in Iceland: Contributions to Mineralogy and Petrology, v. 66, p. 21-28.

Bowers, T. S., and Helgeson, H. C., 1983, Calculation of the thermodynamic and geochemical consequences of nonideal mixing in the system H_2O-CO_2-NaCl on phase relations in geologic systems: Equation of state for H_2O-CO_2-NaCl fluids at high pressures and temperatures: Geochimica et Cosmochimica Acta, v. 47, p. 1247-1275.

Ellis, A. J., 1959, The solubility of calcite in carbon dioxide solutions: American Journal of Science, v. 257, p. 354-365.

Ellis, A. J., 1963, The solubility of calcite in sodium chloride solutions at high temperatures: American Journal of Science, v. 261, p. 259-267.

Ellis, A. J., and Golding, R. M., 1963, The solubility of carbon dioxide above 100°C in water and in sodium chloride solutions: American Journal of Science, v. 261, p. 47-60.

Giggenbach, W. F., 1980, Geothermal gas equilibria: Geochimica et Cosmochimica Acta, v. 44, p. 2021-2032.

Henley, R. W., Truesdell, A. H., and Barton, P. B., Jr., 1984, Fluid-mineral equilibria in hydrothermal systems: Society of Economic Geologists, Reviews in Economic Geology, Volume 1, 267 p.

Holland, H. D., and Malinin, S. D., 1979, On the solubility and occurrence of non-ore minerals; in Barnes, H. L. (ed.), Geochemistry of Hydrothermal Ore Deposits (2d ed.): John Wiley and Sons, New York, p. 461-508.

Keenan, J. H., Keyes, F. G., Hill, P.G., and Moore, J. G., 1969, Steam Tables (international edition-metric units): John Wiley and Sons, New York, 162 p.

Kern, D. M., 1960, The hydration of carbon dioxide: Journal of Chemistry Education, v. 37, p. 14-23.

Kharaka, Y. K., and Barnes, I., 1973, SOLMNEQ: Solution-mineral equilibrium computations: National Technical Information System Technical Report PB 214-899, 82 p.

Malinin, S. D., and Kanukov, A. B., 1971, The solubility of calcite in homogeneous $H_2O-NaCl-CO_2$ systems in the 200°-600°C temperature interval: Geochemistry International, v. 9, p. 410-418.

Miller, J. P., 1952, A portion of the system calcium carbonate-carbon dioxide-water, with geologic implications: American Journal of Science, v. 250, p. 161-203.

Morel, F., and Morgan, J., 1972, A numerical method for computing equilibria in aqueous systems: Environmental Science and Technology, v. 6, p. 58-67.

Plummer, L. N., Parkhurst, D. L., and Kosiur, D., 1975, MIX2: A computer program for modeling chemical reactions in natural waters: U.S. Geological Survey, Water-Resources Investigations, p. 75-61.

Reed, M. H., 1982, Calculation of multicomponent chemical equilibria and reaction processes in systems involving minerals, gases and an aqueous phase: Geochimica et Cosmochimica Acta, v. 46, p. 513-528.

Reed, M. H., and Spycher, N., 1984, Calculation of pH and mineral equilibria in hydrothermal waters with application to geothermometry and studies of boiling and dilution: Geochimica et Cosmochimica Acta, v. 48, p. 1479-1492.

Segnit, E. R., Holland, H. D., and Biscardi, C. J., 1962, The solubility of calcite in aqueous solutions-I. The solubility of calcite in water between 75° and 200° at CO_2 pressures up to 60 atm: Geochimica et Cosmochimica Acta, v. 26, p. 1301-1331.

Sharp, W. E., and Kennedy, G. C., 1965, The system CaO-CO$_2$-H$_2$O in the two-phase region calcite and aqueous solution: Journal of Geology, v. 73, p. 391-403.

Truesdell, A. H., and Jones, B. F., 1974, WATEQ, a computer program for calculating chemical equilibria of natural waters: U.S. Geological Survey, Journal of Research, v. 2, p. 233-248.

Truesdell, A. H., and Singers, W. A., 1971, Computer calculation of downhole chemistry in geothermal areas: New Zealand DSIR Chemistry Division Report CD2136, 145 p.

Truesdell, A. H., and Singers, W. A., 1974, Calculation of aquifer chemistry in hot-water geothermal systems: U.S. Geological Survey, Journal of Research, v. 2, p. 271-278.

Wolery, A. T., 1979, Calculation of chemical equilibrium between aqueous solutions and minerals; the EQ3/6 software package: UCRL-52658, Lawrence Livermore Laboratory.

Chapter 5
FLUID-INCLUSION SYSTEMATICS IN EPITHERMAL SYSTEMS
R. J. Bodnar, T. J. Reynolds, and C. A. Kuehn

INTRODUCTION

Fluid-inclusion analyses have provided some of the most useful information for determining the physical and chemical environments of mineral formation. The purpose of this chapter is to describe those fluid-inclusion characteristics which serve to distinguish relatively near-surface, epithermal formation conditions from deeper and, potentially, higher temperature formation conditions, and to discuss several techniques and problems which are specific to fluid inclusions trapped in the epithermal environment. A detailed summary and critique of fluid-inclusion literature related to epithermal systems has not been attempted. For this information the reader is referred to the recent compilations of Buchanan (1981), Heald-Wetlaufer et al. (1983), Roedder (1984), and Hedenquist and Henley (1985). Moreover, we have not attempted to relate any particular fluid-inclusion characteristic to a specific type or stage of mineralization, because an adequate data base to do so does not presently exist.

This presentation is limited to two subjects--the petrography and petrology of fluid inclusions from the epithermal environment--and is intended to provide the explorationist with a basic understanding of the criteria for recognizing and interpreting inclusions trapped in this environment. Two important topics will be discussed in detail: (1) the identification and interpretation of fluid inclusions trapped from boiling fluids, and (2) the identification of gases (mainly CO_2) in fluid inclusions and the effect of volatiles on calculated pressures and depths of trapping. We will not, however, discuss the important chemical consequences of boiling and dissolved volatiles, as these subjects are covered in detail in other chapters in this volume (see Henley, 1985, this volume; Henley and Brown, 1985, this volume; Fournier, 1985, this volume; and Reed and Spycher, 1985, this volume).

INFORMATION AVAILABLE FROM FLUID-INCLUSION PETROGRAPHY

Characteristics of fluid inclusions trapped in the epithermal environment contrast markedly with those of inclusions formed in deeper environments. This section considers those features which are diagnostic of shallow crustal environments and which are readily observable by anyone with access to a standard petrographic microscope. In addition, owing to the nature of fluid inclusions trapped in the epithermal environment, particular care in the selection of fluid inclusions for detailed microthermometric analysis must be practiced, and such precautions will be discussed. Diagnostic information common to all minerals containing fluid inclusions formed in the epithermal environment will be presented first, followed by a detailed discussion of information available from petrographic observations of fluid inclusions in quartz.

Fluid inclusions in the epithermal environment typically contain only two phases at room temperature--a low-salinity H_2O liquid phase and a vapor bubble. Daughter minerals of halite and sylvite are notably absent in the epithermal environment. Readily observable evidence for gases is lacking also: gases rarely occur as condensed phases in fluid inclusions, and evidence of gases is normally not found by simple crushing tests, for reasons described below. However, low concentrations of gases have been identified by capacitance manometer and mass spectrometric techniques (Sommer et al., 1985; Hedenquist and Henley, 1985).

Exceptions to these generalities may occur when an epithermal system overprints an earlier, higher temperature system, or vice versa. For example, at Summitville, Colorado, which is a rather high-level, fossil hydrothermal system within a volcanic dome (Perkins and Nieman, 1982), some early quartz rarely contains a few isolated healed microfractures defined by vapor-rich $H_2O + CO_2$ (≥ 70 mole % CO_2) inclusions and/or healed microfractures defined by halite-bearing inclusions with small vapor bubbles. These inclusions presumably contain samples of early magmatic fluids trapped before the development of the near-surface epithermal system at this locality. Similarly, high-temperature (360°C), vapor-rich $H_2O + CO_2$ (≥ 70 mole % CO_2) inclusions found in some active geothermal systems (e.g., in certain deep portions of the Geysers geothermal field) could be magmatic fluids. Also, the high concentrations of CO_2 and CH_4 present in fluid inclusions from sediment-hosted gold deposits may not be temporally related to the hydrothermal system attending gold mineralization. Certainly, some epithermal systems must have been subsequently buried and subjected to fluids of deeper origins, and in these cases CO_2-bearing and/or salt-saturated inclusions may postdate the epithermal inclusions.

Fluid inclusions are trapped in many minerals formed in the epithermal environment; quartz, sphalerite, calcite, and fluorite have yielded useful thermometric data. Of these minerals, quartz usually provides the most fertile opportunities for collection of interpretable fluid-inclusion data. Just as the megascopic crustiform banding of quartz is characteristic of void filling in the epithermal environment, so are the microscopic textures in quartz diagnostic. Furthermore, the fluid-inclusion textures in quartz vary systematically in a manner that permits general thermal conditions to be predicted from features observed under the microscope. These systematic variations in epithermal-quartz fluid-inclusion textures may reflect the temperature dependence of the kinetics of dissolution and reprecipitation of quartz during "maturation"* of individual inclusions. Thus, the following discussion of the microscopic textures diagnostic of the epithermal environment are limited to those found in quartz.

---

*The term "maturation" is used to describe the process of dissolution and reprecipitation of a host mineral surrounding a trapped fluid. The initial "immature" inclusion is generally large and very irregularly shaped. With time, this large inclusion will neck down to form numerous smaller, more regularly shaped inclusions, with the most mature inclusion obtaining the negative crystal shape of the host mineral. (See Roedder, 1984; figs. 2-15.)

At the outset, it should be noted that the microscopic textures displayed by quartz discussed below are not unique to the epithermal environment. Similar textures may occur in quartz from deeper, higher-temperature hydrothermal systems. To mitigate possible ambiguities in the final interpretation of the textures observed, many samples must be collected and observed to gain a broad perspective of all features present, and these data must be combined with fundamental geologic knowledge. As is true for most geologic studies, conclusions should not be formulated from a single observation, and this is true for any study involving fluid inclusions.

From a fluid-inclusionist's viewpoint, quartz formed in the epithermal environment can be divided into two groups--quartz that contains fluid inclusions large enough to study ($>1.5\mu$m) and quartz that contains very few or no inclusions large enough to study with a standard petrographic microscope equipped for total magnifications at least as high as 480X. Figure 5.1A shows three different types of quartz commonly found in the epithermal environment that contain few, if any, inclusions large enough to study. The lower part of Figure 5.1A shows finely crystalline, euhedral quartz crystals that grew contemporaneously from many nucleation sites. This texture is often noted as late vug-fillings, but may also be a result of replacement (silicification) of original wallrock. The finely crystalline quartz seldom contains fluid inclusions large enough to study, but on the rare occasion when appropriate inclusions are found (see below), homogenization temperatures are typically <200°C and melting points are usually between 0 and -2°C. Numerous voids, bounded by quartz crystal faces, occur in this type of quartz. Rarely, these spaces may be sealed to trap large fluid inclusions, thus recording either the latest fluids from which the quartz grew, or fluids that circulated at some later period, perhaps even millions of years after the crystals formed. Such large inclusions generally leak during heating runs, indicating that they are structurally weak; thus, attention to this inherent weakness is necessary when attempting homogenization temperature determinations. Furthermore, because the inclusions are weak, mechanical stresses (including the preparation procedure) may permit late fluids to enter the inclusions. If melting points of such angular, intercrystalline inclusions are to be determined, the data should be interpreted with care. More typically, these intercrystalline spaces contain only air, and appear as dark, angular cavities as shown in the early quartz in the lower portion of Figure 5.1A.

The finely banded, botryoidal quartz in the center of Figure 5.1A is chalcedonic quartz. Fluid inclusions have not been observed to date in this type of quartz, but it is suspected that such quartz generally forms at low temperatures.

The coarse-grained, clear, euhedral crystals at the top of Figure 5.1A are also common in the epithermal environment. Generally, inclusions in such clear, euhedral quartz are rare, and if present, are very small ($\leq 3\mu$m in longest dimension). Occasionally, however, clear, euhedral quartz may have inclusions large enough to observe as shown in Figures 5.1B and 5.1C. Figure 5.1C is an enlargement of a growth zone in the center of Figure 5.1B, and shows that the growth zone in the quartz is defined by small fluid inclusions (thus of definite primary origin) of irregular shape and containing variable liquid to vapor volumetric phase ratios. The presence of one-phase, liquid-filled inclusions in this growth zone indicates that healing of the inclusions had continued at low temperatures (≤ 100°C) after the vapor phase had nucleated within the cavity upon cooling, and that the inconsistent liquid to vapor ratios exhibited in the growth zone are probably not a result of inhomogeneous entrapment of liquid and vapor phases during boiling. Similar textural relationships can be seen clearly in Figures 5.1E and 5.1F. In these figures, many vapor-rich inclusions coexist with liquid-filled inclusions in a coarse-grained euhedral quartz growth zone. The presence of many one-phase, liquid-filled inclusions indicates that the vapor-filled inclusions probably resulted from continued healing and "necking" of the long, stringy, dendritic inclusion cavities after a vapor phase had nucleated within the inclusions in a way that separated the vapor phase plus some liquid into some inclusions and only the liquid phase into other inclusions. This same process is responsible for the variable liquid to vapor volumetric ratios observed in healed microfractures crossing clear, euhedral quartz as shown in Figure 5.1D. Again, what distinguishes continued healing of inclusions at low temperatures after nucleation of a vapor phase from entrapment of inclusions during fluid boiling is the observation that one-phase, liquid-filled inclusions coexist with inclusions of variable liquid-to-vapor ratios within the individual plane. Of course, boiling could have occurred during formation of these textures, but continued "necking" after nucleation of another phase would mask evidence for boiling.

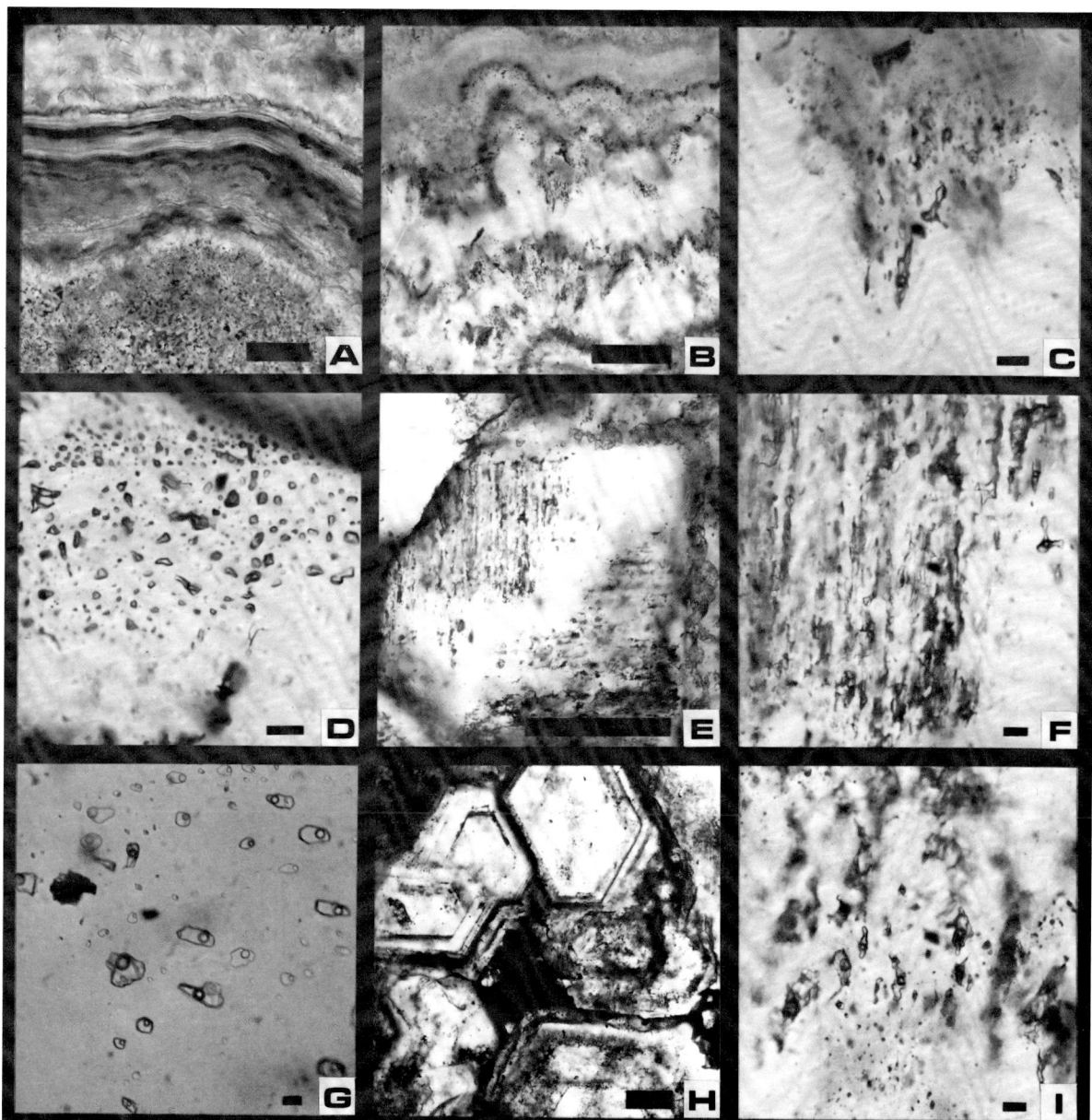

Figure 5.1 A–I.

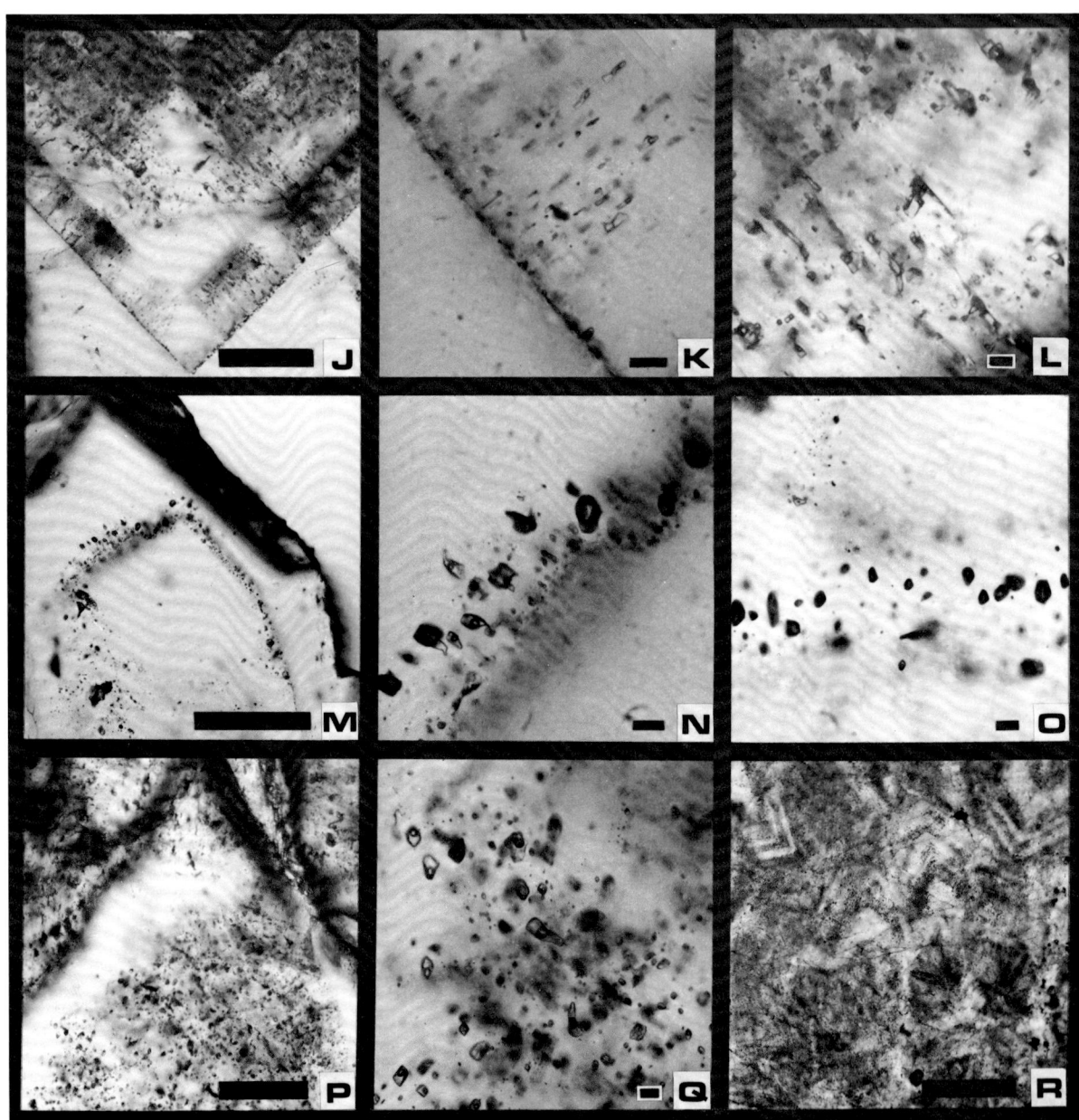

Figure 5.1 J–R.

Figure 5.1. Photomicrographs showing mineralogical and fluid-inclusion textures common in quartz formed in the epithermal environment.

A). Three types of quartz from the epithermal environment that rarely contain fluid inclusions large enough to study. In this example, finely crystalline euhedral quartz containing numerous dark, air-filled voids (bottom) is succeeded by banded, botryoidal quartz (center), and finally by large, clear euhedral quartz crystals (top). Bar scale equals 200 micrometers.

B). Zoned clear, euhedral quartz crystals (center) overgrown by banded chalcedonic quartz (top). Bar scale equals 200 micrometers.

C). Enlargement of central portion of Figure 5.1B showing numerous primary fluid inclusions with variable liquid-to-vapor ratios resulting from necking at low temperature after formation of a vapor phase in the inclusions. Bar scale equals 10 micrometers.

D). Plane of secondary inclusions exhibiting variable liquid-to-vapor ratios (including one-phase liquid inclusions) resulting from necking after the generation of a vapor phase at low temperatures. Bar scale equals 10 micrometers.

E). Clear, euhedral quartz crystal containing numerous primary fluid inclusions oriented perpendicular to growth surfaces. Bar scale equals 200 micrometers.

F). Enlargement of portion of Figure 5.1E showing numerous large, stringy primary inclusions with variable liquid-to-vapor ratios. One-phase liquid inclusions in this zone indicate necking at low temperature after a vapor phase had nucleated in the inclusions. Bar scale equals 10 micrometers.

G). Plane of secondary inclusions with consistent liquid-to-vapor ratios. Bar scale equals 10 micrometers.

H). Clear, euhedral quartz crystals showing growth zones defined by dark bands of primary fluid inclusions. Bar scale equals 200 micrometers.

I). Enlargement of portion of Figure 5.1H showing irregularly shaped, primary fluid inclusions, most with consistent liquid-to-vapor ratios, that homogenize at $230\pm10^{\circ}C$. Bar scale equals 10 micrometers.

J). Portion of a zoned quartz crystal showing primary fluid inclusions defining growth surfaces. Bar scale equals 200 micrometers.

K), L). Enlargements of portions of Figure 5.1J showing irregularly shaped, primary inclusions, most with consistent liquid-to-vapor ratios, that homogenize at $250\pm10^{\circ}C$. Bar scale equals 10 micrometers.

M). Large euhedral quartz crystal showing numerous primary fluid inclusions defining a former crystal surface. Bar scale equals 200 micrometers.

N). Enlargement of growth zone shown in Figure 5.1M containing liquid-rich fluid inclusions with consistent liquid-to-vapor ratios and vapor-rich inclusions with more variable liquid-to-vapor ratios.

O). Healed microfracture in quartz containing all vapor-rich inclusions. Bar scale equals 10 micrometers.

P). Euhedral quartz crystal with primary fluid inclusions defining former growth surfaces. Bar scale equals 200 micrometers.

Q). Enlargement of portion of Figure 5.1P showing regularly shaped primary inclusions that homogenize at $270\pm10^{\circ}C$. Bar scale equals 10 micrometers.

R). Original epithermal quartz that has been subjected to a regional metamorphic event. Although the original epithermal textures are still preserved, the quartz is cut by numerous microfractures defined by small CO_2-bearing inclusions introduced during metamorphism. Bar scale equals 200 micrometers.

Homogenization temperatures collected from groups of primary inclusions or planes of secondary inclusions exhibiting variable liquid-to-vapor volumetric phase ratios due to "necking" after nucleation of the vapor phase will be erroneous, and therefore, misleading. Hence, it is advisable that homogenization temperatures not be collected on such inclusions, although melting points of ice could be valid. The careful inclusionist should painstakingly survey samples to locate the rare, isolated plane(s) or area(s) within a growth zone where the fluid inclusions exhibit consistent liquid-to-vapor volumetric phase ratios and only determine homogenization temperatures for these inclusions. For this type of quartz that contains mostly fluid inclusions of irregular shape and highly variable liquid-to-vapor ratios within individual healed microfractures (planes of secondary inclusions) or within a group of proximal primary inclusions in a growth zone, the rare, isolated inclusions showing consistent liquid-to-vapor volumetric ratios are typically the smallest inclusions. As explained later, more consistent liquid-to-vapor ratios among the smallest inclusions might be a predictable consequence of the healing process. Homogenization temperatures collected from appropriate inclusions in this type of quartz usually are <200°C.

Some quartz crystals may contain growth zones defined by fluid inclusions exhibiting consistent liquid-to-vapor volumetric phase ratios as demonstrated in Figures 5.1H-5.1L and Figures 5.1P and 5.1Q. These three sets of figures show three different examples of growth banding in quartz at low power, and each set has one or two enlargements of individual areas within growth zones with inclusions of consistent liquid-to-vapor ratios. Figure 5.1G shows a plane of secondary inclusions, all with similar liquid-to-vapor phase ratios. In growth zones or planes where consistent liquid-to-vapor volumetric phase ratios predominate, measured homogenization temperatures within a restricted area of a growth zone or within an individual plane will vary mostly within a 10-20°C range, typically around a median temperature ≥ 230°C. The flat, irregularly shaped primary inclusions in Figure 1I homogenize at $\sim 230 \pm 10$°C. The irregularly shaped, though more three-dimensional, inclusions in Figures 5.1K and 5.1L homogenize at $\sim 250 \pm 10$°C, as do the secondary inclusions in Figure 5.1G ($\sim 250 \pm 5$°C). The smooth-surfaced equant to negative crystal-shaped inclusions in Figure 5.1Q homogenize at $\sim 270 \pm 10$°C.

The textures discussed so far should enable a fluid inclusionist with minimal experience to readily distinguish quartz that contains inclusions that will probably homogenize at temperatures ≤ 200°C from quartz that contains inclusions that probably will homogenize at ≥ 230°C; quartz that contains inclusions which homogenize between $\sim 200-230$°C will probably be highly irregular in shape but will have more than a rare plane or area within a growth zone containing consistent liquid-to-vapor volumetric ratios. With a bit more experience, a fluid inclusionist will be able to recognize inclusions which will homogenize consistently around 230°C versus inclusions which will homogenize around 270°C, based on the shapes and surfaces of the inclusions. Of course, during the course of a fluid-inclusion study, such predictions should be supported by homogenization temperatures collected from appropriate inclusions with a heating stage, as exceptions to the above generalities are certain to occur. The general predictability of homogenization temperatures based on consistency of liquid-to-vapor volumetric ratios, shape, and surface smoothness of inclusions may reflect the temperature dependence of the kinetics of dissolution and reprecipitation of quartz during formation and "maturation" of inclusions. The higher the temperature, the faster the healing process, and thus the greater potential for achieving the minimum surface free-energy state (negative crystal morphology); the lower the temperature, the slower the healing process, and, thus, the greater potential for nucleation of a vapor phase prior to complete sealing of an individual fluid inclusion. This would also explain why small inclusions ($\leq 3 \mu m$ in longest dimension) in growth zones exhibiting variable liquid-to-vapor ratios yield more consistent homogenization temperatures. If these small inclusions are "pinched-off" into separate entities early in the healing process, by virtue of being small they are less likely to "neck" further.

From the above discussion, it should be clear that vapor-rich inclusions can and do result from the healing process and in many cases may therefore not be indicative of boiling. Furthermore, we emphasize that boiling may have accompanied formation of any textures shown in Figures 5.1A-5.1L, and 5.1P-5.1Q, but that no fluid-inclusion evidence for boiling was recorded in these examples. Figures 5.1M and 5.1N are photomicrographs of a quartz crystal from a vein collected at the north end of the Creede, Colorado mining district at an elevation of 12,450 ft (ore is at an elevation below 11,000 ft). These inclusions demonstrate excellent evidence for boiling in the hydrothermal system at least once during growth of the quartz crystal at the locality where the crystal was growing. Figure 5.1M shows a well-defined growth zone in a quartz crystal, and Figure 5.1N is an enlargement of a portion of this growth zone showing that it is defined by (primary) fluid inclusions. Within the growth zone shown in Figure 5.1N are liquid-rich inclusions coexisting with vapor-rich inclusions. There are several important features that indicate that the vapor-rich inclusions result from entrapment of a vapor phase during formation of the quartz and are not a result of later healing (necking) processes: (1) most importantly is the absence of one-phase, liquid-filled inclusions; (2) all of the liquid-rich inclusions show consistent liquid-to-vapor ratios and yield consistent homogenization temperatures; and (3) all of the vapor-rich inclusions are larger than the liquid-rich inclusions. Certainly, if the vapor-rich inclusions resulted from necking of larger inclusions after nucleation of a vapor phase, variable liquid-to-vapor volumetric ratios among the liquid-rich inclusions might result; these characteristics are not present in Figure 5.1N. The fact that the vapor-rich inclusions are all of similar size probably reflects the stable size of the vapor bubbles (which should be of a certain diameter at some given P, T condition) in the boiling fluid. Also, it might be expected that some imperfections would be totally filled by liquid (forming the liquid-rich inclusions) and that vapor bubbles might adhere to other imperfections already containing some liquid. As is evident in the photomicrograph in Figure 5.1N, some vapor-rich inclusions do contain some liquid, most probably trapped during formation of the vapor-rich inclusions, which would result in incorrect (too high) homogenization temperatures.

Vapor-rich inclusions can result from continued healing of microfractures after nucleation of vapor bubbles as described above. If a sample contains many such microfractures crisscrossing to form a dense array, a novice inclusionist could easily misinterpret the texture and conclude boiling. Definitive evidence that a vapor phase was truly present in the hydrothermal system would be a healed microfracture containing all vapor-rich inclusions as shown in Figure 5.1 O.

Quartz from deeper environments (e.g., at margins of batholiths, or around and within cupolas of batholiths, or from the surrounding terrains undergoing greenschist to amphibolite grade metamorphism) is characterized by the presence of abundant (millions/mm^3) healed microfractures defined by small ($\leq 5 \mu m$ typically) fluid inclusions resulting in crisscrossing, sweeping, wispy, microscopic textures. These numerous secondary inclusions give "bull quartz" the diagnostic white color commonly observed on the outcrop or in hand specimens. Individual inclusions may contain only one dense, liquid phase at room conditions or may contain a number of phases; volatile components (CO_2, CH_4, N_2) at pressures greater than 1 atm are also typically present.

Growth zoning is not commonly observed in quartz crystals formed at deeper levels in the crust, but does occur. Such quartz is distinguished from growth-banded quartz formed in the epithermal environment by any one of the following possible observations: (1) as the deep quartz forms at higher pressures and potentially higher temperatures, most inclusions in the growth zones will be smooth-surfaced and equant in shape, and most will contain either a single, dense, liquid phase, or two fluid phases which homogenize consistently at temperatures $\leq 270°C$; or (2) the inclusions will contain gases under pressure easily detected by a crushing study; or (3) the inclusions will yield very low melting points ($< -20°C$). Such observations contrast markedly with features of fluid inclusions in epithermal quartz discussed above.

Figure 5.1R is a photomicrograph of quartz thought to have originally formed at epithermal conditions which has been overprinted by a metamorphic event (lower greenschist grade?) from a prospect in British Columbia, Canada. Remnant growth banding in the quartz is evident, as are the ubiquitous crisscrossing microfractures defined by small CO_2-bearing inclusions. Fluid inclusions defining the growth zones in Figure 5.1R display typical epithermal characteristics, most similar to the inclusions shown in Figures 5.1E, 5.1F, 5.1J, and 5.1K. Whether or not the original epithermal inclusions still completely preserve information regarding the epithermal hydrothermal system after a metamorphic event is debatable; more laboratory experiments are necessary to assess the conditions at which inclusions in quartz would begin to be altered.

In summary, petrographic examination of fluid inclusions in quartz can yield valuable information concerning formation conditions, and one need not necessarily be an experienced fluid inclusionist to recognize the important characteristics discussed above. However, we strongly recommend that all fluid inclusionists allow ample time for careful reconnaissance surveys prior to initiating thermometric studies. Unfortunately, many fluid inclusionists tend to eliminate completely or spend too little time on petrographic surveys. A reconnaissance petrographic examination should include many carefully selected samples with consideration of the complexities that may arise from the effects of time to derive the maximum amount of information possible from the inclusions and mineral textures.

IDENTIFICATION OF FLUID INCLUSIONS TRAPPED FROM BOILING SOLUTIONS

A commonly reported feature in studies of the epithermal environment is the occurrence of boiling during one or more stages of mineral deposition. This is not surprising, as boiling is a predictable consequence for fluids circulating at the relatively high-temperature, low-pressure conditions existing in this environment. Moreover, because boiling is a potential mechanism for mineral deposition (Drummond, 1981; Drummond and Ohmoto, 1985), fluid-inclusion evidence for boiling has been proposed as an exploration tool in the search for epithermal precious-metal deposits (Kamilli and Ohmoto, 1977). In this section, we describe various techniques for identifying fluid inclusions that were trapped in a boiling system.

The coexistence of liquid-rich and vapor-rich inclusions is the single, most often cited evidence in support of entrapment from boiling fluids. As shown in the remainder of this section, this is probably also the most important line of evidence that a fluid inclusionist may be able to report conclusively, but several precautions are necessary. First, the sample must actually contain liquid-rich and vapor-rich inclusions. Because fluid inclusions are three-dimensional objects which are being observed in only two dimensions, inclusions having consistent liquid-to-vapor ratios may appear to have variable phase ratios, as shown in Figure 5.2. This problem is easily overcome by testing the inclusions on the heating stage because, if the variable phase ratios are apparent and not real, the inclusions will all exhibit similar modes and temperatures of homogenization. Second, the vapor-rich inclusions must result from entrapment of a vapor phase and not from necking. As noted above, vapor-rich inclusions which result from the healing process may be distinguished from those formed by entrapment of vapor by petrographic observations. Finally, there must be definitive evidence that the liquid-rich and vapor-rich inclusions are contemporaneous, and this evidence is provided when both types of inclusions occur within an individual growth zone as illustrated in Figures 5.1M and 5.1N.

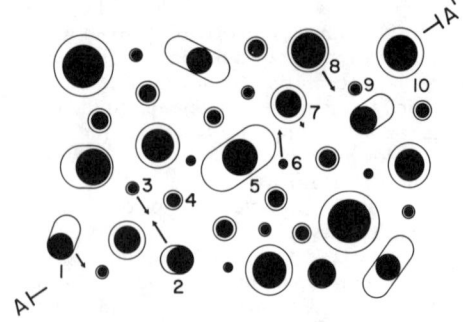

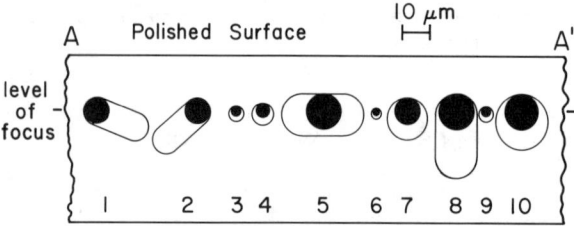

Figure 5.2. (top) Apparent liquid-to-vapor ratios at 25°C of numerous fluid inclusions as seen through the microscope. All fluid inclusions shown contain 25 volume-percent vapor. Inclusions numbered 1-10 along A-A' correspond to inclusions 1-10 below. (bottom) Cross-sectional view of inclusions 1-10 shown above.

As noted in the previous section, additional definitive evidence that a vapor phase existed in a hydrothermal system would be a healed microfracture containing all vapor-rich inclusions (c.f., Figure 5.1 O) as a fracture containing all vapor-rich inclusions could not result from necking alone. Presence of a plane of vapor-rich secondary inclusions means only that boiling occurred somewhere in the system, not necessarily at the point of sample collection and perhaps as much as thousands of feet below. However, in a hydrothermal system which does not intersect a water table, liquid water is potentially always present, and, thus, a mere plane of vapor-rich secondary inclusions may indeed indicate boiling at the sample location. Thus, it may be possible to map zones of boiling conditions by recording relative abundances of vapor-rich inclusions (regardless of their origin, as long as they do not result from necking) to indicate degrees of boiling, ranging from minor boiling to flashing. More research is required to confirm this possibility; experimental studies as well as those conducted in natural settings (active geothermal systems) are likely to be fruitful.

Thus far, we have discussed the characteristics of inclusions from the epithermal environment in qualitative terms, referring to the inclusions simply as liquid-rich and vapor-rich. This qualitative description may, however, be quantified, as it is possible to calculate the phase ratios of inclusions trapped at various PTX conditions using available phase equilibrium and PVT data. Although this information is generally not required for most fluid-inclusion studies, it is nevertheless a worthwhile exercise because it provides the fluid inclusionist with an understanding of the types of inclusions he or she might expect to find, and helps eliminate obviously anomalous inclusions from further study. The calculations are particularly valuable when trying to decide if a vapor-rich inclusion might have trapped all (or almost all) vapor or a mixture of liquid and vapor.

Sample Calculation 1

Predict the room temperature (25°C) phase ratios of the liquid-rich and vapor-rich inclusions trapped in a boiling, 5 wt.-% NaCl fluid at 250°C. Assume that each inclusion traps only a single phase and not mixtures of liquid and vapor. The following data, from Khaibullin and Borisov (1966), Potter and Brown (1977), and Keenan et al. (1978), are required

Composition of liquid phase = 5 wt.-% NaCl

Composition of vapor phase = 0 wt.-% NaCl

Density of liquid phase at trapping
= 0.840 g/cm^3

Density of vapor phase at trapping
= 0.019 g/cm^3

Density of 5 wt.-% NaCl liquid at 25°C
= 1.030 g/cm^3

Density of liquid H$_2$O at 25°C = 0.997 g/cm^3

At the trapping conditions (250°C), the liquid inclusion is filled with a 5 wt.-% NaCl liquid with a density of 0.840 g/cm^3. At 25°C, this inclusion still contains a 5 wt.-% NaCl liquid, but now with a density of 1.030 g/cm^3, as a result of liquid contraction during cooling. The remainder of the inclusion contains a vapor bubble that is essentially a vacuum. The percent decrease in the volume occupied by the liquid, which equals the volume-percent vapor, is therefore

Volume-% vapor

$$= \frac{(1.030 \text{ g/cm}^3 - 0.840 \text{ g/cm}^3)}{1.030 \text{ g/cm}^3} \times 100$$

$$= 18.4\%$$

Thus, the inclusion that trapped only the liquid phase from a boiling, 5 wt.-% NaCl fluid at 250°C will contain at 25°C a liquid occupying 81.6 volume percent of the inclusion and a vapor bubble occupying 18.4 volume percent.

Similarly, for the inclusion that traps only the vapor phase

Volume-% vapor

$$= \frac{(0.997 \text{ g/cm}^3 - 0.019 \text{ g/cm}^3)}{0.997 \text{ g/cm}^3} \times 100$$

$$= 98.1\%$$

This inclusion will contain 98.1 volume-percent vapor and 1.9 volume-percent liquid at 25°C.

Following the calculation procedure outlined above, the room-temperature phase ratios of pure H_2O and 2, 3.5, 5, and 10 wt.-% NaCl inclusions trapped along the liquid-vapor curve have been calculated and the results are listed in Table 5.1. The room temperature appearance of both the liquid-rich and vapor-rich inclusions trapped from a boiling 5 wt.-% NaCl solution are also shown on Figure 5.3.

If inclusions are trapped from boiling fluids, and if the inclusions trap only one or the other of the two fluids present, but not mixtures of the two, the homogenization temperatures are the same as the trapping temperatures. Furthermore, the liquid-rich and vapor-rich inclusions must homogenize at the same temperature. We know, however, both from empirical observations and experimental studies, that the vapor-rich inclusions almost always trap some liquid along with the vapor, whereas liquid inclusions almost never trap any vapor along with the liquid (Bodnar et al., 1985; Robertson, 1968). This being the case, homogenization temperatures of liquid-rich and vapor-rich inclusions trapped from boiling fluids might be distributed as shown on Figure 5.4, with minor variations.

Homogenization temperatures of liquid-rich inclusions should all fall within a narrow range, and, moreover, this range represents the actual range over which boiling occurred. A few scattered temperatures above this value may occur and these represent the few inclusions that have leaked or (less likely) trapped some vapor, but are still liquid rich. Homogenization temperatures of vapor-rich inclusions* will be higher than those of the coexisting liquid-rich inclusions and will be very scattered--a natural consequence of the fact that the vapor-rich inclusions nearly always trap some liquid and that the proportions of liquid and vapor trapped in these inclusions will thus be variable. The maximum possible homogenization temperature of the liquid-rich inclusions is the critical temperature corresponding to the bulk composition of the mixture in the inclusion. Homogenization temperatures of vapor-rich inclusions, on the other hand, may be greater than the critical temperature for the bulk composition in the inclusion, and may extend to the maximum temperature on the solvus for that composition (Bodnar et al., 1985). If a vapor-rich inclusion traps so much liquid that the bulk inclusion density is greater than the critical density, the inclusion will behave like a liquid-rich inclusion. That is, during heating, the vapor bubble will shrink and disappear at some temperature between the boiling temperature and the critical temperature. Many of the anomalously high homogenization temperatures of

Table 5.1--Calculated volume-percent vapor at 25°C of H_2O-NaCl fluid inclusions trapped from boiling solutions

Salinity (wt. %)		Temperature (°C)			
		150	200	250	300
0	L	8.1	13.3	19.8	28.5
	V	99.7	99.2	98.0	95.4
2.0	L	8.1	13.1	19.3	27.5
	V	99.7	99.2	98.0	95.6
3.5	L	7.9	12.9	19.1	26.8
	V	99.7	99.2	98.0	95.7
5.0	L	7.8	12.8	18.6	26.0
	V	99.7	99.2	98.1	95.7
10.0	L	7.7	12.1	17.5	23.5
	V	99.8	99.3	98.2	95.9

L = fluid inclusions that trapped all liquid.
V = fluid inclusions that trapped all vapor.

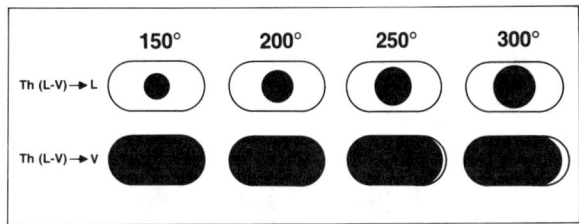

Figure 5.3. Room temperature (25°C) phase relations of H_2O-NaCl fluid inclusions having a salinity of 5 weight-percent NaCl and homogenization temperatures from 150°C to 300°C. Also shown are the room temperature phase relations of inclusions that trapped the vapor phase that would have been in equilibrium with the liquid phase at each temperature. Black corresponds to the vapor phase and the unshaded portion represents the liquid phase.

liquid-rich inclusions in epithermal deposits perhaps are a result of this somewhat special case of heterogeneous entrapment.

---

*Homogenization temperatures of vapor-rich inclusions depicted in Figure 5.4 represent the temperatures that would be observed if such data were obtainable. In practice, homogenization temperatures of vapor-rich inclusions can rarely be measured because the inclusion geometry usually prevents the fluid inclusionist from viewing the final disappearance

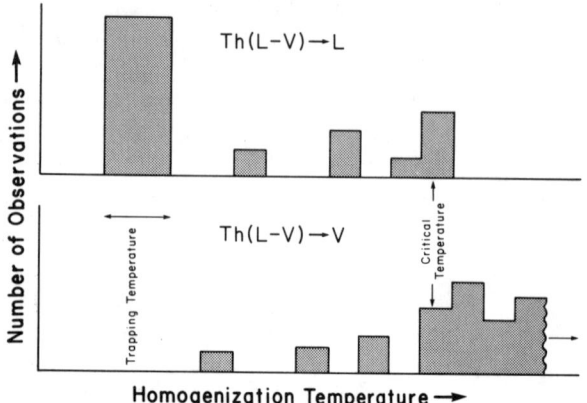

Figure 5.4. Schematic representation of the manner in which liquid-rich and vapor-rich inclusions trapped from a boiling solution would be distributed assuming the inclusions trapped either all liquid or mixtures of liquid and vapor, but never only vapor.

of liquid, that is, assuming a distinct liquid phase is observable at all, and the precision of these measurements is usually very poor (±10-25°C). This problem becomes less severe as (1) the inclusion geometry becomes more irregular, (2) the volume-percent liquid in the inclusion increases, and (3) the homogenization temperature nears the critical temperature.

This qualitative description of the manner in which homogenization temperatures of inclusions trapped in a boiling system should be distributed may be quantified, provided that the necessary PVT and phase-equilibria data for the fluid of interest are available. Moreover, the salinities of inclusions trapping mixtures of liquid and vapor may be calculated, and the variation in salinity with homogenization temperature predicted.

Sample Calculation 2

Using data provided in Sample Calculation 1 above, calculate the bulk-inclusion density and composition of an inclusion that traps 95 volume-percent vapor and 5 volume-percent liquid from a boiling, 5 wt.-% NaCl solution at 250°C. Assuming an inclusion volume of 1 cm^3, the masses of liquid and of vapor trapped in the inclusion are

mass of liquid = 0.05 cm^3 × 0.840 g/cm^3

= 0.042 g

mass of vapor = 0.95 cm^3 × 0.019 g/cm^3

= 0.018 g

$$\text{density} = \frac{0.042 \text{ g} + 0.018 \text{ g}}{1 \text{ cm}^3}$$

= 0.060 g/cm^3

Of the 0.042 g of liquid trapped in the inclusion, 5 weight percent is NaCl and 95 weight percent is H$_2$O. Assume that the trapped vapor is pure H$_2$O. The mass of NaCl trapped in the inclusion is

mass NaCl = 0.05 × 0.042 g

= 0.0021 g

Therefore, the composition of the inclusion is

$$\text{salinity} = \frac{0.0021 \text{ g}}{0.042 \text{ g} + 0.018 \text{ g}} \times 100$$

= 3.50 wt.-% NaCl

Thus, this inclusion, which trapped 5 volume-percent liquid having a composition of 5 wt.-% NaCl and 95 volume-percent vapor having a composition of pure H$_2$O, has a bulk density of 0.060 g/cm^3 and a salinity of 3.50 wt.-% NaCl. Although the exact homogenization temperature of this inclusion cannot be determined owing to lack of PVTX data in this region, we can determine that the homogenization temperature will be greater than the critical temperature of a 3.5 wt.-% NaCl fluid, i.e., greater than 410°C (Khaibullin and Borisov, 1966).

Following the procedure outlined in Sample Calculation 2 and using the data of Khaibullin and Borisov (1966), the compositions and homogenization temperatures of inclusions trapping various proportions of liquid and vapor from a boiling, 5 wt.-% NaCl solution at 250°C have been calculated and are shown on Figure 5.5. Homogenization temperatures of liquid-rich inclusions may vary from the actual trapping temperature (250°C) to the critical temperature for the bulk-fluid composition (417°C), but may not exceed this value. Homogenization temperatures of vapor-rich inclusions may extend to much higher values, but the exact temperatures cannot be predicted because of lack of PVTX data in this region. The salinity may be any value between the liquid-phase composition (5 wt.-% NaCl) and the vapor-phase composition (pure H$_2$O).

If a group of apparently contemporaneous fluid inclusions exhibits variable liquid-to-vapor ratios, it is common practice to not collect homogenization temperature and salinity data from these inclusions as they have obviously trapped mixtures of liquid and vapor (or have leaked or necked). While it is true that the measured values do not represent the actual trapping temperatures or compositions of the fluids in

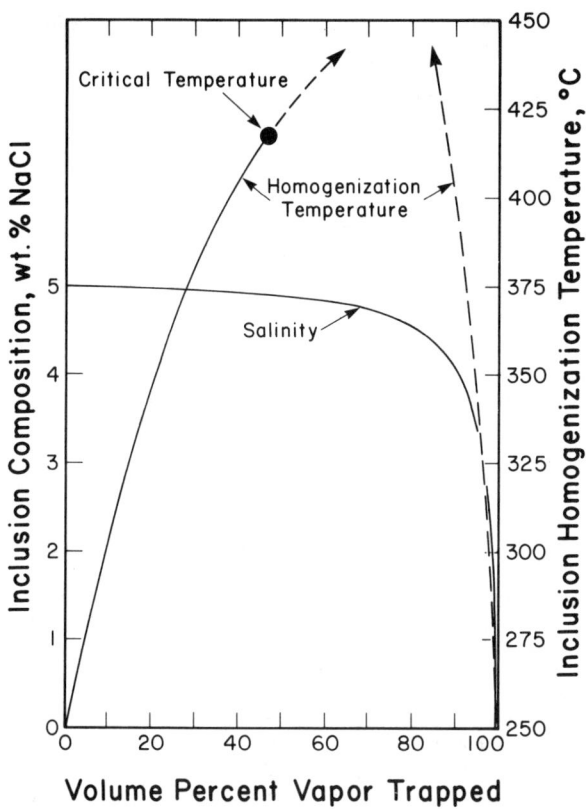

Figure 5.5 Calculated variation in inclusion composition and homogenization temperature as a function of the volume-percent vapor trapped from a boiling, 5 wt.-% NaCl solution at 250°C.

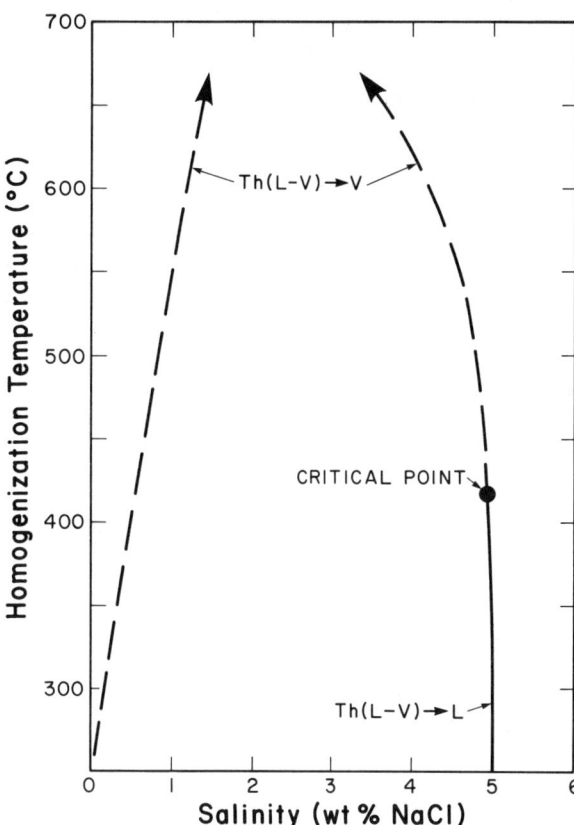

Figure 5.6. Relationship between homogenization temperature and salinity of inclusions trapping various proportions of liquid and vapor from a boiling, 5 wt.-% NaCl solution at 250°C.

equilibrium at trapping, they may provide additional evidence that the inclusions were trapped in a boiling system, owing to the systematic and predictable manner in which homogenization temperature varies as a function of salinity, as shown on Figure 5.6.

In summary, from a practicable standpoint, the occurrence of coexisting, contemporaneously formed liquid-rich and vapor-rich inclusions in a given sample is the best evidence that could normally be observed to indicate that the inclusions were trapped during boiling. However, if homogenization temperature and salinity data can be obtained from such inclusions, additional evidence supporting a boiling intepretation may be determined, owing to the systematic and predictable manner in which these values must vary if the inclusions trapped mixtures of liquid and vapor. It is clear that our understanding of the trapping mechanisms and behavior of inclusions trapped from boiling fluids must be improved, and this will require additional experimental studies as well as detailed studies of inclusions from natural systems where boiling is known to have occurred.

IDENTIFICATION OF GASES IN FLUID INCLUSIONS FROM THE EPITHERMAL ENVIRONMENT

The presence of gases in fluid inclusions from the epithermal environment has been reported by numerous workers, and these volatile components have been shown to play an important role in controlling solution chemistry. Moreover, the presence of gases can significantly affect the pressure, and, therefore, the depth, at which boiling may commence in these systems. In this section we describe the techniques for recognizing gases in fluid inclusions and explain why the presence of gases may not have been recognized in many fluid-inclusion studies of epithermal deposits.

The predominant volatile component in most fluid inclusions from the epithermal environment is CO_2, and we will base our discussion on the PVTX properties of the H_2O-CO_2 system. Phase relations in the system H_2O-CO_2 are well known over a wide range of PTX conditions (Ellis and Golding, 1963; Todheide and Franck, 1963; Takenouchi and Kennedy, 1964; Malinin and Kurovskaya, 1975). An important feature of this system with respect to epithermal deposits is the two-phase field in which H_2O-rich fluids coexist with more CO_2-rich fluids (Fig. 5.7). This field of immiscibility encompasses only a relatively small portion of PTX space at temperatures greater than about 300°C, but expands considerably as temperature decreases. Over the temperature range in which most epithermal deposits formed (100°C-300°C), only very H_2O-rich or very CO_2-rich compositions do not lie in the two-phase field, even up to pressures of several kilobars, except for the small one-phase "vapor" field that spans the entire composition range at very low pressures (Fig. 5.7).

The presence of CO_2 in fluid inclusions may be recognized by (1) the occurrence of three-phase (liquid H_2O, liquid CO_2, and vapor CO_2) inclusions at room temperature, (2) expansion of the vapor bubble when the inclusions are opened by crushing the sample in oil, (3) the nucleation of the CO_2 clathrate compound ($CO_2 \cdot 5\,3/4\,H_2O$) during cooling of a two-phase inclusion from room temperature to lower temperatures, and (4) microanalysis of fluids released from inclusions. The temperature and pressure of trapping and the amount of CO_2 required to generate three-phase inclusions at room temperature, or to produce inclusions which will nucleate a liquid CO_2 phase during cooling, or show expansion of the vapor bubble during crushing tests may be calculated from available phase equilibrium and PVT data for the H_2O-CO_2 system. Because fluid inclusions are constant-volume systems, the only information required to calculate the room temperature phase ratios and temperatures and pressures of various phase changes in the inclusions is the bulk density of the inclusion. The bulk density of an H_2O-CO_2 fluid at a given temperature and pressure of trapping may be determined from a modified Redlich-Kwong equation of state (Connolly and Bodnar, 1983).

Once the bulk-inclusion density is known, the relative amounts of the H_2O-rich phase and the CO_2-rich phase(s) in the inclusion at 25°C may be obtained by determining at which pressure (at T = 25°C) the necessary mass-balance and phase-equilibria requirements imposed by the inclusion density and composition are satisfied. This pressure, obtained using iterative computation techniques shown on the flow chart in Figure 5.8, is unique for a given inclusion composition and density. The initial guess at the internal pressure in the inclusion at 25°C is 64 bars, corresponding to an inclusion containing both liquid and vapor CO_2 at 25°C. That is, 64 bars corresponds to the equilibrium vapor pressure of CO_2 at 25°C. After calculating the amount of H_2O in the inclusion from the estimated bulk composition, the amount of CO_2 that will dissolve in that mass of H_2O at 25°C and 64 bars is determined using the data of Dodds et al. (1956). Then, the volume occupied by this H_2O-rich phase is determined using density data for H_2O-CO_2 solutions at low temperatures from Parkinson and de Nevers (1969).

The amount of "free" CO_2 in the inclusion is the total amount of CO_2, again obtained from the known bulk composition, less that amount of CO_2 dissolved in the H_2O-rich phase.* This amount of CO_2 must fill that portion of the inclusion not occupied by the H_2O-rich phase. The volume and mass of the "free" CO_2 phase may be obtained from mass-balance calculations, because of the assumption that fluid inclusions represent constant-volume, constant-mass, chemical systems; these values provide the density of the "free" CO_2 phase in the inclusion. If this calculated bulk CO_2 density is greater than the density of CO_2 vapor on the CO_2 liquid-vapor curve at 25°C and less than the density of liquid CO_2 at 25°C and vapor saturation, the inclusion will contain both liquid and vapor CO_2 at 25°C and have an internal pressure of 64 bars. The relative amounts of liquid and vapor CO_2 in such an inclusion are obtained from mass balance calculations.

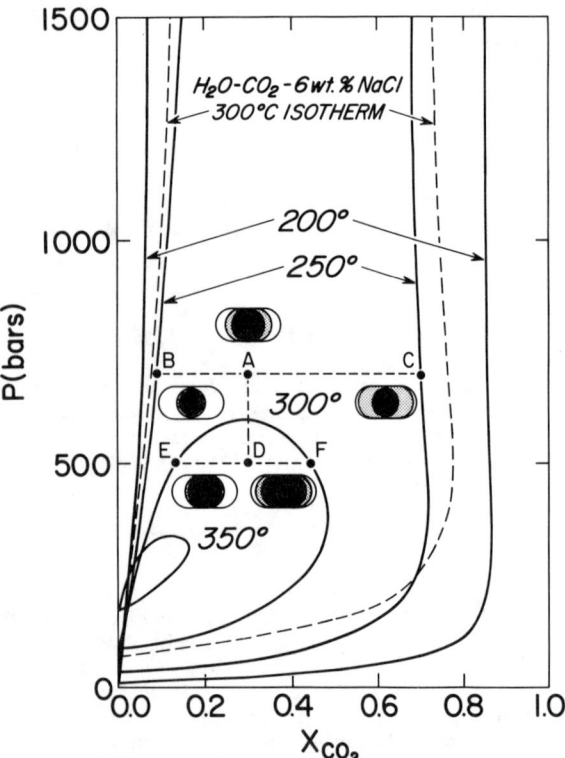

Figure 5.7. Isotherms showing the compositions of coexisting liquid and vapor phases in the system H_2O-CO_2 and H_2O-CO_2-NaCl. Also shown are the 25°C phase relations of H_2O-CO_2 fluid inclusions with a bulk composition of 30 mole-% CO_2 trapped at various temperatures and pressures represented by points A-F. The innermost phase in each inclusion (black) represents a CO_2-rich vapor, the outermost phase (white) represents an H_2O-rich liquid, and the intermediate phase (stippled pattern) represents a CO_2-rich liquid. Room temperature phase relations of these inclusions were calculated as described in the text.

---

*In the calculations of Burruss (1981a,b), Hollister and

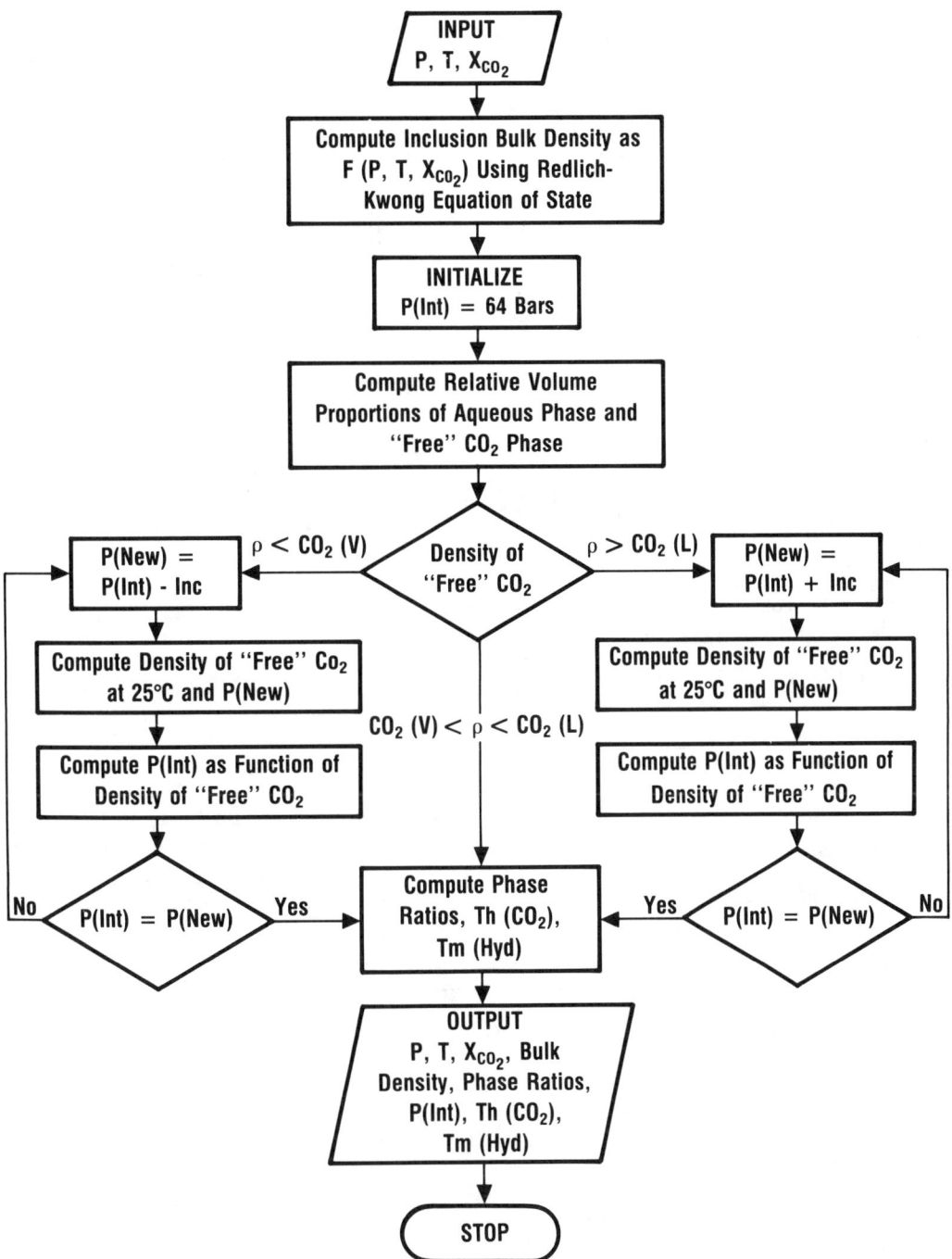

Figure 5.8. Flow chart of the computer program used to calculate room temperature phase ratios and other properties of H_2O-CO_2 fluid inclusions.

Burruss (1976), and Ramboz et al. (1982), the amount of CO_2 dissolved in the H_2O phase is neglected, i.e., considered to be zero. This assumption does not greatly affect the calculated phase relations and temperatures of phase changes for most inclusions in the H_2O-CO_2 system. However, for some inclusions, particularly those with low CO_2 contents or with internal pressures greater than a few tens of bars, neglecting the amount of CO_2 dissolved in H_2O greatly affects the calculated low-temperature properties (Swanenberg, 1980). The amount of H_2O that will dissolve in CO_2 at 25°C and pressures greater than the equilibrium vapor pressure of pure H_2O at this temperature (0.03 bars; Keenan et al., 1978) is approximately an order of magnitude less than the amount of CO_2 that will dissolve in H_2O at these same conditions. Therefore, the CO_2 phases were assumed to be pure in the calculation of room temperature phase ratios of inclusions. This assumption introduces negligible errors into the calculations.

Sample Calculation 3

Assume that H_2O-CO_2 fluid inclusions are trapped in the liquid-vapor field at 250°C and 700 bars. At these conditions, an H_2O-rich fluid with a composition of 91.6 mole-percent H_2O (point B, Fig. 5.7) is in equilibrium with a CO_2-rich fluid with a composition of 38.7 mole-percent H_2O (point C, Fig. 5.7). The density of the H_2O-rich fluid, calculated from a modified Redlich-Kwong equation of state (Connolly and Bodnar, 1983) is 0.831 g/cm^3; the density of the CO_2-rich fluid is 0.638 g/cm^3. Assuming a total inclusion volume of 100 cm^3, what is the density of the "free" CO_2 in the H_2O-rich inclusion? What are the phase ratios of this inclusion at 25°C?

The total mass of fluid (H_2O + CO_2) in the inclusion is

Total mass = 100 cm^3 × 0.831 g/cm^3
= 83.1 g

The mass of H_2O in the inclusion is

Mass H_2O

$$= \frac{(0.916 \text{ moles} \times 18 \text{ g/mole})}{(0.916 \text{ moles} \times 18 \text{ g/mole}) + (0.084 \text{ moles} \times 44 \text{ g/mole})} \times 83.1 \text{ g}$$

= 67.88 g

The number of moles of H_2O in the inclusion is

$$\text{Moles } H_2O = \frac{67.88 \text{ g}}{18 \text{ g/mole}} = 3.77 \text{ moles}$$

The total mass of CO_2 in the inclusion is

Mass CO_2 = 83.1 g − 67.88 g = 15.22 g

The total number of moles of CO_2 in the inclusion is

$$\text{Moles } CO_2 = \frac{15.22 \text{ g}}{44 \text{ g/mole}} = 0.346 \text{ moles}$$

Now, for the initial iteration in the calculation procedure, assume that at 25°C the inclusion contains three phases—liquid H_2O and liquid and vapor CO_2. At 25°C, the equilibrium vapor pressure of CO_2 is 64 bars (Quinn and Jones, 1936). Thus, our initial assumption is that the internal pressure of the inclusion is 64 bars. At 25°C and 64 bars, liquid H_2O has a density of 0.9998 g/cm^3 (Keenan et al., 1978). Thus, the volume of liquid H_2O in the inclusion is

$$\text{Volume } H_2O = \frac{67.88 \text{ g}}{0.9998 \text{ g/cm}^3} = 67.89 \text{ cm}^3$$

Now, the aqueous phase is not pure H_2O but, rather, is a CO_2-saturated, H_2O-rich phase. The solubility of CO_2 in H_2O at 25°C and 64 bars is 2.433 mole percent (Fig. 5.9). So, the total number of moles of CO_2 dissolved in the H_2O phase in our model inclusion is

$$\text{Moles } CO_2 = \frac{0.02433 \times 3.77 \text{ moles}}{1.0 - 0.02433}$$

$$= 0.094 \text{ moles } CO_2$$

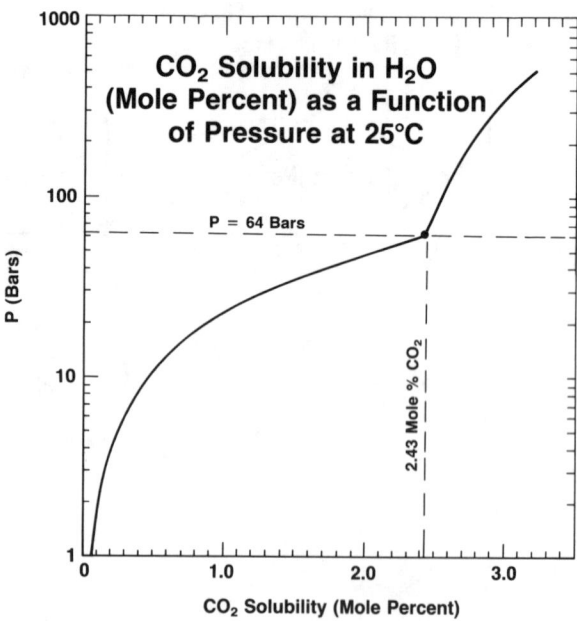

Figure 5.9. Solubility of CO_2 in H_2O as a function of pressure at 25°C.

The volume of CO_2 in the CO_2-saturated aqueous solution at 25°C and 64 bars, assuming a molal volume of CO_2 in solution of 29 cm^3/mole (Wiebe and Gaddy, 1939), is

Volume of CO_2 in solution

$$= (0.094 \text{ moles} \times 29 \text{ cm}^3/\text{mole}) \times \frac{67.88 \text{ g}}{1000 \text{ g}}$$

$$= 0.185 \text{ cm}^3$$

The total volume of the CO_2-saturated aqueous liquid phase is

Volume aqueous phase = 67.89 cm^3 + 0.185 cm^3

$$= 68.075 \text{ cm}^3$$

The mass of CO_2 remaining ("free" CO_2) in the inclusion is

Mass "free" CO_2

$$= (0.346 \text{ moles} - 0.094 \text{ moles}) \times 44 \text{ g/mole}$$

$$= 11.088 \text{ g}$$

The volume occupied by "free" CO_2 is

Volume "free" CO_2 = 100 cm^3 - 68.075 cm^3

$$= 31.92 \text{ cm}^3$$

The bulk density of "free" CO_2 is

Density of "free" CO_2 = $\dfrac{11.08 \text{ g}}{31.92 \text{ cm}^3}$

$$= 0.347 \text{ g/cm}^3$$

Because the density of "free" CO_2 is greater than the density of CO_2 vapor at 25°C and 64 bars (0.2375 g/cm^3), and less than the density of CO_2 liquid at 25°C and 64 bars (0.71 g/cm^3) (Fig. 5.10), the inclusion will contain both liquid and vapor CO_2 at 25°C.

The volume of CO_2 liquid in this inclusion is

Volume of CO_2 liquid

$$= \frac{11.08 \text{ g} - (0.2375 \text{ g/cm}^3 \times 31.92 \text{ cm}^3)}{0.479 \text{ g/cm}^3}$$

$$= 7.3 \text{ cm}^3$$

The volume of CO_2 vapor is

Volume of CO_2 vapor = 31.9 cm^3 - 7.3 cm^3

$$= 24.6 \text{ cm}^3$$

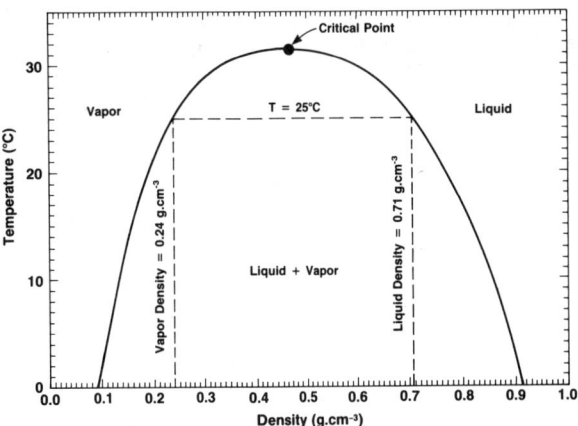

Figure 5.10. Density of liquid and vapor CO_2 at phase equilibrium from 0°C to the critical temperature (31.1°C).

Therefore, assuming a total inclusion volume of 100 cm^3, the inclusion will contain

68.1 volume-percent aqueous phase
24.6 volume-percent CO_2 vapor
7.3 volume-percent CO_2 liquid

100.0 volume percent

Homogenization temperatures of the CO_2 phases ("free" CO_2) in H_2O-CO_2 inclusions may be estimated using the calculated density of "free" CO_2 and the known densities of liquid and vapor on the CO_2 solvus (Fig. 5.10). The homogenization temperature of the "free" CO_2 phases is that temperature on the liquid-vapor curve at which CO_2, either liquid or vapor, that has the same density as that of "free" CO_2 at 25°C and a pressure equal to the calculated internal pressure of the inclusion at this temperature. Obviously, if the inclusion has an internal pressure of 64 bars and contains both liquid and vapor CO_2 at 25°C, the homogenization temperature is between 25°C and the critical temperature (31.0°C). Similarly, if the inclusion contains only one CO_2 phase at 25°C, either liquid or vapor, the CO_2 phases are already homogenized and the homogenization temperature is less than 25°C. From Figure 5.10, determine the homogenization temperature, and the phase to which the "free" CO_2 phases will homogenize. (Answer: $T_h \approx 30.3$°C, to vapor phase).

Following the calculation procedure outlined above, the room temperature phase ratios of H_2O-CO_2 inclusions trapped at the PTX conditions shown on Figure 5.11 have been determined. The appearance of these inclusions at 25°C is shown on Figure 5.12 and the data are listed in Table 5.2. Also listed in Table 5.2 are the calculated homogenization temperatures of the CO_2 phases, the dissociation temperature of the clathrate, and the internal pressure at 25°C. All inclusions were trapped from a fluid having a bulk composition of 4 mole-percent CO_2. At 450 bars (point 1, Fig. 5.11), all temperatures from 200°C to 300°C are in the one-phase field, and fluid inclusions that trap this fluid would contain two phases, liquid water and CO_2 vapor at 25°C (Fig. 5.12, Table 5.2). At 350 bars (point 2, Fig. 5.11) the fluid is also in the one-phase field at 250°C and 300°C and the resulting inclusions would be similar to those trapped at 450 bars. However, at 200°C and 350 bars, a 4 mole-% composition is in the two-phase field, and would split into two immiscible fluids. At 25°C fluid inclusions of the H_2O-rich phase (96.4 mole-% H_2O; Table 5.2) would contain two phases, liquid water and CO_2 vapor; those of the CO_2-rich phase (16.2 mole-% H_2O) would contain three phases, liquid water, and liquid and gaseous CO_2 (Fig. 5.12). With slight heating, the CO_2 phase in the CO_2-rich inclusion would homogenize to the vapor phase at 30.9°C (Table 5.2).

At 250 bars (point 3, Fig. 5.11), a 4 mole-% CO_2 bulk composition would be in the two-phase field at 200°C and 250°C, but would be in the one-phase field at 300°C. Inclusion pairs trapped at 200°C and 250°C would be of the H_2O-rich and CO_2-rich varieties when observed at 25°C, and the CO_2-rich inclusion trapped at 200°C would contain a small amount of liquid CO_2 (Fig. 5.12; Table 5.2).

At pressures of 200 and 100 bars (points 4 and 5, Fig. 5.11), all temperatures between 200°C and 300°C are in the two-phase field for a bulk composition of 4 mole-% CO_2. Inclusions that would result from trapping either the H_2O-rich or the CO_2-rich phase at these conditions would be liquid rich and vapor rich, respectively, and all would contain only two phases, liquid water and CO_2 vapor at 25°C (Fig. 5.12, Table 5.2).

Several important features are revealed by the phase ratios shown on Figure 5.12 and the data listed in Table 5.2. Most importantly, three-phase inclusions are rarely produced at these PTX trapping conditions. This in spite of the fact that we chose a CO_2 concentration (4 mole %) that is much higher than normally encountered in the epithermal environment. Had we used a more realistic CO_2 content for the epithermal environment, such as 1.0 mole-% CO_2 or less, three-phase inclusions would be completely absent. Moreover, only the vapor-rich inclusions contain three phases and, owing to optical problems, the presence of liquid and vapor CO_2 in these inclusions might easily be overlooked. Secondly, the liquid-to-vapor ratio of the two-phase inclusions is very similar to that of simple H_2O-salt inclusions trapped in the same temperature range (compare with Fig. 5.3). As a result, even though these inclusions contain significant amounts of CO_2, the presence of CO_2 would not be revealed by simple petrography.

The discussions above show that the presence of CO_2 in fluid inclusions from epithermal deposits will generally not be discernible from room temperature phase relations. However, one relatively simple test for CO_2, or more appropriately for the presence of noncondensed gases, is the crushing test. If a fluid inclusion contains water or a simple salt solution it would have an equilibrium vapor pressure of about 0.03 bars at 25°C (Keenan et al., 1978). Therefore, if we immerse the sample containing this inclusion in oil and break the inclusion open, the oil will be forced into the inclusion because the external pressure ($\sim$1 bar) is considerably higher than the internal pressure of the inclusion ($\sim$0.03 bars). If we observe this procedure under the microscope, the bubble will appear to collapse and disappear at the instant the inclusion is opened. If the inclusion contains some dissolved, noncondensed gases, the internal pressure will be greater than 0.03 bars. Under these conditions the vapor bubble will either only partially collapse if $0.03 < P_{int} < 1.0$, or will expand if $P_{int} > 1.0$ bar.

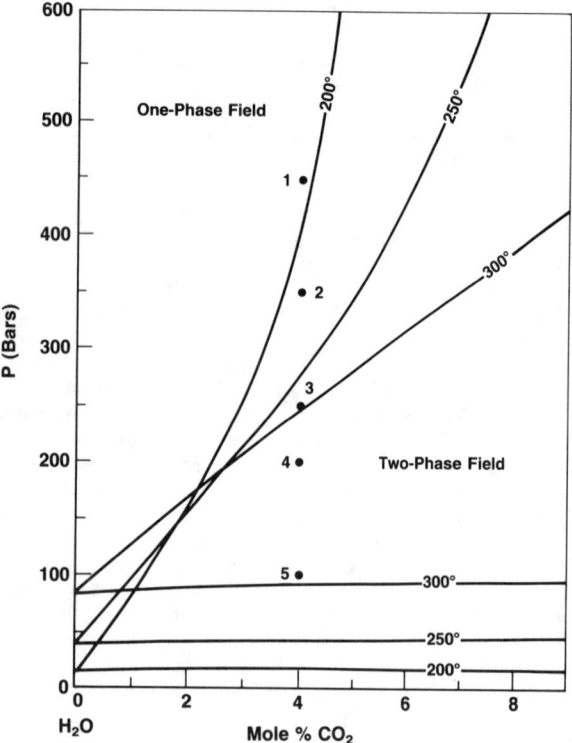

Figure 5.11. The 200°C, 250°C, and 300°C isotherms in the low pressure (600 bars), water-rich portion of the H_2O–CO_2 phase diagram. The isotherms represent the boundary between the one-phase field (upper-left portion of the figure) and the two-phase field (lower-right portion of the figure) at that particular temperature. Points 1-5 correspond to points 1-5 on Figure 5.12.

Sample Calculation 4

By what amount will the CO_2 phases in the inclusion in Sample Calculation 3 above increase in

Table 5.2--Calculated room temperature properties of H_2O-CO_2 fluid inclusions

1	2	3	4	5	6	7	8	9
200	450	0.960	80	0	20	17.0	10.0	56
200	350	0.964	80	0	20	11.3	10.0	51
200	350	0.162	3	34	63	30.9	10.0	64
200	250	0.971	82	0	18	-0.7	9.0	41
200	250	0.157	2	11	87	28.5	10.0	64
200	200	0.975	82	0	18	-8.9	7.6	34
200	200	0.180	2	0	98	24.0	10.0	62
200	100	0.988	81	0	19	-33.4	1.7	17
200	100	0.211	1	0	99	0.8	9.2	42
250	450	0.960	75	0	25	9.3	9.9	49
250	350	0.960	74	0	26	8.0	9.8	48
250	250	0.963	73	0	27	3.0	9.4	44
250	250	0.328	4	0	96	22.9	10.0	62
250	200	0.972	75	0	25	-9.4	7.6	34
250	200	0.364	4	0	96	15.1	10.0	55
250	100	0.989	74	0	26	-40.0	0.1	14
250	100	0.535	3	0	97	-14.2	6.7	30
300	450	0.960	68	0	32	1.4	9.2	42
300	350	0.960	66	0	34	-0.3	9.0	41
300	250	0.960	64	0	36	-2.3	8.6	39
300	200	0.972	67	0	33	-16.1	6.3	29
300	200	0.610	6	0	94	-1.1	8.8	40
300	100	0.996	69	0	31	-56.6	-19.0	6
300	100	0.890	5	0	95	-56.6	-12.0	8

1. Trapping temperature (°C)
2. Trapping pressure (bars)
3. Composition (mole-fraction H_2O)
4. Volume-percent aqueous liquid phase at 25°C
5. Volume-percent liquid CO_2 at 25°C
6. Volume-percent CO_2 vapor at 25°C
7. Homogenization temperature of CO_2 phases (°C)
8. Clathrate dissociation temperature (°C)
9. Internal pressure at 25°C (bars)

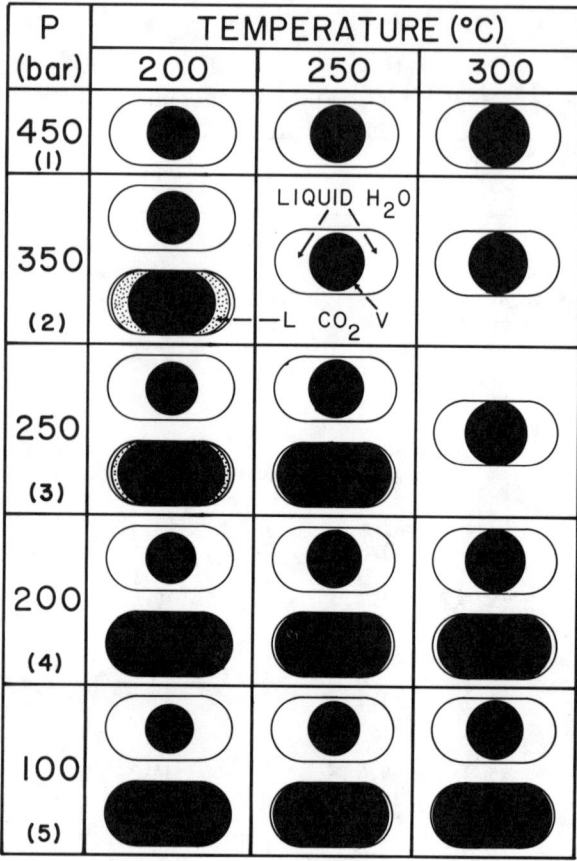

Figure 5.12. Room temperature (25°C) phase relations of H_2O-CO_2 fluid inclusions trapped at temperatures of 200°C, 250°C, and 300°C and pressures of 100, 200, 250, 350, and 450 bars, corresponding to points 1-5 on Figure 5.11. The innermost phase in each inclusion (black) represents a CO_2-rich vapor and the outer phase (white) represents an H_2O-rich liquid. The phase between the CO_2-rich vapor and the H_2O-rich liquid in the inclusions trapped at 200°C and 250 and 350 bars (stippled pattern) represents a CO_2-rich liquid. Room temperature phase relations of these inclusions were calculated using the method described in the text and are listed along with calculated temperatures of various phase changes in Table 5.2.

volume (expand) when the inclusion is opened by crushing the sample in oil? The bulk density of CO_2 in the unopened inclusion is 0.347 g/cm³, and the density of CO_2 at 25°C and 1 bar is 0.00172 g/cm³. Neglect the CO_2 dissolved in the H_2O liquid phase which, under normal conditions, would exsolve from the H_2O phase when the inclusion is opened and contribute to the expansion of the vapor bubble.

$$\text{volume increase} = \frac{0.347 \text{ g/cm}^3}{0.00172 \text{ g/cm}^3} = 202 \text{ times}$$

Thus, the volume of the CO_2 phase after crushing will be 202 times larger than the volume occupied by the CO_2 phases in the unopened inclusion. If the CO_2 phases in the unopened inclusion had a diameter of 3 microns, assuming a spherical geometry, the CO_2 bubble formed after crushing would have a diameter of approximately 17.5 microns.

Figure 5.13A shows a quartz chip from the Carlin sediment-hosted disseminated-gold deposit in the crushing stage before the crushing test was begun. During the crushing test, numerous small bubbles appeared as the quartz flaked off the edge of the chip (Fig. 5.13B, C, D). These bubbles represent the contents of fluid inclusions that have expanded upon opening of the inclusions, indicating that the inclusions contained a gas at an internal pressure greater than ambient atmospheric pressure (1 bar). Because of the small size of the inclusions in the sample shown in Figure 5.13, as is typical of epithermal deposits, individual inclusions could not be observed as they opened; each bubble in Figure 5.13 contains the contents of numerous fluid inclusions. This is suggested by the observation that during the crushing process several tens of small bubbles rapidly coalesced into one large, stable bubble (Figs. 5.13C, D). Other small bubbles that were not incorporated into these larger bubbles quickly dissolved into the oil.

The inclusions contained in the Carlin sample shown in Figure 5.13 are not typical of inclusions from most epithermal deposits because they contain on the order of 5-20 mole-% CO_2, whereas most inclusions from bonanza-type vein deposits contain on the order of a few tenths of a mole-% CO_2 (Fig. 5.14). Moreover, when crushing studies are conducted on inclusions from bonanza-type deposits, evidence of gases (i.e., expanding vapor bubbles) is rarely found. The question, then, is how much CO_2 is required before we will see the bubble expand during crushing? This value may be easily calculated from PVTX data for the H_2O-CO_2 system.

Sample Calculation 5

Assume we have an H_2O-CO_2 inclusion containing a vapor bubble (all CO_2) occupying 25 volume percent of the inclusion at 25°C and that the internal pressure is 1 bar. How much CO_2 does the inclusion contain?

At 25°C and 1 bar, the solubility of CO_2 in H_2O is 0.15 wt.% or 0.06 mole % (Fig. 5.9; Dodds et al., 1956). Assuming an inclusion volume of 100 cm³, the CO_2-saturated liquid, with a density of approximately

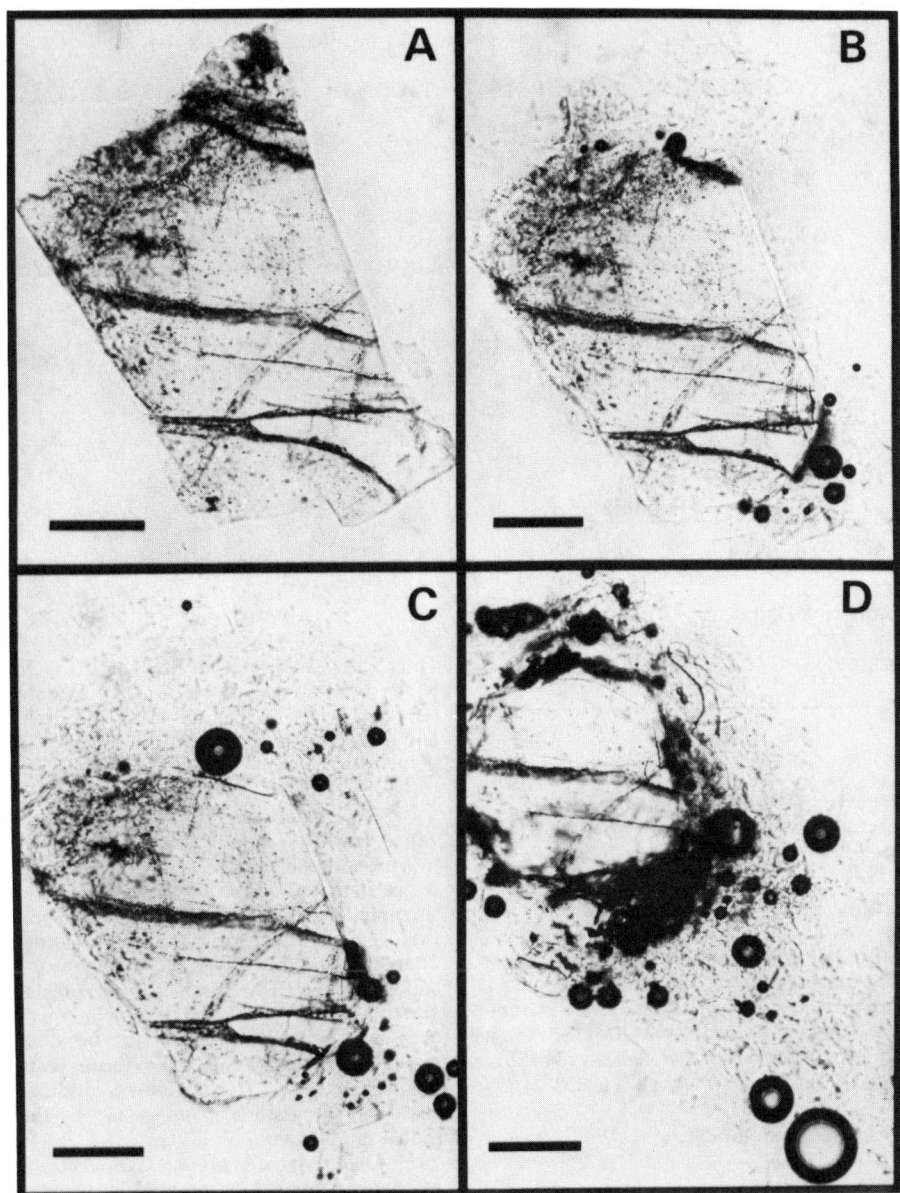

Figure 5.13. Crushing test on a quartz chip containing high-density, CO_2-rich inclusions from the Carlin deposit.

A). Quartz chip in oil in crushing stage prior to beginning of crushing test. Fluid inclusions in this sample are small (<10 micrometers) and appear as dark specks and diffuse bands where the inclusions occur along fractures.

B). Initial stages of crushing test. As quartz fragments begin to flake off the edges of the chip, numerous inclusions are opened, releasing gases into the oil (lower and upper right edges of the chip).

C). With continued crushing more inclusions are opened and smaller bubbles representing gases released from one or a few inclusions coalesce to from larger bubbles.

D). Final stages of crushing test. Several large bubbles occur away from chip. The interface between crushed and uncrushed sample is composed of a mixture of gas-saturated oil and finely ground quartz and appears as a dark "meniscus" surrounding the uncrushed quartz.

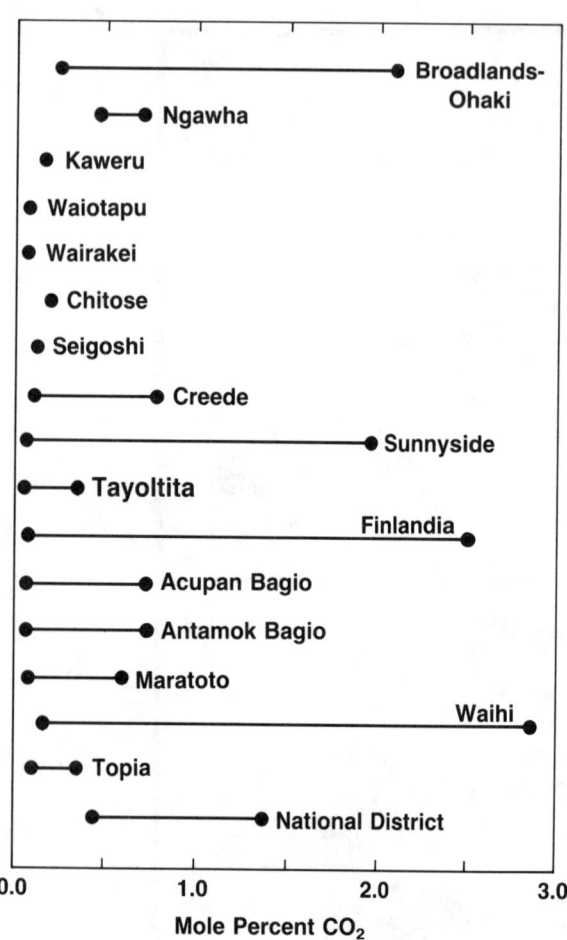

Figure 5.14. Summary of reported CO_2 concentrations of hydrothermal fluids from several terrestrial geothermal systems and epithermal precious-metal deposits. Compiled from data given by Hedenquist and Henley (1985), Sommer et al. (1985) and Vikre (1985).

0.997 g/cm^3 (Parkinson and de Nevers, 1969), has a total mass of

Mass solution = $(0.997$ g/cm$^3)$ $(75$ cm$^3)$

= 74.775 g

Of this total mass, 0.15 wt.-% is CO_2 or

Mass CO_2 in solution = $(.0015)$ $(74.775$ g$)$

= 0.1122 g

The vapor bubble (assumed to be all CO_2) has a density of 0.00172 g/cm^3 (Quinn and Jones, 1936) and contains a total mass of CO_2 of

Mass CO_2 in vapor = $(0.00172$ g/cm$^3)$ $(25$ cm$^3)$

= 0.043 g

The total mass of CO_2 in the inclusion (CO_2 vapor plus CO_2 in solution) equals

Total CO_2 in inclusion = 0.1122 g + 0.043 g

= 0.1552 g

The total mass of fluid in the inclusion (H_2O plus CO_2) equals

Total fluid in inclusion = 0.1552 g + 74.6628 g

= 74.818 g

Therefore, the bulk composition of the inclusion equals

$$\text{Bulk composition} = \frac{0.1552 \text{ g}}{74.818 \text{ g}}$$

= 0.207 wt.-% CO_2

= 0.047 molal

= 0.086 mole-%

Thus, the vapor bubble of an H_2O-CO_2 fluid inclusion with a bulk composition 0.047 molal (0.086 mole percent) will not expand when the inclusion is crushed and observed under the microscope.

The minimum amount of CO_2 required before the vapor bubble will expand during crushing studies (~0.1 mole percent) is within the range of CO_2 concentrations commonly reported from modern terrestrial geothermal systems and their fossil analogs, the epithermal precious-metal deposits (Fig. 5.14). Therefore, if inclusions from bonanza-type epithermal deposits are crushed in oil, the explosive release of gas and formation of large vapor bubbles in the oil (as shown in Fig. 5.13) will not be observed. This is exactly what is found when these tests are made, and the presence of CO_2 is never indicated by crushing tests on inclusions from most epithermal precious-metal deposits.

Using slightly higher CO_2 contents of 0.2 and 0.5 mole percent, which is still in the range of reported values from the epithermal environment (Fig. 5.14), the 25°C internal pressures of inclusions trapped at epithermal P-T conditions have been calculated and are shown in Figure 5.15. For a CO_2 content of 0.2 mole percent, the internal pressures of inclusions from the epithermal environment will be 4-6 bars at 25°C.

Consider an H_2O-CO_2 fluid inclusion with a bulk CO_2 concentration of 0.2 mole percent trapped at 300°C and 300 bars. Using the calculation procedure described above, this inclusion will contain two phases at 25°C--an aqueous liquid phase and a CO_2 vapor bubble occupying 25.2 volume percent of the inclusion. Furthermore, the internal pressure of this inclusion at 25°C will be 5 bars and the CO_2 vapor phase will have a density of 0.010 g/cm^3. What will happen to the vapor bubble in this inclusion if we crush the sample and open the inclusion in oil? From the calculated density of the CO_2 vapor phase in the unopened inclusion (0.010 g/cm^3) and the known density of CO_2 gas at 25°C and 1 bar (0.00172 g/cm^3), we can predict that the vapor bubble volume after crushing will be 5.8 times its volume before crushing. If the vapor bubble diameter was initially 2 microns, its diameter after crushing would be 3.6 microns. As a result, explosive release of CO_2 into the surrounding oil to produce large, easily observable vapor bubbles, as shown in Figure 5.13, will not be observed. Rather, the bubble will expand only slightly during crushing and achieve a volume approximately equal to the total inclusion volume. Therefore, if individual inclusions are not monitored during the crushing test, and this is generally not possible with inclusions from epithermal deposits because of the small size of the inclusions and because of the randomness of the crushing procedure, the slight expansion of the vapor bubble to fill the inclusion will be missed. This probably explains why evidence of gases is almost never revealed during crushing tests on inclusions from epithermal deposits. Moreover, if an inclusion has an internal CO_2 pressure of less than 10 bars at room temperature, the clathrate compound does not form on cooling (Hedenquist and Henley, 1985). Because fluid inclusions trapped in the epithermal environment generally have internal CO_2 pressures below this value (Fig. 5.15), the presence of CO_2 will not be revealed by recognition of the CO_2 clathrate.

The fourth technique for identifying CO_2, or other gases, in inclusions from the epithermal environment is microanalysis of the inclusion fluids. Virtually all reported occurrences of volatiles in inclusions from the epithermal environment are a result of microanalytical techniques, and the majority of these are from mass spectrometric analyses. The major problem with this approach is that these analyses generally are bulk analyses of the fluids released from a large number of inclusions, either by crushing or thermal decrepitation. Because most samples from epithermal deposits contain numerous growth zones and/or planes of inclusions representing fluids of potentially different compositions trapped at different times, the resulting analyses probably are not representative of any one fluid type, but rather mixtures of fluids of different compositions and, perhaps, different origins. Recently, however, Sommer et al. (1985) described a laser-decrepitation-capacitance manometer technique which permits individual inclusions to be sampled. Using this technique, these workers determined that the CO_2/H_2O ratio of individual inclusions from the Topia, Mexico epithermal precious-metal deposit was 0.0305 ± 0.004.

INTERPRETATION OF FLUID INCLUSIONS FROM THE EPITHERMAL ENVIRONMENT

The purpose of most fluid-inclusion studies is to obtain information on the temperatures, pressures (depths), and compositions of the fluids responsible for mineral deposition, and the variation in these properties in time and space in individual hydrothermal systems. With currently available optical equipment

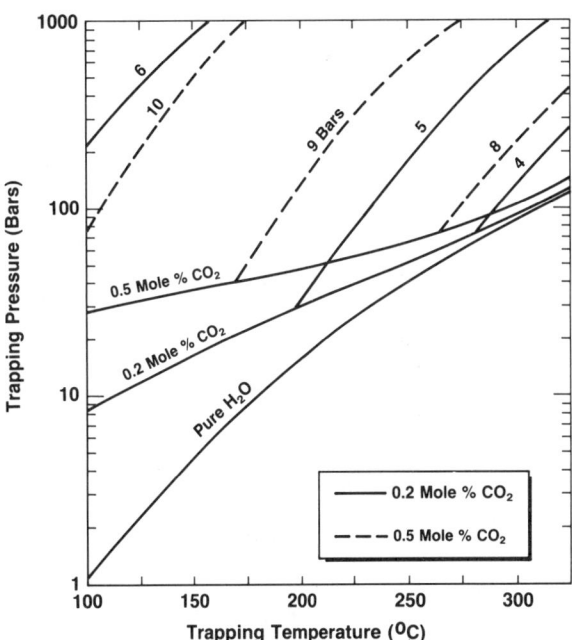

Figure 5.15. Calculated internal pressures (at 25°C) of H_2O-CO_2 inclusions with compositions of 0.2 molal (solid lines) and 0.5 molal (dashed lines) CO_2 trapped at various epithermal P-T conditions. Also shown are the boiling curves for the fluids of these compositions, as well as the pure H_2O.

and heating/freezing stages, it is possible to collect a large amount of highly accurate data from fluid inclusions in a relatively short period of time, and generally these data consist of homogenization temperatures and melting temperatures of inclusion fluids. Mineralization conditions predicted from these data are, however, often suspect, owing to incorrect assumptions made during data interpretation. In the relatively low-pressure epithermal environment, fluid-inclusion homogenization temperatures require little or no pressure correction to obtain the trapping temperature. Thus, homogenization temperatures provide a good approximation of the mineralization temperature and are not usually subject to errors in interpretation. The major errors in data interpretation are consequently related to composition and pressure determinations.

Generally, the inclusion composition is reported as a salinity in terms of wt.-% NaCl equivalent and is obtained from the melting temperature of ice in the inclusion. The assumption that is made in obtaining this salinity is that the depression of the freezing point is due solely to dissolved solids, mainly chlorides of sodium, potassium, and calcium. However, as discussed above, some inclusions from the epithermal environment may also contain small amounts of gases that are not usually detectable by normal petrographic, microthermometric, or crushing studies, and these gases also contribute to the freezing point depression. Thus, in the pure H_2O-CO_2 system, dissolved CO_2 can result in lowering of the ice-melting temperature to -1.48°C, without formation of liquid CO_2 or the CO_2 hydrate ($CO_2 \cdot 5\ 3/4\ H_2O$) during cooling.

Hedenquist and Henley (1985) have shown that this behavior may account for the apparently too high salinities determined from fluid-inclusion freezing temperatures for many active geothermal systems. For example, inclusions from well BR25 at Broadlands have a freezing point of -0.6°C, which corresponds to an NaCl equivalent salinity of 1.02 wt.%. Hedenquist and Henley (1985) have shown, however, that 87% of the freezing point lowering is due to dissolved CO_2, and the actual dissolved salt concentration is only 0.13 wt.%. The considerable effect that this error in calculated chloride concentration has on pH and metal solubility estimates is discussed by these workers.

Failure to recognize small amounts of gases in fluid inclusions can also seriously affect the pressures and corresponding depths of formation calculated from inclusion data. Most calculations of depths of formation from fluid-inclusion data use the boiling point-depth relationship for H_2O-NaCl calculated by Haas (1971). Using these data, pressures calculated for epithermal fluid inclusions in the temperature range 200-300°C are on the order of a few to several tens of bars. Depths obtained from these relatively low pressures are on the order of a few hundred meters to a kilometer at most for homogenization temperatures of 200-300°C, as shown by the temperature-depth curve for a 2 wt.-% NaCl solution on Figure 5.16. The addition of even small amounts of CO_2 to H_2O-NaCl, however, significantly raises the vapor pressure and, concomitantly, the calculated depth of formation. For example, addition of 0.5 molal CO_2 to a 0.5 molal (2.84 wt.%) NaCl solution at 250°C raises the vapor pressure from 69 bars to 108 bars, which increases the depth to the boiling curve from 450 meters to 1120 meters (Loucks, 1984). Thus, this small amount of CO_2, which is in the range of CO_2 concentrations generally reported for active geothermal systems and fluid inclusions from epithermal deposits (Fig. 5.14), increases the calculated depth of formation almost 2 1/2 times as compared to depths calculated assuming only H_2O-NaCl.

APPLICATION OF FLUID INCLUSIONS IN EXPLORATION FOR EPITHERMAL PRECIOUS-METAL DEPOSITS

Imaginative application of conceptual models of ore genesis is of paramount importance to the contemporary exploration geologist. Over the last decade, through a synthesis of detailed investigations of actual deposits, experimental studies, fluid-inclusion research, and studies of active geothermal systems, a generalized model for the formation of epithermal ore deposits has evolved. According to some authors, boiling is an integral part of this model, and is thought to be intimately associated with mineralization in relatively shallow, near-surface deposits. Theoretical and experimental studies (Drummond and Ohmoto, 1985) have shown that under certain conditions, changes in the fluid chemistry as a result of boiling could be an effective mechanism for the precipitation of metals from solution. In addition, the characteristic low-pH alteration assemblages commonly associated with these epithermal systems are consistent with the "boiling off" and subsequent recondensation of acidic gases, mostly CO_2 and H_2S, in the presence of cooler groundwater above the veins.

If boiling and mineralization are genetically related, then, from an exploration geologist's viewpoint, fluid-inclusion evidence of boiling is a favorable characteristic in the search for epithermal deposits, and the pressures and corresponding depths calculated from inclusions trapped in a "boiling system" define that portion of the earth's crust that can host these deposits. The modification of depth estimates to include the effects of dissolved gases greatly expands the crustal depth range of "epithermal" precious-metal systems. However, the actual computation of depth estimates becomes highly questionable without knowledge of gas contents of

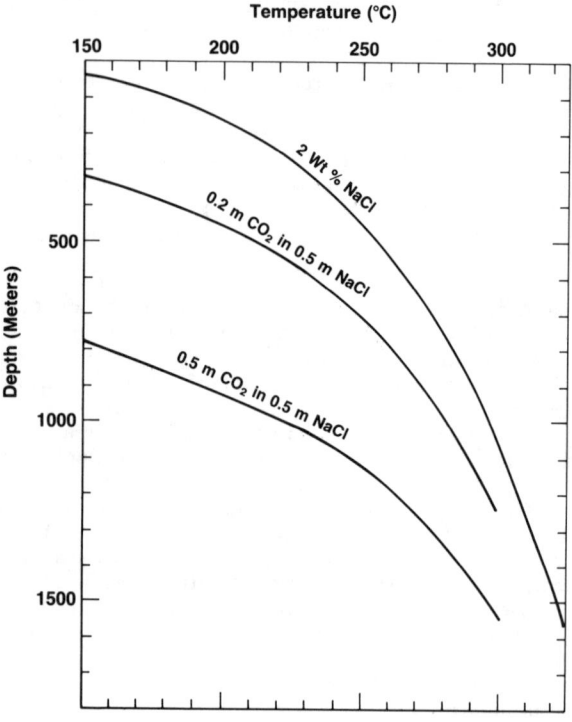

Figure 5.16. Relationship between temperature and the depth at which boiling will commence for an H_2O-NaCl solution (2 wt.-% NaCl) and H_2O-NaCl-CO_2 solutions containing 0.2 and 0.5 molal CO_2 and 0.5 molal NaCl. (Modified from Loucks, 1984.)

inclusions and the effects of various gases on water-gas immiscibility.

In some deposits, it can be demonstrated unquestionably through geologic reconstructon that mineralization occurred very near the surface (Vikre, 1985). For these surficial "hot-spring" deposits either the ore fluid had a very low gas content or, if the ore fluid did contain dissolved gases and consequently began "boiling" at some much deeper level, then near-surface deposition did not result from changes in fluid chemistry as a result of boiling, but by some other mechanism such as cooling, oxidation, or mixing.

A consideration of the effects of dissolved gases on the depths at which boiling will occur helps explain some apparent inconsistencies in the fluid-inclusion literature. If these deposits all formed very near the surface, at or near the one-phase/two-phase (boiling) interface, then preservation of these deposits through geologic time is unlikely. However, bonanza-type vein deposits and sediment-hosted disseminated deposits are not uncommon, and often occur in tectonically active areas where uplift and erosion have taken place. Moreover, the vertical range over which mineralization occurs in individual deposits commonly extends to depths considerably greater than those which could be achieved by simple H_2O-NaCl or H_2O hydrostatic boiling. For example, at Guanajuato, Mexico, evidence for boiling can be found in samples collected from 650 m below the present-day surface (Buchanan, 1980). However, average fluid-inclusion temperatures of 230°C suggest a pressure of 28 bars (Haas, 1976), which converts to only 340 m of boiling hydrostatic head (avg. = 0.83 g/cm^3). This "deep-ore" problem is compounded if denudation since the time of mineralization is accounted for, but is explicable if the effects of dissolved gases are considered.

SUGGESTIONS FOR FUTURE FLUID-INCLUSION RESEARCH

Fluid inclusions from the epithermal environment exhibit distinct petrographic and chemical characteristics which serve to distinguish them from those formed in deeper, higher temperature systems. At the present time, however, it is not possible to relate these inclusion properties to a specific type or stage of mineralization because a sufficient data base does not exist, and the data that are available are often inconclusive or contradictory. Many more detailed fluid-inclusion studies of epithermal systems (especially ore deposits, but including barren systems as well) will be required before these differences can be reconciled. These studies should concentrate on relating observed fluid inclusions to relative stages in a system's evolution, and variations in fluid-inclusion characteristics within a system must be mapped in detail. Such careful, detailed studies of the temporal and spatial variations in fluid-inclusion characteristics in the porphyry-copper deposits have contributed greatly to our understanding of the genesis of these deposits, and should prove equally valuable in understanding the epithermal environment.

One area in which additional work would be timely is in defining the position of boiling in space and time in epithermal systems and the relationship of boiling to ore deposition in mined ore deposits. It is generally accepted that boiling does occur in most epithermal systems, but in those systems which have been studied in detail, the temporal and spatial relationship of boiling and ore deposition is highly variable. For example, at Finlandia, Kamilli and Ohmoto (1977) found that boiling only occurred at the location and paragenetic stage in which precious-metal deposition occurred. At Sunnyside, boiling did not occur during gold mineralization, but was observed during the later quartz-rhodochrosite-fluorite stage of mineralization (Casadevall and Ohmoto, 1977). Similarly, Radtke et al. (1980) report that boiling at Carlin was associated with late acid leaching and hypogene oxidation, but did not occur during earlier quartz-pyrite-gold mineralization. At Buckskin Mountain, Vikre (1985) found that boiling was associated with precious-metal deposition at shallow levels, but did not occur during precious-metal deposition at deeper levels in the deposit. From these and other examples, it is possible that boiling may simply be a characteristic of epithermal systems in general, but may or may not be related to precious-metal deposition. Thus, the presence of boiling may be useful for identifying epithermal systems but cannot, at this time, be used to predict whether or not precious-metal deposition might have occurred.

A second, but related, area in which more work might be useful is to document crystal habits and other characteristics of various minerals formed from boiling fluids, as evidence of boiling may not be recorded by fluid inclusions. There probably are certain conditions where fluid inclusions would not be trapped, or if formed, would not record all fluid phases resulting from boiling. For example, the vapor phase resulting from near-surface boiling may not be trapped in the minerals precipitating, but may simply escape through the system plumbing to the surface. Hedenquist and Henley (1985) report that calcite deposited in the casing of well 80 at Wairakei during discharge of a high-temperature, two-phase mixture contained only liquid-rich inclusions. Also, the minerals precipitating at near-surface conditions may be so fine grained as to preclude observation of possibly sub-micron-sized inclusions.

Robertson (1968) has shown that 91% of the inclusions trapped in sodium nitrate crystals grown in a boiling solution had liquid-to-vapor ratios suggesting that the inclusions trapped only the liquid phase. Only 9% of the inclusions had variable liquid-to-vapor ratios indicating entrapment of varying amounts of liquid and vapor. In a natural sample in which 10% of the inclusions contain variable liquid-to-vapor ratios, these inclusions might easily be overlooked or dismissed as being inclusions which have necked or leaked, the conclusion being that the inclusions were trapped in a non-boiling system. Roedder (1984) describes several other examples in which inclusions have preferentially trapped only one of two coexisting fluid phases, resulting in inclusion characteristics indicative of formation in a single-phase fluid. Experimental studies may help to define those conditions under which only one type of inclusion may result during boiling, and should address the problem of inclusion formation in "slightly" boiling or effervescing systems, as might occur when a fluid moves down a pressure gradient and slowly exsolves small amounts of vapor, as compared to inclusions formed during violent boiling or "flashing," as might occur when an overpressured system is breached by development of fractures. The effect of gases should also be considered in these studies because the vapor phase produced during boiling is the non-wetting phase and may inhibit entrapment as inclusions.

REFERENCES

Bodnar, R. J., Burnham, C. W., and Sterner, S. M., 1985, Synthetic fluid inclusions in natural quartz. III. Phase equilibria in the H_2O-NaCl system to 1000°C and 1500 bars: Geochimica et Cosmochimica Acta v. 49, p. 1861-1873.

Buchanan, L. J., 1980, The Las Torres Mine, Guanajuato, Mexico--Ore controls of a fossil geothermal system: Unpublished Ph.D. thesis, Colorado School of Mines, 138 p.

Buchanan, L. J., 1981, Precious-metals deposits associated with volcanic environments in the southwest; in Dickinson, W. R., and Payne, W. D. (eds.), Relations of Tectonics to Ore Deposits in the Southern Cordillera: Arizona Geological Society Digest, v. 14, p. 237-262.

Burruss, R. C., 1981a, Analysis of phase equilibria in C-O-H-S fluid inclusions; in Hollister, L. S., and Crawford, M. L. (eds.), Short Course in Fluid Inclusions--Applications to Petrology: Mineralogical Association of Canada, Calgary, Alberta, p. 39-74.

Burruss, R. C., 1981b, Analysis of fluid inclusions--Phase equilibria at constant volume: American Journal of Science, v. 281, p. 1104-1126.

Casadevall, T., and Ohmoto, H., 1977, Sunnyside mine, Eureka mining district, San Juan County, Colorado--Geochemistry of gold and base-metal ore deposition in a volcanic environment: Economic Geology, v. 72, p. 1285-1320.

Connolly, J. A. D., and Bodnar, R. J., 1983, A modified Redlich-Kwong equation of state for H_2O-CO_2 mixtures--Application to fluid inclusion studies (abs.): EOS, v. 64, p. 350.

Dodds, W. S., Stutzman, L. F., and Sollami, B. J., 1956, Carbon dioxide solubility in water: Industrial and Engineering Chemistry, v. 1, no. 1, p. 92-95.

Drummond, S. E., Jr., 1981, Boiling and mixing of hydrothermal fluids--Chemical effects on mineral precipitation: Unpublished Ph.D. thesis, The Pennsylvania State University, 380 p.

Drummond, S. E., and Ohmoto, H., 1985, Chemical evolution and mineral deposition in boiling hydrothermal systems: Economic Geology, v. 80, p. 126-147.

Ellis, A. J., and Golding, R. M., 1963, The solubility of carbon dioxide above 100°C in water and sodium chloride solutions: American Journal of Science, v. 261, p. 47-60.

Fournier, R. O., 1985, The behavior of silica in hydrothermal solutions; in Berger, B. R., and Bethke, P. M. (eds.), Geology and Geochemistry of Epithermal Systems: Society of Economic Geologists, Reviews in Economic Geology, v. 2.

Haas, J. L., Jr., 1971, The effect of salinity on the maximum thermal gradient of a hydrothermal system at hydrostatic pressures: Economic Geology, v. 66, p. 940-946.

Haas, J. L., Jr., 1976, Physical properties of the coexisting phases and thermochemical properties of H_2O component in boiling NaCl solutions: U.S. Geological Survey, Bulletin 1421-A, 73 p.

Heald-Wetlaufer, P., Hayba, D. O., Foley, N. K., and Goss, J. A., 1983, Comparative anatomy of epithermal precious- and base-metal districts hosted by volcanic rocks: U.S. Geological Survey, Open-File Report 83-710, 16 p.

Hedenquist, J. W., and Henley, R. W., 1985, The importance of CO_2 on freezing point measurements of fluid inclusions: Evidence from active geothermal systems and implications for epithermal ore deposition: Economic Geology, v. 80, p. 1379-1406.

Henley, R. W., 1985, The geothermal framework of epithermal deposits; in Berger, B. R., and Bethke, P. M. (eds.), Geology and Geochemistry of Epithermal Systems: Society of Economic Geologists, Reviews in Economic Geology, v. 2.

Henley, R. W., and Brown, K. L., 1985, A practical guide to the thermodynamics of geothermal fluids and hydrothermal ore deposits; in Berger, B. R., and Bethke, P. M. (eds.), Geology and Geochemistry of Epithermal Systems: Society of Economic Geologists, Reviews in Economic Geology, v. 2.

Hollister, L. S., and Burruss, R. C., 1976, Phase equilibria in fluid inclusions from the Khtada Lake metamorphic complex: Geochimica et Cosmochimica Acta, v. 40, p. 163-175.

Kamilli, R. J., and Ohmoto, H., 1977, Paragenesis, zoning, fluid inclusion, and isotopic studies of the Finlandia Vein, Colqui district, central Peru: Economic Geology, v. 72, p. 950-982.

Keenan, J. H., Keyes, F. G., Hill, P. G., and Moore, J. G., 1978, Steam Tables--Thermodynamic Properties of Water, including Vapor, Liquid, and Solid Phases: John Wiley and Sons, New York, 156 p.

Khaibullin, I. K., and Borisov, N. M., 1966, Experimental investigation of the thermal properties of aqueous and vapor solutions of sodium and potassium chlorides at phase equilibrium: High Temperature, v. 4, p. 489-494.

Loucks, R. R., 1984, Zoning and ore genesis at Topia, Durango, Mexico: Unpublished Ph.D. thesis, Harvard, 416 p.

Malinin, S. D., and Kurovskaya, N. A., 1975, Solubility of CO_2 in chloride solutions at elevated temperatures and CO_2 pressures: Geokhimiya, no. 4, p. 547-550 (in Russian); translated in Geochemical International, v. 12, no. 2, p. 199-201.

Parkinson, W. J., and de Nevers, N., 1969, Partial molal volumes of carbon dioxide in water solutions: Industrial and Engineering Chemistry Fundamentals, v. 8, p. 709-713.

Perkins, R. M., and Nieman, G. W., 1982, Epithermal

gold mineralization in the South Mountain volcanic dome, Summitville, Colorado; in Symposium of the Genesis of Rocky Mountain Ore Deposits: Changes with Time and Tectonics; Proceedings of Denver Region Exploration Geologists Society, p. 165-172.

Potter, R. W., II, and Brown, D. L., 1977, The volumetric properties of aqueous sodium chloride solutions from $0°$ to $500°C$ at pressures up to 2000 bars based on a regression of available data in the literature: U.S. Geological Survey, Bulletin 1421-A, 36 p.

Quinn, E. L., and Jones, C. L., 1936, Carbon dioxide: American Chemical Society Monograph 72, Rheinhold, New York.

Radtke, A. S., Rye, R. O., and Dickson, F. W., 1980, Geology and stable isotope studies of the Carlin gold deposit, Nevada: Economic Geology, v. 75, no. 5, p. 641-672.

Ramboz, C., Pichavant, M., and Weisbrod, A., 1982, Fluid immiscibility in natural processes--Use and misuse of fluid inclusion data. II. Interpretation of fluid inclusion data in terms of immiscibility: Chemical Geology, v. 37, p. 29-48.

Reed, M. H., and Spycher, N. F., 1985, Boiling, cooling, and oxidation in epithermal systems: Numerical modeling approach; in Berger, B. R., and Bethke, P. M. (eds.), Geology and Geochemistry of Epithermal Systems: Society of Economic Geologists, Reviews in Economic Geology, v. 2.

Robertson, J. M., 1968, Crystal growth from boiling solutions--an experimental study: Unpublished M.S. thesis, University of Michigan, 53 p.

Roedder, E., 1984, Fluid Inclusions: Mineralogical Society of America, Reviews in Mineralogy, v. 12, 644 p.

Sommer, M. A., II, Yonover, R. N., Bourcier, W. L., and Gibson, E. K., 1985, Determination of H_2O and CO_2 concentrations in fluid inclusions in minerals using laser decrepitation and capacitance manometer analysis: Analytical Chemistry, v. 57, p. 449-453.

Swanenberg, H. E. C., 1980, Fluid inclusions in high-grade metamorphic rocks from southwest Norway: Unpublished Ph.D. thesis, University of Utrech., Geologica Vetraiectine, v. 23, 146 p.

Takenouchi, S., and Kennedy, G. C., 1964, The binary system H_2O-CO_2 at high temperatures and pressures: American Journal of Science, v. 262, p. 1055-1074.

Todheide, K., and Franck, E. U., 1963, Das Zweiphasengebiet und die Kritische Kurve in system Kohlendioxid-Wasser bis zu Drucken von 3500 bar: Z. Phys. Chem. Neuefolge, v. 37, p. 388-401.

Vikre, P. G., 1985, Precious-metal vein systems in the National District, Humboldt County, Nevada: Economic Geology, v. 80, p. 363-393.

Wiebe, R., and Gaddy, V. L., 1939, The solubility of carbon dioxide in water at 50, 75 and $100°C$ at pressures to 700 atmosphere: Journal of the American Chemical Society, v. 61, p. 315-318.

Chapter 6
LIGHT STABLE-ISOTOPE SYSTEMATICS IN THE EPITHERMAL ENVIRONMENT
Cyrus W. Field and Richard H. Fifarek

INTRODUCTION

Stable-isotope geochemistry has made important contributions to the widely acknowledged "renaissance" in the earth sciences for more than three decades. This status may be ascribed both to theoretical and practical considerations. First, the isotopic species of an element may be fractionated (partitioned unequally) between two or more coexisting phases because of mass-dependent differences in their chemical and physical behaviors, and the amount of such fractionation normally varies inversely with temperature and independently of pressure. Accordingly, the isotopic abundances of an element may serve to define the mechanisms of formation, thermal environment, and provenance of rocks, minerals, and fluids. Second, the analytical procedures now available render most geologic materials well suited for routine and rapid isotopic measurements. Some important milestones of the 1930's and 1940's leading to our present understanding include the discovery of deuterium and formulation of the theoretical basis for stable-isotope fractionation by Harold C. Urey and colleagues at the University of Chicago and the development of improved mass spectrometers by Alfred O. Nier at the University of Minnesota. The subsequent construction of laboratory facilities elsewhere was commonly directed by graduates and associates of these pioneers and their respective institutions.

As of today, the literature relevant to stable-isotope geochemistry is voluminous and far beyond the scope of this topical overview. Most investigations, apart from those concerned with theory or laboratory experimentation, have been focused on one or more of the following objectives: (1) the conditions and mechanisms of rock or mineral formation; (2) the sources of magma, sediment, ore, petroleum, or water and of metamorphic, ore-forming, or geothermal fluids; (3) the effects of contamination related to magma-country rock, magma-water, or water-rock reactions or to natural versus man-induced pollutants; and (4) the geothermometry of rocks, ores, aqueous systems, and paleoclimates. Although many studies have dealt with the application of stable isotopes to mineral deposits and problems of ore genesis, relatively few have been concerned specifically and in detail with epithermal systems. This meager data base, given the current overriding interest in deposits of the precious metals, undoubtedly will be enlarged both in detail and scope during the forthcoming decade. Our emphasis in this review is confined exclusively to the stable isotopes of carbon, hydrogen, oxygen, and sulfur. Prior to our discussion of the data for epithermal deposits, we provide a background synopsis of the conventions of data presentation, fractionations that may accompany equilibrium isotope-exchange reactions between various fluid and mineral phases, and the distributions of the stable isotopes in major geologic environments. Those who require more precise elaboration should consult recent textbooks by Faure (1977) and Hoefs (1980) and reviews by Friedman and O'Neil (1977), O'Neil (1977), Ohmoto and Rye (1979), and Taylor (1979), and appropriate references cited therein.

CONVENTIONS, SYSTEMATICS, AND RATIONALE

According to Lange's Handbook of Chemistry (Dean, 1979), the relative abundances of the stable isotopes of hydrogen, carbon, oxygen, and sulfur are as follows:

$^{1}_{1}H = 99.985\%$ $^{32}_{16}S = 95.0\%$

$^{2}_{1}H = D = 0.015$ $^{33}_{16}S = 0.76$

$^{12}_{6}C = 98.892$ $^{34}_{16}S = 4.22$

$^{13}_{6}C = 1.108$ $^{36}_{16}S = 0.014$

$^{16}_{8}O = 99.759$

$^{17}_{8}O = 0.037$

$^{18}_{8}O = 0.204$

Although these abundances are not known to the accuracy implied, this uncertainty is not important because a precision of 0.02 percent or better is obtained routinely from analyses based on comparative measurements of the heavy-to-light isotopic ratio of an element in a sample with respect to that in a standard. The analyses are performed on a gas-source mass spectrometer, and they require that the isotopic element measured in both the sample and standard be converted quantitatively to a gas (carbon and oxygen to CO_2, hydrogen to H_2, and sulfur to SO_2). Depending on the element under consideration, the instrumental record provides a comparison of sample and standard ratios in terms of D/H, $^{13}C/^{12}C$, $^{18}O/^{16}O$, and $^{34}S/^{32}S$. Differences in the ratio of the sample (R_x) with respect to that of the standard (R_s) are normally expressed as deviations from the standard by del (δ) values in parts per thousand (permil, or $^o/oo$), as given by the equation

$$\delta_x \, ^o/oo = \left(\frac{R_x}{R_s} - \frac{R_s}{R_s}\right) 10^3 = \left(\frac{R_x}{R_s} - 1\right) 10^3 \quad (1)$$

Positive or negative del values signify the permil enrichment or depletion, respectively, of the heavy isotope (D, ^{13}C, ^{18}O, or ^{34}S) in the sample relative to the standard. Conventional standards are a belemnite from the Cretaceous Peedee Formation (PDB) for carbon, Standard Mean Ocean Water (SMOW) for hydrogen and oxygen, and troilite from the Canyon Diablo meteorite (CD) for sulfur. All standards are 0 permil, by definition (eq. 1).

Fractionation

Isotopes of an element may be fractionated during equilibrium and unidirectional reactions that accompany chemical and physical processes of either inorganic or biogenic origin. The fractionation effects are generally directly proportional to mass differences between the isotopic species and inversely proportional to temperature, provided other parameters such as valence and bond strength are the same. Equilibrium isotope-exchange reactions are probably the most important, and many of the fractionations first predicted from theory have been corroborated by subsequent laboratory studies and analyses of natural minerals, liquids, and gases. Fractionations that accompany unidirectional reactions are more difficult to evaluate. They evolve from kinetic effects and result in the reaction products being depleted in the heavy isotope.

The isotopic fractionation factor (α) between two compounds, A and B, is equal to the quotient of the heavy-to-light isotope ratios in the compounds. Moreover, under conditions of equilibrium, the fractionation factor and ratio quotient are related to the equilibrium constant (K) of isotope exchange as follows,

$$\alpha_{A-B} = \frac{R_A}{R_B} = K^{1/n} \quad (2)$$

where n is the maximum number of exchangeable isotopes in any of the compounds. However, this generalization with respect to K does not hold true for compounds of hydrogen or for those containing two or more isotopes of an element that do not occupy equivalent molecular positions. Because instrumental measurements of the stable-isotope abundances are normally reported as del (δ) values in permil rather than as ratios (R), other identities must be used to equate these deviations from the standard ($\delta \, ^o/oo$) to the fractionation factor (α). The interrelationships given by

$$\alpha_{A-B} = \frac{R_A}{R_B} = \frac{1 + \delta_A \, ^o/oo/1000}{1 + \delta_B \, ^o/oo/1000} = \frac{1000 + \delta_A \, ^o/oo}{1000 + \delta_B \, ^o/oo} \quad (3)$$

are derived from equations (1) and (2).

Isotopic effects deduced from experimental and theoretical investigations are usually reported in terms of the fractionation factor. The α values are close to unity and commonly range from $1 \pm 0.0x$ to $1 \pm 0.00x$. In contrast, isotopic effects obtained from applied studies of minerals and rocks are generally reported as delta (Δ) values, which is the difference of measured del values, in permil, between two phases. Delta values may be readily determined from inspection of the analytical data (given as $\delta \, ^o/oo$ values) and are similar to values derived from the logarithmic transformation of their associated fractionation factors, as provided by the useful approximation

$$\Delta_{A-B} = \delta_A \, ^o/oo - \delta_B \, ^o/oo \simeq 1000 \ln \alpha_{A-B} \quad (4)$$

For most practical purposes, such as geothermometry, the delta value is essentially identical to the more precise, but less easily calculated, $1000 \ln \alpha$ value. Errors introduced by this approximation (eq. 4) become significant, relative to analytical uncertainties ($\pm 0.2 \, ^o/oo$ or more), only as the delta values exceed 10 permil and as their component del values become unsymmetrically distributed with respect to the reference standard ($0 \, ^o/oo$).

Illustrative of the foregoing considerations, the equation

$$^{32}SO_2 + H_2^{34}S = {^{34}SO_2} + H_2^{32}S \quad (5)$$

represents the equilibrium isotope-exchange reaction for ^{32}S and ^{34}S between sulfur dioxide and hydrogen sulfide gases. It follows, from equations (1), (2), and (3), that

$$\alpha_{SO2-H2S} = K_{SO2-H2S} = \frac{R_{SO2}}{R_{H2S}} = \frac{(^{34}S/^{32}S)_{SO2}}{(^{34}S/^{32}S)_{H2S}}$$

$$= \frac{1 + \delta^{34}S_{SO2} \, ^o/oo/1000}{1 + \delta^{34}S_{H2S} \, ^o/oo/1000} = \frac{1000 + \delta^{34}S_{SO2} \, ^o/oo}{1000 + \delta^{34}S_{H2S} \, ^o/oo} \quad (6)$$

and from equation (4) that

$$\Delta_{SO_2-H_2S} = \delta^{34}S_{SO_2} \text{ °/oo} - \delta^{34}S_{H_2S} \text{ °/oo}$$
$$= 1000 \ln \alpha_{SO_2-H_2S} \quad (7)$$

The equilibrium fractionation factors for this reaction are 1.0339 and 1.0074 at 100°C and 500°C, respectively, according to the equation for SO_2-H_2S equilibration given by Ohmoto and Rye (1979, p. 516). Equivalent values of 1000 ln α (and Δ, see eq. (7)) are 33.3 and 7.4 permil at 100°C and 500°C, respectively. Thus, under equilibrium conditions, SO_2 is enriched in the heavy ^{34}S isotope by 33.3 and 7.4 permil relative to coexisting H_2S at these two temperatures. Had equation (5) been written in terms of the ^{32}S and ^{36}S isotopic species of SO_2 and H_2S, the associated fractionation factors would have increased to 1.0678 and 1.0148 at 100°C and 500°C, respectively. Thus, as previously stated, fractionation effects relate inversely to temperature and directly to differences in mass between the isotopic species involved. However, small fractionations between coexisting phases of a mineral group formed at similar temperatures, such as variations in $\delta^{34}S$ among sulfides or $\delta^{18}O$ among silicates, relate to differing strengths of the metal-sulfur or metal-oxygen bonds (Bachinski, 1969; O'Neil, 1977). The heavy isotope is preferentially concentrated (enriched) in that phase having the strongest bond.

Equilibrium Reactions

The temperature dependency of stable-isotope fractionation provides geologists with a potentially large number of geothermometers. However, successful application to the study of rocks and minerals requires that the isotopic element be geologically common, and that the fractionation factors for the isotopes between two or more phases be known over the appropriate range of temperatures. Although the most reliable equilibrium fractionation factors are obtained from laboratory investigations of isotope exchange, experimental problems involving disequilibrium and metastability, particularly at low temperatures, commonly necessitate that they be estimated indirectly by less certain methods. These estimates are derived either from theory, using statistical mechanics and various spectral and thermodynamic data, or from empirical consideration of measured fractionation trends in natural mineral assemblages. Fractionation equations for isotopic reaction pairs of many geologically important compounds of carbon, hydrogen, oxygen, and sulfur are presented in Tables 6.1 through 6.5. The general form of these equations

$$1000 \ln \alpha = a(10^6/T^2) + b \quad (8)$$

indicates that the fractionation effect, given by 1000 ln α, varies curvilinearly with the inverse square of temperature (in °K). Coefficients a and b represent the slope and intercept, respectively, of these linear curves. Variations in the fractionation effect (1000 ln α) over the temperature interval 0 to 600°C for many of these reaction pairs are graphically illustrated in Figures 6.1 through 6.5. Positive or negative values of 1000 ln α (≃ Δ values) denote permil enrichment or depletion, respectively, of the heavy isotope in the first-named member of the pair. For purposes of illustration, some of the fractionation curves have been extrapolated to temperatures beyond that of the experimental data and sometimes beyond that of the stability range for at least one member of the reaction pair.

Fractionation equations for reaction pairs of isotopic carbon compounds are listed in Table 6.1 and the resultant curves over the temperature range from 0°C to 600°C are shown in Figure 6.1. They are based entirely on the sources of data compiled by Friedman and O'Neil (1977), although Ohmoto and Rye (1979) provide more complex equations for these and other carbon compounds. As previously noted, the heavy isotope is preferentially concentrated in that member of the reaction pair to which the element is most strongly bonded. Equilibrium values of 1000 ln α (Fig. 6.1) for the carbon reaction pairs show that ^{13}C tends to be enriched in the more condensed and(or) oxidized member. The latter redox effect, also common to isotopic compounds of sulfur, is confirmed by data from sedimentary, metamorphic, and geothermal environments wherein ^{13}C is found to be concentrated in carbonates relative to graphite or organic carbon, and in CO_2 relative to CH_4.

The fractionation equations for isotopic compounds of hydrogen are given in Table 6.2 and the portrayal of these effects with temperature in Figure 6.2 are largely from the compilation of Friedman and O'Neil (1977). However, more recent experimental data for some mineral-H_2O equilibria include those for

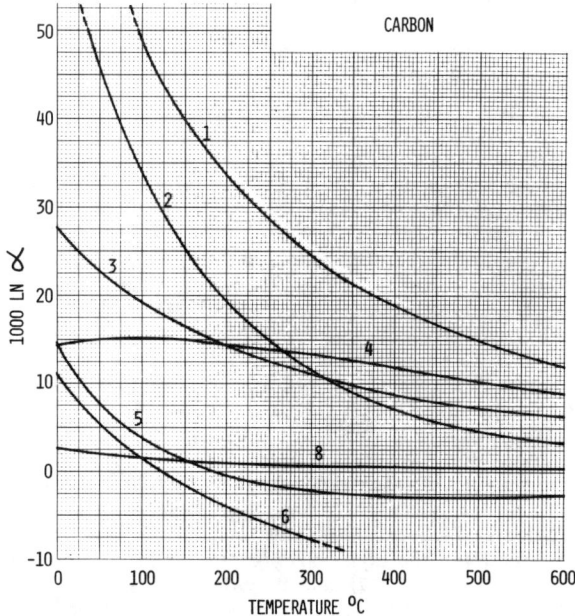

Figure 6.1. Fractionation curves for isotopic compounds of carbon.

Table 6.1 Fractionation equations for isotopic compounds of carbon

	Carbon Pair	1000 ln α	T (°C) Range	Reference
1	CO_2-CH_4	not given	0-700	Friedman and O'Neil (1977, Fig. 29)
2	gph-CH_4	not given	0-700	Friedman and O'Neil (1977, Fig. 32)
3	cal-gph	$\approx 1.74(10^6/T^2) + 5.22$	0-700	Friedman and O'Neil (1977, Fig. 29)
4	CO_2-gph	not given	0-700	Friedman and O'Neil (1977, Fig. 30)
5	cal-CO_2	$2.988(10^6/T^2) - 7.666(10^3/T) + 2.461$	0-700	Friedman and O'Neil (1977, Fig. 31)
6	HCO_3^--CO_2	$9.552(10^3/T) - 24.10$	5-125	Friedman and O'Neil (1977, Fig. 27)
7	$CO_3^=$-CO_2	not given ($\approx$ eq. 6)	0-100	Friedman and O'Neil (1977, Fig. 28)
8	dol-cal	$0.18(10^6/T^2) + 0.17$	100-650	Friedman and O'Neil (1977, Fig. 33)

Comments

1. Although Friedman and O'Neil (1977) and sources cited therein do not provide fractionation equations for some of the carbon pairs listed above, their graphical portrayal of fractionation trends for these pairs are the basis of those curves illustrated in Figure 6.1 and our approximation of the calcite-graphite equation (eq. 3) above.

2. Ohmoto and Rye (1979, p. 551, Table 10-3) give more complex polynomial equations for the equilibrium fractionation factors of these and other isotopic compounds of carbon.

3. Fractionation equations and isotopic trends for the carbon pairs have been calculated largely from empirical and theoretical considerations, and not from experimental determinations, over the temperature ranges listed (see Friedman and O'Neil, 1977, and Ohmoto and Rye, 1979).

4. Mineral abbreviations and compound states are as follows: cal, calcite; CH_4, gaseous; CO_2, gaseous; $CO_3^=$, aqueous; dol, dolomite; gph, graphite; HCO_3^-, aqueous.

epidote (Graham et al., 1980), goethite (Yapp and Pedley, 1985), and hornblende (Graham et al., 1984). Less certain and(or) complete are the data for gibbsite and kaolinite (Taylor, 1979) and for montmorillonite (Savin and Epstein, 1970a; O'Neil and Kharaka, 1976), and there are virtually none for chlorite. Equations and curves are not given for equilibrium reactions of H_2-$H_2O_{(v)}$, H_2-CH_4, and H_2S-H_2O. However, their fractionations are enormous, as summarized by Friedman and O'Neil (1977), with 1000 ln values ranging from -900 and less to -400 and less within the temperature interval from 0°C to 300°C, respectively. Presumably reactions such as those with H_2 are not important in most geologic environments (Taylor, 1974a; 1979). Values of 1000 ln are largely negative, for the equations (Table 6.2) as written and their curves (Fig. 6.2), and they demonstrate that D is enriched in H_2O relative to coexisting gas or mineral phases. The parallelism of fractionation curves for the various mica-water systems (Fig. 6.2, eqs. 5, 6, and 7) is noteworthy. According to Suzuoki and Epstein (1976), the fractionation of D between the micas and water is not only a function of temperature, but also relates to the molar fraction of cations (Al, Fe, and Mg) in six-fold coordination. From their general equation for mica-water fractionation given by

1000 ln α

$= -22.4(10^6/T^2) + 28.2 + (2X_{Al} - 4X_{Mg} - 68X_{Fe})$ (9)

where X is the molar fraction of each cation, it may be inferred that values of 1000 ln α at constant temperature become progressively more depleted (negative) in D with the compositional succession from Al-rich through Mg-rich to Fe-rich micas. In terms of theoretical mineral assemblages, Al-pure "muscovite" would be enriched in D by 6 permil relative to coexisting Mg-pure "phlogopite", and by 70 permil relative to coexisting Fe-pure "annite"; regardless of temperature. We have selected compositionally reasonable molar fractions of these cations

($X_{Al}:X_{Mg}:X_{Fe}$) for the fractionation equations and curves (6 and 7) illustrative of phlogopite (0.10:0.80:0.10) and biotite (0.10:0.45:0.45) in Table 6.2 and Figure 6.2. These and(or) possibly other compositional variations may similarly affect the distributions of hydrogen isotopes in chlorites and smectites.

Fractionation equations and curves for equilibrium isotope-exchange reactions involving oxygen compounds are presented in Table 6.3 and Figure 6.3, respectively. Although many of these reactions are from sources in Friedman and O'Neil (1977), they include data for reactions of quartz and the feldspars from Matsuhisa et al. (1979), kaolinite from Kulla and Anderson (1978), and biotite and chlorite from Bottinga and Javoy (1973) and Javoy (1977) based on empirical and theoretical considerations. Limited data from Sheppard et al. (1969), Savin and Epstein (1970a), and Taylor (1979) suggest that the equation for montmorillonite-water fractionation may be grossly analagous to that for barite water. Note that the fractionations of ^{18}O and given by 1000 lnα values for the various mineral-water systems (Fig. 6.3) are substantially less than those for deuterium. With the exceptions of biotite, chlorite, and magnetite, most minerals concentrate ^{18}O relative to coexisting water, which is a reversal of the isotopic effect for deuterium. The order of enrichment among the silicates is largely the reverse of Bowen's reaction series, with quartz enriched in ^{18}O relative to most other common rock-forming minerals. Fractionation equations (Table 6.3) for albite and potassium feldspar are identical, as are their respective 1000 lnα values (Fig. 6.3) for a given

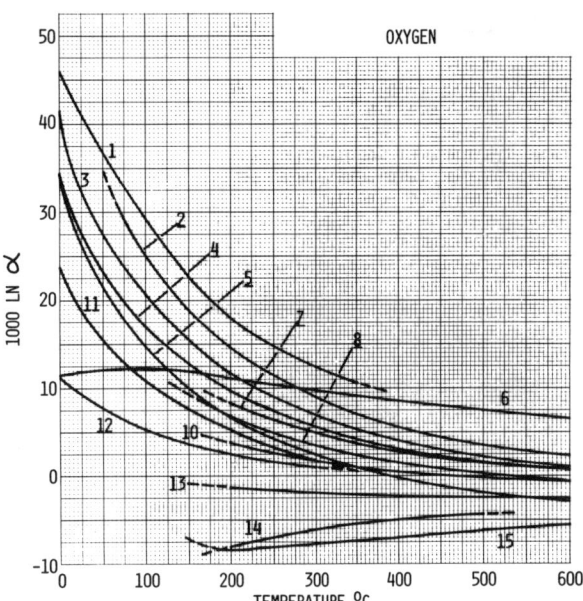

Figure 6.3. Fractionation curves for isotopic compounds of oxygen.

temperature. In contrast, anorthite is slightly depleted in ^{18}O relative to the alkali feldspars at all temperatures, which reflects a progressive weakening of the Si-O bond with increasing substitution of aluminum for silicon and sodium for calcium in the solid solution sequence of the plagioclase feldspars.

Equations and curves pertaining to equilibrium fractionations between isotopic compounds of sulfur are listed in Table 6.4 and are graphically illustrated in Figure 6.4. The equation for sulfate-H_2S fractionation (eq. 1) is from Ohmoto and Lasaga (1982), whereas all others are those provided by Ohmoto and Rye (1979). Large fractionations that characterize the sulfate-H_2S and SO_2-H_2S reactions are related primarily to the different states of sulfur oxidation among the members of these pairs, with ^{34}S being preferentially enriched in the more oxidized member of the pair (Fig. 6.4, inset curves 1 and 2). They are analogous to the large isotopic effect previously described for carbon that attends the CO_2-CH_4 redox reaction (Fig. 6.1, curve 1). The smaller 1000 lnα values for sulfide-H_2S reactions as contrasted with those for silicate-H_2O reactions is partly attributable to the smaller percentage mass difference between ^{34}S and ^{32}S as compared to that between ^{18}O and ^{16}O. It is apparent from the signs of the 1000 lnα values for various sulfide-H_2S reactions (Fig. 6.4) that molybdenite, pyrite, and sphalerite are variably but weakly enriched in ^{34}S relative to coexisting H_2S, whereas galena and the other sulfides are depleted. The relative enrichment of ^{34}S between different sulfide minerals (mo>py>sl, etc.) is governed by the relative strengths of the metal-sulfur bonds, as first predicted from theory by Sakai (1968) and Bachinski (1969).

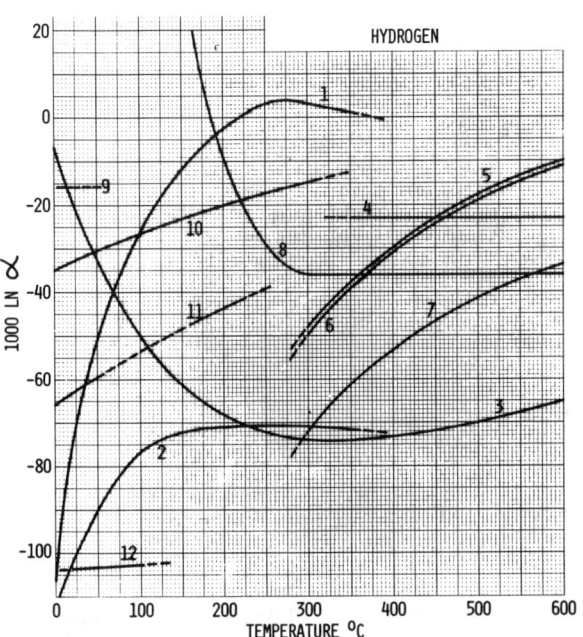

Figure 6.2. Fractionation curves for isotopic compounds of hydrogen.

Table 6.2 Fractionation equations for isotopic compounds of hydrogen

	Hydrogen Pair	1000 lnα	T (°C) Range	Reference
1	$H_2O_{(v)}$-$H_2O_{(l)}$	$-24.844(10^6/T^2) + 76.248(10^3/T) - 52.612$	0-50	Friedman and O'Neil (1977, Figs. 34 and 35)
2	CH_4-$H_2O_{(l)}$	not given	0-350	Friedman and O'Neil (1977, Figs. 34, 35, and 36)
3	CH_4-$H_2O_{(v)}$	not given	0-700	Friedman and O'Neil (1977, Fig. 36)
4	hbd-H_2O	-23.1	350-805	Graham et al. (1984)
5	mus-H_2O	$-22.1(10^6/T^2) + 19.1$	450-850	Suzuoki and Epstein (1976)
6	phl-H_2O	$-22.4(10^6/T^2) + 28.2 + C$, where* $C = -9.8$	450-850	Suzuoki and Epstein (1976)
7	bio-H_2O	$-22.4(10^6/T^2) + 28.2 + C$, where* $C = -32.2$	450-850	Suzuoki and Epstein (1976)
8	epi-H_2O	$22.9(10^6/T^2) - 138.8$ -35.9	<300 300-650	Graham et al. (1980)
9	gib-H_2O	-16.1	at 20	Taylor (1979)
10	kal-H_2O	not given	20-300	Taylor (1979, Fig. 6.2)
11	mon-H_2O	-64 and -43.9	15 and 200	Savin and Epstein (1970a) and O'Neil and Kharaka (1976)
12	goe-H_2O	-104 and -103	0 and 100	Yapp and Pedley (1985)

Comments

1. Fractionation curves illustrated in Figure 6.2 for hydrogen pairs lacking fractionation equations are based on, or determined from, the graphical portrayal of these isotopic trends given by Friedman and O'Neil (1977) and Taylor (1979).

2. Mineral abbreviations and compound states are as follows: bio, biotite; CH_4, gaseous; epi, epidote; gib, gibbsite; goe, goethite; H_2O, liquid unless otherwise specified; hbd, hornblende; kal, kaolinite; mon, montmorillonite; mus, muscovite; and phl, phlogopite.

* Where C is calculated using the general equation: $C = 2X_{Al} - 4X_{Mg} - 68X_{Fe}$. See text or Suzuoki and Epstein (1976) for an elaboration.

Applications

Geologists who are unfamiliar with the innuendos of stable-isotope geochemistry may question our apparently excessive concern with experimental and(or) theoretical systems defined by mineral-H_2O and mineral-H_2S reactions. Justification for this approach comes from the fact that the results of such laboratory investigations offer guides and constraints to the interpretation of stable-isotope data, and ideally, they provide the means by which some relative genetic parameters of rock and mineral systems may be quantified. Two prominent applications will now be described in order to stress the importance of laboratory research. These and other uses will be implicit to subsequent discussions and interpretations of the stable-isotope data.

Geothermometry--Although the temperature dependency of fractionation serves as the rationale for the application of stable isotopes to geothermometry, the acquisition of reliable data from geologic materials requires that several prerequisites be

Table 6.3 Fractionation equations for isotopic compounds of oxygen

	Oxygen Pair	1000 ln α	T (°C) Range	Reference
1	$CO_{2(g)}-H_2O_{(l)}$	$-0.021(10^6/T^2) + 17.994(10^3/T) - 19.97$	0-100	Friedman and O'Neil (1977, Figs. 5 and 6)
2	anh-H_2O	$3.88(10^6/T^2) - 2.90$	100-575	Friedman and O'Neil (1977, Fig. 14)
3	qtz-H_2O	$3.34(10^6/T^2) - 3.31$ $2.05(10^6/T^2) - 1.14$	250-500 500-800	Matsuhisa et al. (1979) Matsuhisa et al. (1979)
4	cal-H_2O	$2.78(10^6/T^2) - 2.89$	0-500	Friedman and O'Neil (1977, Fig. 13)
5	bar-H_2O	$3.0(10^6/T^2) - 6.79$	<180-350	Friedman and O'Neil (1977, Fig. 15)
6	$CO_{2(g)}$-cal	$-1.803(10^6/T^2) + 10.611(10^3/T) - 2.780$	0-750	Friedman and O'Neil (1977, Fig. 12)
7	Kf-H_2O or Ab-H_2O	$2.39(10^6/T^2) - 2.51$ $1.59(10^6/T^2) - 1.16$	400-500 500-800	Matsuhisa et al. (1979) Matsuhisa et al. (1979)
8	mus-H_2O	$2.38(10^6/T^2) - 3.89$	400-650	Friedman and O'Neil (1977, Fig. 19)
9	mon-H_2O	$\simeq$ eq. 5	15-300	Sheppard et al. (1969), Savin and Epstein (1970a), and Taylor (1979)
10	An-H_2O	$1.49(10^6/T^2) - 2.81$ $1.04(10^6/T^2) - 2.01$	400-500 500-800	Matsuhisa et al. (1979) Matsuhisa et al. (1979)
11	kal-H_2O	$2.05(10^6/T^2) - 3.85$	<150-300	Kulla and Anderson (1978)
12	$H_2O_{(l)}-H_2O_{(v)}$	$0.766(10^6/T^2) + 1.205(10^3/T) - 3.493$	0-300	Friedman and O'Neil (1977, Fig. 8)
13	bio-H_2O	$0.41(10^6/T^2) - 3.10$ (?)	500-800	Bottinga and Javoy (1973) and Javoy (1977)
14	chl-H_2O	$-1.34(10^6/T^2) - 2.07$ (?)	500-800	Bottinga and Javoy (1973) and Javoy (1977)
15	mag-H_2O	$-1.47(10^6/T^2) - 3.70$	500-800	Bottinga and Javoy (1973) and Javoy (1977)

Comment

1. Mineral abbreviations and compound states are as follows: Ab, albite; An, anorthite; anh, anhydrite; bar, barite; bio, biotite; cal, calcite; chl, chlorite; CO_2, gaseous; H_2O, liquid unless otherwise specified; kal, kaolinite; Kf, K-feldspar; mag, magnetite; mon, montmorillonite; mus, muscovite; and qtz, quartz.

fulfilled. These include not only a knowledge of variations in the fractionation factor with temperature between two or more mineral phases, as determined preferably from experiment, but also that the minerals formed contemporaneously and in isotopic equilibrium, and that this equilibrium was subsequently preserved. These fundamental tenets notwithstanding, because the fractionation of D and ^{18}O in silicates is

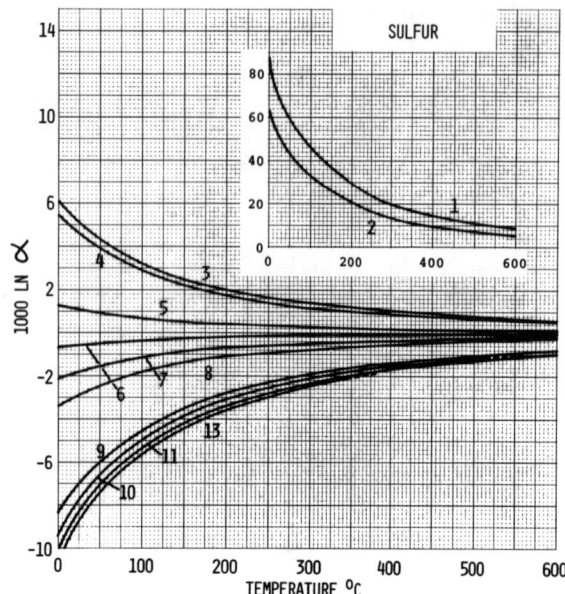

Figure 6.4. Fractionation curves for isotopic compounds of sulfur.

Table 6.4 Fractionation equations for isotopic compounds of sulfur

Sulfur Pair	1000 lnα	T (°C) Range	Reference
1 $SO_4^=$(a,s)–H_2S	$6.463(10^6/T^2) + 0.56$	≈200 to >600	Ohmoto and Lasaga (1982)
2 SO_2–H_2S	$4.70(10^6/T^2) - 0.5$	350–1050	Ohmoto and Rye (1979)
3 mo–H_2S	$0.45(10^6/T^2)$	uncertain	Ohmoto and Rye (1979)
4 py–H_2S	$0.40(10^6/T^2)$	200–700	Ohmoto and Rye (1979)
5 sl–H_2S	$0.10(10^6/T^2)$	50–705	Ohmoto and Rye (1979)
6 cp–H_2S	$-0.05(10^6/T^2)$	200–600	Ohmoto and Rye (1979)
7 S^o–H_2S	$-0.16(10^6/T^2)$	200–400	Ohmoto and Rye (1979)
8 bn–H_2S	$-0.25(10^6/T^2)$	uncertain	Ohmoto and Rye (1979)
9 gn–H_2S	$-0.63(10^6/T^2)$	50–700	Ohmoto and Rye (1979)
10 cb–H_2S	$-0.70(10^6/T^2)$	uncertain	Ohmoto and Rye (1979)
11 sb–H_2S	$-0.75(10^6/T^2)$	uncertain	Ohmoto and Rye (1979)
12 cc–H_2S	≈ eq. 11	uncertain	Ohmoto and Rye (1979)
13 ag–H_2S	$-0.80(10^6/T^2)$	uncertain	Ohmoto and Rye (1979)

Comment

1. Mineral abbreviations and compound states are as follows: ag, argentite; bn, bornite; cb, cinnabar; cc, chalcocite; cp, chalcopyrite; gn, galena; H_2S, gaseous; mo, molybdenite; py, pyrite; S^o, native sulfur; sb, stibnite; sl, sphalerite; SO_2, gaseous; $SO_4^=$, either aqueous or solid sulfate.

referenced to silicate-H_2O reactions and that of ^{34}S in sulfides to sulfide-H_2S reactions, it follows that more geologically useful equations for the fractionation of these isotopes between mineral phases, such as between silicate-silicate and sulfide-sulfide pairs, may be derived from algebraic identities. For example, the ^{18}O fractionation equations given for the quartz-H_2O and barite-H_2O equilibrations (Table 6.3, eqs. (3) and (5)) are likely to be of interest only to some geothermalists, whereas their algebraic difference

$$1000 \ln\alpha_{qtz-H_2O} - 1000 \ln\alpha_{bar-H_2O}$$
$$= 1000 \ln\alpha_{qtz-bar}$$
$$= 0.34(10^6/T^2) + 3.48 \quad (10)$$

yields the fractionation equation for the quartz-barite reaction, which is an assemblage of interest to many explorationists. By having analyses of $\delta^{18}O$ performed on the members of this pair, and applying the approximation given by equation (4)

$$1000 \ln\alpha_{qtz-bar} \simeq \Delta_{qtz-bar}$$
$$= \delta^{18}O_{qtz} \text{ o/oo} - \delta^{18}O_{bar} \text{ o/oo} \quad (11)$$

the depositional temperature (in °K) of the quartz-barite assemblage is readily calculated by substitution and solving for T in equation (10).

Listed in Table 6.5 are the equilibrium fractionation equations, and their analogues rearranged in terms of T, for isotopic mineral pairs of oxygen and sulfur in potentially useful assemblages. Variations of their respective $1000 \ln\alpha$ ($\simeq \Delta$) values over the temperature range from 0°C to 600°C are illustrated in Figure 6.5. The equations listed have either been calculated from those in Tables 6.1-6.4 by the method described (eq. 10), or they have been taken from sources previously noted such as Friedman and O'Neil (1977), Matsuhisa et al. (1979), Ohmoto and Rye (1979), etc. Many other equations for mineral-mineral reactions can be derived from those provided in Tables 6.1-6.4, but those given in Table 6.5 are restricted to mineral pairs commonly found in hydrothermal deposits.

The use of these and other fractionation equations in isotope geothermometry requires further elaboration. Because of problems with disequilibrium that are especially prevalent at low reaction temperatures, not all of the equations in Tables 6.1-6.5 are equally reliable for isotope geothermometry. For example, the quartz-H_2O reaction has been repeatedly investigated (see Friedman and O'Neil, 1977; O'Neil, 1977; Matsuhisa et al., 1979), because the exchange of oxygen isotopes between quartz and water is particularly slow, and probably most other systems will be reexamined eventually. For similar reasons, isotopic equilibrium is unlikely to be attained with sulfate-sulfide reactions below 350°C, although it may be enhanced by conditions such as long residence time and(or) low rate of cooling, low pH, and high sulfur content of the fluids (Ohmoto and Rye, 1979; Ohmoto and Lasaga, 1982). Pyrite-chalcopyrite pairs commonly give unreasonable isotopic temperatures

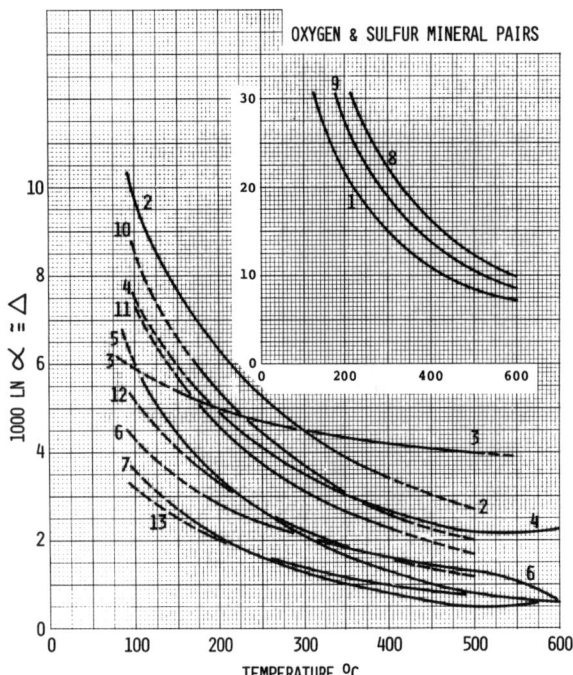

Figure 6.5. Fractionation curves for isotopic mineral pairs of oxygen and sulfur.

indicative of disequilibrium (Field and Gustafson, 1976), whereas sphalerite-galena pairs generally do not (Ohmoto and Rye, 1979). Furthermore, the carbon-bearing minerals are not only few in number, but the flat slope of the dolomite-calcite fractionation curve with respect to temperature renders this geologically common mineral pair of little value as a geothermometer; in contrast to the steeper slope of the calcite-graphite fractionation curve (Fig. 6.1, curves 8 and 3). Application of hydrogen isotopes to geothermometry is severely limited because of uncertainties caused by compositional variations of the minerals, post-depositional re-equilibration, and for some reactions the insensitivity of fractionation to changes in temperature (Fig. 6.2). The ^{18}O compositions of the silicate minerals are subject to re-equilibration by aqueous fluids, and this post-depositional effect is largest for the feldspars and biotite, and least for quartz and magnetite (Brigham, 1984, and references cited therein). Finally, the geologist must exercise care in the selection of samples for analysis. Although minerals that exhibit textural evidence of noncontemporaneity are unlikely to be in isotopic equilibrium, the converse is not always true.

Provenance--Geochemical investigations commonly have as their objective the identification of the source or provenance of fluids and(or) minerals by means of chemical tracers. Stable isotopes have been successfully applied to many of these endeavors. In particular, the isotopes of hydrogen and oxygen are used routinely to identify the source of aqueous fluids,

Table 6.5 Fractionation and temperature equations for isotopic mineral pairs of oxygen and sulfur

	Mineral Pair	$1000 \ln \alpha \simeq \Delta$	$T^\circ K$
1	qtz-mag	$4.81(10^6/T^2) + 0.39$ (250–500°C) $3.52(10^6/T^2) + 2.56$ (500–800°C)	$2.19(10^3)/(\Delta - 0.39)^{1/2}$ $1.88(10^3)/(\Delta - 2.56)^{1/2}$
2	qtz-kal	$1.29(10^6/T^2) + 0.54$ (250–500°C)	$1.13(10^3)/(\Delta - 0.54)^{1/2}$
3	qtz-bar	$0.34(10^6/T^2) + 3.48$ (250–500°C) $0.95(10^6/T^2) + 5.65$ (500–800°C)	$0.58(10^3)/(\Delta - 3.48)^{1/2}$ $-0.97(10^3)/(\Delta - 5.65)^{1/2}$
4	qtz-mus	$0.96(10^6/T^2) + 0.58$ (250–500°C) $-0.33(10^6/T^2) + 2.75$ (500–650°C)	$0.98(10^3)/(\Delta - 0.58)^{1/2}$ $-0.57(10^3)/(\Delta - 2.75)^{1/2}$
5	qtz-Kf or qtz-Ab	$0.95(10^6/T^2) - 0.80$ (400–500°C) $0.46(10^6/T^2) + 0.02$ (500–800°C)	$0.97(10^3)/(\Delta + 0.80)^{1/2}$ $0.68(10^3)/(\Delta - 0.02)^{1/2}$
6	anh-qtz	$0.54(10^6/T^2) + 0.41$ (250–500°C) $1.83(10^6/T^2) - 1.76$ (500–800°C)	$0.73(10^3)/(\Delta - 0.41)^{1/2}$ $1.35(10^3)/(\Delta + 1.76)^{1/2}$
7	qtz-cal	$0.56(10^6/T^2) - 0.42$ (250–500°C) $-0.73(10^6/T^2) + 1.75$ (500–800°C)	$0.74(10^3)/(\Delta + 0.42)^{1/2}$ $-0.85(10^3)/(\Delta - 1.75)^{1/2}$
8	$SO_4^={}_{(a,s)}$-gn	$7.093(10^6/T^2) + 0.56$ ($\simeq$200 to >600°C)	$2.66(10^3)/(\Delta - 0.56)^{1/2}$
9	$SO_4^={}_{(a,s)}$-py	$6.063(10^6/T^2) + 0.56$ ($\simeq$200 to >600°C)	$2.46(10^3)/(\Delta - 0.56)^{1/2}$
10	py-ag	$1.20(10^6/T^2)$ (uncertain)	$1.10(10^3)/(\Delta)^{1/2}$
11	py-gn	$1.03(10^6/T^2)$ (200–700°C)	$1.01(10^3)/(\Delta)^{1/2}$
12	sl-gn	$0.73(10^6/T^2)$ (50–700°C)	$0.85(10^3)/(\Delta)^{1/2}$
13	py-cp	$0.45(10^6/T^2)$ (200 to >600°C)	$0.67(10^3)/(\Delta)^{1/2}$

Comments

1. These equations are derived from those listed in Tables 6.3 and 6.4 by methods described in the text.

2. Mineral and compound abbreviations are as follows: Ab, albite; ag, argentite; anh, anhydrite; bar, barite; cal, calcite; cp, chalcopyrite; gn, galena; kal, kaolinite; Kf, K-feldspar; mag, magnetite; mus, muscovite; py, pyrite; qtz, quartz; sl, sphalerite; and $SO_4^=$ either aqueous or solid sulfate.

whether or not now present, in many rock- and ore-forming environments. The viability of this method stems from the fact that the isotopic domains of magmatic, meteoric, and ocean waters are generally distinct, and that the composition of fluids (hydrothermal, metamorphic, etc.) may be calculated from the analytical data for one or more minerals using the appropriate reaction equations (Tables 6.2 and 6.3). For example, given the δD and $\delta^{18}O$ values for a sample of hydrothermal sericite, and assuming equilibrium and a temperature of formation based on independent criteria (fluid inclusions, mineral assemblages, sulfur isotopes, etc.), the compositions of the fluid may be determined by substitution of the data and assumed temperature (in °K) in the muscovite-H_2O reaction equations (Table 6.2, eq. (5); Table 6.3, eq. (8)). Similarly, the $\delta^{18}O$ composition of the fluid may be checked from an additional analysis of coexisting quartz and use of the quartz-H_2O reaction (Table 6.3, eq. (3)). In addition, having the $\delta^{18}O$ values of both quartz and sericite allows use of the quartz-muscovite isotopic geothermometer (Table 6.5, eq. (4)), which thereby serves to check the validity of the assumed temperature.

There are variants to the foregoing procedures, and where applicable they may offer advantages in terms of time and expense. Provided it can be assumed that the hydrothermal fluid was predominantly of meteoric origin and that the system was characterized by large water-to-rock ratios, it is possible to estimate both the hydrogen- and oxygen-isotopic composition of the fluid from a single $\delta^{18}O$ analysis of quartz or other oxygen-bearing mineral. This approximation is feasible, to the extent that all preceding assumptions are valid, because the D and $\delta^{18}O$ contents of meteoric waters, although extraordinarily variable, change sympathetically and linearly according to the equation

$$\delta D\ ^o/oo = 8\ \delta^{18}O\ ^o/oo + 10 \qquad (12)$$

cited by Craig (1966) and Taylor (1974a). However, because of the qualification with respect to water:rock ratios, which if low may perturb the $\delta^{18}O$ compositions of fluids and minerals precipitated therefrom, it is normally customary to calculate the $\delta^{18}O$ values of the fluids from the δD values of minerals using equation (12), or from measured δD values of fluids extracted from mineral inclusions. Meteoric waters are characteristically depleted in D and ^{18}O by values that may exceed -150 and -20 permil, respectively, relative to their oceanic source (0 $^o/oo$ for δD and $\delta^{18}O$, by definition). These depletions take place as a result of equilibrium fractionations at low temperatures that accompany the progressive condensation and crystallization of rain and snow from a finite quantity of atmospheric water vapor (see Figs. 6.2 and 6.3, curves 1 and 12, respectively). Their magnitudes vary directly with altitude, latitude, and the relative amount of water vapor removed from the system. Because isotopic studies have demonstrated that waters of meteoric origin dominate geothermal and most hydrothermal systems, especially those of epithermal character, the present-day compositions of these waters over much of North America is shown in Figure 6.6 (after Taylor, 1979, p. 243). Note that precipitation over Nevada, and most of the Basin and Range province, is characterized by δD and $\delta^{18}O$ values of -130 to -80 and -80 to -11 permil, respectively. These depletions are similar to those of "fossil" waters calculated from the mineral data for Tertiary hydrothermal systems of this region. However, it is appropriate to conclude this discussion of "calculated" fluid compositions with a note of caution. Studies by Truesdell (1974) and others and the data and discussions of Friedman and O'Neil (1977) and Taylor (1967, 1979) indicate that the calculated compositions of water from saline hydrothermal fluids may be in error, and usually depleted in D and ^{18}O with respect to the true values. This solute effect is caused by the tendency of some cations, particularly those of large ionic potential, to hydrate in solution with the isotopically heavier molecules of water, which thus leaves the free water that equilibrates with minerals proportionately depleted in D and ^{18}O. Although difficult to evaluate, this solute-controlled isotopic effect is probably small in epithermal systems having fluids of low salinity.

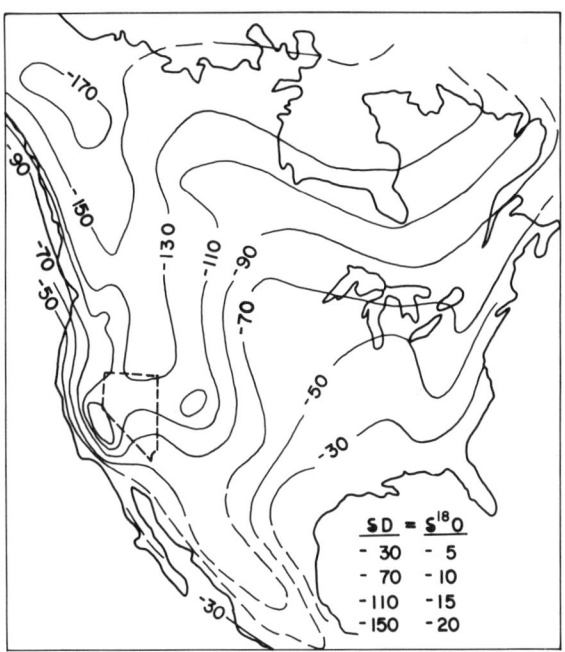

Figure 6.6. Distributions of δD and $\delta^{18}O$ in meteoric waters over part of North America (after Taylor, 1979).

Other applications of the stable isotopes to investigations of provenance warrant brief mention. It is well known that authigenic minerals formed in sedimentary environments are conspicuously enriched in ^{18}O relative to their magmatic counterparts because of the larger fractionations permitted at low temperatures. Smaller, yet significant, enrichments of ^{18}O found in some igneous rocks and minerals have been used in conjunction with other geochemical and geologic data to document examples and sources of magma contamination and the sedimentary component of S-type granites (Magaritz et al., 1978, and references therein). The isotopes of carbon and sulfur also undergo appreciable fractionations at low to intermediate temperatures, and these may be enhanced when the reaction pairs involve both oxidized and reduced forms of these elements (Fig. 6.1, curves 1, 2, and 3; Fig. 6.4, curves 1 and 2). As a consequence, the carbonates are commonly enriched in ^{13}C relative to graphite, organic carbon, and hydrocarbons, and marine sulfates are enriched in ^{34}S relative to magmatic sulfur ($\simeq 0\ ^o/oo$) and in contrast to ^{34}S-depleted sedimentary sulfides. These enrichments and depletions originally served as a qualitatively convenient means by which to interpret the isotopic data in terms of oceanic, magmatic, and biogenic sources and processes. However, Sakai (1968) and Ohmoto (1972) demonstrated the fallibility of subjective interpretations by showing that the isotopes of carbon and sulfur could undergo large fractionations at relatively high temperatures by way of inorganic redox reactions controlled by Eh and pH. In spite of errors inherent with isotopic generalizations, Thode et al. (1954) and Feely and Kulp (1957) deduced from carbon- and sulfur-isotopic data the biogenic origin of calcite and native sulfur in the cap rock of Gulf Coast salt domes from sources of evaporitic anhydrite at depth. In addition to the scientific merits of their contribution, the practical corollary was that salt domes lacking an anhydrite-calcite cap rock are unlikely to contain economic accumulations of native sulfur and petroleum!

GEOLOGIC DISTRIBUTIONS

Abundances of the stable isotopes in geologically important environments are now summarized and briefly described. The purposes are twofold: first, these data offer background and perspective for discussions of the epithermal environment that follow; and second, they exhibit trends that are largely consistent with those derived from theory and experiment as previously described. For some readers it may be disconcerting to observe that the isotopic range for any of the four elements may be broad and overlapping from one environment to another. However, this apparent lack of isotopic definition is principally an artifact of these compilations. The isotopic signature of an element for a particular member of an environment, such as a single mineral deposit, plutonic phase, sedimentary formation, etc., is usually narrow and well defined. Nonetheless, the environment may exhibit a larger isotopic range because its members may be compositionally dissimilar owing to differences of provenance, depositional conditions, and age.

Hydrogen and Oxygen

It is evident from previous discussions that D and ^{18}O form an isotopic couplet that is uniquely suited to the study of fluids and minerals in aqueous systems. However, because oxygen is a major constituent of the crust (≈ 46.6 wt%), whereas hydrogen (≈ 0.1 wt%) is not, the ^{18}O content of fluids is more likely to be perturbed by water-rock reactions than is the D content. Accordingly, it is useful to first review the distributions of ^{18}O in various geologic environments. These data, as portrayed in Figure 6.7, are taken largely from the summaries of Taylor (1967, 1974a, 1979), Garlick (1972), Faure (1977), Hoefs (1980), and other sources. The $\delta^{18}O$ values for ultramafics (>5 and <7 $^o/oo$) are similar to those of meteorites (≈ 3 to $7^o/oo$) and are consistent with the presumed mantle origin of the former. More siliceous igneous clans are progressively enriched in ^{18}O for the sequence from basalts and gabbros (≈ 5.5 to <8 $^o/oo$), through andesites and granodiorites (≈ 5.5 to >12 $^o/oo$), to rhyolites and granites (≈ 6 to 13 $^o/oo$). This generalized isotopic trend is compatible with observed fractionations between the common rock-forming minerals wherein $\delta^{18}O$ values are largest in quartz, carbonates, and alkali feldspars; intermediate in plagioclase feldspars, micas, and ferromagnesian minerals; and smallest in the Fe-Ti oxides. A similar trend of diminishing $\delta^{18}O$ values is evident among the authigenic minerals of marine sedimentary rocks that include cherts (≈ 20 to 39 $^o/oo$: Garlick, 1972; Kolodny and Epstein, 1976), carbonates (≈ 15 to 36 $^o/oo$: Garlick, 1972; Veizer and Hoefs, 1976), shales (≈ 11 to 29 $^o/oo$: Savin and Epstein, 1970b), and ferromanganese nodules (≈ 10 to 14 $^o/oo$; Field et al., 1983). The large ^{18}O enrichment of sedimentary rocks and minerals as compared to those of magmatic origin results from the larger fractionations permitted at the low temperatures prevailing in the hydrosphere, and in spite of the fact that magmas are enriched in ^{18}O (≈ 6 to 10 $^o/oo$) relative to sea water (≈ 0 $^o/oo$). This distinction between isotopically heavy authigenic sedimentary minerals and their lighter magmatic counterparts gives support to hypotheses of assimilation or anatectic melting of sedimentary rocks to account for the few documented examples of ^{18}O-enriched (up to 16 $^o/oo$) plutonic and volcanic rocks (Taylor and Turi, 1976; Magaritz et al., 1978). By analogy, the sources of fluids and mineral constituents in hydrothermal systems may also be constrained by similar isotopic differences. The data for hydrothermal deposits in Figure 6.7 are subdivided between Cordilleran and volcanogenic massive-sulfide types, with the principal differences being that the former are associated with epizonal plutons whereas the latter are deposited in a submarine environment on or short distances below the sea floor (Sawkins, 1972). For purposes of comparison, we have excluded the data for epithermal deposits from the Cordilleran subgroup as summarized in Figures 6.7, 6.9, and 6.10. Values of $\delta^{18}O$ for quartz (≈ -4 to 13 $^o/oo$), carbonates (<6 to 14$^o/oo$), and sulfates (-6 to 20 $^o/oo$) of the Cordilleran subgroup are chiefly from Sheppard et al. (1971), Fuex and Baker (1973), and Watanabe and Sakai (1983). Although there are abundant data for hydrothermal silicates other than quartz (feldspars, micas, clays, oxides, etc.), they are not illustrated in Figure 6.7 to preserve clarity among the dominant mineral phases. Their isotopic distributions would largely mimic those of quartz, but they would be more depleted in ^{18}O because of smaller fractionation factors at any given temperature (Table 6.3 and Fig. 6.3). Distributions of $\delta^{18}O$ values in quartz (≈ 7 to 14 $^o/oo$), carbonates (≈ 9 to 20$^o/oo$), and sulfates (≈ 5 to 15 $^o/oo$ and more) of the volcanogenic massive-sulfide deposits are taken from Kusakabe and Chiba (1983), Watanabe and Sakai (1983), and Fifarek (1985). Although the data base for massive-sulfide deposits is less extensive than that for the Cordilleran types, their minerals appear to be isotopically less variable and more enriched in ^{18}O. Such differences, if real, probably relate to the larger range of depositional temperatures and differing sources of the aqueous component (magmatic versus meteoric) in Cordilleran hydrothermal systems.

Data relevant to the foregoing discussion are given by the distributions of δD and $\delta^{18}O$ in various waters and minerals, as displayed in Figure 6.8, provide data relevant to the foregoing discussion, and are illustrative of several useful applications of the hydrogen-oxygen isotope pair. The isotopic locations of standard mean ocean water (SMOW) and the meteoric water line (MWL) as derived from equation 12 (Craig, 1966; Taylor, 1974a), plotted at the top center and diagonally down the left-hand margin,

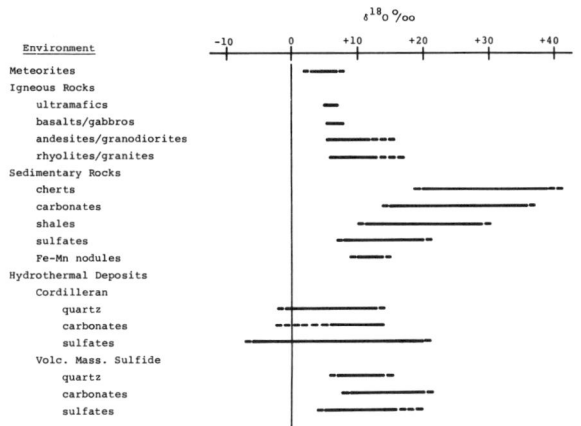

Figure 6.7. Distributions of $\delta^{18}O$ in geologic environments.

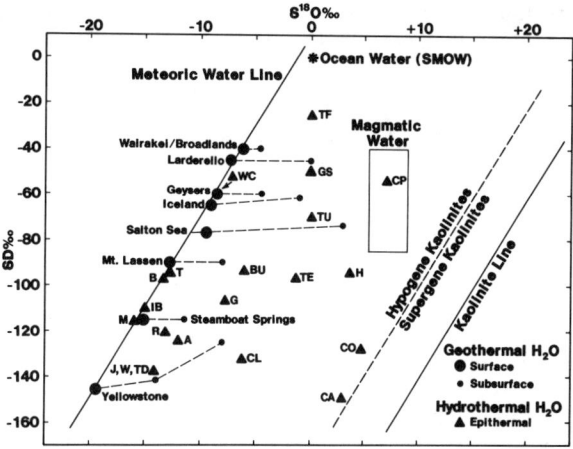

Figure 6.8. Distributions of δD and $\delta^{18}O$ in various waters, minerals, and hydrothermal fluids of epithermal deposits.

respectively, are useful points of reference. The diagonal at the right-hand margin represents the kaolinite line (KL) cited by Sheppard et al. (1969) and based on the work of Savin and Epstein (1970a), which marks a continuum of δD and $\delta^{18}O$ values in kaolinites from modern soils. Parallelism of lines MWL and KL documents equilibrium isotope exchange of D and ^{18}O between meteoric waters and kaolinites formed during the weathering and transformation of rocks to clay-rich soils. The compositional domain of magmatic waters, given by the rectangular box at center-right of Figure 6.7, has been calculated from isotopic analyses of hydrous magmatic silicates. Most magmatic waters have a relatively confined range of values between -85 and -40 permil D and 5.5 to 9 permil $\delta^{18}O$ (Taylor, 1974a; 1979). Not illustrated, for reasons of clarity and possible lack of relevance, are the broad isotopic fields of metamorphic waters and the saline brines of sedimentary basins. According to Taylor (1979), the metamorphic waters inherit their compositional variability ($\approx$-65 to -20 $^o/oo$ δD, and 5 to 25 $^o/oo$ $\delta^{18}O$) from dehydration and fluid-mineral reactions with isotopically variable igneous and sedimentary rocks, whereas the data for the brines are broadly scattered to the right of the MWL and suggest varying mixtures of both connate and meteoric waters that have been modified by other fluid-sediment interactions at depth within the basins.

Compositional variations of waters associated with a number of well-known geothermal systems, as modified from the data of Craig (1966), White et al. (1973), and Truesdell and Hulston (1980), are plotted on Figure 6.8. Surface waters (large closed circles) have δD and $\delta^{18}O$ values located on or near the MWL, whereas the values for related subsurface waters (small closed circles) trend variably and horizontally to the right (dashed lines) from the MWL. Such trends, known as the "oxygen isotope shift", are characterized by increasing values of $\delta^{18}O$ at nearly constant δD and are common to the subsurface fluids of many geothermal systems. The shift to larger $\delta^{18}O$ values is caused by equilibrium isotope-exchange reactions between ^{18}O-depleted meteoric waters and ^{18}O-enriched rocks during water-rock reactions. In general, the size of the ^{18}O-shift correlates directly with temperature and salinity of the fluids, and inversely with the mass ratios of water to rock. Because the effects of the ^{18}O-shift and water:rock ratios are interrelated and may influence interpretations of the analytical data for fluids and minerals, these phenomena will be discussed at greater length in forthcoming considerations of the epithermal deposits. However, the reader should note that meteoric waters similar to those within the compositional range from Wairakei/Broadlands to the Salton Sea ($\delta D \approx$ -40 to -80 $^o/oo$) could be driven by an ^{18}O-shift into the isotopic domain of magmatic waters by exchange reactions involving small to subequal amounts of water relative to that of rock (low to intermediate water:rock mass ratios). Values of δD, in contrast to those of $\delta^{18}O$, are not significantly affected by water:rock exchange because at these water:rock mass ratios the principal source of hydrogen is in the aqueous fluids. Thus, the geothermal fluids retain the δD value of their meteoric source, and water-rock isotope-exchange reactions are manifest as the subhorizontal trend lines of Figure 6.8. The slight positive slope to a few trend lines, such as for the Salton Sea and Yellowstone geothermal areas, may result from deuterium enrichment that is unrelated to water-rock reactions. Concentration of deuterium is caused by evaporation of surface waters prior to recharge at the Salton Sea (Craig, 1966) and by the boiling of subsurface waters at Yellowstone (Truesdell et al., 1977) and perhaps elsewhere.

As previously noted, the isotopic compositions of hydrothermal and other fluids may be calculated from the measured δD and $\delta^{18}O$ values of associated minerals by means of the appropriate mineral-water fractionation curves (Tables 6.2 and 6.3) and using an estimated temperature of deposition. The "calculated" fluid compositions derived from many biotites and a few sericites of porphyry-type deposits plot within the magmatic water box (Sheppard et al., 1971; Osatenko and Jones, 1976; Taylor, 1979). However, the data for a few biotite, most sericites, and nearly all other hydrous minerals of hydrothermal origin are located within the broad range of δD and $\delta^{18}O$ values between the MWL and KL boundaries. Compositions of the fluids calculated from the isotopic data for these minerals, or from the fluids of inclusions contained therein, suggest waters of either meteoric origin, or of mixed meteoric-magmatic or magmatic-oceanic parentage.

According to Sheppard et al. (1969) and Taylor (1974a), minerals such as the kaolinites that may form either by hypogene or supergene processes may be distinguished on the basis of temperature-controlled fractionations, which determine their isotopic positions relative to the hypogene/supergene kaolinite line of Figure 6.8. Supergene kaolinites formed at low temperatures and in equilibration with meteoric waters plot to the right of this line and up to the KL, whereas the hypogene kaolinites formed at higher temperatures plot to the left.

Also illustrated on Figure 6.8 are the "calculated" compositions of hydrothermal fluids responsible for the deposition of many epithermal precious-metal deposits (closed triangles) of the western U.S. and a few elsewhere. These data are from the results of other investigators and will be cited subsequently. They are based on the δD values obtained from fluid inclusions and(or) hydrous minerals, and on $\delta^{18}O$ values determined from quartz and(or) other associated minerals. The broad isotopic distribution of these "calculated" compositions largely precludes a significant input of magmatic water to these hydrothermal systems, but it does record a major contribution from meteoric sources. Further detail and elaboration about these fluids will be deferred to our concluding discussion of the epithermal deposits.

Carbon

Abundances of ^{13}C in various geologic environments are illustrated in Figure 6.9, and they are based largely on the data cited by Craig (1953), Bender (1972), Fuex and Baker (1973), Ohmoto and Rye (1979), and Hoefs (1980). Values of $\delta^{13}C$ are surprisingly variable in the products of high-temperature environments such as meteorites ($\simeq$-12 to 9 $^o/oo$), igneous rocks ($\simeq$-10 to 3 $^o/oo$), diamonds (-6 to 3 $^o/oo$), and carbonatites ($\simeq$-10 to 2 $^o/oo$); particularly because the compositional extremes have been omitted from these ranges. The causes of such isotopic variability are uncertain, but possibly relate to equilibrium or kinetic redox reactions, contamination, inhomogeneities of source, and(or) differing proportions of compositionally distinct carbon in the samples. For example, the data given for meteorites and igneous rocks is that of total carbon, which consists both of ^{13}C-enriched and oxidized (carbonate, up to 66 $^o/oo$) and ^{13}C-depleted and reduced (graphite, "organic," carbonyl, etc.; up to -30 $^o/oo$) forms of carbon. In accordance with fractionation theory (Table 6.1 and Fig. 6.1), similar relative enrichments and depletions of ^{13}C are present between the oxidized ($CO_2 \simeq$ -10 to 2 $^o/oo$) and reduced ($CH_4 \simeq$ -31 to -16 $^o/oo$) components of volcanic/geothermal gases. Narrower compositional variations characterize marine limestones ($\simeq$-5 to 4 $^o/oo$), sea water HCO_3^- (-5 to -2 $^o/oo$), and atmospheric CO_2 ($\simeq$-8 to -6 $^o/oo$), and the relative ^{13}C enrichments among these compounds ($CaCO_3 > HCO_3^- > CO_2$) are consistent with experimental-theoretical fractionation trends. The cluster of values around and near 0 permil is expectable because the carbon PDB standard is calcite of marine derivation. Fractionations that accompany kinetic photosynthetic reactions of atmospheric and hydrospheric CO_2 to form organic carbon lead to marked depletions of ^{13}C in marine and land plants ($\simeq$-34 to -12 $^o/oo$), and this isotopic record of biogenic processes is preserved in carbonaceous sediments, coal, and petroleum ($\simeq$-35 to -10 $^o/oo$). Although the pronounced isotopic differences between organic carbon and inorganic carbonates seem academic in view of obvious visible differences between these compounds, they do serve as a tracer of biogenic precursors where reduced forms of carbon become oxidized and remobilized in some magmatic, metamorphic, and hydrothermal (?) environments. The data for hydrothermal carbonates provided by Sheppard et al. (1971), Fuex and Baker (1973), Ohmoto and Rye (1979), and Fifarek (1985) show remarkably little isotopic variability ($\simeq$-11 to 1 $^o/oo$), regardless of textural variety or genetic occurrence. According to Ohmoto and Rye (1979), the $\delta^{13}C$ value of carbon in mantle-derived igneous rocks is -5 ± 2 permil, and this value is not demonstrably different from that of carbon in average sedimentary or crustal rocks based on considerations of mass balance. Thus, carbon isotopes do not offer promise as a means for distinguishing between mantle and crustal sources of magma and igneous rock. The carbon in minerals and fluids of hydrothermal systems may be derived from mantle sources, as is permissive from the isotopic evidence, or from diverse sources in country rocks either by oxidation of reduced forms or by dissolution and(or) decarbonation reactions of carbonates (Ohmoto and Rye, 1979).

Sulfur

Isotopic abundances of sulfur portrayed in Figure 6.10 are taken principally from the data and summaries of Field (1972), Field et al. (1976, 1983, 1984), Ohmoto and Rye (1979), Claypool et al. (1980), Hoefs (1980), and Sakai et al. (1984). Compositions of orthomagmatic total sulfur in igneous rocks ($\simeq$-3 to 3 $^o/oo$) are predictably close to that of meteoritic sulfide-sulfur ($\simeq$0 $^o/oo$). However, those of the component oxidized and ^{34}S-enriched sulfate (up to 10 $^o/oo$) and reduced and ^{34}S-depleted sulfide (up to -10 $^o/oo$) forms are more variable, because of redox reactions. The $\delta^{34}S$ values of magmatic Cu-Fe-Ni sulfides in layered mafic intrusions ($\simeq$-6 to 14 $^o/oo$) exhibit larger variations attributable both to contamination from sedimentary sources in nearby country rocks and to redox reactions within the host

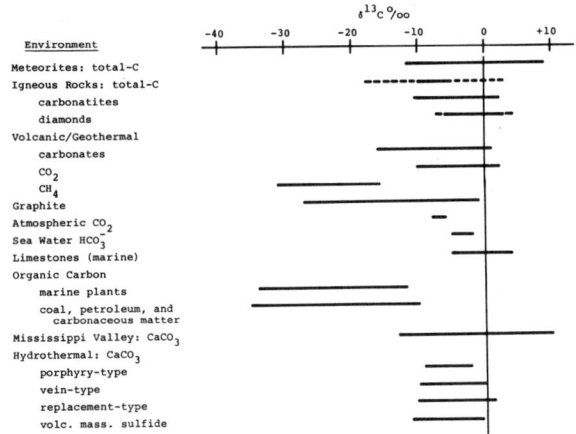

Figure 6.9. Distributions of $\delta^{13}C$ in geologic environments.

magma chambers. Sulfur-bearing products of volcanic/geothermal emanations are broadly isotopically similar to, but more variable, than magmatic sulfur ($\simeq$-3 to 3 ^{o}/oo), and they are increasingly depleted in ^{34}S with some overlap in the redox sequence from SO_2 ($\simeq$-8 to 18 ^{o}/oo), through native sulfur ($\simeq$-15 to 16 ^{o}/oo), to H_2S ($\simeq$-9 to 6 ^{o}/oo). Sea water sulfate has a value of 20 ± 1 ^{o}/oo at present, but has ranged from about 10 to 30 ^{o}/oo over Phanerozoic time as deduced from studies of marine evaporites (Claypool et al., 1980). Thus, the ^{34}S-age curve serves within broad limits to date marine sedimentary strata (and volcanogenic massive-sulfide deposits). The ^{34}S-enriched sulfate ($\simeq$ 20 to 35 ^{o}/oo) and sulfide ($\simeq$-6 to 25 ^{o}/oo) minerals of the and(or) evaporites. Sedimentary sulfides of diagenetic-syngenetic or later epigenetic origin have an extraordinarily large range of $\delta^{34}S$ values ($\simeq$-50 to 50 ^{o}/oo). Such large variations, especially the ^{34}S depletions, have been documented by experimental investigations, and they are caused by kinetic fractionations of -20 to -50 permil and more that may accompany the biogenic reduction of $SO_4^=$ to H_2S (Ohmoto and Rye, 1979). Although sulfides variably depleted in ^{34}S are considered to be typical of those formed by biogenic processes, others may have ^{34}S enrichments that relate to factors such as source, reservoir of sulfate, and other conditions within the system. Regardless of environment or locale, hydrothermal sulfates of hypogene origin are enriched in ^{34}S relative to associated sulfides, which is consistent with fractionation theory and the inferred range of depositional temperatures from about 200°C to 600°C (Figs. 6.4 and 6.5). Contrary to the impression given by the data in Figure 6.10, variations in the $\delta^{34}S$ values for either sulfates or sulfides of individual deposits rarely exceed 5 to 7 ^{o}/oo, although some deposits may exhibit a compositionally distinct cluster of absolute values that may relate to source, age, and(or) unique conditions within the system (Field et al., 1983; Fifarek, 1985). Both the sulfates and sulfides of Cordilleran-type hydrothermal deposits are generally depleted in ^{34}S relative to their counterparts in volcanogenic massive-sulfide deposits. The Cordilleran subgroup includes a large spectrum of porphyry, vein, and replacement types of deposits that apparently are devoid of a distinctive isotopic signature regardless of differences in host rock, metals, minerals, and depositional textures of ore and gangue. The ^{34}S data for sulfates and sulfides of a few deposits have values suggestive of country rock contamination, but the majority are consistent with derivation from a $\simeq$0 permil source of deep-seated "magmatic" sulfur. In contrast, sulfates and sulfides of the volcanogenic massive-sulfide subgroup are normally ^{34}S-enriched because they derive their sulfur largely or entirely from isotopically heavy sea-water sulfate. Data for the sulfates ($\simeq$12 to 39 ^{o}/oo) are essentially equivalent to those of temporally similar marine evaporitic sulfates ($\simeq$10 to 30 ^{o}/oo), and those for associated sulfides ($\simeq$8 to 22 ^{o}/oo) are correlative to the age-trend but depleted in ^{34}S by about 15 to 18 ^{o}/oo relative to oceanic sulfate-sulfur as a consequence of temperature dependent fractionation (Sangster, 1968; Franklin et al., 1981; Fifarek, 1985). The data for a single mid-ocean hydrothermal vent (21° North, East Pacific Rise, Baja, California) are illustrative of such isotopic effects under contemporary oceanic conditions (Styrt et al., 1981).

EPITHERMAL DEPOSITS

Distributions of the stable isotopes in epithermal deposits are now considered. Also included are the data for several geothermal systems because they are regarded by many to be contemporary analogues of the epithermal environment (White, 1981). For the purposes of comparison and discussion, we have subdivided the epithermal deposits into sediment-hosted, volcanic-hosted, and zoned polymetallic vein occurrences. Our subdivisions differ partly from those of Hayba et al. (1985, this volume) in that most or all deposits of our zoned polymetallic vein subtype are, or may be, grouped in their Adularia-Sericite subtype of

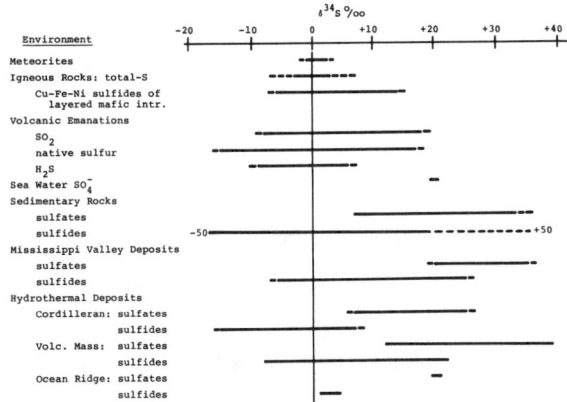

Figure 6.10. Distributions of $\delta^{34}S$ in geologic environments.

volcanic-hosted epithermal deposits. These deposits may exhibit characteristics suggestive of a deeper level of hydrothermal mineralization, such as fluid inclusions having higher salinities and homogenization temperatures; a not uncommon magmatic component to the hydrothermal fluids as deduced from hydrogen- and oxygen-isotope data; host rocks that include plutonic and sedimentary lithologies; alteration that lacks widespread and pervasive zones of advanced argillic and alunitic assemblages; and ores that contain relatively abundant sulfides and sulfosalts of the base metals. Presumably such differences are not genetic, but instead represent variations in style and content of mineralization that have developed from chemical and thermal gradients in deeply convecting hydrothermal systems.

Carbon

Data for ^{13}C in various geothermal systems and epithermal deposits are illustrated in Figure 6.11. The δ^{13}C values for all occurrences, except the Geysers and Pueblo Viejo, are remarkably uniform within the narrow range of -10 to 1 permil, and suggest that the carbon was derived either from magmatic or marine limestone sources (Fig. 6.9). Extreme ^{13}C depletions (-25 to -24 °/oo) of carbonaceous material in volcaniclastic sedimentary rocks at Pueblo Viejo (Kesler et al., 1981) are typical of reduced forms of organic carbon. Oxidized carbon compounds such as CO_2, aqueous HCO_3^-, host-rock carbonate, and vein calcite at the Geysers geothermal area exhibit large variations in δ^{13}C (-19 to 1 °/oo) according to the work of White et al. (1973) and Sternfeld (1981), and other references cited by these authors. Detailed fluid-inclusion, isotopic, and mineralogical studies by Sternfeld (1981) suggest that much of the ^{13}C variability may be attributed to multiple sources of biogenic, marine, and magmatic carbon in the igneous and sedimentary host rocks that were remobilized and redeposited during subsequent and temporally separate metamorphic and geothermal events. Vein calcites diminish in δ^{13}C (~4 °/oo/1000 m) with increasing depth, as a result of temperature-induced fractionation, and a distinct population of ^{13}C-depleted calcites is attributed to late-stage re-equilibration with CO_2-rich steam. In contrast to the Geysers, isotopic compositions of carbon in carbonate clasts of host rocks at the Cerro Prieto (Williams and Elders, 1984) and Salton Sea (Clayton et al., 1968) geothermal areas and in vein carbonates from these and the Broadlands (Blattner, 1975) and Wairakei (Clayton and Steiner, 1975) areas of New Zealand are less variable. Values of δ^{13}C in host rock and vein carbonates decrease with increasing depth at Broadlands and Cerro Prieto. The cause of these isotopic trends is uncertain. They may possibly relate to diminishing equilibrium fractionations with increasing temperatures at depth, and (or) to other complexities such as boiling, decarbonation and (or) dissolution reactions, influx of ^{13}C-depleted organic carbon, and possibly other processes (see Blattner, 1975; Williams and Elders, 1984). Data for carbonates from Wairakei (Clayton and Steiner, 1975) and the Salton Sea (Clayton et al., 1968) do not show isotopic trends related to depth or reservoir temperatures. However, ^{13}C-depleted carbonates from the Salton Sea field correlate inversely with the total carbonate (wt.%) content of the host, which suggests they are the residuals of decarbonation reactions accompanied by the loss of ^{13}C-enriched CO_2 (Clayton et al., 1968).

Carbonates from unaltered and altered host rocks and veins of the sediment-hosted epithermal deposits at Carlin (Radtke et al., 1980) and Cortez (Rye et al., 1974) exhibit a narrow and overlapping range of δ^{13}C values (Fig. 6.11; ~-6 to 1 °/oo). These and other data suggest that most of the carbon in hydrothermal calcites was probably extracted by dissolution reactions from the carbonate host rocks at depth. However, the fluids at Carlin also must have contained a component of oxidized ^{13}C-depleted organic carbon from the host rocks, provided isotopic equilibrium prevailed, to account for the relatively light compositions of one main-stage and several late low-temperature vein calcites (Radtke et al., 1980).

The various carbonate minerals formed in zoned polymetallic veins also show a narrow range of δ^{13}C values (~-10 to 0.1 °/oo), but as a group they are slightly depleted in ^{13}C relative to those of sediment-hosted deposits. Data for rhodochrosites, manganosiderites, and siderites from Creede, Colorado (-8.2 to -4.0 °/oo; Bethke and Rye, 1979), calcites and other carbonates (?) from Tui, New Zealand (-7.8 to 0.1 °/oo; Robinson, 1974), and calcites from Casapalca, Peru (-10.0 to -2.6 °/oo; Rye and Sawkins, 1974), are collectively interpreted as being indicative of magmatic carbon (~-5 ± 2 °/oo; Ohmoto and Rye, 1979). This conclusion is also supported by hydrogen and oxygen-isotope data of inclusion fluids and host minerals at Creede and Casapalca. However, the source of carbon in calcites and rhodocrosites at the Sunnyside mine (-7.9 to -1.8 °/oo), near Creede, is not definitive, and may have been derived in part from dissolution of marine limestones or magmatic sources at depth, or from atmospheric CO_2 dissolved in circulating meteoric water (Casadevall and Ohmoto, 1977).

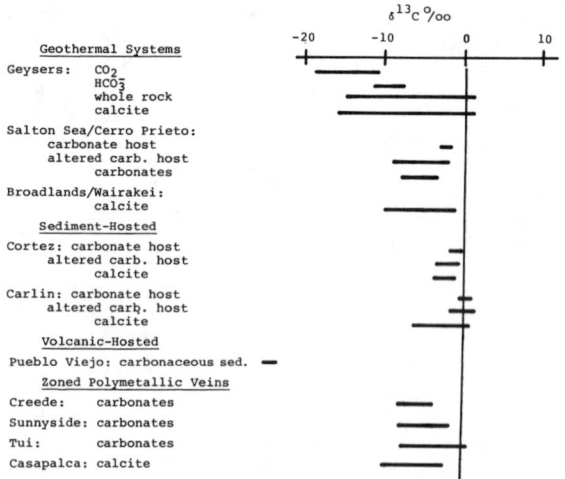

Figure 6.11. Distributions of δ^{13}C in epithermal deposits.

Thus, in the absence of contributing information from other isotopic elements, fluid inclusions, and assemblages of vein and alteration minerals, the ^{13}C data alone are unlikely to provide unique interpretations to geochemical problems. Moreover, vertical ^{13}C gradients present in some geothermal systems have not been reported in the epithermal deposits. However, early calcites (-10.1 to -6.1 °/oo) are depleted in ^{13}C relative to late calcites (-6.4 to -2.6 °/oo) at Casapalca (Rye and Sawkins, 1974), and rhodochrosites (-7.9 to -6.3 °/oo) are depleted in ^{13}C relative to calcites (-3.8 to -2.8 °/oo) at the Sunnyside mine (Casadevall and Ohmoto, 1977).

Sulfur

Abundances of ^{34}S in geothermal systems and epithermal deposits are portrayed in Figure 6.12. As expected, sulfide-sulfur from whole-rock samples of post-glacial basalts near the geothermal fields in Iceland yield δ^{34}S values (-1.8 to 0.4 °/oo) similar to those of 0 permil magmatic sulfur (Sakai et al., 1980). The data for aqueous $SO_4^=$ and sulfate minerals occupy two distinct isotopic populations. One group represented by samples from Iceland (Sakai et al., 1980), Yellowstone (Schoen and Rye, 1970; Truesdell et al., 1977, 1978), and Wairakei (Steiner and Rafter, 1966) is enriched in ^{34}S ($\approx$15 to 23 °/oo) and consists of hypogene sulfates that isotopically equilibrated with H_2S at depth and at relatively high temperatures ($\geq$300°C). The other group, also from Iceland, Yellowstone, and Wairakei, is isotopically more variable and relatively depleted in ^{34}S (-6 to 12 °/oo). It consists of mixed proportions of the deep ^{34}S-enriched hypogene sulfate and shallow ^{34}S-depleted "supergene sulfate formed by near-surface, non-equilibrium, and quantitative oxidation of ^{34}S-depleted H_2S (Truesdell et al., 1977, 1978). Isotopically light aqueous and mineral sulfates are not only common to the acid-sulfate springs of geothermal areas, but they are also typical of some alunites and barites of volcanic-hosted epithermal deposits; particularly those associated with advanced-argillic alteration (the acid-sulfate type of Hayba et al., 1985, this volume). Native sulfur at Yellowstone has formed chiefly by near-surface inorganic oxidation of H_2S, as is consistent with the compositional similarities between elemental (-8.4 to 3.2 °/oo) and reduced (-5.0 to 4.0 °/oo) sulfur compounds (Schoen and Rye, 1970). However, the relatively broad spread of δ^{34}S values (8.4 to 4.0 °/oo for S° and H_2S ($\sim$0 °/oo) undergoes slight fractionation upon separation into discrete ^{34}S-depleted vapor and ^{34}S-enriched aqueous phases (Truesdell et al., 1978). The sulfide minerals normally exhibit narrow compositional ranges that are similar to those of associated H_2S, although there may be some variability attributable to differing sources of sulfur both within and between the geothermal fields. For example, the pyrites from Iceland consist of two distinct isotopic populations: one is enriched in ^{34}S ($\sim$2.9 to 7.9 °/oo) and has been derived through reduction of isotopically heavy seawater sulfate, and the other is depleted in ^{34}S (-4.6 to 0.9 °/oo) and has originated from a magmatic source (Sakai et al., 1980).

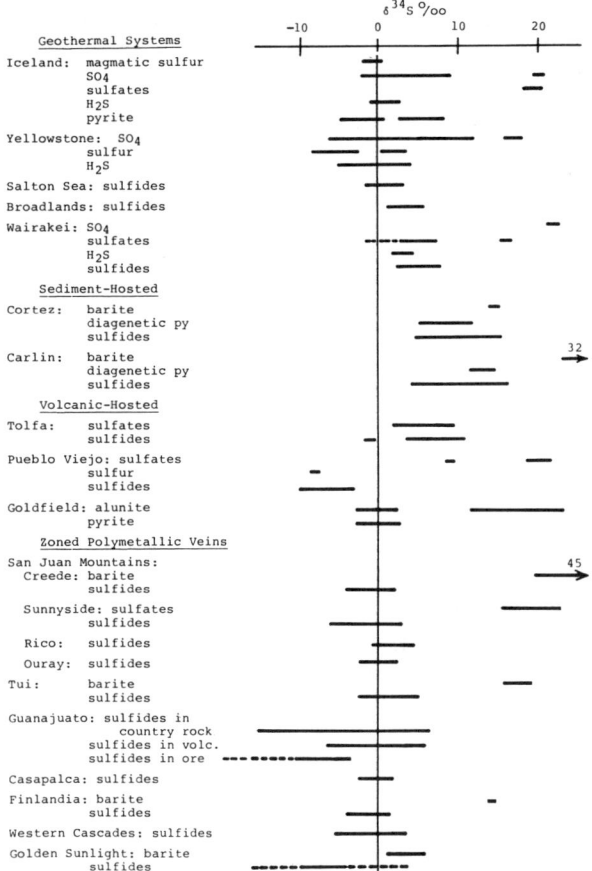

Figure 6.12. Distributions of δ^{34}S in epithermal deposits.

Ranges of δ^{34}S are narrow for pyrite and Ag-Cu sulfides (-1.4 to 3.0 °/oo) of the Salton Sea; pyrite, pyrrhotite, galena, and sphalerite (1.4 to 5.1 °/oo) of Broadlands; and pyrite and pyrrhotite (2.7 to 6.8 °/oo) of Wairakei. The sulfide compositions at the Salton Sea area are compatible with, but not proof of, a magmatic source of sulfur (White, 1968, 1974), whereas those at the Broadlands and Wairakei are slightly more enriched in ^{34}S and suggest a crustal provenance either by leaching and partial reduction of sulfate from basement rocks (Browne et al., 1975) or by magma generation in the upper crust (Steiner and Rafter, 1966). Although the data base is meager, and except for two sphalerite-galena pairs from Broadlands, there is little evidence of complete isotopic equilibrium between sulfate-sulfide and sulfide-sulfide assemblages in the geothermal environment.

All sulfide and most sulfate minerals in and near the sediment-hosted gold deposits at Carlin (Radtke et al., 1980) and Cortez (Rye et al., 1974) are enriched in ^{34}S relative to their counterparts in geothermal systems and other epithermal deposits. The evidence, particularly from Carlin, suggests that most of the sulfide-sulfur in pyrite-galena-realgar-sphalerite-stibnite ore (4.2 to 16.1 $^o/oo$) was derived from the hydrothermal remobilization of diagenetic pyrite (11.7 to 14.3 $^o/oo$ at Carlin; 5.1 to 11.4 $^o/oo$ at Cortez) in the sedimentary host rocks. However, the source of sulfate-sulfur in barites (27.8 to 31.7 $^o/oo$ at Carlin) is uncertain, and may have originated either by hydrothermal solution of sedimentary barite in the country rocks, or by fractionation that may have accompanied partial oxidation of sedimentary sulfide-sulfur incorporated in the fluids (Radtke et al., 1980). Regardless of origin of the sulfate-sulfur, it is remarkable that isotopic temperatures calculated from barite-pyrite values (250°C to 305°C) are reasonably consistent with those derived from the homogenization of fluid inclusions (180°C to 365°C).

The volcanic-hosted deposits of Tolfa (Italy), Pueblo Viejo and Goldfield have geologic features common to many geothermal systems in that their volcanic host rocks have been pervasively altered to alunite-kaolinite + pyrophyllite assemblages of advanced argillic alteration. According to Field and Lombardi (1972) and Cortecci et al. (1981), isotopic similarities between alunite and barite (1.9 to 9.6 $^o/oo$) and hypogene pyrite, cinnabar, and galena (3.4 to 10.3 $^o/oo$) at Tolfa suggest that these sulfates, and possibly marcasite (-1.5 and -0.6 $^o/oo$), are of supergene origin. However, the sulfate-sulfide assemblages of Tolfa may consist of two isotopically distinct populations because the ^{34}S-depleted marcasites are associated with the lightest alunites (1.9 and 2.5 $^o/oo$). Perhaps these minerals formed in surficial acid-sulfate pools from ^{34}S-depleted H_2S that had separated with boiling of reservoir fluids at depth, as proposed by Truesdell et al. (1978) for some hot springs of the Yellowstone area. In contrast, most sulfates at Pueblo Viejo are considered to be of hypogene origin by Kesler et al. (1981), because the barite and alunite are enriched in ^{34}S relative to pyrite and sphalerite (18.8 to 21.6 $^o/oo$ versus -10.1 to -3.5 $^o/oo$). However, Jensen et al. (1971) have documented alunites of both hypogene (11.6 to 23.3 $^o/oo$) and supergene (-2.5 to 1.7 $^o/oo$) origin at Goldfield; the latter group being isotopically similar to hypogene pyrite (-2.8 to 2.4 $^o/oo$) from which they derived their sulfur. Thus, isotopic and geologic evidence suggest that the sulfates (aqueous and mineral) of many geothermal and volcanic-hosted environments are of supergene origin, and acquired their distinctive ^{34}S-depleted sulfate-sulfur (as contrasted to ^{34}S-enriched hypogene sulfates) by near-surface oxidation of ^{34}S-depleted hypogene sulfide-sulfur (H_2S and mineral; see Field, 1966; Schoen and Rye, 1970; Jensen et al., 1971; Field and Lombardi, 1972). The source of hydrothermal sulfur at Goldfield is considered to be magmatic (Jensen et al., 1971). In contrast, the source at Pueblo Viejo is thought to be a mixture of pristine and biogenically reduced sulfate (15 $^o/oo$) from Cretaceous sea water (Kesler et al., 1981), and that at Tolfa may have been derived from gypsiferous Miocene-Pliocene mudstones that underlie this young Pliocene-Pleistocene volcanic complex (Field and Lombardi, 1972). This latter interpretation is supported by the work of Taylor and Turi (1976) who attribute the exceedingly high $\delta^{18}O$ values (15.3 to 16.4 $^o/oo$) of Tolfa quartz latites and rhyolites to magmatic assimilation of ^{18}O-enriched argillaceous sedimentary rocks at depth. Although fractionation effects between hypogene sulfide and sulfate-sulfide minerals of the volcanic-hosted group appear to be consistent with the attainment of at least partial isotopic equlibrium (with $\delta^{34}S$ values of sulfates > sulfides and those of sulfides mostly in the order py > sl > gn), a more precise evaluation of these apparent trends is difficult because the data for contemporaneous or spatially associated mineral pairs and triplets are not available.

The isotopic data base for zoned polymetallic vein deposits (Fig. 6.12) is large and detailed, particularly for those of the Creede district (Bethke et al., 1973; Bethke and Rye, 1979; Foley et al., 1982; Hayba et al., 1985, this volume) and the Sunnyside mine of the San Juan Mountains, Colorado (Casadevall and Ohmoto, 1977); Tui, New Zealand (Robinson, 1974); Casapalca, Peru (Rye and Sawkins, 1974); Finlandia vein, Colqui district, Peru (Kamilli and Ohmoto, 1977); mining districts of the Western Cascades, Oregon (Taylor, 1971, 1974b; Power, 1985; Field and Power, 1985); and Golden Sunlight mine, Montana (Porter and Ripley, 1985). Sulfides in most of these deposits have $\delta^{34}S$ values near 0 permil that contrast markedly with associated ^{34}S-enriched hypogene sulfates, such as at Creede (-4.1 to 1.7 $^o/oo$ versus 19.8 to 45 $^o/oo$), Sunnyside (-6.3 to 2.7 $^o/oo$ versus 15.3 to 22.9 $^o/oo$), Tui (-2.4 to 4.9 $^o/oo$ versus 16.0 to 19.5 $^o/oo$), and Finlandia (-4.0 to 1.5 $^o/oo$ versus 14.0 to 14.1 $^o/oo$). Moreover, compositions of sulfides that are unassociated with sulfates, such as those given by Jensen et al. (1960) for Rico (-0.6 to 4.0 $^o/oo$) and Ouray (-2.0 to 1.9 $^o/oo$) in the San Juan Mountains of Colorado, and by Rye and Sawkins (1974) for Casapalca, are also closely grouped near 0 permil. Because of the overall isotopic similarity of most sulfides in these deposits to the 0 permil value of magmatic sulfur, and other geologic and geochemical considerations, a magmatic source of sulfur is advocated by the authors of investigations at Rico, Ouray, Casapalca, and the Western Cascades, and possibly at Creede and Finlandia. However, the commonly assumed "genetic" equivalence of 0 permil sulfides and magmatic (0 $^o/oo$) sources of sulfur is a hazardous generalization when applied to hydrothermal environments. It is valid for sulfides only to the extent it may be assumed that concentrations of reduced sulfur (H_2S) are approximately equal to those of total sulfur in these systems. However, under conditions of high f_{O_2} and (or) low pH, as may be inferred from the presence of oxide and sulfate minerals in vein and alteration assemblages, part of the H_2S becomes oxidized to $SO_4^=$, and total sulfur must then consist of both oxidized and reduced components. Because of the large fractionation of ^{34}S between sulfates and sulfides and of mass balance considerations between oxidized and reduced forms of aqueous sulfur in the fluids, sulfide minerals become increasingly depleted in ^{34}S relative to the $\delta^{34}S$ of total sulfur in the system, as ratios of $SO_4^=:H_2S$ increase with increasing f_{O_2} and (or) decreasing pH. Redox changes as described, such as increasing states of oxidation with evolution of hydrothermal systems, have been proposed by Robinson (1974) to account for the progressive ^{34}S depletion observed in the paragenetic order of sulfide deposition at Tui. This so-called Eh-pH control of mineral ^{34}S compositions was first defined by Sakai (1968) and Ohmoto (1972), and has been subsequently refined by Rye and Ohmoto (1974) and Ohmoto and Rye (1979). It is for reasons of this Eh-pH control, as deduced from mineralogical and geochemical evidence, that isotopically heavy marine sulfates have been proposed as sources of sulfur in the hydrothermal deposits at Sunnyside (upper Paleozoic evaporites of ~12 °/oo; Casadevall and Ohmoto, 1977) and Tui (Jurassic sea water of ~16 °/oo; Robinson, 1974), and in spite of the fact that compositions of associated sulfides (-6.3 to 2.7 °/oo and -2.4 to 4.9 °/oo, respectively) bracket that of 0 °/oo magmatic sulfur. Moreover, this control is implicit to the interpretation of a magmatic source of sulfur at the Golden Sunlight deposit (Porter and Ripley, 1985), although both sulfates (1.2 to 5.9 °/oo) and sulfides (-15.8 to -4.0 °/oo) are appreciably depleted in ^{34}S relative to the majority of data for hypogene equivalents (Fig. 6.12; also see Field and Gustafson, 1976, Fig. 3, for a graphical portrayal of this control). In contrast, Gross (1975) has proposed a crustal source for sulfur and metals contained in vein sulfides (-19.5 to -3.4 °/oo) of pyrite-sphalerite-galena-argentite ore from Guanajuato, Mexico, which are considered to have been derived from metal-rich and sulfide-bearing (-16.6 to 6.3 °/oo) Mesozoic sedimentary country rocks by heated ground waters during Oligocene volcanic activity. Fractionation trends for sulfate-sulfide and sulfide-sulfide mineral pairs from the zoned polymetallic veins are largely consistent with those predicted from theory and experiment (Tables 6.4 and 6.5, and Figs. 6.4 and 6.5). However, with the exception of a few sphalerite-galena pairs, from Sunnyside, Tui, Casapalca, Finlandia, and the Western Cascades, the ^{34}S temperature estimates are rarely consistent with those obtained by fluid-inclusion homogenization methods. This disparity implies that isotopic equilibrium largely did not prevail in these systems for reasons that may include low depositional temperatures, relatively rapid accent of fluids and deposition of minerals, and possibly abrupt changes in $SO_4^=:H_2S$ ratios with the ascent of fluids into more oxidizing environments. The lack of equilibrium between sulfates and sulfides is clearly expectable based on the discussions and data presented by Ohmoto and Rye (1979) and Ohmoto and Lasaga (1982). Moroever, biogenic processes (Hayba et al., 1985, this volume), as previously suggested by Kesler et al. (1981) to account for the ^{34}S-depleted compositions of bedded sulfides at Pueblo Viejo, in addition to the effects of chemical and (or) isotopic disequilibrium (Bethke et al., 1973) are not considered to be responsible for the large ^{34}S enrichments (19 to 45 °/oo) of sulfates at Creede, Colorado.

Hydrogen and Oxygen

Distributions of ^{18}O in the host rocks and minerals of many geothermal and epithermal occurrences previously described are portrayed in Figure 6.13. Because these data are voluminous and based on numerous investigations of variable detail, our discussion will focus principally on the major isotopic trends and reasons thereof. More detailed information may be obtained as needed from the references cited. Although these data exhibit a large

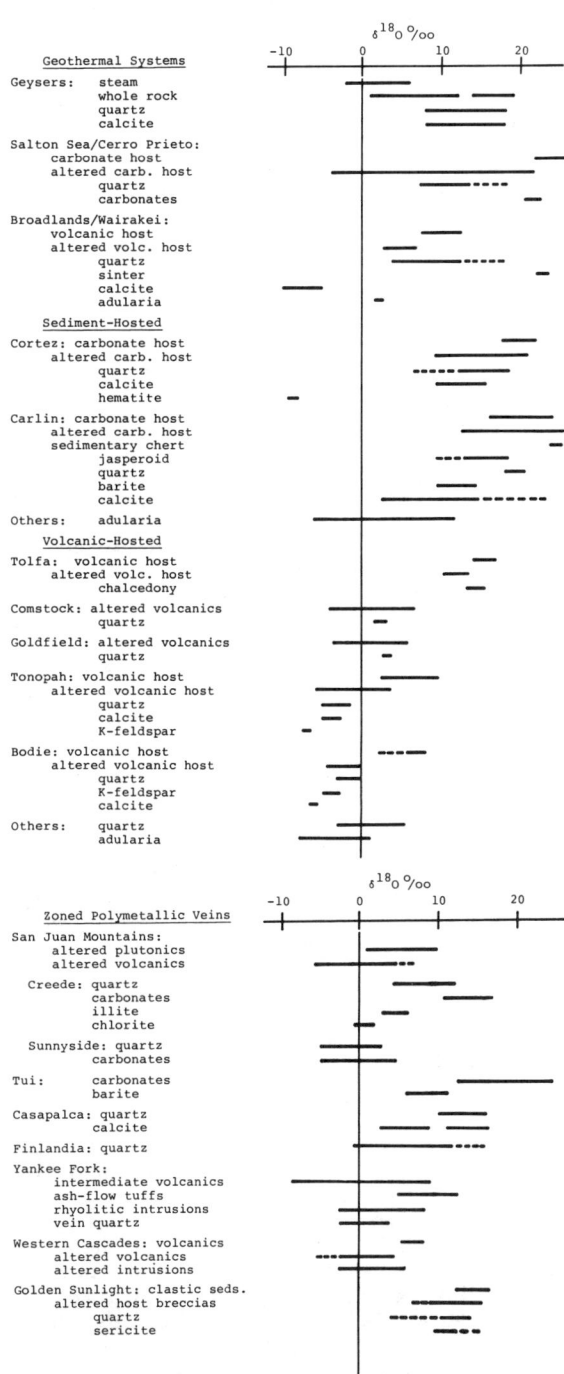

Figure 6.13. Distributions of $\delta^{18}O$ in epithermal deposits.

amount of compositional scatter, they reveal a number of systematic trends when examined on the basis of individual occurrences. The large isotopic variability results from (1) differing ^{18}O compositions of the host rocks, (2) differing ^{18}O compositions of the hydrothermal fluids, and (3) interactions between the host rocks and fluids over a range of low temperatures (100°C to 300°C) and attainments of equilibrium that produced hydrothermal minerals of variable and generally intermediate ^{18}O compositions.

As previously noted, most of the common marine sedimentary rocks are isotopically more variable and enriched in ^{18}O ($\simeq$10 to 40 °/oo) than are the common igneous rocks ($\simeq$5 to 12 °/oo; see Fig. 6.7). These petrologic distinctions account for the obvious isotopic differences between the ^{18}O-enriched "unaltered" sedimentary host rocks of the Geysers and Salton Sea geothermal areas and the Cortez and Carlin epithermal deposits, as compared to the less ^{18}O-enriched "unaltered" volcanic host rocks of nearly all other epithermal deposits.

The derivation of nearly all geothermal and epithermal fluids from meteoric sources of water has also been mentioned previously. This conclusion is based on the D and ^{18}O compositions of these fluids, which although variable plot in close proximity to or by lateral "oxygen isotope shift" away from the meteoric water line (Fig. 6.8). The δD and $\delta^{18}O$ compositions of hydrothermal fluids that formed the epithermal deposits of this discussion and those for many of their nearby present-day meteoric waters are listed in Table 6.6. Sources of data for most deposits of the Basin and Range are from O'Neil and Silberman (1974), except for those of Carlin (Radtke et al., 1980) and Cortez (Rye et al., 1974). Data for epithermal deposits of other geographic localities include Tolfa (Lombardi and Sheppard, 1977), Yankee Fork-Idaho Batholith (Criss and Taylor, 1983; Criss et al., 1985), and others previously mentioned. On the basis of isotopic similarities to present-day meteoric waters and of widespread D and ^{18}O depletions, the sources of most epithermal fluids must have been local meteoric waters (Table 6.6). Although magmatic waters may constitute a small proportion of these hydrothermal fluids, they have been detected with confidence from D analyses of fluid inclusions only at Casapalca (Rye and Sawkins, 1974), for the early and intermediate stages of carbonate mineralization at Creede (Bethke and Rye, 1979), and at Golden Sunlight (Porter and Ripley, 1985). Based on isotopic and other considerations, the fluids of Tui (Robinson, 1974), Finlandia (Kamilli and Ohmoto, 1977), and possibly the Comstock Lode (Taylor, 1973) may be partly of magmatic origin. Thus, the compositions of hydrothermal fluids depicted by closed triangles on Figure 6.8, with the exception of those of Casapalca and Golden Sunlight, are illustrative of meteoric water-dominated epithermal systems. Compositionally complex fluids such as those of Creede (Bethke and Rye, 1979), Sunnyside (Casadevall and Ohmoto, 1977), Finlandia (Kamilli and Ohmoto, 1977) and parts of several other hydrothermal systems are not portrayed in Figure 6.8. Fluids from these deposits have exceedingly variable δD and $\delta^{18}O$ values (Figs. 6.6 to 6.8) and thus enclose large isotopic domains as summarized by Taylor (1979). This variability is attributable either to paragenetically distinct and (or) mixed sources of magmatic and meteoric waters, as deduced from the analyses of many samples, or to subsequent contamination by deuterium-depleted ground waters that were trapped in pseudosecondary inclusions (Foley et al., 1982).

Distributions of the isotopic data (Fig. 6.13) show that altered host rocks, regardless of igneous or sedimentary parentage, are variably depleted in ^{18}O relative to their unaltered counterparts. These trends are evident from comparisons of the ^{18}O data for unaltered and altered host rocks at Salton Sea-Cerro Prieto (Clayton et al., 1968; Williams and Elders, 1984), Broadlands-Wairakei (Blattner, 1975; Clayton and Steiner, 1975), Cortez (Rye et al., 1974), Carlin (Radtke et al., 1980), Tolfa (Taylor and Turi, 1976), Lombardi and Sheppard, 1977), Tonopah (Taylor, 1973), Bodie (O'Neil et al., 1973), San Juan Mountains, Colorado (Taylor, 1974b; Lea et al., 1984), Western Cascades, Oregon (Taylor, 1971; 1974a), and Golden Sunlight (Porter and Ripley, 1985). Although comparisons to unaltered equivalents are lacking, altered sedimentary host rocks at the Geysers (Sternfeld, 1981) and volcanic host rocks at the Comstock Lode and Goldfield (Taylor, 1973; O'Neil and Silberman, 1974), and at Yankee Fork (Criss and Taylor, 1983; Criss et al., 1985) are similarly depleted in ^{18}O relative to normal magmatic compositions. These ^{18}O depletions of volcanic and sedimentary host rocks are a corollary of the "oxygen isotope shift" previously noted for many geothermal fluids (Fig. 6.8). They arise during geothermal-hydrothermal activity from isotope-exchange reactions between ^{18}O-depleted meteoric waters and ^{18}O-enriched host rocks, and result in the ^{18}O enrichment of the fluids and ^{18}O depletion of the rocks as a consequence of water-rock reactions. Progressive depletions of ^{18}O with increasing depth in host rocks of the Cerro Prieto (Williams and Elders, 1984), Geysers (Sternfeld, 1981, and references therein), and Salton Sea (Clayton et al., 1968) areas cannot be related to temperature-induced fractionation trends, and thus must be caused by changes in fluid compositions with depth that result from recharge and (or) water-rock reactions. Isotopic trends and permutations resulting from these phenomena range from subtle to dramatic, and they will be discussed extensively in the section that follows.

Compositions of the fracture-controlled hydrothermal gangue minerals from geothermal systems and epithermal deposits (Fig. 6.13) are largely compatible with isotopic effects described in the foregoing discussion and from previous considerations of fractionation. The data are from sources previously cited, except for the "other" categories of sediment- and volcanic-hosted deposits that are from O'Neil and Silberman (1974; also see Table 6.6) and represent miscellaneous mines and prospects from the Basin and Range province. Values of $\delta^{18}O$ for these minerals generally occupy a range that is intermediate between the compositions of associated altered-unaltered host rocks (Fig. 6.13) and those of the geothermal or

Table 6.6 Compositions of δD and $\delta^{18}O$ in hydrothermal fluids of Tertiary epithermal deposits and some local meteoric waters

	Epithermal Fluids		Meteoric Water	
	δD	$\delta^{18}O$	δD	$\delta^{18}O$*

Basin and Range

Volcanic-Hosted Deposits

Bullfrog (BU)	−94	−6.0	−100	−13.8
Aurora (A)	−124	−12.0	−130	−17.5
Trade Dollar (TD)	−136	−14.2	−135	−18.1
Wonder (W)	−139	−13.5	−135	−18.1
Jarbidge (J)	−139	−14.4	−135	−18.1
Rawhide (R)	−120	−14.2	−120	−17.5
Gilbert (G)	−111	−8.0	−120	−17.5
Tonopah (T)	−90	−12.8	−110	−15.0
Bodie (B)	−98	−13.5	−130	−17.5
Comstock Lode (CL)	−133	−6.2	−115	−15.6

Sediment-Hosted Deposits

Tenmile (TE)	−97	−1.4	−130	−17.5
Humboldt (H)	−95	+3.7	−130	−17.5
Cortez (CO)	−128	+4.8	−130	−17.5
Carlin (CA)	−153	+5.4	−135	−18.1
Manhattan (M)	−116	−15.9	−125	−16.9

Elsewhere

Volcanic-Hosted Deposits

Tolfa (TF)	−20	−1	−37	−6.7

Zoned Polymetallic Vein

Creede		complex		
Sunnyside		complex		
Tui (TU)	−70	0	−30	−5.5
Casapalca	−54	7		
Finlandia		complex		
Yankee Fork (Idaho Batholith, IB)	−120	−16	−140	−19
Western Cascades (WC)	−63	−9		
Golden Sunlight (GS)	−50	0		

*calculated from equation for the meteoric water line (eq. 12)

hydrothermal fluids (Table 6.6 and Fig. 6.8). The $\delta^{18}O$ values of hydrothermal minerals formed by mineral-H_2O exchange reactions are determined by the temperature of deposition, which controls the extent of isotopic fractionation, and fluid compositions, which in turn are controlled by the source of fluids, types of host rock, and water-host rock reactions. Isotopic effects related to one or more of these determinants are qualitatively apparent from the data of Figure 6.13. The temperature control, with mineral-H_2O fractionation increasing with decreasing temperature, must account for the large ^{18}O enrichments of siliceous hot-spring sinters (22.2 and 23.6 °/oo) relative to vein quartz (mostly 3.9 to 12.3 °/oo) formed at higher temperatures and depths in the Broadlands-Wairakei areas (Clayton and Steiner, 1975; Blattner, 1975). Moreover, this temperature control of fractionation is largely the cause of progressive ^{18}O depletions of vein quartz and (or) calcite with increasing depth at Broadlands (Blattner, 1975), Cerro Prieto (Williams and Elders, 1984), Geysers (Sternfeld, 1981), Salton Sea (Clayton et al., 1968), and Wairakei

(Clayton and Steiner, 1975). However, the parallel trend of decreasing $\delta^{18}O$ values with depth in host rocks of several geothermal areas, as previously noted, cannot be attributed to temperature effects, but instead must result from changes in fluid composition. In addition, compositions of the gangue minerals appear to be influenced by those of their host rocks. For example, hydrothermal calcite and quartz associated with ^{18}O-enriched sedimentary host rocks of Carlin and Cortez are largely isotopically heavier than their counterparts in relatively ^{18}O-depleted volcanic host rocks of the other epithermal deposits; particularly in the Basin and Range province where compositions of the meteoric waters are least variable (Table 6.6). The isotopic composition of hydrothermal fluids, as partly determined by source, may also leave its imprint on the minerals. Quartz and calcite deposited from ^{18}O-enriched magmatic waters (Casapalca, Golden Sunlight, and possibly carbonates at Creede) or from relatively undepleted meteoric waters (Tolfa and Tui) are isotopically heavier than their equivalents formed in hydrothermal systems having more depleted fluids of meteoric origin such as those of the Basin and Range (see Table 6.6 and Fig. 6.13). The sequence of relative ^{18}O enrichments among the vein minerals (qtz>cal>bar>Kf) is generally consistent with that obtained from experimentally derived fractionation data (Fig. 6.3). However, analyses for coexisting mineral pairs and triplets are conspicuously few, and they are largely suggestive of isotopic disequilibrium. This result is not surprising because of the relatively low temperatures (100°C to 300°C), changes in fluid chemistry, and varied sequences of mineral paragenesis that may characterize geothermal-epithermal systems, and which collectively render isotopic equilibrium unlikely. Investigations of fluid-mineral isotopic equilibria in geothermal systems by Clayton et al. (1968), Blattner (1975), and Clayton and Steiner (1975) have demonstrated that quartz is most resistent to isotopic exchange, whereas calcite and alkali feldspars may rapidly undergo re-equilibration and thus be susceptible to compositional change during post-depositional stages of hydrothermal activity.

The principal trends for ^{18}O in host rocks and hydrothermal gangue minerals associated with epithermal activity are summarized by the composite illustration given in Figure 6.14 (after Taylor, 1971; 1973). Volcanic rocks of the Western Cascades in Oregon host numerous zoned polymetallic vein deposits of the base and precious metals. These deposits are mostly clustered within larger district-sized areas of hydrothermal alteration (Field and Power, 1985) that are cored by small granodiorite intrusions of Tertiary age. According to Taylor (1971), the volcanic country rocks (originally 5.5 to 8 °/oo) are progressively depleted in ^{18}O with increasing proximity to the intrusions (Fig. 6.14A; after Taylor, 1971), as a consequence of hydrothermal alteration imposed by reactions between heated meteoric ground waters ($\approx$ -9 °/oo, Table 6.6) and the volcanic host rocks (now $\approx$-5.5 to 5.5 °/oo). This trend of ^{18}O depletion, which increases with intensity of alteration, is analogous to the "oxygen isotope shift" of geothermal waters, but opposite in direction. Volcanic host rocks of the Tonopah district exhibit similar depletions of ^{18}O with progressive alteration (Fig. 6.14B; after Taylor, 1973), and evidence for this having been a meteoric water-dominated hydrothermal system is additionally strengthened by the pronounced ^{18}O depletions ($\approx$-7 to 1 °/oo) of associated quartz, calcite, and adularia vein minerals. The areal distribution of ^{18}O-depleted country rocks are large at Bohemia in the Western Cascades (Taylor, 1971), Tonopah (Taylor, 1973), and at Yankee Fork and other hydrothermally altered areas of the Idaho Batholith (Criss and Taylor, 1983; Criss et al., 1985), and these "negative" isotopic anomalies form well-defined targets appropriate to the reconnaissance stage of mineral exploration.

Water:rock ratios--In this section we examine the systematics of oxygen- and hydrogen-isotopic exchange between fluid and rock and present an

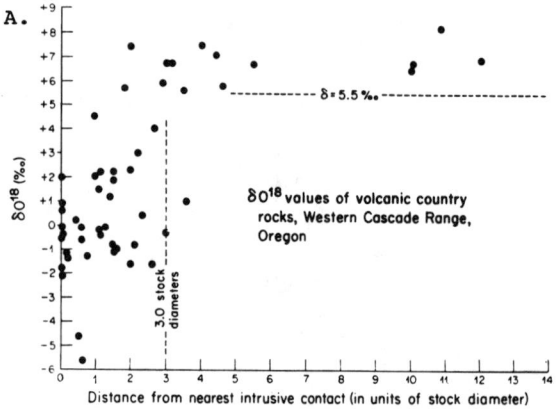

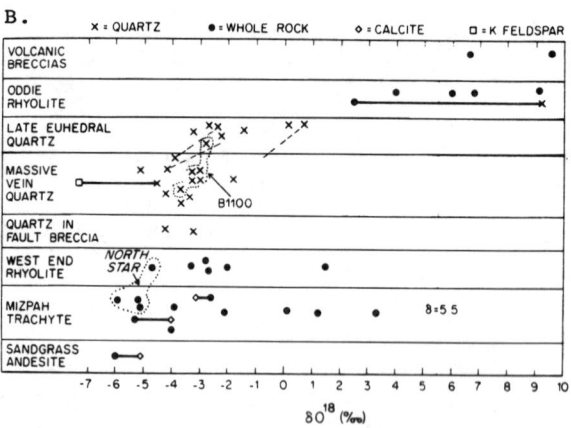

Figure 6.14. Variations of $\delta^{18}O$ in (A) volcanic host rocks adjacent to Tertiary intrusions in the Western Cascades of Oregon (after Taylor, 1971, p. 7867, Fig. 8) and (B) volcanic host rocks and magmatic and hydrothermal minerals in the Tonopah district of Nevada (after Taylor, 1973, p. 755, Fig. 6).

application of these principles to epithermal systems. The application serves not only to impose constraints on the hydrothermal environment, but also as an illustration of the assumptions, problems, and uncertainties that may be involved in modeling isotopic data. The development of the principles of isotopic exchange in water-rock systems generally follows that of Ohmoto and Rye (1974).

The final isotopic composition of water (δ_w^f) after equilibration with rock is a function of: (a) the initial (unexchanged) composition of the water (δ_w^i) and rock (δ_r^i), (b) the temperature of equilibration, which determines the fractionation factor between rock and water (Δ_{r-w}), and (c) the ratio of exchanged oxygen and hydrogen atoms in the water to those in rock (w/r). This relationship, after Ohmoto and Rye (1974), is expressed as

$$\delta_w^f = \frac{\delta_r^i - \Delta_{r-w} + [(w/r)(\delta_w^i)]}{1 + (w/r)} \quad (13)$$

Epithermal precious-metal deposits are commonly hosted either by volcanic rocks of intermediate to felsic composition or by clastic or chemically precipitated sedimentary rocks. Unaltered andesites, dacites, and rhyolites typically have $\delta^{18}O$ and δD values of about 7 °/oo (Fig. 6.7) and -70 °/oo, respectively. Rocks of this compositional range contain approximately 50 weight-percent oxygen and as much as 0.11 weight-percent hydrogen (1 wt.-% H_2O), according to the average analyses reported by Nockolds et al. (1978). Thus, the oxygen- and hydrogen-isotopic composition of a fluid that has equilibrated with volcanic rock of these characteristics can be determined from

$$\delta^{18}O_w^f = \frac{7 - \Delta_{r-w} + (1.8R)(\delta^{18}O_w^i)}{1 + (1.8R)} \quad (14)$$

and

$$\delta D_w^f = \frac{-70 - \Delta_{r-w} + (100R)(\delta D_w^i)}{1 + (100R)} \quad (15)$$

where the coefficients 1.8 and 100 represent ratios of the weight-percent oxygen in water (88.8%) to that in rock (50%) and the weight-percent hydrogen in water (11.2%) to that in rock (0.112%), respectively, and R is the water:rock mass ratio. Because values of R represent proportions of water and rock that have isotopically equilibrated, they are easier to relate to natural systems than are values of the atomic ratio w/r.

Isotopic compositions of sedimentary rocks are more variable than those of igneous rocks (Fig. 6.7). However, the stratigraphic sequences that host epithermal deposits typically consist of silty argillaceous limestone, dolomite, quartzite, and minor shale. If a "typical" host sequence contains subequal amounts of siliciclastic and carbonate components, then it would average approximately 16 °/oo $\delta^{18}O$ and -60 °/oo δD and contain about 50 weight-percent oxygen and perhaps 0.28 weight-percent hydrogen (2.5 wt.-% H_2O). The final oxygen and hydrogen isotopic composition of a fluid that has equilibrated with a sedimentary rock of this composition can be calculated from

$$\delta^{18}O_w^f = \frac{16 - \Delta_{r-w} + (1.8R)(\delta^{18}O_w^i)}{1 + (1.8R)} \quad (16)$$

and

$$\delta D_w^f = \frac{-60 - \Delta_{r-w} + (40R)(\delta D_w^i)}{1 + (40R)} \quad (17)$$

The fractionation of oxygen isotopes between fluids and rock has been variously assumed to be similar to those of smectite-H_2O (Cathles, 1983), plagioclase feldspar (An_{30})-H_2O (Taylor, 1974a, 1979; Ohmoto and Rye, 1974; Green et al., 1983) and muscovite-H_2O (Spooner et al., 1977). The assumptions are based on comparisons of these mineral-H_2O fractionations with experimental rock-H_2O fractionations and with the oxygen-isotopic compositions of naturally altered rocks and secondary minerals. For the illustrations that follow, we have used fractionation factors derived from the plagioclase feldspar (An_{30})-H_2O equation of O'Neil and Taylor (1967)

$$1000 \ln\alpha = 2.68 (10^6/T^2) - 3.29 \quad (18)$$

Use of this equation rather than the more recent one of Matsuhisa et al., 1979 (from eqs. 7 and 10 in Table 6.3) was done to maintain continuity between our results and those of other investigators (see Taylor, 1979, and above). These fractionations are intermediate between those derived from the muscovite-H_2O and smectite-H_2O curves over the temperature range from 100°C to 300°C. The fractionation of hydrogen isotopes between fluids and rock is similarly assumed to be equivalent to that of biotite or chlorite-H_2O (Taylor, 1974a). Our computations are based on chlorite-H_2O fractionation factors which were taken from Taylor (1979, Fig. 6.2).

Using the equations, rock compositions, and mineral-H_2O systems described above, the final isotopic composition of meteoric water ($\delta^{18}O_w^i$ = -16 °/oo and δD_w^i = -120 °/oo) and magmatic waters ($\delta^{18}O_w^i$ = 7.5 °/oo and δD_w^i = -60 °/oo) were computed at temperatures of 100°C to 300°C and R values of 0.01 to 10. The initial isotopic composition of the meteoric water is typical of present-day meteoric water in western Nevada (Table 6.6) and the selected temperatures and R values are considered relevant to epithermal systems. The results of the computations, portrayed in Figure 6.15, illustrate the systematics of isotopic exchange between meteoric or magmatic waters and rocks in the Basin and Range province. First, meteoric water is enriched in ^{18}O and D under most geologic conditions, whereas magmatic water is enriched in deuterium but may be either depleted in ^{18}O through exchange with igneous rocks or enriched in ^{18}O via exchange with sedimentary rocks. Second, the magnitude of the isotopic enrichment or depletion varies inversely with the water:rock mass ratio. Third, the change in fluid δ^{18}O values exceeds that of δD values at high water:rock mass ratios ($\gtrsim$1) but the reverse is true at low ratios ($\lesssim$1). This effect results from the small H:O mass ratio in rocks relative to that in water and from the generally larger fractionation factors for hydrogen than for oxygen. Accordingly, the isotopic influence of rock hydrogen is significant only at low water:rock mass ratios, but is potentially larger than that of rock oxygen. A corollary to this effect is that the generally small "^{18}O shift" observed for geothermal waters (Fig. 6.8) must be a product of isotopic exchange at relatively high water:rock mass ratios. Fourth, at equivalent temperatures and water:rock mass ratios, fluids that have equilibrated with sedimentary rock are isotopically heavier than those that have equilibrated with igneous rock because the former is enriched in ^{18}O and D and contains a greater quantity of hydrogen relative to the latter.

The calculated and analyzed isotopic compositions of the hydrothermal fluids associated with epithermal deposits fall in the range -15 to 5 ^{o}/oo δ^{18}O and -150 to -90 ^{o}/oo δD (Table 6.6 and Fig. 6.8). Consequently, these fluids are isotopically depleted relative to magmatic water or evolved magmatic water, but are similar to meteoric water and its evolved counterparts (Fig. 6.15). Because it is unlikely that magmatic fluid will be isotopically depleted in deuterium through equilibration with rock (most mineral-H$_2$O fractionation factors are negative for hydrogen), it is concluded that meteoric water is the predominant fluid in epithermal systems. However, if fluid mixing is commonplace, then a minor component of magmatic water cannot be precluded on the basis of isotopic considerations.

The isotopic composition of a convected fluid may differ from that of a static fluid because, among other factors, the convected fluid has equilibrated over a range of temperatures and water:rock mass ratios. Therefore, to model the evolution of western Nevada meteoric water during convection, the isotopic composition was calculated at 20°C intervals from 100°C to 300°C. The final composition of the water (δ_w^f) computed at each temperature became the initial composition of the water (δ_w^i) for the calculations at the next higher temperature. Since water:rock mass ratios may vary according to the convection path, the series of calculations were performed at R values of 0.01, 0.1, 1, and 10.

The results are presented in Figure 6.16 and compared to compositions of the fluids responsible for those volcanic-hosted (Fig. 6.16A) and sediment-hosted (Fig. 6.16B) epithermal deposits listed in Table 6.6. The computations imply that convecting meteoric water is progressively enriched in ^{18}O and D during heating and exchange with unaltered wall rock. This water will be isotopically heavier than non-convecting meteoric water (e.g. pore fluid), under identical conditions of equilibration, because of its "history" of exchange (compare Figs. 6.15 and 6.16). Moreover,

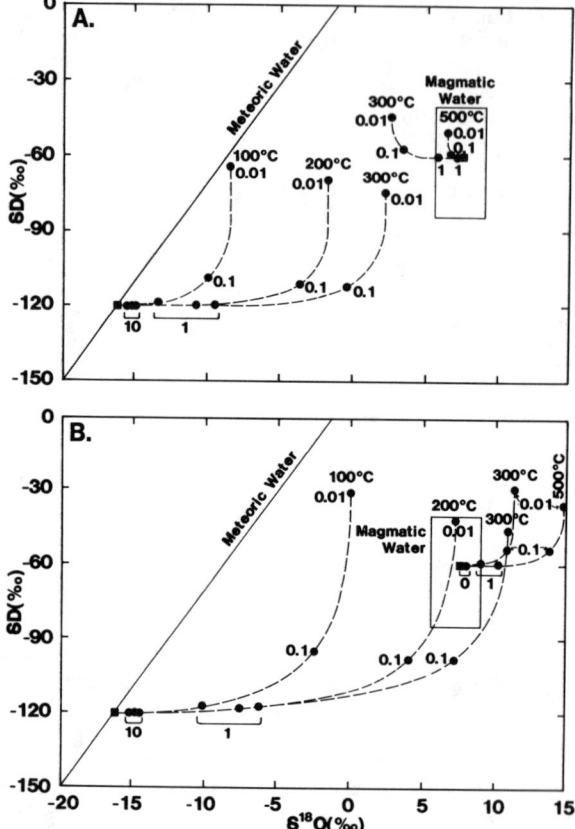

Figure 6.15. Variations of δD and δ^{18}O in fluids that equilibrate with (A) volcanic and (B) sedimentary rock as a function of the initial fluid composition, temperature, and water:rock ratio.

isotopically light meteoric water may attain compositions similar to those of magmatic or evolved magmatic water through exchange at low water:rock mass ratios ($\lesssim$0.1). Therefore, the distinction between a magmatic or meteoric source of fluids and the demonstration of mixing between magmatic and meteoric waters cannot be made solely on the basis of the calculated composition of a hydrothermal fluid.

The oxygen-isotopic composition of meteoric water in the Basin and Range province has remained essentially unchanged since the Early Tertiary, whereas the hydrogen-isotopic composition may have decreased slightly (10-20 ^{o}/oo) in response to the climatic cooling (Sheppard et al., 1969; Taylor, 1973). Thus, the isotopic composition of meteoric water in this region may be regarded as an approximation of the initial composition of Tertiary meteoric hydrothermal fluids. Meteoric water compositions near the Aurora, Rawhide, Gilbert, Tenmile, Humboldt, Cortez and Manhattan epithermal deposits have δD values between -130 and -120 ^{o}/oo (Table 6.6). Accordingly, the isotopic compositions of the hydrothermal fluids for these deposits can be interpreted in terms of the calculated curves displayed in Figure 6.16. Isotopic data for the remaining deposits in the Basin and Range province (Table 6.6) are more appropriately compared to calculated curves that have been shifted, relative to those in Figure 6.16, in the direction and to the extent that the initial compositions of the fluids differed from -16 ‰ $\delta^{18}O$ and -120 ‰ δD. Assuming that temperatures of water-rock equilibration were typically between 150°C and 300°C (hachured fields in Fig. 6.16), as indicated by fluid-inclusion studies, then most of the epithermal systems were characterized by high water:rock mass ratios (>0.5). However, the model calculations imply that the Bodie, Tonopah, and Tenmile fluids evolved at unusually low water:rock mass ratios (∼0.01-0.2) and temperatures (≤100°C). Furthermore, the model cannot account for the apparent decrease in the δD composition of the Carlin and Comstock Lode fluids.

Most of the water:rock ratios determined above should be regarded as tentative (particularly the anomalously low values) because processes other than fluid-rock exchange may have influenced the isotopic composition of the fluids. For example, boiling in the convection system would enrich the evolved meteoric fluid in ^{18}O and D along a trend subparallel to the MWL (Truesdell et al., 1977; see Radtke et al., 1980). Unexchanged meteoric water may become involved at the site of mineralization through mixing or with the complete equilibration (alteration) of the host rocks. This effect would drive the fluid composition towards the initial $\delta^{18}O$ and δD values. Inaccurate water:rock ratios may also result if the isotopic composition of modern meteoric water is used as an approximation of Tertiary meteoric water in areas that have undergone considerable uplift or subsidence. Similarly, δD analyses of extracted inclusion fluids that contained a significant fraction of secondary fluids could lead to erroneous conclusions (see Foley et al., 1982).

The final isotopic composition of the rock (δ_r^f) that equilibrated with the meteoric fluid at each interval in the convection model was calculated from the relationship

$$\delta_r^f = \delta_w^f + \Delta_{r-w} \qquad (19)$$

and portrayed as curves in Figure 6.17. These results indicate that rocks become progressively depleted in ^{18}O and D with an increase in the water:rock mass ratio. The depletion in ^{18}O is negligible at low ratios because of the overwhelming abundance of rock oxygen. With increasing temperature, both the $\delta^{18}O$ and δD values increase at water:rock mass ratios of 1 or less, whereas at a ratio of 10 the $\delta^{18}O$ value decreases and the δD value increases.

Decreases in the $\delta^{18}O$ values of hydrothermally altered rocks near centers of hydrothermal activity have been documented for the Tonopah epithermal Au-Ag deposit (Taylor, 1973), zoned polymetallic veins of the Bohemia district in the Western Cascades (Taylor, 1971), Yankee Fork and other altered and mineralized localities of the Idaho Batholith (Criss and Taylor, 1983; Criss et al., 1985), and several volcanogenic massive sulfide deposits (see Franklin et al., 1981). The whole rock $\delta^{18}O$ compositions at Tonopah range from values typical of unaltered igneous rocks (5.5 to 10.0 ‰) to nearly -6 permil and whole rock δD compositions range from -150 to -135 permil. The field represented by six Tonopah samples for which both δD and $\delta^{18}O$ analyses have been reported is shown in Figure 6.17. Compositions of these samples are most consistent with isotopic exchange at high water:rock mass ratios (>1), although the δD values are somewhat larger than predicted. On the basis of whole-rock ^{18}O depletions observed at Tonopah and elsewhere, water:rock mass ratios for many epithermal districts throughout the western U.S. range from about 0.2 to 2 (Taylor, 1974a) and these ratios are broadly similar to those of geothermal systems that range from about 0.15 at the Geysers (Sternfeld, 1981) through 0.45 and 1.3 at Salton Sea (Clayton et al., 1968) and Cerro Prieto (Williams and Elders, 1984), to as large as 4.3 at Wairakei (Clayton and Steiner, 1975). The outer haloes of ^{18}O depletion extend beyond the megascopically identifiable effects of alteration and anomalies of trace and minor elements (Taylor, 1971; Green et al., 1983).

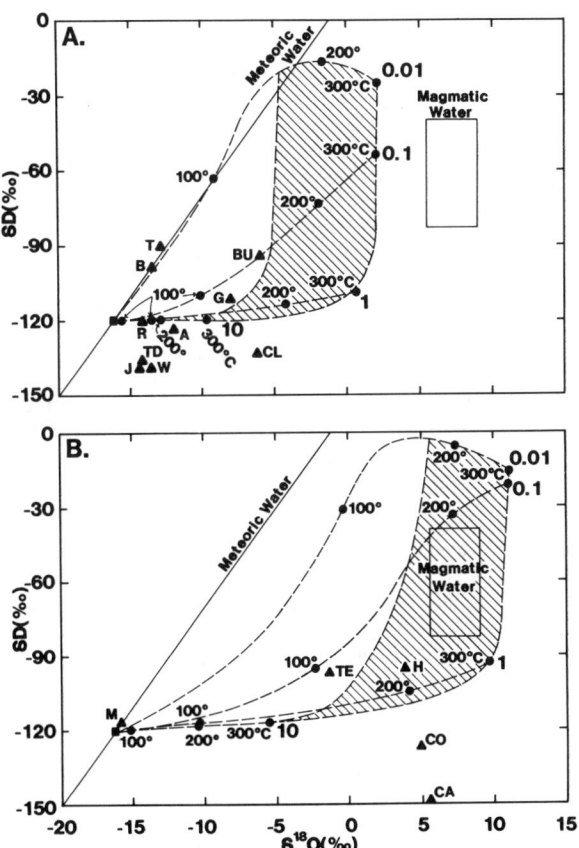

Figure 6.16. Variations of δD and $\delta^{18}O$ in fluids during convection through (A) volcanic and (B) sedimentary rock as a function of temperature and water:rock ratio.

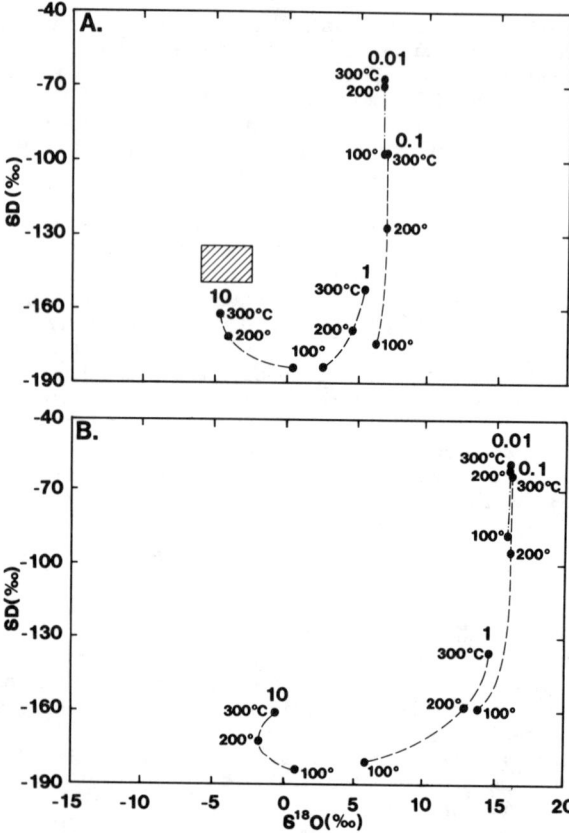

Figure 6.17. Variations of δD and $\delta^{18}O$ in (A) volcanic and (B) sedimentary rocks that have equilibrated with convected meteoric water over a range of temperatures and water:rock ratios.

Accordingly, these cones of ^{18}O depletion may be potentially useful in the exploration for hidden mineral deposits (see Criss and Taylor, 1983; Criss et al., 1985).

Two main conclusions pertaining to epithermal systems may be summarized from this discussion on the exchange of oxygen and hydrogen isotopes between fluids and rock. First, meteoric water is the predominant, and possibly only, source of fluid in most of the epithermal gold-silver deposits studied to date. Second, at reasonable temperatures of 150°C to 300°C, the isotopic data for hydrothermal fluids and their associated altered rocks both imply that high (≥ 1) water:rock mass ratios prevailed during mineral deposition. Because typical porosities limit bulk water:rock mass ratios to generally less than 0.1 (e.g. 10% porosity is equivalent to a ratio of 0.04), then fluid-rock equilibration and mineral deposition must have occurred in open systems through which masses of fluid circulated that were equivalent to or larger than those of the rocks.

SUMMARY

This overview of light-stable isotopes in the epithermal environment is prefaced by a review of the general principles of equilibrium isotope-exchange fractionation, including the equations and graphical portrayal of fractionation effects between common isotopic compounds, and a summary of D, ^{13}C, ^{18}O, and ^{34}S distributions in geologically important habitats. The available data for these isotopes in geothermal and epithermal systems are largely consistent with trends that might be inferred from experimental-theoretical considerations and known distributions in common rock types and mineral deposits. Nonetheless, the epithermal data exhibit isotopic characteristics that are intermediate between those of near-surface geothermal systems and deeper hydrothermal deposits. Depletions of D and ^{18}O in the host rocks, minerals, and inclusion fluids of the epithermal deposits, and compositional similarities of these fluids to present-day meteoric waters, suggest that the hydrothermal fluids were predominantly ground waters of meteoric origin; although several deposits (Casapalca, Golden Sunlight, and possibly Creede) may have had varying amounts of a magmatic component. Sources of carbon in the carbonate minerals are not uniquely defined, and probably have originated from nearby sedimentary host rocks (Carlin and Cortez), magmas, and possibly from biogenic and other provenances. Those of sulfur are considered to have been derived from magmas or igneous rocks, sulfates (Sunnyside and Tolfa) and sulfides (Carlin, Cortez, and Guanajuato) of sedimentary rocks, and sea water (Pueblo Viejo and Tui). Temperature-controlled fractionations over lateral or vertical gradients ranging from 300°C to 100°C outward and upward in hydrothermal systems should result in heavy-isotope enrichments of about 12 ‰ ^{13}C for calcite, 14 ‰ ^{18}O for quartz, 12 ‰ ^{18}O for calcite, 26 ‰ ^{34}S for sulfates, and 1.7 ‰ ^{34}S for pyrite. Although these "hypothetical" fractionation trends have not been reported from any of the epithermal deposits studied to date, they have been noted for ^{13}C and ^{18}O in calcite and quartz of geothermal systems (Broadlands, Cerro Prieto, Geysers, Salton Sea, and Wairakei). Their absence from epithermal deposits may possibly be the result of inadequate sample representation, or of isotopic changes in fluid composition caused by boiling, redox and (or) water-rock reactions, disequilibrium, and different sources of the isotopic elements. Near-surface occurrences of supergene sulfates, which have formed by oxidation of ascending H_2S or of preexisting sulfides in the host rocks, are markedly depleted in ^{34}S relative to the deeper hypogene sulfates. The altered sedimentary and volcanic host rocks of both epithermal deposits and geothermal systems are conspicuously depleted in ^{18}O relative to their peripheral and unaltered equivalents. These ^{18}O depletions are a corollary of the well-known "oxygen isotope shift" of geothermal fluids, and they form in hydrothermal systems characterized by relatively high water:rock mass ratios (≥ 1). The resultant "negative" ^{18}O anomalies may serve as a useful guide to mineral exploration because they are isotopically unambiguous and areally extensive. Isotopic investigations of epithermal deposits should be continued, especially in conjunction with geologic and other topical studies, and particular emphasis should be given to systematic and three-dimensional sampling of host rocks beyond ore-bearing structures.

REFERENCES

Bachinski, D. J., 1969, Bond strength and sulfur isotope fractionation in coexisting sulfides: Economic Geology, v. 64, p. 56-65.

Bender, M. L., 1972, Carbon isotope fractionation; in Fairbridge, R. W. (ed.), The Encyclopedia of Geochemistry and Environmental Sciences: Van Nostrand Reinhold Company, New York, p. 133-136.

Bethke, P. M., Barton, P. B., and Rye, R. O., 1973, Hydrogen, oxygen, and sulfur isotopic compositions of ore fluids in the Creede district, Mineral County, Colorado abs.: Economic Geology, v. 68, p. 1205.

Bethke, P. M., and Rye, R. O., 1979, Environment of ore deposition in the Creede mining district, San Juan Mountains, Colorado--Part IV, Source of fluids from oxygen, hydrogen, and carbon isotope studies: Economic Geology, v. 74, p. 1832-1851.

Blattner, P., 1975, Oxygen isotopic composition of fissure-grown quartz, adularia, and calcite from Broadlands geothermal field, New Zealand: American Journal of Science, v. 275, p. 785-800.

Bottinga, Y., and Javoy, M., 1973, Comments on oxygen isotope geothermometry: Earth and Planetary Science Letters, v. 20, p. 250-265.

Brigham, R. H., 1984, K-feldspar genesis and stable isotope relations of the Papoose Flat Pluton, Inyo Mountains, California: Unpublished Ph.D. dissertation, Stanford University, 172 p.

Browne, P. R. L., Rafter, T. A., and Robinson, B. W., 1975, Sulphur isotope ratios of sulphides from the Broadlands geothermal field, New Zealand: New Zealand Journal of Science, v. 18, p. 35-40.

Casadevall, T., and Ohmoto, H., 1977, Sunnyside mine, Eureka mining district, San Juan County, Colorado--Geochemistry of gold and base-metal ore deposition in a volcanic environment: Economic Geology, v. 72, p. 1285-1320.

Cathles, L. M., 1983, An analysis of the hydrothermal system responsible for massive sulfide deposition in the Hokuroku Basin of Japan; in Ohmoto, H., and Skinner, B. J. (eds.), The Kuroko and Related Volcanogenic Massive Sulfide Deposits: Economic Geology, Monograph 5, p. 439-487.

Claypool, G. E., Holser, W. T., Kaplan, I. R., Sakai, H., and Zak, I., 1980, The age curves of sulfur and oxygen isotopes in marine sulfate and their mutual interpretation: Chemical Geology, v. 28, p. 199-260.

Clayton, R. N., Muffler, L. J. P., and White, D. E., 1968, Oxygen isotope study of calcite and silicates of the River Ranch No. 1 well, Salton Sea geothermal field, California: American Journal of Science, v. 266, p. 968-979.

Clayton, R. N., and Steiner, A., 1975, Oxygen isotope studies of the geothermal system at Wairakei, New Zealand: Geochimica et Cosmochimica Acta, v. 39, p. 1179-1186.

Cortecci, G., Lombardi, G., Reyes, E., and Turi, B., 1981, A sulfur isotopic study of alunites from Latium and Tuscany, central Italy: Mineralium Deposita, v. 16, p. 147-156.

Craig, H., 1953, The geochemistry of stable carbon isotopes: Geochimica et Cosmochimica Acta, v. 3, p. 53-92.

Craig, H., 1966, Isotopic composition and origin of the Red Sea and Salton Sea geothermal brines: Science, v. 154, p. 1544-1548.

Criss, R. E., Champion, D. E., and McIntyre, D. H., 1985, Oxygen isotope, aeromagnetic, and gravity anomalies associated with hydrothermally altered zones in the Yankee Fork mining district, Custer County, Idaho: Economic Geology, v. 80, p. 1277-1296.

Criss, R. E., and Taylor, H. P., Jr., 1983, An $^{18}O/^{16}O$ and D/H study of Tertiary hydrothermal systems in the southern half of the Idaho Batholith: Geological Society of America Bulletin, v. 94, p. 640-663.

Dean, J. A. (ed.), 1979, Lango's Handbook of Chemistry, Twelfth Edition, McGraw-Hill Book Company.

Faure, G., 1977, Principles of Isotope Geology: John Wiley and Sons, New York, 464 p.

Feely, H. W., and Kulp, J. L., 1957, The origin of Gulf Coast salt dome sulfur deposits: Bulletin of the American Association of Petroleum Geologists, v. 41, p. 1802-1853.

Field, C. W., 1966, Sulfur isotopic method for discriminating between sulfates of hypogene and supergene origin: Economic Geology, v. 61, p. 1428-1435.

Field, C. W., 1972, Isotope geology--stable; in Fairbridge, R. W. (ed.), The Encyclopedia of Geochemistry and Environmental Sciences: Van Nostrand Reinhold Company, p. 618-622.

Field, C. W., Dymond, J. R., Heath, G. R., Corliss, J. B., and Dasch, E. J., 1976, Sulfur isotope reconnaissance of epigenetic pyrite in ocean-floor basalts, Leg 34 and elsewhere; in Yeats, R. S., Hart, S. R. et al. (eds.), 1976: Initial Reports of the Deep Sea Drilling Project, v. XXXIV, Washington, D.C. (U.S. Government Printing Office), p. 381-384.

Field, C. W., and Gustafson, L. B., 1976, Sulfur isotopes in the porphyry copper deposit at El Salvador, Chile: Economic Geology, v. 71, p. 1533-1548.

Field, C. W., and Lombardi, G., 1972, Sulfur isotopic evidence for the supergene origin of alunite deposits, Tolfa district, Italy: Mineralium Deposita, v. 7, p. 113-125.

Field, C. W., and Power, S. G., 1985, Metallization in the Western Cascades, Oregon and southern Washington abs.: Geological Society of America, Abstracts with Programs, v. 17, no. 4, p. 218.

Field, C. W., Rye, R. O., Dymond, J. R., Whelan, J. F., and Senechal, R. G., 1983, Metalliferous sediments of the East Pacific; in Shanks, W. C., III (ed.), Cameron Volume on Unconventional Mineral Deposits: Society of Mining Engineers, New York, p. 133-156.

Field, C. W., Sakai, H., and Ueda, A., 1984, Isotopic constraints on the origin of sulfur in oceanic igneous rocks; in Wauschkuhn, A., Kluth, C., and Zimmerman, R. A. (eds.), Syngenesis and Epigenesis in the Formation of Mineral Deposits: Springer-Verlag, Berlin-Heidelberg, p. 573-589.

Fifarek, R. H., 1985, Alteration geochemistry, fluid inclusion, and stable isotope study of the Red Ledge volcanogenic massive sulfide deposit, Idaho: Unpublished Ph.D. dissertation, Oregon State University, 187 p.

Foley, N. K., Bethke, P. M., and Rye, R. O., 1982, A reinterpretation of δD_{H_2O} values of inclusion fluids in quartz from shallow ore bodies abs.: Geological Society of America, Abstracts with Programs, v. 14, no. 7, p. 489-490.

Franklin, J. M., Sangster, D. M., and Lydon, J. W., 1981, Volcanic-associated massive sulfide deposits: Economic Geology, Seventy-Fifth Anniversary Volume, p. 485-627.

Friedman, I., and O'Neil, J. R., 1977, Compilation of stable isotope fractionation factors of geochemical interest; in Fleischer, M. (ed.), Data of Geochemistry, Sixth Edition: U.S. Geological Survey, Professional Paper 440-KK, p. KK1-KK12.

Fuex, A. N., and Baker, D. R., 1973, Stable carbon isotopes in selected granitic, mafic, and ultramafic rocks: Geochimica et Cosmochimica Acta, v. 37, p. 2509-2521.

Garlick, G. D., 1972, Oxygen isotope geochemistry; in Fairbridge, R. W. (ed.), The Encyclopedia of Geochemistry and Environmental Sciences: Van Nostrand Reinhold Company, New York, p. 864-874.

Graham, C. M., Harmon, R. S., and Sheppard, S. M. F., 1984, Experimental hydrogen isotope studies--hydrogen isotope exchange between amphibole and water: American Mineralogist, v. 69, p. 128-138.

Graham, C. M., Sheppard, S. M. F., and Heaton, T. H. E., 1980, Experimental hydrogen isotope studies--I. systematics of hydrogen isotope fractionation in the systems epidote-H_2O, zoisite-H_2O, and AlO(OH)-H_2O: Geochimica et Cosmochimica Acta, v. 44, p. 353-364.

Green, G. R., Ohmoto, H., Date, J., and Takahashi, T., 1983, Whole-rock oxygen isotope distribution in the Fukazawa-Kosaka area, Hokuroku district, Japan, and its potential application to mineral exploration: Economic Geology, Seventy-Fifth Anniversary Volume, p. 395-411.

Gross, W. H., 1975, New ore discovery and source of silver-gold veins, Guanajuato, Mexico: Economic Geology, v. 70, p. 1175-1189.

Hayba, D. O., Bethke, P. M., Heald, P., and Foley, N. K., 1985, Geologic, mineralogic, and geochemical characteristics of volcanic-hosted epithermal precious-metal deposits; in Berger, B. R., and Bethke, P. M. (eds.), Geology and Geochemistry of Epithermal Systems: Society of Economic Geologists, Reviews in Economic Geology, v. 2.

Hoefs, J., 1980, Stable Isotope Geochemistry, Second Edition: Springer-Verlag, Berlin-Heidelberg, New York, 208 p.

Javoy, M., 1977, Stable isotopes and geothermometry: Journal of the Geological Society of London, v. 133, p. 609-636.

Jensen, M. L., Ashley, R. P., and Albers, J. P., 1971, Primary and secondary sulfates at Goldfield, Nevada: Economic Geology, v. 66, p. 618-626.

Jensen, M. L., Field, C. W., and Nakai, N., 1960, Sulfur isotopes and the origin of sandstone-type uranium deposits: Biennial Progress Report for 1959-1960, U.S. Atomic Energy Commission Contract AT(30-1)-2261, 281 p.

Kamilli, R. J., and Ohmoto, H., 1977, Paragenesis, zoning, fluid inclusion, and isotope studies of the Finlandia vein, Colqui district, central Peru: Economic Geology, v. 72, p. 950-982.

Kesler, S. E., Russell, N., Seaward, M., Rivera, J., McCurdy, K., Cumming, G. L., and Sutter, J. F., 1981, Geology and geochemistry of sulfide mineralization underlying the Pueblo Viejo gold-silver oxide deposit, Dominican Republic: Economic Geology, v. 76, p. 1096-1117.

Kolodny, Y., and Epstein, S., 1976, Stable isotope geochemistry of deep sea cherts: Geochimica et Cosmochimica Acta, v. 40, p. 1195-1209.

Kulla, J. B., and Anderson, T. F., 1978, Experimental oxygen isotope fractionation between kaolinite and water; in Zartman, R. E. (ed.), Short Papers of the Fourth International Conference, Geochronology, Cosmochronology, Isotope Geology 1978: U.S. Geological Survey, Open-File Report 78-701, p. 234-235.

Kusakabe, M., and Chiba, H., 1983, Oxygen and sulfur isotope composition of barite and anhydrite from the Fukazawa deposit, Japan; in Ohmoto, H., and Skinner, B. J. (eds.), The Kuroko and Related Volcanogenic Massive Sulfide Deposits: Economic Geology, Monograph 5, p. 292-301.

Lea, D. W., Larson, P. B., and Taylor, H. P., Jr., 1984, Oxygen isotope and fluid inclusion study of veins and wallrocks in the Eureka Graben, San Juan Mountains, Colorado abs.: Geological Society of America, Abstracts with Programs, v. 16, no. 6, p. 572.

Lombardi, G., and Sheppard, S. M. F., 1977, Petrographic and isotopic studies of the altered acid volcanics of the Tolfa-Cerite area, Italy--the genesis of the clays: Clay Minerals, v. 12, p. 147-162.

Magaritz, M., Whitford, D. J., and James, D. E., 1978, Oxygen isotopes and the origin of high $^{87}Sr/^{86}Sr$ andesites: Earth and Planetary Science Letters, v. 40, p. 220-230.

Matsuhisa, Y., Goldsmith, J. R., and Clayton, R. N., 1979, Oxygen isotopic fractionation in the system quartz-albite-anorthite-water: Geochimica et Cosmochimica Acta, v. 43, p. 1131-1140.

Nockolds, S. R., Knox, R. W. O'B., and Chinner, G. A., 1978, Petrology: Cambridge University Press, New York, 435 p.

Ohmoto, H., 1972, Systematics of sulfur and carbon isotopes in hydrothermal ore deposits: Economic Geology, v. 67, p. 551-578.

Ohmoto, H., and Lasaga, A. C., 1982, Kinetics of reactions between aqueous sulfates and sulfides in hydrothermal systems: Geochimica et Cosmochimica Acta, v. 46, p. 1727-1745.

Ohmoto, H., and Rye, R. O., 1974, Hydrogen and oxygen isotopic compositions of fluid inclusions in the Kuroko deposits, Japan: Economic Geology, v. 69, p. 947-953.

Ohmoto, H., and Rye, R. O., 1979, Isotopes of sulfur and carbon; in Barnes, H. L. (ed.), Geochemistry of Hydrothermal Ore Deposits, Second Edition: John Wiley and Sons, New York, p. 509-567.

O'Neil, J. R., 1977, Stable isotopes in mineralogy: Physics and Chemistry of Minerals, v. 2, p. 105-123.

O'Neil, J. R., and Kharaka, Y. K., 1976, Hydrogen and oxygen isotope exchange reactions between clay minerals and water: Geochimica et Cosmochimica Acta, v. 40, p. 241-246.

O'Neil, J. R., and Silberman, M. L., 1974, Stable isotope relations in epithermal Au-Ag deposits: Ecoomic Geology, v. 69, p. 902-909.

O'Neil, J. R., Silberman, M. L., Fabbi, B. P., and Chesterman, C. W., 1973, Stable isotope and chemical relations during mineralization in the Bodie mining district, Mono County, California: Economic Geology, v. 68, p. 765-784.

O'Neil, J. R., and Taylor, H. P., Jr., 1967, The oxygen isotope and cation exchange chemistry of feldspars: American Mineralogist, v. 52, p. 1414-1437.

Osatenko, M. J., and Jones, M. B., 1976, Valley Copper; in Brown, A. S. (ed.), Porphyry Deposits of the Canadian Cordillera: The Canadian Institute of Mining and Metallurgy, Special Volume 15, p. 130-143.

Porter, E. W., and Ripley, E., 1985, Petrologic and stable isotope study of the gold-bearing breccia pipe at the Golden Sunlight deposit, Montana: Economic Geology, v. 80, p. 1689-1706.

Power, S. G., 1985, The "tops" of porphyry copper deposits--mineralization and plutonism in the Western Cascades, Oregon: Unpublished Ph.D. dissertation, Oregon State University, 234 p.

Radtke, A. S., Rye, R. O., and Dickson, F. W., 1980, Geology and stable isotope studies of the Carlin gold deposit, Nevada: Economic Geology, v. 75, p. 641-672.

Robinson, B. W., 1974, The origin of mineralization at the Tui mine, Te Aroha, New Zealand, in the light of stable isotope studies: Economic Geology, v. 69, p. 910-925.

Rye, R. O., Doe, B. R., and Wells, J. D., 1974, Stable isotope and lead isotope study of the Cortez, Nevada, gold deposit and surrounding area: U.S. Geological Survey, Journal of Research, v. 2, p. 13-23.

Rye, R. O., and Ohmoto, H., 1974, Sulfur and carbon isotopes and ore genesis--a review: Economic Geology, v. 69, p. 826-842.

Rye, R. O., and Sawkins, F. J., 1974, Fluid inclusion and stable isotope studies on the Casapalca Ag-Pb-Zn-Cu deposit, central Andes, Peru: Economic Geology, v. 69, p. 181-205.

Sakai, H., 1968, Isotopic properties of sulfur compounds in hydrothermal processes: Geochemical Journal (Japan), v. 2, p. 29-49.

Sakai, H., Des Marais, D. J., Ueda, A., and Moore, J. G., 1984, Concentrations and isotope ratios of carbon, nitrogen, and sulfur in ocean-floor basalts: Geochimica et Cosmochimica Acta, v. 48, p. 2433-2441.

Sakai, H., Gunnlaugsson, E., Tomasson, J., and Rouse, J. E., 1980, Sulfur isotope systematics in Icelandic geothermal systems and influence of seawater circulation at Reykjanes: Geochimica et Cosmochimica Acta, v. 44, p. 1223-1231.

Sangster, D. F., 1968, Relative sulphur isotope abundances of ancient seas and stratabound sulphide deposits: Proceedings of the Geological Association of Canada, v. 17, p. 79-91.

Savin, S. M., and Epstein, S., 1970a, The oxygen and hydrogen isotope geochemistry of clay minerals: Geochimica et Cosmochimica Acta, v. 34, p. 25-42.

Savin, S. M., and Epstein, S., 1970b, The oxygen and hydrogen isotope geochemistry of ocean sediments and shales: Geochimica et Cosmochimica Acta, v. 34, p. 43-63.

Sawkins, F. J., 1972, Sulfide ore deposits in relation to plate tectonics: Journal of Geology, v. 80, p. 377-397.

Schoen, R., and Rye, R. O., 1970, Sulfur isotope distribution in solfataras, Yellowstone National Park: Science, v. 170, p. 1082-1084.

Sheppard, S. M. F., Nielsen, R. L., and Taylor, H. P., Jr., 1969, Oxygen and hydrogen isotope ratios of clay minerals from porphyry copper deposits: Economic Geology, v. 64, p. 755-777.

Sheppard, S. M. F., Nielsen, R. L., and Taylor, H. P., Jr., 1971, Hydrogen and oxygen isotope ratios in minerals from porphyry copper deposits: Economic Geology, v. 66, p. 515-542.

Spooner, E. T. C., Beckinsale, R. D., England, P. C., and Senior, A., 1977, Hydration, ^{18}O enrichment and oxidation during ocean floor hydrothermal metamorphism of ophiolite metabasic rocks from E. Liguria, Italy: Geochimica et Cosmochimica Acta, v. 41, p. 857-871.

Steiner, A., and Rafter, T. A., 1966, Sulfur isotopes in pyrite, pyrrhotite, alunite, and anhydrite from steam wells in the Taupo Volcanic Zone, New Zealand: Economic Geology, v. 61, p. 1115-1129.

Sternfeld, J. N., 1981, The hydrothermal petrology and stable isotope geochemistry of two wells in the Geysers geothermal field, Sonoma County, California: Unpublished M.S. thesis, University of California (Riverside), 202 p.

Styrt, M. M., Brackman, A. J., Holland, H. D., Clark, B. C., Pisutha-Arnond, V., Eldridge, C. S., and Ohmoto, H., 1981, The mineralogy and the isotopic composition of sulfur in hydrothermal sulfide/sulfate deposits of the East Pacific Rise, $21°N$ latitude: Earth and Planetary Science Letters, v. 53, p. 382-390.

Suzuoki, T., and Epstein, S., 1976, Hydrogen isotope fractionation between OH-bearing minerals and waters: Geochimica et Cosmochimica Acta, v. 40, p. 1229-1240.

Taylor, H. P., Jr., 1967, Oxygen isotope studies of hydrothermal mineral deposits; in Barnes, H. L. (ed.), Geochemistry of Hydrothermal Ore Deposits: Holt, Rinehart and Winston, Inc., p. 109-142.

Taylor, H. P., Jr., 1971, Oxygen isotopic evidence for large-scale interaction between meteoric ground waters and Tertiary granodiorite intrusions, Western Cascade Range, Oregon: Journal of Geophysical Research, v. 76, p. 7855-7874.

Taylor, H. P., Jr., 1973, O^{18}/O^{16} evidence for meteoric-hydrothermal alteration and ore deposition in the Tonopah, Comstock Lode, and Goldfield mining districts, Nevada: Economic Geology, v. 68, p. 747-764.

Taylor, H. P., Jr., 1974a, The application of oxygen and hydrogen isotope studies to problems of hydrothermal alteration and ore deposition: Economic Geology, v. 69, p. 843-883.

Taylor, H. P., Jr., 1974b, Oxygen and hydrogen isotope evidence for large-scale circulation and interaction between ground waters and igneous intrusions, with particular reference to the San Juan volcanic field, Colorado; in Hoffman, A. W., Giletti, B. J., Yoder, H. S., Jr., and Yund, R. A. (eds.), Geochemical Transport and Kinetics: Carnegie Institution of Washington Publication 634, p. 299-323.

Taylor, H. P., Jr., 1979, Oxygen and hydrogen isotope relationships in hydrothermal mineral deposits; in Barnes, H. L. (ed.), Geochemistry of Hydrothermal Ore Deposits, Second Edition: John Wiley and Sons, New York, p. 236-277.

Taylor, H. P., Jr., and Turi, B., 1976, High ^{18}O igneous rocks from the Tuscan magmatic province, Italy: Contributions to Mineralogy and Petrology, v. 55, p. 33-54.

Thode, H. G., Wanless, R. K., and Wallouch, R., 1954, The origin of native sulfur deposits from isotope fractionation studies: Geochimica et Cosmochimica Acta, v. 5, p. 288-298.

Truesdell, A. H., 1974, Oxygen isotope activities and concentrations in aqueous salt solutions at elevated temperatures--consequences for isotope geochemistry: Earth and Planetary Science Letters, v. 23, p. 387-396.

Truesdell, A. H., and Hulston, J. R., 1980, Isotopic evidence on environments of geothermal systems; in Fritz, P., and Fontes, J. Ch. (eds.), Handbook of Environmental Isotope Geochemistry: Elsevier Scientific Publishing Company, New York, p. 179-226.

Truesdell, A. H., Nathenson, M., and Rye, R. O., 1977, The effects of subsurface boiling and dilution on the isotopic compositions of Yellowstone thermal waters: Journal of Geophysical Research, v. 82, p. 3694-3704.

Truesdell, A. H., Rye, R. O., Whelan, J. F., and Thompson, J. M., 1978, Sulfate chemical and isotopic patterns in thermal waters of Yellowstone Park, Wyoming; in Zartman, R. E. (ed.), Short Papers of the Fourth International Conference, Geochronology, Cosmochronology, Isotope Geology: U.S. Geological Survey, Open-File Report 78-701, p. 435-436.

Veizer, J., and Hoefs, J., 1976, The nature of O^{18}/O^{16} and C^{13}/C^{12} secular trends in sedimentary carbonate rocks: Geochimica et Cosmochimica Acta, v. 40, p. 1387-1395.

Watanabe, M., and Sakai, H., 1983, Stable isotope geochemistry of sulfates from the Neogene ore deposits in the green tuff region, Japan; in Ohmoto, H., and Skinner, B. J. (eds.), The Kuroko and Related Volcanogenic Massive Sulfide Deposits: Economic Geology, Monograph 5, p. 282-291.

White, D. E., 1968, Environments of generation of some base-metal ore deposits: Economic Geology, v. 63, p. 301-335.

White, D. E., 1974, Diverse origins of hydrothermal ore fluids: Economic Geology, v. 69, p. 954-973.

White, D. E., 1981, Active geothermal systems and hydrothermal ore deposits: Economic Geology, Seventy-Fifth Anniversary Volume, p. 392-423.

White, D. E., Barnes, I., and O'Neil, J. R., 1973, Thermal and mineral waters of nonmeteoric origin, California coast ranges: Geological Society of America Bulletin, v. 84, p. 547-560.

Williams, A. E., and Elders, W. A., 1984, Stable isotope systematics of oxygen and carbon in rocks and minerals from the Cerro Prieto geothermal anomaly, Baja, California, Mexico: Geothermics, v. 13, p. 49-63.

Yapp, C. J., and Pedley, M. D., 1985, Stable hydrogen isotopes in iron oxides--II. D/H variations among natural goethites: Geochimica et Cosmochimica Acta, v. 49, p. 487-495.

Chapter 7
GEOLOGIC, MINERALOGIC, AND GEOCHEMICAL CHARACTERISTICS OF VOLCANIC-HOSTED EPITHERMAL PRECIOUS-METAL DEPOSITS

Daniel O. Hayba, Philip M. Bethke
Pamela Heald, and Nora K. Foley

INTRODUCTION

In Chapter 1, R. W. Henley summarized our understanding of the chemical and hydrodynamic structure and the transport properties of active hydrothermal systems, with particular emphasis on terrestrial magmatic-hydrothermal systems. Such an overview is especially valuable because active geothermal systems are modern "archetypes" of the ancient systems which concentrated metals in their upper portions to form epithermal ore deposits. More than any other factor, the study of active systems has provided the framework on which the observations on epithermal deposits have been arranged in the relatively recent development of comprehensive models of epithermal ore formation. The Principle of Uniformitarianism has served us well in this instance.

In this chapter, we focus on observations on epithermal ore deposits in continental silicic to andesitic volcanic terranes. Volcanic-hosted deposits offer the most direct comparison with many of the well-studied modern geothermal systems. We first compare the attributes from a number of epithermal ore deposits and show how they may be used to identify two important, and distinct volcanic-related hydrothermal environments. We then examine the best-studied deposit of each type: Creede and Summitville, both of which are located in the San Juan Mountains in southwest Colorado. In so doing, we are able to examine epithermal deposits for evidences of processes that are now occurring in geothermal systems. Finally, we use the observational base and interpretations derived from each deposit type to develop generalized "geothermal" models of mineralization. The models have been taken, in large part, from the excellent synthesis by Henley and Ellis (1983). We feel that their models are soundly based on a myriad of direct observations on active geothermal systems, and are consistent, for the most part, with the observations made at Creede and Summitville.

SUMMARY OF THE CHARACTERISTICS OF VOLCANIC-HOSTED EPITHERMAL ORE DEPOSITS

In 1981, Buchanan published a valuable compilation of selected observations from over 60 gold-silver vein deposits in unmetamorphosed volcanic-to-subvolcanic environments. These data and the integrated model derived from them have formed a useful basis for numerous subsequent analyses of the characteristics of epithermal deposits, some of which include Giles and Nelson (1982), Ashley (1982), Bonham and Giles (1983), Heald-Wetlaufer et al. (1983), Berger and Eimon (1983), Sillitoe and Bonham (1984), Heald et al. (1986), and Bonham (1986). Because of the large number of deposits in Buchanan's 1981 study, a detailed evaluation of the data base was not possible. Heald et al. (1986) undertook such a detailed study of 16 carefully selected, well-studied, Tertiary volcanic-hosted epithermal deposits. Using this data base, Heald et al. (1986) showed that two types of epithermal deposits could be distinguished and that Buchanan's data base supplemented and supported this conclusion. The two types, distinguished primarily on the basis of vein and alteration mineralogies, are the Adularia-Sericite type and the Acid-Sulfate type. The characteristic features of these two main types are shown in Table 7.1 and are discussed below. Adularia-Sericite-type deposits are far more numerous than the Acid-Sulfate-type deposits (Table 7.2), and so predictably, there is more information on the former.

Major telluride deposits (e.g., Cripple Creek, Colorado) were excluded in the study by Heald et al. (1986) because their unique mineral assemblage suggests that these deposits make up a distinct class (Heald-Wetlaufer et al., 1983; Bonham and Giles, 1983; Bonham, 1986). Hot-spring-type deposits (see Silberman and Berger, 1985, this volume; Berger and Silberman, 1985, this volume) were not well represented in their compilation because of a paucity of published descriptive data.

Characteristics of Adularia-Sericite-type Deposits

Structural setting--The most common regional structural setting for Adularia-Sericite-type deposits is along the margins of calderas, although other tectonic settings (typically structurally complex volcanic environments) are not uncommon. The importance of the caldera setting lies in the excellent plumbing system it provides for hydrothermal circulation (Lipman et al., 1976; Steven and Lipman, 1976). It is important to note, however, that relatively few calderas in the western United States have been mineralized (McKee, 1979; Rytuba, 1981), and in the San Juan Mountains, Colorado, only about 1/3 of the known calderas have had significant mineral production (Steven and Lipman, 1976).

Size of deposit--There is a large range in the size of the Adularia-Sericite-type deposits. Guanajuato, a silver- and base-metal-rich district, covers a surface area of approximately 190 sq km. Districts with relatively low base-metal contents, such as Oatman, tend to be considerably smaller (12 sq km); Comstock is an important exception. The length-to-width ratio

Table 7.1--Characteristics of the adularia-sericite type and acid-sulfate type deposits (compiled from Heald et al., 1986).

	Acid-Sulfate	Adularia-Sericite
Structural setting	Intrusive centers, 4 out of the 5 studied related to the margins of calderas	Structurally complex volcanic environments, commonly in calderas
Size length:width ratio	relatively small equidimensional	variable; some very large usually 3:1 or greater
Host rocks	rhyodacite typical	silicic to intermediate volcanics
Timing of ore and host	similar ages of host and ore (<0.5 m.y.)	ages of host and ore distinct (>1 m.y.)
Mineralogy	enargite, pyrite, native gold, electrum, and base-metal sulfides Chlorite rare no selenides Mn-minerals rare sometimes bismuthinite	argentite, tetrahedrite, tennantite, native silver and gold, and base-metal sulfides chlorite common selenides present Mn gangue present no bismuthinite
Production data	Both gold- and silver-rich deposits noteworthy Cu production	Both gold- and silver-rich deposits variable base-metal production
Alteration	Advanced argillic to argillic (± - sericitic) Extensive hypogene alunite Major hypogene kaolinite No adularia	Sericitic to argillic supergene alunite occasional kaolinite Abundant adularia
Temperature	$200°$ to $300°C^2$	$200°$ to $300°C$
Salinity	1 to 24 wt% NaCl eq.[3]	0 to 13 wt% NaCl eq.
Source of fluids	Dominantly meteoric, possibly significant magmatic component	Dominantly meteoric
Source of sulfide sulfur	Deep-seated, probably magmatic	Deep-seated, probably derived by leaching wallrocks deep in system
Source of lead	Volcanic rocks or magmatic fluids	Precambrian or Phanerozoic rocks under volcanics

[1] Could be secondary in some districts.
[2] Limited data, possibly unrelated to ore.
[3] Salinities of 5 to 24 wt% NaCl eq. are probably related to the intense acid-sulfate alteration which preceded ore deposition.

of the surface projection of mineralized veins in this type of deposit is generally on the order of 3:1. The vertical range of mineralization is usually on the order of 400 to 700 meters compared to strike lengths of several kilometers.

Host rocks--The composition of the host rocks of Adularia-Sericite-type deposits ranges from rhyolitic to andesitic, and ore is commonly hosted by several different compositional units within a district. In a few of the districts, the ore fluids also mineralized associated sediments (Creede, Guanajuato) or intrusive rocks (Silver City, Idaho), but ore is mainly confined to the volcanic rocks. The occurrence of ore in several lithologies in Adularia-Sericite deposits implies that composition of the host rock(s) is not a controlling factor. The lack of a genetic tie to the host is further implied by the fact that ore deposition in Adularia-Sericite-type deposits almost always occurred more than 1 m.y. subsequent to the formation of the host rock.

Mineralogy--The mineralogy of the Adularia-Sericite-type deposit is characterized by the presence of vein adularia and sericite and by the absence of both hypogene alunite and the assemblage: enargite + pyrite +/- covellite (Table 7.2). Chlorite is also characteristically present. Although alunite is found in some of these deposits, in each case it appears to be a near-surface supergene occurrence, unrelated to the primary ore-forming system.

Metal ratios--The high silver-to-gold production ratios of most of the Adularia-Sericite-type deposits reflect the abundance of native silver, silver sulfides, and sulfosalts. Districts such as Round Mountain, Nevada, and Oatman, Arizona, have low silver-to-gold ratios. The precious metals in these two districts are present mainly as native gold, native silver, and electrum; silver sulfides and silver sulfosalts are rare (Table 7.1). Base-metal production is also usually low for the more gold-rich deposits. In most cases this reflects the lack of base-metal sulfides and sulfosalts, but base-metals are sometimes not recovered in milling operations and therefore production data may not always accurately reflect the ore mineralogy. The fairly large sample size of the Adularia-Sericite-type deposits in Table 7.2 permits speculation that there may be a continuum from the base-metal-rich, silver-rich districts, such as Colqui, Peru to the base-metal-poor, gold-rich districts, such as Oatman.

Wallrock alteration--The alteration patterns of Adularia-Sericite-type deposits are not yet well defined, due in part to the lack of detailed alteration studies. In general, Adularia-Sericite-type deposits are characterized by the predominance of sericitic alteration* that often borders a silicified zone near the vein. Also near the vein, fine-grained potassium feldspar and/or chlorite are often disseminated in the wallrock. The sericitic zone typically grades outward into a propylitic zone. An argillic zone between the sericitic and propylitic zones is sometimes present. At Creede, Pachuca (Mexico) and Oatman, the alteration over the ore body has been described as a sericitic "cap", interpreted to be the result of the condensation of acid volatiles released at depth during boiling (Barton et al., 1977; Buchanan, 1981). In many, if not most districts, the outermost propylitic alteration zone (typical of both Acid-Sulfate- and Adularia-Sericite-type deposits) appears to have formed prior to ore deposition and may be unrelated to the ore-forming hydrothermal system.

---

*The term "sericitic" is used in this chapter in the sense of alteration consisting of a mica-type mineral (e.g., illite) + quartz + pyrite, including mixed-layer illite-smectite in which illite layers are predominant.

Thermal history--For the Adularia-Sericite-type deposits, fluid-inclusion studies tied to a detailed paragenetic sequence (compiled in Hayba, 1983) show that most ore deposition occurred at temperatures between 200°C and 300°C, with late-stage fluids typically depositing only gangue minerals between 140°C and 200°C. Of the 11 Adularia-Sericite deposits evaluated in Heald et al. (1986), boiling has unequivocally been demonstrated to be associated with precious-metal deposition only at Colqui (Kamilli and Ohmoto, 1977). Precious-metal deposition due to boiling has been proposed for other districts based on less definitive evidence (cf. Guanajuato (Buchanan, 1979), Tonopah (Fahley, 1981)). Roedder (1984, p. 426) notes that some of the evidence for boiling at Guanajuato and Tonopah is ambiguous and that the occurrence of boiling in these deposits may have been overstated. Boiling also has been noted at Eureka (Casadevall and Ohmoto, 1977), Pachuca (Drier, 1976), and Creede (Barton et al., 1977) during times other than precious-metal deposition. Although boiling is an effective mechanism for depositing ores (Henley et al., 1984; Henley, 1985, this volume; Drummond and Ohmoto, 1985; Reed and Spycher, 1985, this volume), it is not the only mechanism. In the study of any ore deposit, the evidence for boiling should be critically evaluated before assuming it contributed to precious-metal deposition. Bodnar et al. (1985, this volume) discuss problems in using fluid-inclusion evidence to document boiling in ore deposits. In addition to boiling, the mixing of fluids from two or more sources has played a role in the thermal history of Adularia-Sericite-type deposits as has been documented by fluid-inclusion and stable-isotope studies at Creede (discussed in detail later).

Composition of fluids--Salinities, determined from freezing point depression measurements, range from 0 to 13 wt.-% NaCl equivalent for fluid inclusions from those Adularia-Sericite-type deposits evaluated. Most of the deposits have consistently low salinities, usually less than 3 wt.-% (Hayba, 1983), but Creede and Colqui have unusually high salinity fluids, ranging from 5 to 12 wt.-% NaCl equivalent (Woods et al., 1982; Kamilli and Ohmoto, 1977). The high salinities in these two deposits may have a bearing on their relatively high base-metal contents because chloride complexes are an effective means of transporting base metals (Barnes, 1979; Henley, 1985, this volume). On the other hand, fluid inclusions from the Sunnyside mine, Eureka mining district, Colorado, which is also very base-metal rich, have low salinities (0-3.6 percent). Hedenquist and Henley (1985) and Henley

Table 7.2--Selected mineralogical and production data for 54 epithermal districts (from Heald et al., 1986). Both types of districts are ordered from higher to lower base-metal production. The last 14 districts may represent a distinctively low base-metal subtype of the adularia-sericite-type district. AD, adularia; AL, alunite; SS, sulfosalts; AGS, silver sulfides; SP, sphalerite; GN, galena; EN, enargite; BA, barite; RC, rhodochrosite; FL - fluorite; X - denotes presence.

District	Ag:Au	Base Metal[1] percent	AD	AL	SS	AGS	SP	GN	EN	BA	RC	FL
ACID-SULFATE TYPE												
*Red Mtn., Colo.	68	high	X	X	X	X	X	X	X	X	X	X
*Julcani, Peru	467	mod	X	X	X	X	X	X	X			
*Lake City II, Colo.[5]	23	mod	X	X	X	X	X	X	X			
*Summitville, Colo.	1.2	5?	X			X	X	X	X			
*Goldfield, Nev.	0.3	1	X	X		X		X	X			
ADULARIA-SERICITE TYPE												
*Lake City (I), Colo.[5]	259	high	X		X		X	X		X	X	X
*Colqui, Peru[6]	26	17.5			X	X	X	X		X	X	X
Tovar, Mex.	45	10			X	X	X	X		X		
Parral, Mex.	150	4-20			X	X	X	X		X		
Bohemia, Oreg.	6	9	X			X	X	X				
Namiquipa, Mex.	Ag only	8.5				X	X	X		X		
*Eureka, Colo.[7]	47	7.5	X		X		X	X		X	X	X
*Creede, Colo.	400	5	X	X[2]	X	X	X	X		X	X	X
*Guanajuato, Mex.	200	0-10	X		X	X	X	X				
Guanacevi, Mex.	100-500	6-12	X		X	X	X	X		X		
Yoquivo, Mex.	74	mod	X		X	X	X	X				
Calico, Calif.	1200	mod			X	X				X		
Fresnillo, Mex.	64	4	X		X	X	X	X				
*Pachuca, Mex.	200	3.5	X		X	X	X	X		X	X	X
El Tigre, Mex.	162	3			X	X	X	X				
Great Barrier, New Zea.	4-30	3	X		X	X	X	X		X	X	
Silver Peak, Nev.	243	1-4	X		X	X	X	X		X		
Piz Piz, Nicaragua	2	2	X			X	X	X			X	X
*Tonopah, Nev.	80-110	2	X		X	X	X	X		X	X	X
Zalcualpan, Mex.	Ag only	1.4			X	X	X	X		X		
Tayoltita, Mex.	51	1	X		X	X	X	X				
Temalscaltepec, Mex.	267	low	X		X	X	X	X				
Golden Plateau, Aus.	0.8	minor	X			X	X	X				
Mohave, Calif.	2-12	minor	X		X	X	X	X				
Divide, Nev.	101	v low	X		X?	X		X		X		
Bodie, Calif.	5	v low	X		X	X	X					
Stateline, Utah	4	v low	X									
Guadalupe, Mex.	40	rare				X	X	X				
Mogollon, N. Mex.	58	1	X		X	X	X	X				
*Comstock, Nev.	23-40	1	X	X[3]	X	X	X	X				
*Silver City, Idaho	23	1	X	X	X	X	X	X				X
*De Lamar, Idaho	30	1	X		X	X	X	X				
Gold Circle, Nev.	15	1	X		X	X	X					
National, Nev.	1	1	X		X		X	X				
Aurora, Nev.	14	1	X		X	X						
Ramsey-Tala, Nev.	1	1	X			X						
Republic, Wash.	6.3	1	X		X	X						
Searchlight, Nev.	1	0.3	X					X				
Rawhide, Nev.	14	0.2	X		X	X						
Ocampo, Mex.	60	0.03			X	X	X	X				

Table 7.2--Selected mineralogical and production data for 54 epithermal districts (from Heald et al., 1986)--(continued)

District	Ag:Au	Base Metal[1] percent	AD	AL	SS	AGS	SP	GN	EN	BA	RC	FL
ADULARIA-SERICITE TYPE (continued)												
Rochester, Nev.	113	0.02	X	X[4]	X	X	X	X				
Tuscarora, Nev.	44-100	0.02	X		X	X	X	X	X			
*Round Mtn. Nev.	0.2	0.01	X	X[2]							X	X
Cornucopia, Nev.	68	0			X	X	X	X		X		
Wonder, Nev.	94	0	X		X	X	X	X				
Seven Troughs, Nev.	5.4	0	X		X	X						
El Oro, Mex.	7	0			X	X						
Jarbidge, Nev.	3	0	X		X	X						
Gilbert, Nev.	Au only	0	X		X	X						
Hayden Hill, Calif.	1.5	0	X		X							
Katherine, Ariz.	3	0	X									
*Oatman, Ariz.	0.4	0	X									

*Denotes the 16 districts studied in detail in Heald et al., 1986; the others were modified from Buchanan (1981).

[1] Base-metal production, usually as sulfides, rarely as metals, as a percent of the total tonnage (Buchanan, 1981).

[2] Alunite younger than mineralization.

[3] Alunite shallow only, may be supergene.

[4] Alunite older than mineralization.

[5] Slack (1980) grouped the Galenea (Henson Creek) and Lake (Lake San Cristobal) districts, Colo. into the Lake City district; the earlier ores (Lake City I) occur primarily in the Galena district and the later ones (Lake City II) in the Lake district.

[6] Data for the Finlandia vein.

[7] Data mainly for the Sunnyside vein.

(1985, this volume) have noted that the interpretation of salinity from measurements of fluid inclusions in some deposits can be skewed to higher values if the fluids have relatively high gas contents, because CO_2 and H_2S contribute to the freezing point depression. However, the presence of CO_2 in these epithermal deposits has not been documented. Although high concentrations of CO_2 can easily be detected by the crushing test techniques described by Roedder (1970), Bodnar et al. (1985, this volume) show that inclusions trapped under "epithermal" conditions may contain significant amounts of CO_2 which go undetected. A few analyses of inclusion fluids from Creede (Roedder, compiled in Hayba, 1983), Eureka (Casadevall and Ohmoto, 1977), and Colqui (Kamilli and Ohmoto, 1977) indicate CO_2 concentrations of less than 0.5 molal in almost all samples.

Paleodepth--Paleodepth estimates from both geologic reconstructions and pressures estimated from fluid-inclusion studies show that most of the Adularia-Sericite-type deposits appear to have formed at paleodepths of 300-600 m, although both methods have rather large uncertainties (Roedder and Bodnar, 1980). Colqui, and possibly Eureka, are thought to have formed at greater depths, approximately 1000 and 1400 m, respectively. Based only on geologic reconstruction estimates, Round Mountain, Oatman, and DeLamar seem to have formed at shallower depths, possibly as shallow as 100 m. The shallow paleodepths and high gold/silver ratios of Round Mountain and Oatman are consistent with models for gold deposition presented by Henley (1985, this volume), Silberman and Berger (1985, this volume), and Hedenquist and Henley (1985).

Sources of fluids--The predominance of meteoric water has been documented in several epithermal deposits, but only Creede (Bethke and Rye, 1979), Colqui (Kamilli and Ohmoto, 1977), and Sunnyside (Casadevall and Ohmoto, 1977) have had isotopic studies related to paragenetic sequences. However, even with these more comprehensive studies, the systematics of hydrogen and oxygen isotopes may be more complex than previously thought. Very detailed sampling of quartz crystals at Creede by Foley et al. (1982) has shown that the previously reported δD values for quartz (Bethke and Rye, 1979) do not truly represent primary fluids, but rather represent a mixture, during extraction and analysis, of hydrothermal fluids trapped in primary fluid inclusions, and isotopically lighter, overlying ground waters, trapped in pseudosecondary inclusions (discussed more thoroughly later). Without such detailed data, it is difficult to draw any more specific conclusions other than that the mineralizing solutions were deeply circulating, dominantly meteoric waters. It should be noted, however, that a magmatic component of up to ten percent could be hidden in the uncertainties and cannot be ruled out.

Source of sulfur and lead--For the 16 deposits considered by Heald et al. (1986), the sulfur-isotopic data for the Adularia-Sericite deposits are limited to measurements made on Creede (R. O. Rye, U.S.G.S., personal communication, 1985), Sunnyside (Casadevall and Ohmoto, 1977), and Colqui (Kamilli and Ohmoto, 1977). Although the $\delta^{34}S$ values for the sulfides for all three deposits cluster very close to 0 permil, quite different interpretations of the sources of the sulfur have been proposed for the different deposits. This is due in part to geologic considerations, but mainly to the $\delta^{34}S$ values for the sulfate minerals; if equilibrium conditions are assumed, there must be a material balance between the isotopically light sulfides and heavy sulfates. For Sunnyside and Colqui, the sulfates range from +10 to +25 permil, but at Creede they vary from +17 to +45 permil. The significance of the unusually heavy sulfate sulfur at Creede will be discussed later.

At Sunnyside, Casadevall and Ohmoto (1977) suggest that the sulfur was derived from evaporite-bearing sedimentary rocks which have a bulk sulfur value of $\delta^{34}S$ = +12 permil as sulfate and are located outside of the San Juan and Silverton calderas. They base their conclusion on the assumptions of sulfide-sulfate isotopic equilibrium (based on a pyrite-anhydrite pair having a $\delta^{34}S$ difference of 22.3 permil) and that total sulfate concentration in the fluid was greater than or equal to total sulfide concentration (indicated from mineralogical data). In order to produce the 0 permil sulfides in equilibrium with the heavier sulfates at the presumed equilibration temperature of 300°C, a source of $\delta^{34}S$ = +12 permil is required. Kamilli and Ohmoto (1977) also prefer a sedimentary sulfate source for the sulfur at Colqui, but suggest that an igneous origin is possible.

At Creede, it is clear that sulfate-sulfide isotope relationships were not governed by equilibrium fractionation, and that aqueous sulfide and sulfate were essentially independent systems (Bethke et al., 1973). Although the uniquely heavy sulfates at Creede require extensive isotopic evolution, Bethke et al.'s proposal for non-equilibrium, essentially independent sulfide-sulfate systems is also tenable for the other deposits. It is probable that non-equilibrium sulfide-sulfate conditions existed for most epithermal deposits (Ohmoto and Lasaga, 1982). Rather than assuming sulfide-sulfate equilibrium and relying on scarce sedimentary units, sulfate/sulfide ratios, and material balance considerations to produce 0 permil sulfides, it is perhaps better to assume that the isotopic composition of the sulfide sulfur was controlled by the thick volcanic piles associated with each of these deposits.

For those Adularia-Sericite districts where lead-isotopic studies have been done, Tonopah (Zartman, 1974), Pachuca (Cumming et al., 1979) Creede, Lake City, Colorado, and Sunnyside (Doe et al., 1979), the relatively radiogenic initial-lead ratios of galena suggest that a significant component of the lead may have been derived from Precambrian or Phanerozoic rocks underlying the volcanic rocks. This indicates that the ore components of Adularia-Sericite deposits may have been derived, in large part, by leaching of wallrocks deep in the system.

Characteristics of Acid-Sulfate-type Deposits

Structural setting--Four of the five Acid-Sulfate deposits listed in Table 7.2 are spatially related to the margins of calderas. The other Acid-Sulfate deposit, Julcani, is associated with silicic domes at the intersection of major faults. The association of these deposits with intrusive centers, particularly ring-fracture volcanic domes on the margins of calderas, appears to be a critical genetic factor (Heald et al., 1986) in contrast to the Adularia-Sericite-type deposit where the role of calderas is one of ground preparation for later hydrothermal fluids (Lipman et al., 1976; Steven and Lipman, 1976).

Size of deposit--On the whole, Acid-Sulfate deposits are smaller in terms of tonnage than Adularia-Sericite-type deposits, although this may be a result of a limited data base. In addition, the surface projection of the productive areas is relatively equidimensional rather than elongated. For example, the areal extent of the most productive part of Goldfield, Nevada, is only approximately 2 km long by 1.5 km wide. For all the Acid-Sulfate-type deposits, the vertical extent of the mineralized area is much smaller than the horizontal extent, usually less than 500 meters.

Host rock--The primary host rock for the Acid-Sulfate-type deposit is almost exclusively rhyodacite which is commonly porphyritic. At Julcani, some ore also occurs in dacite, and at Goldfield, trachyandesite and rhyolite, in addition to rhyodacite, host the ore. The timing of ore deposition relative to the emplacement of the host rock is of particular importance. Age dates show that ore deposition very closely (0.5 m.y.) followed emplacement of the host rock (Table 7.3) indicating a possible genetic relationship. An exception is the Lake City II deposit (i.e., the Lake district; see footnote 5 on Table 7.2) where the ores occur in a quartz latite ash-flow tuff which is considerably older than the ore. However, the

Table 7.3--Age and type of host rock and age of mineralization for five Acid-Sulfate districts (modified from Heald et al., 1986).

District	Host Rock	Age of Host Rock (m.y.)	Age of Mineralization	Reference
Red Mtn., Colo.	rhyodacite dome	21.3 to 23.6	23.1 ± 0.6	Lipman et al., 1976; Mehnert et al., 1973; Gilzean et al., 1984
Julcani, Peru	rhyodacite dome; dacite dome	9.67 ± 0.05 to 10.13 ± 0.03	9.83 ± 0.08	Petersen et al., 1977; Noble and Silberman, 1984
Lake City II, Colo.	quartz latite ash-flow	<28.4, >27.8	22.5[1]	Slack, 1980; Lipman et al., 1976
Summitville, Colo.	rhyodacite dome	22.8 ± 0.6	22.4 ± 0.5	Mehnert et al., 1973
Goldfield, Nev.	rhyodacite dome; trachyandesite; rhyolite	21.3 ± 0.3; 21.5 ± 0.5; 28-33	21.0 ± 0.4	Silberman and Ashley, 1970

[1] Slack (1980) proposes that these later ores were generated during emplacement of the nearby Red Mountain rhyodacite dome, approximately 22.5 m.y. ago.

ore at Lake City II fills fractures that form a radial pattern around a rhyodacite flow dome that Slack (1980) suggested is genetically related to the Lake City II ores and has a similar age to the ores. Hon et al. (1985) have shown that some veins included in the Lake City II district are older than the rhyodacite flow dome and probably formed from an earlier hydrothermal system(s).

Mineralogy--Acid-Sulfate-type deposits are characterized by the occurrence of the vein mineral assemblage enargite + pyrite +/- covellite. Adularia and chlorite are absent or rarely present. Ore occurs primarily as native gold and electrum with sulfides, sulfosalts, and tellurides. Bismuthinite has been identified in 3 of the 5 Acid-Sulfate deposits in Table 7.2, but selenides, rhodochrosite, and fluorite are rare.

Metal ratios--Summitville and Goldfield have low silver-to-gold ratios (2:1) which reflect the high proportion of free gold and gold-bearing minerals. Julcani, Red Mountain and Lake City II have high silver-to-gold ratios (10:1) and are characterized by more abundant silver mineralization, primarily in the form of silver sulfides and sulfosalts. Copper constitutes a major proportion of the base-metal production for Acid-Sulfate-type deposits, especially at Goldfield and Summitville, where copper accounts for more than 85% of the base-metal production. The silver-rich districts have proportionally more base metals, but copper production is secondary to lead (1:2 Cu:Pb ratio).

Wallrock alteration--A definitive characteristic of Acid-Sulfate-type deposits is the association of advanced-argillic alteration with the ore. Kaolinite, usually accompanied by alunite, occurs close to the vein and is often coextensive with silicification. Farther from the vein, argillic alteration, sometimes intermixed with sericitic alteration, surrounds the zone of advanced-argillic alteration. The argillic alteration zone is commonly mineralogically zoned, with kaolinite nearer the vein and smectite farther from the vein. The outermost alteration zone consists of propylitic alteration.

Thermal history--The limited fluid-inclusion data for Acid-Sulfate-type deposits indicate that ore deposition occurred primarily at temperatures similar to those of Adularia-Sericite-type deposits (200°C to 300°C). The salinities of Acid-Sulfate-type deposits, which are not yet well documented, show a wide range. Secondary inclusions in quartz phenocrysts which are thought to contain the fluids responsible for the intense quartz + alunite +/- pyrite alteration which preceded ore deposition have salinity ranges of 5-24, 7-21 and 5-18 wt.-% NaCl equivalent at Julcani, Summitville and Goldfield, respectively (Bruha and Noble, 1983). However, limited data from primary inclusions in quartz and sphalerite associated with main stage ore deposition from Lake City II (Slack, 1980), Red Mountain (Nash, 1975), and Summitville (Stoffregen, 1985) indicate salinities on the order of 1 to 6 wt.-% NaCl equivalent. These limited data are examined more closely in a later section, but clearly, much remains to be learned from fluid-inclusion studies of Acid-Sulfate-type deposits.

Paleodepth--Paleodepth estimates for Acid-Sulfate-type deposits appear to be similar to Adularia-Sericite deposits (300-600 m), although Nash (1975) suggests that the Red Mountain deposit may have formed at depths greater than 1200 m. Based solely on geologic reconstruction, the gold-rich deposits, Goldfield and Summitville, may have formed at somewhat shallower depths than the silver-rich deposits.

Source of constituents--The stable-isotope data on Acid-Sulfate deposits is limited to some $\delta^{18}O$ numbers for Goldfield (Taylor, 1973) and $\delta^{34}S$ data for Julcani (Goodell, 1970), Goldfield (Jenson et al., 1971), and Summitville (R.O. Rye, U.S.G.S., personal communication, 1985). Without any deuterium data, the source of the hydrothermal fluid is equivocal. The possibility of a significant magmatic component is discussed later. The paucity of sulfur-isotope data permits only generalities. The sulfides from these three districts plot between -7 and +3 permil suggesting a magmatic source for the sulfur. Similarly, lead-isotopic studies at Summitville and Red Mountain show that the galena is very similar isotopically to the enclosing volcanic rock, implying that either the adjacent rocks or related magmatic fluids were the primary source of the lead (Doe et al., 1979).

Summary of Characteristics

The primary characteristics which distinguish the Adularia-Sericite-type deposits from the Acid-Sulfate-type deposits are the vein mineralogy and the alteration assemblages (Heald et al., 1986). As shown in Table 7.1, there are also many other important, but less definitive, characteristics of each type which need to be considered in developing genetic models. A comparative study, such as the one done by Heald et al. (1986) helps determine which characteristics are the salient features of a deposit-type model and which features are just local variations. In order to examine each of these two types of epithermal deposits in greater detail, a characteristic deposit from each group will be reviewed in the next sections.

THE ADULARIA-SERICITE ENVIRONMENT:
CREEDE AS AN EXAMPLE

In the foregoing section the characteristics of 16 well-documented Tertiary volcanic-hosted epithermal ore deposits were summarized based on the comparative study done by Heald et al. (1986). It was concluded that two distinctive types of deposits could be distinguished: 1) those deposits characterized by an alteration assemblage dominated by adularia and sericite, and 2) those deposits characterized by an alteration assemblage containing kaolinite and alunite. The most thoroughly studied of the Adularia-Sericite-type deposits is the Creede mining district, Colorado. Therefore, Creede is useful as the exemplar for the Adularia-Sericite-type deposits.

Creede as an Exemplar

In basing our discussion of the Adularia-Sericite environment on Creede we recognize that some features of the Creede district are not representative of the group as a whole. For example, the chemical and isotopic evolution of the Creede ore fluids in a closed-basin lake prior to incorporation into the hydrothermal system is an aspect that may be uncommon, although a similar geologic setting may have existed at Calico, California. The presence of an evaporative lake is obviously not a requirement for ore formation in these deposits, although the high salinities developed by evaporation probably made the Creede system highly efficient in the transport of base-metals (cf. Henley, 1985, this volume). We also recognize that within this group of deposits there is considerable diversity in specific characteristics. For example, Round Mountain and Oatman are both distinctively gold-rich and sulfide-poor relative to the other districts discussed in the previous section. Other

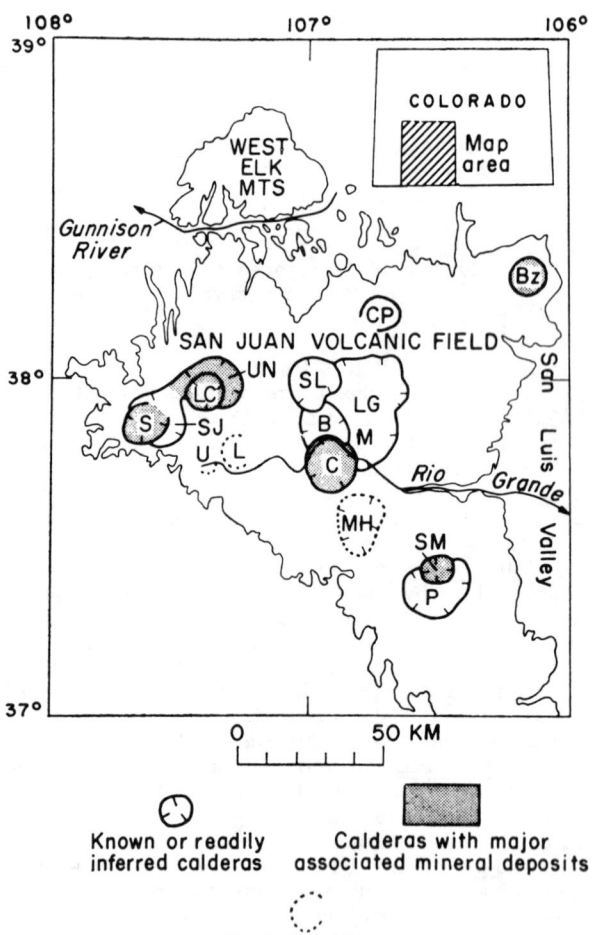

Figure 7.1. Calderas of the San Juan volcanic field: S, Silverton; LC, Lake City; CP, Cochetopa Park; Bz, Bonanza; LG, La Garita; SL, San Luis; B, Bachelor; C, Creede; MH, Mount Hope; P, Platoro; SM, Summitville; L, Lost Lake; U, Ute Creek; SJ, San Juan; UN, Uncompahgre; M, general location of the Mammoth Mountain caldera. After Steven and Eaton (1975).

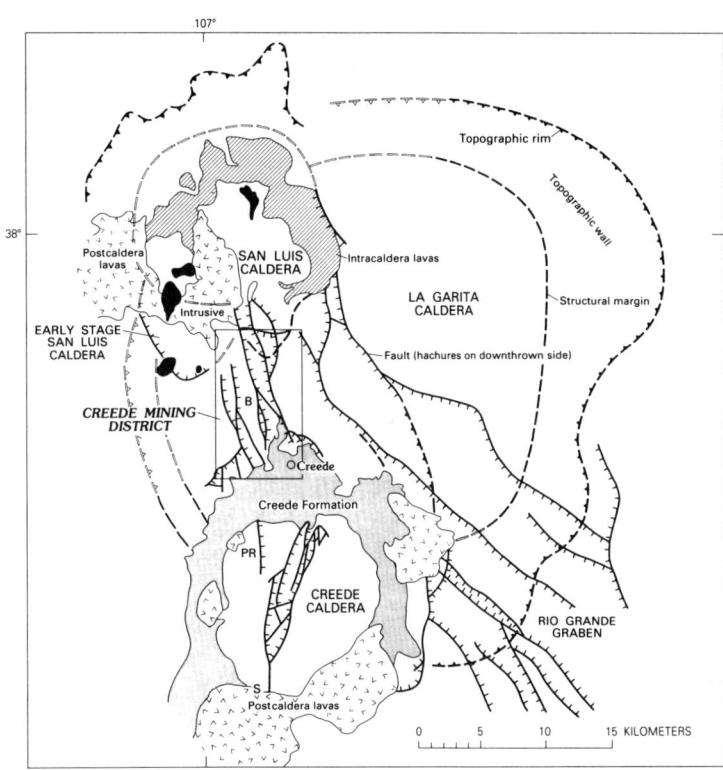

Figure 7.2. Generalized geology of the Creede and San Luis calderas in relation to remnants of the Bachelor (B) and La Garita calderas and to the Creede mining district (shown in box). Control is moderate to good where boundaries are shown by solid symbols, and conjectural where shown by open symbols. Pr, Point of rocks volcano; S, Spar City. After Steven and Lipman (1976).

differences could be enumerated. In spite of the diversity in specific characteristics, we would argue that the differences between the various districts represent variations on a theme, whose main characteristics are illustrated by the similarities.

We are aware that the "Hot-Spring-type" deposits such as McLaughlin, California, and Hasbrouck Mountain and Sulphur, Nevada, are not specifically treated when Creede is used as an exemplar because there is no evidence of surface discharge of the ore fluids at Creede. It is our current opinion, however, that such hot-spring-type deposits form in the surficial parts of hydrothermal systems similar to those which deposited the Creede ores. Berger and Eimon (1983), Henley (1985, this volume), Silberman and Berger (1985, this volume), and Berger and Silberman (1985, this volume) treat hot-spring-type deposits separately because of important structural attributes that, in terms of ore controls, somewhat distinguish these deposits from the deeper classical vein deposits. However, Silberman and Berger (1985, this volume) also present evidence from Bodie, California, that links hot-spring-type deposits directly to the Creede-type veins.

Summary of Important Studies

The Creede mining district has been the focus of both extensive and intensive study for the past 30 years. Studies by T. A. Steven and co-workers of the U.S.Geological Survey unravelled the volcanic history of the district, related the ore deposits to that history, and documented the structural control of ore deposition (Steven and Ratté, 1965; Ratté and Steven, 1967; Steven and Ratté, 1973; Steven and Lipman, 1973; Steven, 1967; Steven and Eaton, 1975). Continued studies by T. A. Steven and P. W. Lipman and co-workers have worked out the volcanic stratigraphy and volcano-tectonic evolution of the entire San Juan volcanic field, providing a particularly well-developed regional context for the Creede district (Steven and Lipman, 1976; Lipman et al., 1970; Lipman et al., 1978; and Doe et al., 1979). Detailed mineralogical-geochemical studies have served to develop a well-documented, comprehensive model of ore deposition (Roedder, 1960; Roedder, 1965; Roedder, 1977; Bethke et al., 1976; Barton et al., 1977; Wetlaufer, 1977; Bethke and Rye, 1979; Woods et al., 1982; Foley et al., 1982; Heald-Wetlaufer and Plumlee, 1984; Hayba, 1984; Robinson and Norman, 1984; Plumlee and Hayba, 1985). These studies have been complemented, and the models developed from them improved, by detailed, unpublished geologic studies by the Homestake Mining Company, Pioneer Nuclear Corporation, Chevron Resources, Freeport Mining Company, Minerals Engineering Company, and by a number of theses (Cannaday, 1955; Chaffee, 1967; Hull, 1970; Giudice, 1980; Robinson, 1981; Battory, 1982; McCrink, 1982; Wason, 1983; Rice, 1984; Horton, 1983; Vergo, 1984; and Misantoni, 1985).

Geologic and Mineralogic Characteristics

<u>Volcanic-hydrothermal history</u>--The San Juan Mountains are the main erosional remnant of a volcanic field which covered most of the southern

Rocky Mountains in mid-Tertiary time. Approximately 2/3 of the volume of the present field is composed of a series of early lavas and volcaniclastic aprons, mainly of andesitic composition, related to a number of stratovolcanic centers. These "Early Intermediate" composition lavas are overlain by a series of quartz latitic to rhyolitic ash-flow sheets. Fifteen calderas have been identified as sources of these ash-flows (Fig. 7.1). Many of these calderas are nested to form complexes such as the central San Juan caldera complex which hosts the Creede mining district.

The Creede mining district is located along the western edge of the central San Juan caldera complex, a series of 7 nested calderas from which quartz-latitic to rhyolitic ash-flow sheets erupted over the brief interval of 27.6 to 26.2 m.y. (Fig. 7.2). Only 5 of the 7 calderas can be accurately located, the remaining two having caved into the younger Creede caldera. The Creede ores are hosted by the intracaldera fill of the resurgent Bachelor caldera, the second caldera in the series. The ores are contained in a set of fractures comprising a graben running between the Creede caldera, the youngest of the series, and the slightly older San Luis caldera.

Radiometric dating of vein adularia and mixed-layer clay alteration minerals (Bethke et al., 1976; Vergo, 1984) indicates that the ores were deposited approximately 1 million years after the youngest dated volcanic event in the district. These dates are consistent with the observation of Steven and Ratte (1960b) that mineralization was confined to structures young enough to cut the sediments of the Creede formation which fill the moat of the resurgent Creede caldera.

Studies of the Creede Formation by Steven and Ratté (1965), Steven and Van Lonen (1970), McCrink (1982), and Battory (1982) have shown that the sediments filling the moat of the Creede caldera consist of landslide debris and stream channel fill on the margins of the moat, and lacustrine deposits of airfall and water-lain tuffs, interbedded with a few thin ash-flow tuffs, in the center of the moat. These sediments accumulated in a shallow, playa lake environment, and the lacustrine sediments were strongly zeolitized during diagenesis.

Steven and Eaton (1975) proposed that the convecting hydrothermal system was driven by an unexposed pluton beneath the Creede district. They observed that the maximum displacement along the Amethyst and Bulldog fault systems occurs in the central, most heavily mineralized part of the district, where subtle magnetic and gravity anomalies are suggestive of a buried intrusion.

<u>Creede ore deposits</u>--The base- and precious-metal mineralization at Creede fills open fractures. The ore zone, which occupies a narrow vertical range from 250 to 400 m, has been mined nearly continuously for approximately 3 km along strike on the Amethyst-OH vein system, and for over 2 km on the Bulldog Mountain vein system (Fig. 7.3). This horizontally-dominated aspect ratio is characteristic of deposits of the Adularia-Sericite-type. Mineralization is confined to the fracture system comprising the Creede graben bounded on the east by the Solomon-Holy Moses fault system and on the west by the Alpha-Corsair system (Fig. 7.3). This graben system follows closely the structures of an older graben interpreted to be the keystone graben of the resurgent Bachelor caldera. Detailed ore petrology studies of material from the major producing structures (Amethyst, OH, P and Bulldog Mountain vein systems) indicate that all were part of a single hydrologic system and were filled during the same mineralizing event (Barton et al., 1977; Bethke and Rye, 1979; Heald-Wetlaufer and Plumlee, 1984).

A significant, low-grade, bulk-tonnage silver resource also exists as disseminated replacements in the wallrocks of the upper parts of the Amethyst vein system near its intersection with the OH vein (Giudice, 1980). Additional bulk-tonnage resources are present as mineralized stream sediments in the stream channel facies of the Creede Formation adjacent to its

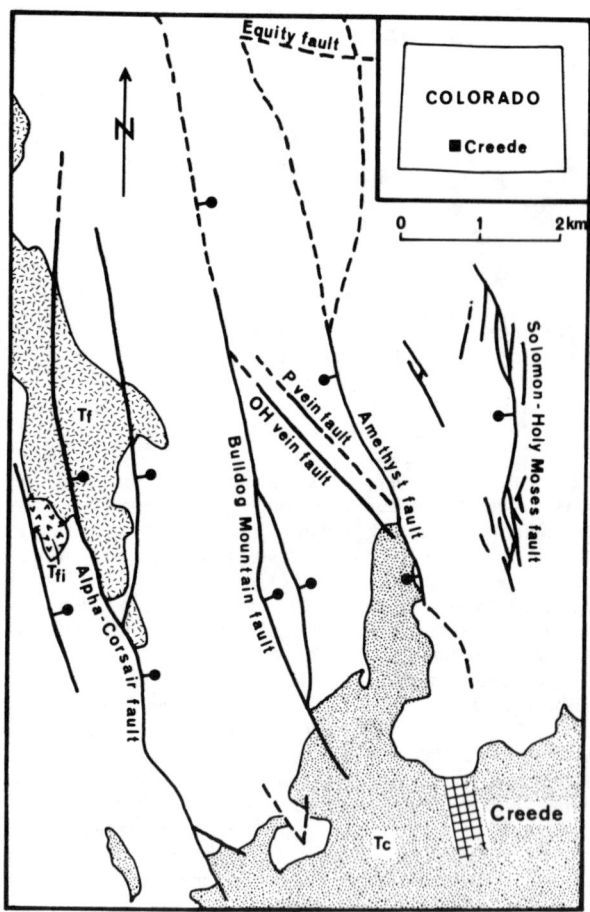

Figure 7.3. Generalized geology of the Creede mining district, modified from Steven and Eaton (1975). Area of map is shown in Figure 7.2. Faults are dashed where uncertain or inferred; bar and ball show the downdropped side. Tc, Creede Formation; Tf, Fisher quartz latite flow; Tfi, volcanic neck of Fisher quartz latite.

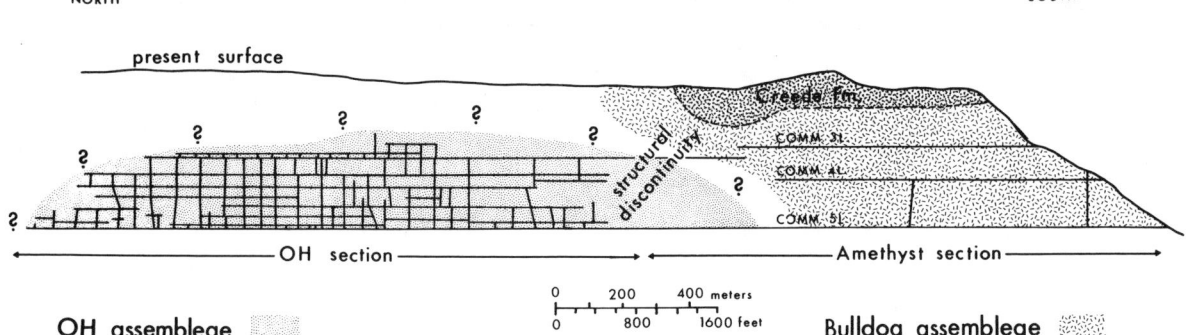

Figure 7.4. Generalized longitudinal projection onto a vertical plane of the OH vein and the southern end of the Amethyst vein. The main workings and the general distribution of the OH and Bulldog assemblages are shown. The "structural discontinuity" marks the position that would have been the intersection of the OH and Amethyst veins had fracturing been more continuous. The Amethyst vein has been mined northward for about 2 km beyond the "structural discontinuity" but is not shown here. Adapted from Barton et al. (1977).

truncation by the Amethyst vein (Wason, 1983; Rice, 1984). The metals have been introduced into the clastic facies of the Creede Formation by leakage of the ore fluid from underlying or adjacent veins.

Mineralogic characteristics--Mineralization in the Creede veins is strongly zoned from an association in the north (OH Assemblage) of chlorite + hematite + quartz + adularia + sphalerite + galena + chalcopyrite + pyrite +/- fluorite and tetrahedrite to a barite + rhodochrosite + quartz + adularia + galena + sphalerite + fluorite + tetrahedrite + silver sulfosalt + native silver association to the south (Bulldog Assemblage) (Fig. 7.4). Heald-Wetlaufer and Plumlee (1984) have demonstrated that the Bulldog Mountain and OH-P-Amethyst vein systems each contain both the OH and Bulldog assemblages, and that the two are contemporaneous and related to each other through facies changes along the fluid-flow path. The mineralization in each of these facies may be divided into 5 stages on the basis of mineralogy and texture, and each of the stages can be correlated between the two assemblages. Details of the mineralogy and the correlation of stages between the two assemblages are given in Table 7.4.

A late association of covellite + chalcocite + acanthite has been described by Giudice (1980) and Robinson and Norman (1984) in the disseminated ore from the upper and southern Amethyst vein system. This assemblage is a volumetrically minor, but economically important, component of the Bulldog Assemblage. It has not yet been incorporated into the geochemical model for Creede ore deposition, because its limited occurrence and fine-grained nature makes it difficult to study.

Many minerals of the Creede ores (sphalerite, rhodochrosite, siderite, Fe-chlorite, tetrahedrite-tennantite, proustite-pyrargyrite, late gel-pyrite, and, probably, other sulfosalt minerals) exhibit marked compositional variations in both time and space. Of these minerals, only sphalerite (Barton, et al., 1977), the Mn-Fe-carbonates (Wetlaufer, 1977) and Fe-chlorite (Horton, 1983) have been studied extensively. The compositional variations in the Mn-Fe carbonates and Fe-chlorite have proven too complex to be very useful to developing a genetic model of Creede, but those in the sphalerite (primarily iron content) have provided the main basis for documenting the evolution of the ore fluid in time and space, and for constraining the chemical environment. The sphalerite is beautifully banded, reflecting, mainly, the variations in iron content. This banding records a wealth of information about crystal growth and dissolution. Even more importantly, the growth zones can be correlated along the OH vein and between the OH and other veins. Figure 7.5 shows a composite microprobe tracing of samples of sphalerite from the OH vein that is representative of most of the B and D stage paragenesis.

Hydrothermal leaching (intense for barite, fluorite and rhodochrosite, and less intense for sphalerite and galena) demonstrates that at times the solutions entering the ore zone were undersaturated with respect to selected components. The increased intensity of leaching and number of leaching horizons at the north end of the OH vein suggest that the solutions entered the ore zone in the north and traversed the vein system from north to south.

Supergene oxidation has affected the Creede ores to a limited extent producing extensive manganese oxide (presumably after rhodochrosite) in the southern Amethyst system and local occurrences of kaolinite and halloysite. Some of the oxidized rocks contain distinctive fracture coatings of dendritic native silver. Veinlets of fine-grained alunite which crosscut mineralization in the upper levels of the Amethyst vein system have been dated by potassium-argon techniques and range in age from 3.5 to 5.0 million years (M. Lanphere, U.S.G.S., personal communication, 1981). This age corresponds closely to the regional uplift of the southern Rocky Mountains which, presumably, raised the Creede orebodies above the water table. It is not clear whether any or all of

Table 7.4--Mineralogy and paragenetic stages of the OH and Bulldog assemblages of the Creede district based on the work of P. B. Barton and P. M. Bethke as reported in Bethke and Rye (1979) for the OH, P, and Amethyst vein systems, on that of Robinson (1981) and Giudice (1980) for the upper and southern Amethyst vein, and on the detailed study by G. S. Plumlee and P. Heald on the Bulldog Mountain vein system as reported in Heald-Wetlaufer and Plumlee (1984).

Stage	OH Assemblage (OH, P northern Amethyst veins and northern 1/3 Bulldog vein system)	Bulldog Assemblage (southern 2/3 Bulldog vein system, southernmost OH vein and southern Amethyst vein)	Stage
E (youngest)	Fibrous pyrite with some marcasite and stibnite	Fibrous pyrite with minor marcasite, stibnite and sulfosalts. Late Ag (may be supergene)	V
D	Relatively coarse-grained sphalerite, galena, chalcopyrite, and quartz, some hematite; silver minerals notably absent; subdivided into three substages on basis of color banding in sphalerite: inner light, middle dark, and outer light	Coarse-grained sphalerite and galena with minor late chalcopyrite, tetrahedrite and Ag-Cu sulfides and sulfosalts	IV
C	Volumetrically minor siderite-manganosiderite and fluorite on quartz; sits on deep etch of earlier B stage; most fluorite deeply etched - commonly completely removed	Volumetrically minor (Mn,Fe)-carbonate and fluorite on white and amethystine quartz; most fluorite etched - commonly completely removed	III
B	Relatively fine-grained sphalerite, galena, chalcopyrite, chlorite, hematite, pyrite, and some tetrahedrite-tennantite	Interbanded barite and fine-grained sphalerite, galena, and tetrahedrite	II
A (oldest)	Primarily quartz with minor chlorite and sulfide	Early massive rhodochrosite, later barite; minor disseminated sphalerite and galena	I

the wire and leaf native silver, which is characteristic of the Bulldog assemblage, is of supergene origin. Our current interpretation is that it is primary.

Wallrock alteration--Wallrock alteration at Creede consists of a feldspar-destructive, mixed-layer illite/smectite alteration present along the top of the orebodies.* Alteration intensity ranges from weak, where only the pumice fragments are altered, to intense, where the entire rock has been altered and where all evidence of primary volcanic texture has been obliterated (Horton, 1983). The alteration is strongly fracture-controlled, and the intensity of argillization decreases away from the veins.

---

*Heald et al. (1986) included such mixed-layer illite-smectite alteration in the sericite category, and we will use the term "sericite" in their sense to include both illite and illite-smectite mixed-layer clay minerals.

The most intense alteration forms a "clay cap" that marks the upper limit of mining. The vertical extent of the clay cap has not been well established, but surface outcrop above highly altered areas often shows no alteration effects, and drilling above the OH

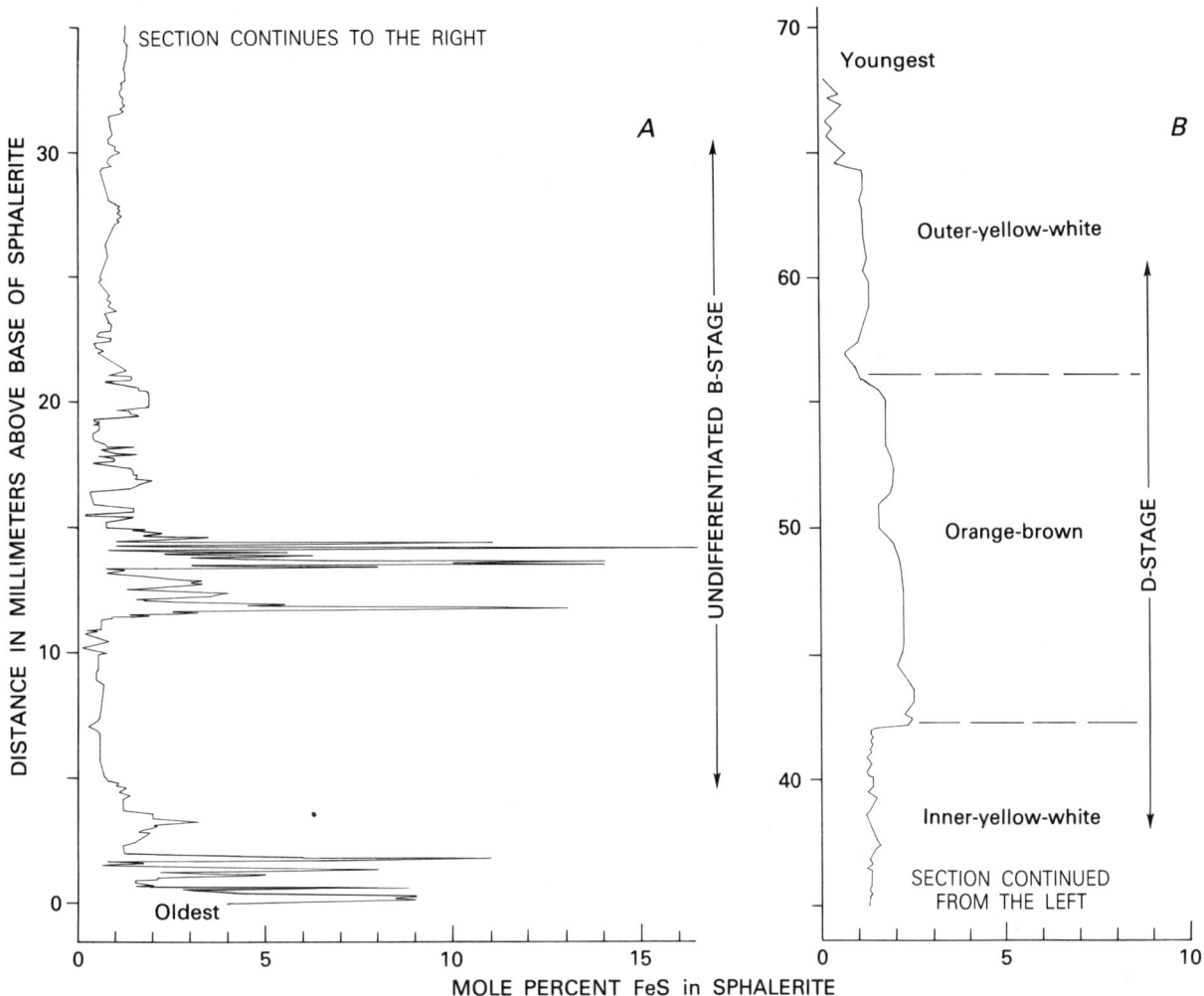

Figure 7.5. Coarse-scale plot of the iron content across a growth-banded aggregate of sphalerite crystals, modified from Barton et al. (1977). Data are taken from composite microprobe tracings of samples from the north end of the OH vein and are replotted to eliminate instrumental scatter. A). tracing of the iron content across finely banded, early (B-stage), undifferentiated sphalerite; B). tracing of the iron content across coarse, late (D-stage) sphalerite that is subdivided on the basis of color and iron and cadmium contents.

vein is reported to have encountered only fresh rock approximately 100 m above the base of alteration. Along the OH vein, the clay cap is continuous and regularly distributed (Fig. 7.6). In other parts of the district, the distribution of the clay cap is much more irregular; the base of alteration undulates over a vertical interval of several hundred feet, and parts of the veins show no alteration at all at any exposed level. The continuity and regularity of the base of the clay cap above the OH vein is interpreted to reflect a more uniform hydrologic regime in the vertical, simple tension fracture that contains the vein, in contrast to the hydrology in the more complex structures of the other vein systems in the district. Barton et al. (1977) interpreted the clay or sericitic alteration to mark a region of recondensation of acid-rich steam derived by boiling of the ore fluid.

In the upper parts of the vein systems, particularly near the junction of the Amethyst and OH veins, the wallrock adjacent to the veins is highly silicified and sometimes bleached. The silicification appears to pre-date the sericitic alteration and to have protected the affected wallrocks from it. In areas of intense silicification, the sericitic alteration often borders the silicified zone, although the silicification may pass into fresh, unsilicified wallrock, particularly in areas lying beneath the sericitic cap. It is suggested, but not demonstrated, that the silicification predated the sericitic alteration and protected the walls from the altering fluids.

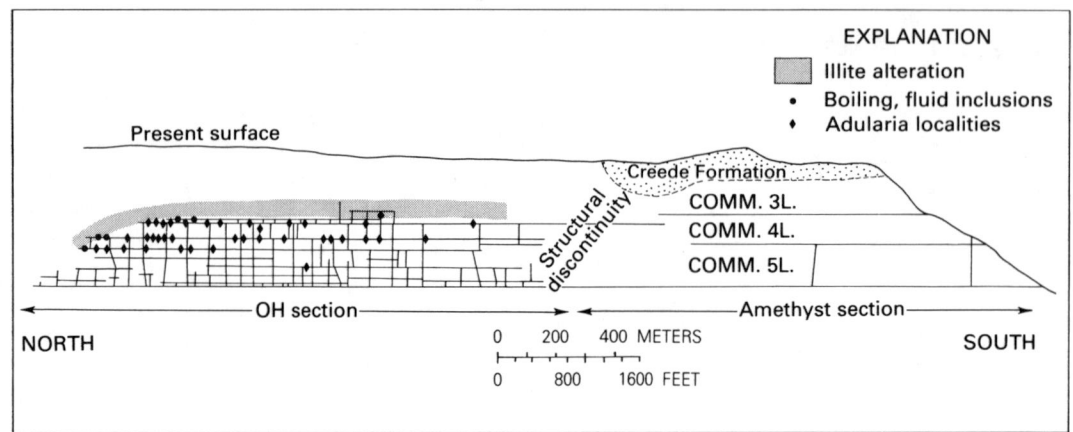

Figure 7.6. Generalized longitudinal projection onto a vertical plane of the OH vein and the southern end of the Amethyst vein. The base of the intense sericitic alteration is indicated; the position of the top is unknown.

Detailed studies of the variation in structure and composition of the mixed-layer clay minerals by Horton (1983) and Vergo (1984) show that the illite component of the clays ranges from greater than 95% to less than 60%. In the area just north of the Amethyst-OH intersection, where the ground between the two is cut by many smaller veins, the proportion of illite in the mixed-layer clays increases toward the Amethyst vein. The degree of ordering of the stacking sequence of the illite and smectite layers, which ranges from random to long-range ordered (Reichweite = 0 to 3), also increases toward the Amethyst vein. Using empirical relationships between temperature and the structure and composition of mixed-layer illite-smectite clays, Horton (1983) estimated the position of isotherms surrounding the upper portions of the Amethyst vein near its intersection with the OH vein. Vergo (1984) found the mixed-layer clays along the Bulldog vein system to be more smectite-rich than most of Horton's samples (presumably because they formed at lower temperatures). Vergo did not find any systematic variation in the illite-smectite ratio with distance from the vein for distances up to 70 m from the vein, and calculated that the system must have been active for a period of at least 10,000 years to produce such an apparently flat thermal gradient.

Below the zone of intense sericitic alteration, the wallrocks enclosing the veins are essentially fresh, but were enriched in potassium by a period of potassium feldspar-stable, metal-barren hydrothermal alteration that occurred approximately 2 million years prior to ore deposition (Ratté and Steven, 1967; Bethke et al., 1985). This intense potassium metasomatism was presumably related to hydrothermal circulation in the keystone graben of the Bachelor caldera, and is similar in most respects to the potassic alteration at Bodie (O'Neil et al., 1973; Silberman and Berger, 1985, this volume). It does not appear to have been associated with any earlier period of mineralization.

Geochemical Environment

An extremely detailed knowledge of the geochemical environment of the Creede ore-forming system has been developed through fluid-inclusion, light-stable isotope, and lead-isotope studies, and through thermochemical analysis of the ore and gangue mineral assemblages. A summary of the general characteristics of the geochemical environment are shown in Table 7.5.

Temperature-salinity ranges--Much of the evidence used in defining the depositional parameters of the Creede system has come from the study of fluid inclusions. Numerous homogenization measurements have defined a temperature range of 120° to 280°C for the Creede ore-forming fluids. Most of the measurements have been on sphalerite, quartz and fluorite from the OH-Amethyst vein system (Woods et al., 1982; Robinson and Norman, 1984; J. Goss, U.S.G.S., personal communication, 1985) and on barite from the mineralized stream channel of the Creede Formation near its truncation by the Amethyst vein (Rice, 1984). There is a clear trend from higher temperatures in the north (OH data, as summarized by Woods et al., 1982) to lower temperatures in the south (southern Amethyst data, Robinson and Norman, 1984, and Creede Formation data, Rice, 1984) although the temperature ranges for minerals and veins overlap (Fig. 7.7).

Fluid-inclusion studies have also shown that the Creede ores were deposited from relatively concentrated NaCl brines and that these brines mixed with overlying ground water in the ore zone (discussed below). From freezing and crushing and leaching studies, Roedder (1963, 1965) showed that the ore fluid ranged in salinity from about 4 to 12 wt.-% NaCl equivalent and had average atomic ratios of Na:K:Ca of 9:1:2. These ratios are in excellent agreement with those calculated for a temperature of 260°C using the alkali geothermometer of Fournier and Truesdell (1973). As noted in the previous section, with the exception of Colqui, Peru, the salinity of the Creede ore fluids is much higher than all the other Adularia-Sericite-type deposits evaluated by Heald et al. (1986).

Depth-pressure ranges--In addition to the temperature and salinity data, fluid-inclusion studies

Table 7.5--General environmental parameters for the OH vein, Creede, Colorado (modified from Barton et al., 1977)

Parameter	Range observed	Reference environment	Source of information
Temperature	190 - 285°C	250°C	Fluid inclusions[1]
	40 - 50 bars	50 bars	Evidence of boiling in fluid inclusions
Depth	450 - 600 m	500 m	Estimated from pressure and geologic reconstr.
Salinity	4 - 12 wt%	6 wt%	Fluid inclusions
Na:K	7.4 - 9.9	9	Analyses of fluid inclusions
Total S	0.018-0.30 molal	0.02 molal[2]	Analyses of fluid inclusions
pH		5.4	Calculated

[1]Most of the fluid inclusion evidence is from the later half of the mineralization which is much coarser grained.

[2]Because of the problem of sulfur contributed by oxidation of sulfides during sample handling, the lower total sulfur values are considered more reliable.

have shown that the ore fluid was, at times, boiling near the top of the OH vein (the evidence for boiling is discussed later). Using the tables from Haas (1971), the pressure at the top of the ore body was approximately 40 bars, based on the boiling point of a 250°C, 1 molal NaCl fluid. For most of ore deposition, the pressure must have been slightly greater than this, because boiling appears to have occurred infrequently. A hydrostatic head of approximately 450 m is necessary to produce the requisite pressure, which is in excellent agreement with the geologic reconstruction by Steven and Eaton (1975).

Chemical parameters--In 1977, Barton et al. did a thermochemical analysis of the Creede system and were able to put limits on such parameters as activities of S_2 and O_2, pH, and total sulfur. In terms of the activities of S_2 and O_2 (Fig. 7.8), the Creede environment is located near the junctures of the hematite, pyrite and Fe-chlorite stability fields based on the common occurrence of that mineral assemblage and on the iron content of sphalerite which usually ranged between 1 and 4 mole-% FeS (Fig. 7.5).

Because of the common occurrence of adularia associated with minor amounts of sericite in the wall rock near the vein, Barton et al. (1977) estimated the pH of the fluid during mineral deposition in the Creede system based on the feldspar hydrolysis reaction

3 K-feldspar + 2 H^+ = Kmica + 6 qtz + 2 K^+ (1)

A pH of 5.4 is estimated from the thermodynamic data of Montoya and Hemley (1975) for a 1 molal solution with a Na/K ratio of 9. This is a nearly neutral pH at 250°C.

Roedder et al. (1963) measured total sulfur concentrations of about 0.02 molal in inclusion fluids

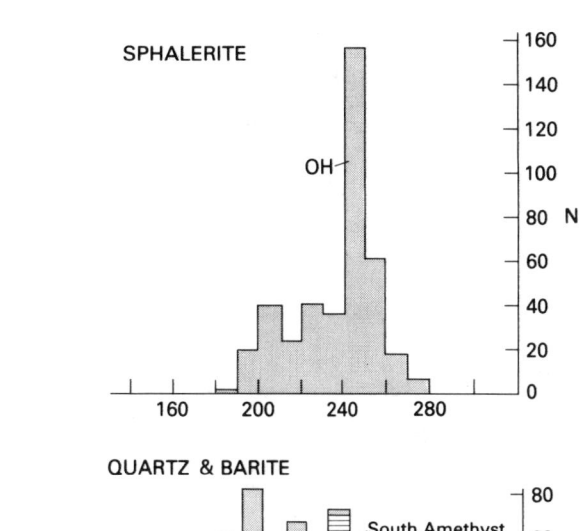

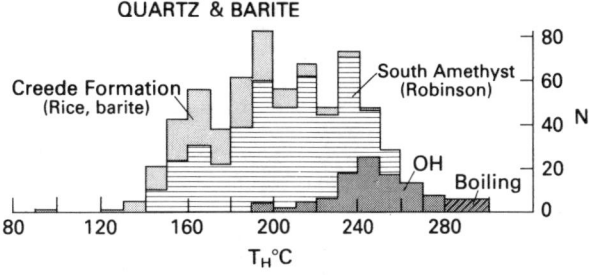

Figure 7.7. Histograms of homogenization temperatures for fluid inclusions in quartz and sphalerite from the OH vein (Woods et al., 1982), quartz from the southern Amethyst vein (Robinson and Norman, 1984), and barite from the Creede Formation (Rice, 1984).

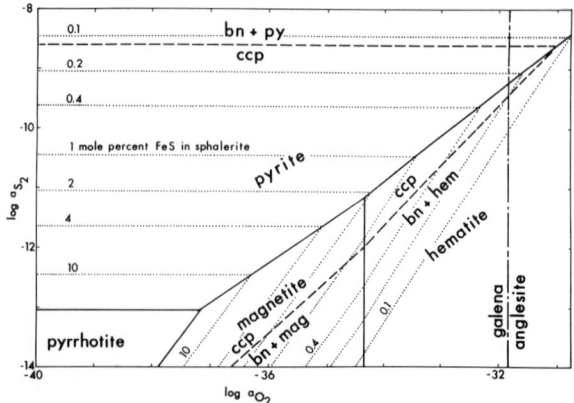

Figure 7.8. Log a_{S_2}-a_{O_2} diagram at 250°C showing the mineral stability fields for the significant minerals in the Creede ores. The shaded field of magnetite is completely preempted by chlorite. The contour for 20 mole percent FeS in sphalerite coincides with the pyrrhotite field boundary. Quartz is present throughout the diagram. Abbreviations: py = pyrite, ccp = chalcopyrite, bn = bornite, hem = hematite, chl = chlorite. The standard state for S_2 and O_2 is the ideal diatomic gas at 1 atmosphere and 250°C. The data for the iron-copper sulfides and oxides are summarized in Barton and Skinner (Table 7.2, 1979); thick solid lines: boundaries between pyrite, pyrrhotite, chlorite, and hematite; long light dashes: iron content of sphalerite. After Barton et al. (1977).

from Creede, but other concentrations in different samples are possible. However, at this total sulfur concentration (0.02 m), the pH of the pyrite + hematite + Fe-chlorite triple point is near 5.4, which lends credence to the thermodynamic estimate of pH (Fig. 7.9).

Chemical buffering of the ore fluids--Barton et al. (1977) proposed that reactions between iron-rich chlorite, hematite, pyrite, quartz and water controlled the redox conditions for the OH vein. A means of examining that suggestion arises from Woods et al.'s (1982) observation that the iron contents of growth-banded sphalerite from the OH vein show a positive correlation with the homogenization temperatures (Th) of primary fluid inclusions contained in the growth bands (Fig. 7.10). Previous investigators (Barton and Toulmin, 1964; Scott and Barnes, 1971) have shown that the iron content of sphalerite in a pyrite-saturated system is a function solely of temperature and the activity of sulfur (the role of pressure may safely be neglected for these shallow deposits), based on the definitive equation for the iron in sphalerite in pyrite-saturated system given below

$$\text{FeS (in sph)} + 1/2\, S_2 = FeS_2 \text{ (py)} \qquad (2)$$

In order to determine if the systematic variation of Th with iron content is due to a mineralogically reasonable buffer, we will examine the following reaction which could buffer the activity of sulfur in the ore-forming solution

$$3\ \text{daphnite} + 3\ \text{K-feldspar} + 5\ S_2$$
$$= 5\ \text{hematite} + 5\ \text{pyrite} + 3\ \text{Kmica}$$
$$+ 9\ \text{quartz} + 9\ H_2O \qquad (3)$$

This reaction differs from that of the proposed buffer of Barton et al. (1977) in that the Fe-chlorite in this reaction (3) is daphnite ($Fe_5Al_2Si_3O_{10}(OH)_8$) rather than an aluminum-free end member. Daphnite has a more realistic composition which closely approximates that of the Creede chlorite (Emmons and Larsen, 1923). However, using daphnite in the reaction requires that other aluminum-bearing phases be considered to balance the reaction. The two most logical choices for Creede are K-feldspar and sericite, the two phases which Barton et al. (1977) suggest controlled the pH of the system (discussed above).

In order to see if reaction (3) predicts the observed correlation of temperature with sphalerite iron content, it is necessary to (1) know the change in the free energy of the reaction with temperature so that the activity of sulfur can be estimated, and (2) use that sulfur activity to calculate the iron content of sphalerite at that temperature. Estimating the change in the free energy of the reaction is limited by the lack of free energy data on the daphnite component. However, since we are only trying to predict the change in the activity of sulfur with temperature for this buffer reaction, the accuracy of the free energy data is not as important as knowing how it changes with temperature. Hemingway et al. (1984) have recently measured the heat capacity of two natural chlorites (one Fe-rich, the other Mg-rich), and their data agree well with Helgeson et al.'s (1978) heat capacity data for daphnite between 200°-300°C. Although the heat-capacity data allow us to calculate the change in the free energy with temperature, in order to use those data it is still necessary to estimate the chemical potential of daphnite at one temperature. Therefore, we have used the estimate of Barton et al. (1977) of the chlorite + pyrite + hematite triple point at 250°C at approximately -11.0 for log S_2 activity and -34.2 for log O_2 activity. Using these values, the chemical potential at 250°C for the daphnite component of the Creede chlorite is estimated to be -1462 kcal/mole.

Using the activity of sulfur estimated from the above thermodynamic calculation at a given temperature, the iron content of the sphalerite can be calculated using the following equation

$$\log X_{(py)} = 6.809 - 7340/T - 0.5 \log a_{S_2} \qquad (4)$$

This equation which is a numerical expression of reaction (2) has the same slope as Scott and Barnes (1971) equation, but the intercept has been changed to agree with Czamanske's (1974) measurements on the iron content of sphalerite in equilibrium with pyrite +

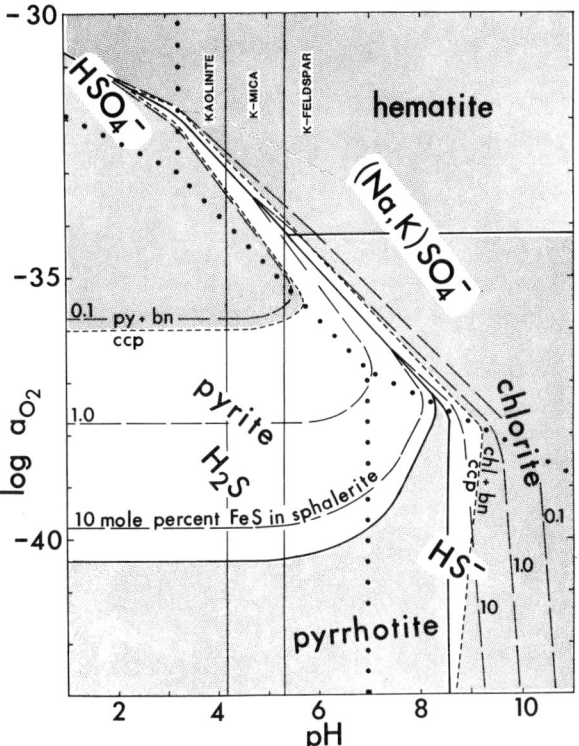

Figure 7.9. Log a_{O_2} - pH diagram at 250°C and total sulfur = 0.02 molal showing the mineral stability fields pertinent to the Creede ores (modified from Barton et al., 1977). The salinity is 1 molal, with Na+/K+ = 9. Dotted lines: boundaries between aqueous sulfur species; thick solid lines: boundaries between pyrite, pyrrhotite, chlorite, and hematite; short dashes: limit of stability of chalcopyrite; long light dashes: iron content of sphalerite; medium solid lines: boundaries between kaolinite, muscovite, and potassium feldspar. Abbreviations: py = pyrite, bn = bornite, ccp = chalcopyrite, chl = chlorite.

bornite + chalcopyrite, as discussed by Barton et al. (1977, p. 10).

In order to directly relate the iron content of the sphalerite to the buffer, we can combine equations (2) and (3) to get the following reaction

3 daphnite + 3 K-feldspar + 5 pyrite

= 10 $FeS_{(in\ sph)}$ + 5 hematite + 3 Kmica

+ 9 quartz + 9 H_2O (5)

Figure 7.10 shows that the predicted iron content between 200-280°C fairly closely matches the measured data, indicating that the ore-forming system may have been indeed buffered by reaction (3). It is important to note that the line on Figure 7.10 was forced to go through the data at 250°C by our estimate of the free energy of daphnite, but that the slope of the line was determined by the heat-capacity data. The scatter in the data in Figure 7.10, other than can be attributed to analysis and correlation error, is presumably due to perturbations of the chemical system away from the buffered environment, especially at the higher iron concentrations, as discussed by Barton et al. (1977).

Sources of metals--Lead-isotope studies reported by Doe et al. (1979) and unpublished data of Foley (U.S.G.S., 1985) show that the lead-isotopic composition of galena from the OH, Amethyst and Bulldog Mountain vein systems is remarkably uniform and is more radiogenic than that of any volcanic rock in the San Juan Mountains, or for that matter, than any Mesozoic or Cenozoic rock from the entire Rocky Mountain region measured to date. Doe et al. (1979) conclude that this requires the bulk of the lead in the Creede system to be derived by leaching of 1.4 to 1.7 billion-year-old Precambrian rock underlying the district. Surprisingly, the lead-isotopic composition of galenas from the Alpha-Corsair fault system and from the mineralized Creede Formation at Monon Hill, nearly on strike of the Alpha-Corsair fault system, are less radiogenic than the galenas from the main part of the district. These galenas appear to have a much larger component of lead from the volcanic rocks, suggesting that the lead deposited along the Alpha-Corsair system was derived from a different, possibly more shallow, lead reservoir than that of the Amethyst-OH-Bulldog Mountain system.

Sources of sulfur--Sulfur-isotope studies at Creede, only partially reported to date (Bethke et al., 1973), indicate a complex sulfur history not yet fully understood. Sulfur-isotopic equilibrium between sulfide minerals and reduced aqueous sulfur species was apparently closely approached. It is clear, however, that there was little, if any, sulfur-isotopic exchange between oxidized and reduced aqueous sulfur species in the ore zone, and they appear to have operated as separate isotopic (and probably chemical) systems in the upper part of the hydrothermal system. The narrow range of sulfur-isotopic composition of the sulfide minerals is interpreted to reflect a deep, nearly 0 permil, sulfide reservoir, whereas the extremely heavy values of $\delta^{18}O$ and $\delta^{34}S$ from sulfate minerals (up to 19 and 45 permil, respectively) require that at least a large part of the sulfate underwent biogenic reduction in the playa lake, the presumed reservoir for the ore fluids. Some of the sulfate was carried deep into the roots of the Creede hydrothermal system where temperatures were high enough (~350°C) for it to equilibrate with the wallrock silicates and the 0 permil sulfur reservoir. Most of the sulfate did not penetrate deeply enough to attain a high enough temperature to undergo substantial isotopic exchange. This partially-to-unexchanged sulfate did, however, mix into the hydrothermal system, presumably along the margins of the upwelling brine plume, to produce the exceptionally wide range of sulfur- and oxygen-isotopic compositions measured on the barites.

Hydrologic Environment

The hydrologic environment is a major factor in the localization of epithermal ore bodies. Unfortunately, it is also the environment which we are at present least able to treat quantitatively. However, a number of geologic, isotopic, and fluid-inclusion evidences can be used to place some restraints on the hydrologic environment of the Creede system, and to develop a qualitative model which can be compared with observed hydrologies of active geothermal systems.

Figure 7.11 shows a generalized hydrologic model for the Creede mining district as it is presently understood. It is similar in most aspects to the model orginally proposed by Steven and Eaton (1975) based upon geologic grounds, and to that proposed by Barton et al. (1977) based upon mineralogical and geochemical evidence. It is also similar in overall aspects to models of geothermal systems in the Taupo Volcanic Zone, New Zealand (cf. Henley and Ellis, 1983).

The Creede ores were deposited along the top of a saline, deeply circulating hydrothermal system, charged primarily with meteoric waters, that displaced the regional ground water regime. An intrusion underlying the district at a depth of 3 to 5 km is speculated to have been the heat source which provided the buoyancy of the brine plume. In the upper part of the system, fluid movement was fracture-controlled and nearly horizontal, from north to south. The ore zone was overlain by a zone of fresh ground water, approximately 500 meters thick, , which flowed southward down the regional hydraulic gradient. The base of this ground water zone was heated to temperatures of about 160°C by heat transfer from the underlying brine. Precipitation of the ores at the interface between the deep brine and the overlying ground water appears to have resulted from the dual processes of boiling and mixing. The various evidences supporting this general hydrologic model are presented below.

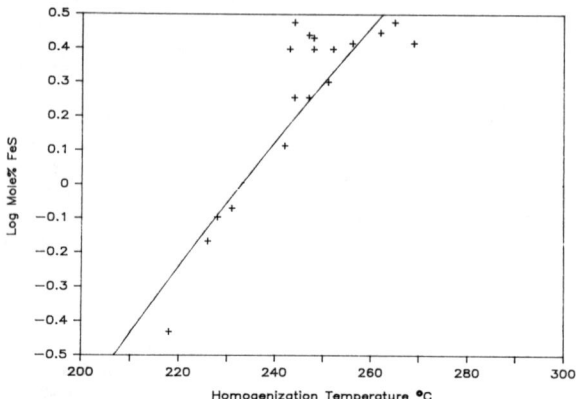

Figure 7.10. Diagram showing the temperature of homogenization of fluid inclusions vs. the iron content of the host sphalerite growth zone for sample locality NJP-X on the OH vein. The line shows the predicted iron content of the sphalerite if the sulfur fugacity of the system had been buffered by the triple point - Fe-chlorite (daphnite), pyrite, hematite.

Geologic constraints--Steven and Eaton (1975) suggested that the circulating hydrothermal system responsible for mineralization was influenced by two major lithologic factors. The first was the location of soft, non-welded to poorly-welded, relatively impermeable, tuffs along the top of the ore zone. This

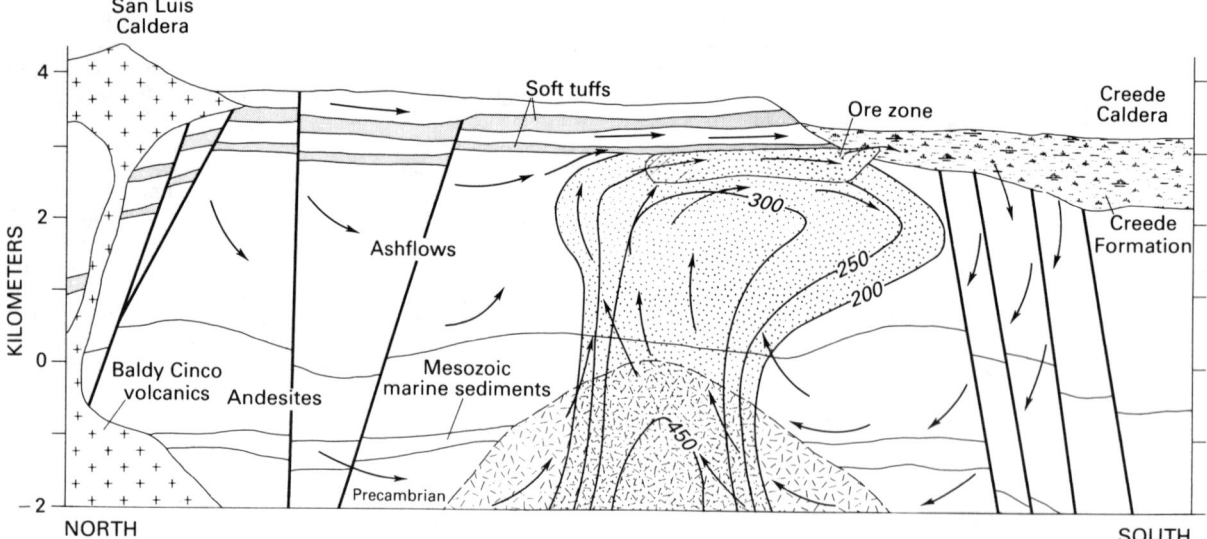

Figure 7.11. Schematic representation of the Creede hydrothermal system. Upwelling plume (stippled pattern), outlined by the 200°C isotherm, displaces regional ground water flow from highlands in the north to the low area of the Creede caldera moat to the south. Heat source responsible for buoyancy of plume is shown as stock beneath district (hatched pattern).

aquitard largely blocked the upward movement of the hydrothermal solutions and forced the solutions to flow laterally to the south. The second lithologic factor was the location of the permeable talus-regolith and fanglomerate deposits of the Creede Formation at the southern end of the vein system (Figs. 7.3 and 7.11). Where cut by the flow-controlling fractures, these coarse clastics provided an outlet to the south for the hydrothermal solutions.

The mineralogically and geochemically based hydrologic model of Barton et al. (1977) is nearly identical to that of Steven and Eaton (1975). Barton et al. describe the system as a freely convecting hydrothermal cell that extracted metals and sulfur from sources at depth and deposited gangue and ore minerals near the top of the system. They attribute ore deposition to cooling and a slight pH rise due to boiling of the hydrothermal fluid. Recondensation of the CO_2 and H_2S, which were strongly fractionated into the vapor phase during boiling, in the cooler, overlying rocks led to the formation of the intensely sericitized cap above the ore (Fig. 7.6). The development of this zone of intense alteration above the orebodies undoubtedly served to increase the efficiency of the aquitard along the top of the system.

Influence of topography--In addition to the influence of the soft-tuff aquitard and sericitic alteration zone, it is probable that the topography at the time of ore-formation played an important role in maintaining a "cap" of cooler ground water above the deeply circulating ore-forming brine. According to Steven and Eaton (1975), at the time of ore-formation the area of the San Juan Mountains comprised a widespread volcanic plateau punctuated by regions of rough topography in caldera areas. Local relief in the vicinity of Creede approached 1.8 km over horizontal distances of 10 km. The low point was the playa lake in the moat of the Creede caldera, and maximum elevations were attained along a string of volcanoes located along the present Continental Divide, about 10 km north of the geographic center of the mining district. This high relief imposed a strong regional hydraulic gradient on the ground water from north to south across the ore zone. As noted by Henley (1985, this volume), such a regional hydraulic gradient will tend to establish a lateral ground-water flow across the top of a geothermal system. Such an effect has been evaluated by Hanaoka (1980) using numerical modeling techniques. Hanaoka's analysis is simplified in that: (1) it is calculated for pure water; (2) the conditions were chosen to obviate boiling; (3) the permeability and thermal conduction-dispersion coefficient were both uniform; and (4) the effect of pressure on viscosity was ignored. Changes in these variables will modify the quantitative aspects of Hanaoka's results, but, within reasonable bounds, are not likely to affect the general topology of his models. His calculations showed that in areas of moderate relief, a two-tiered flow regime may be produced, consisting of a deep, hot, convecting cell overlain by a cooler region of ground-water flow forming a "hydrologic" cap to the hydrothermal system. Flow is parallel and lateral along the interface between the two regimes, and heat is transferred from the deep to the shallow system by conduction and dispersion. Hanaoka's model is for moderate relief consistent with the conditions observed in a number of geothermal systems (Healy and Hochstein, 1973; Healy, 1975; Ellis and Mahon, 1977; Hedenquist, 1983). It is also tantalizingly consistent with the evidence for the interaction between a hot, deep, saline fluid, and a cooler, overlying fresh ground water at Creede.

Isotopic constraints--Light-stable-isotope studies have shown that in addition to the deep, saline fluids and the shallow groundwaters discussed above, fluids from a third, isotopically distinct, probably magmatic, source were involved in the Creede mineralization. The evidence for the episodic introduction of magmatic waters into the Creede hydrothermal system comes from carbon- and oxygen-isotopic studies of the vein carbonates (Bethke and Rye, 1979). These studies indicate that both the early (A-stage) rhodochrosite and the younger (C-stage) siderite-manganosiderite were deposited from fluids that equilibrated with silicates at very high temperatures (presumably magmatic fluids). This supposition was based primarily on the large $\delta^{18}O$ values obtained, but it is also consistent with the carbon-isotopic composition and with the hydrogen-isotopic composition of the inclusion fluids in a few rhodochrosite samples. The interpretation remains open to question, however, because it requires either that the vein system was occupied by magmatic fluids at two different times, separated by a period when it was occupied by the meteoric waters from the lake sediment reservoir, or that CO_2 given off from the magma did not exchange oxygen with the meteoric waters that filled the vein system. Neither seems a reasonable proposition. Carbonate minerals in many hydrothermal ore deposits appear to have been deposited from fluids substantially heavier or lighter in $\delta^{18}O$ than those from which other apparently coeval minerals were deposited (R.O. Rye, U.S.G.S., personal communication, 1985). Perhaps we do not fully understand the isotope systematics of CO_2 in hydrothermal systems.

Based on the hydrogen- and/or oxygen-isotopic composition of the alteration minerals, and chlorite and the inclusion fluids in sphalerite, Bethke and Rye (1979) postulated that the deep, saline fluids responsible for ore deposition originated as pore-waters in the Creede Formation which accumulated in the evaporative, closed-basin playa lake in the moat of the Creede caldera. Such an interpretation is consistent with the unusually high salinities of the Creede ore fluids, and with sulfur- and oxygen-isotopic studies on barite (R. Rye, P. Barton and P. Bethke, U.S.G.S., unpublished data, 1985) which suggest that most of the sulfate in the Creede system must have undergone considerable isotopic evolution in the playa lake sediments.

The recent work of Foley et al. (1982) demonstrated the existence of a zone of heated, fresh meteoric water overlying the ore zone. They showed that fluids in primary inclusions in quartz were similar in salinity, homogenization temperature, and hydrogen-isotopic composition (and therefore, presumably, source) to primary inclusions in sphalerite. This interpretation was contrary to that originally proposed by Bethke and Rye (1979), which suggested that the quartz was deposited from fluids that originated as meteoric waters in the high country north of the district. Very painstaking sampling and analysis showed that the hydrogen-isotopic analyses of fluid inclusions in quartz reported by Bethke and Rye (1979) were biased by fluids released from pseudo-secondary inclusions during analysis. These pseudo-secondary inclusions homogenized over a similar but slightly lower temperature range as did the primaries, but contained fresh waters whose hydrogen isotopic composition was at least 30 permil lighter than that of the brine in the primary inclusions. Foley et al. (1982) proposed that these light, fresh waters were unevolved meteoric waters that constituted the regional ground waters. These fluids overlay the ore zone, but episodically entered it due to hydrologic fluctuations during vein filling. Presumably, the thermal shock of these heated, but somewhat cooler ground waters, caused fractures in growth-strained quartz crystals which, on rehealing, trapped some of the fluids.

The lead-isotope studies by Doe et al. (1979) and unpublished data of Foley (U.S.G.S., 1985), discussed in the section on sources of metals, also provide evidence that the Creede hydrothermal system circulated to depths of several kilometers. Their data suggest that the bulk of the lead in the Creede system was derived by selective leaching of radiogenic lead from the 1.4 to 1.7 billion-year-old Precambrian rock underlying the district. The depth to the Precambrian at Creede (and hence the minimum depth of hydrothermal circulation) can only be estimated, but must be at least 2.5-3.0 km below the ore zone (T. Steven, U.S.G.S., written communication, 1982).

Boiling and Mixing in the Ore Zone

The most recent modification to the interpretation of the hydrology is the evidence for mixing of the hydrothermal fluid with overlying ground water developed by Hayba (1984) for the OH vein and by Robinson and Norman (1984) for the southern Amethyst vein. Although boiling has been documented in the upper levels of the OH vein (Roedder, 1970; Woods et al., 1982), it appears that mixing was responsible for at least some of the vein mineral deposition. These two processes, boiling and mixing, are each important mechanisms of mineral deposition; evidences for their roles in the Creede hydrothermal system will be discussed below.

Boiling--One of the major factors influencing the hydrologic interpretation of the district is the fact that boiling occurred during at least some of the depositional history. As noted earlier, there are very few deposits exhibiting unequivocal evidence that boiling occurred during precious-metal deposition. At Creede, there is good evidence for boiling, but it appears to have taken place during stage D, a base-metal stage, rather than during the precious-metal stage B (Table 7.4). The case for boiling at Creede is based primarily on two lines of evidence: (1) fluid inclusions, and (2) the distributions of vein adularia and sericitic alteration.

Over 25 years ago, Roedder (1960) discovered some coeval fluid inclusions in D-stage quartz that had widely varying liquid/vapor ratios, and thus a wide range of homogenization temperatures, many over 280°C (about the maximum temperature for Creede mineral deposition based on data from over 2,500 fluid inclusion measurements). He interpreted these inclusions as having resulted from the trapping of varying proportions of liquid and vapor from a heterogeneous, two-phase system. This was the first such demonstration of boiling in any ore deposit. His interpretation is further bolstered by the presence of "empty" or "steam" inclusions in some of these same quartz samples (e.g., Roedder 1970, Fig. 7.7), and by the occasional occurrence of ore stalactites (indicative of growth in a two-phase regime) protruding into open vugs along the OH and Bulldog Mountain vein systems.

To date, evidence for boiling has been found at six localities along the top of the OH vein (Fig. 7.6) in D-stage quartz. It is notable that Robinson and Norman (1984) found no direct evidence of boiling in quartz samples from the southern Amethyst vein although they carefully looked for indications of boiling. No such evidence has yet been found in the limited number of fluid-inclusion measurements on material from the Bulldog Mountain vein system, but the material so far examined has come from areas on that vein which lie to the south of the area on the OH from which boiling has been documented. The implication is that boiling was primarily confined to the northern part of the district, at least in the later stages of ore deposition during which most of the material studied to date was deposited. This surmise is consistent with the extensive evidence for mixing (discussed below) of the ore fluids with cooler overlying fresh waters along the top of the system. To date, fluid-inclusion evidence for boiling at Creede has been found only in quartz crystals, even though it has been sought carefully in sphalerite and fluorite. It is possible that boiling also occurred during the "B-stage" deposition of the precious-metal ores, but evidence for boiling during this staged has been carefully looked for and has not been found. The fine-grained nature of B-Stage material stage makes recognition of fluid-inclusion evidence for boiling difficult, but the lower homogenization temperatures for B-Stage inclusions than for D-Stage inclusions is consistent with a lack of boiling.

Also shown on Figure 7.6 is the distribution of vein adularia and intense sericitic alteration. Adularia occurs mainly below the zone of boiling documented by fluid inclusions, while the zone of intense sericitic alteration, which caps the OH vein, is above the zone of documented boiling. The sericitic cap, which is spectacular in its contrast and sharpness of contact with the unaltered wallrock below (the productive part of the vein), may have resulted from the recondensation of volatiles in the overlying, probably fresh waters and from the subsequent hydrolysis reactions. At Broadlands, adularia and calcite deposition has been related to the rise in pH of the fluid on boiling, due to the strong fractionation of acid-forming volatiles such as CO_2 and H_2S into the vapor phase (Browne and Ellis, 1970). This pattern of intense sericitic alteration overlying a zone of vein adularia (+ carbonate in some mining districts) provides indirect, but potentially useful evidence for boiling in epithermal systems. It may be a particularly useful indicator of boiling in deposits where no material

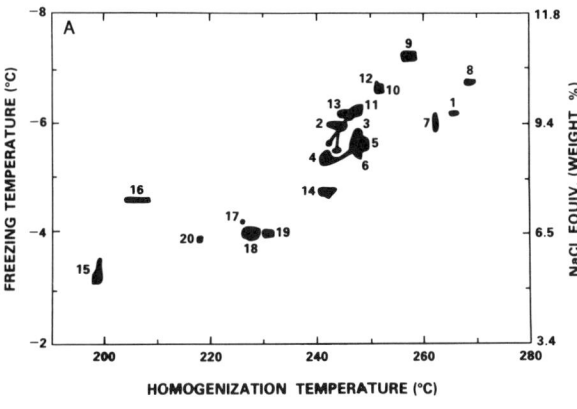

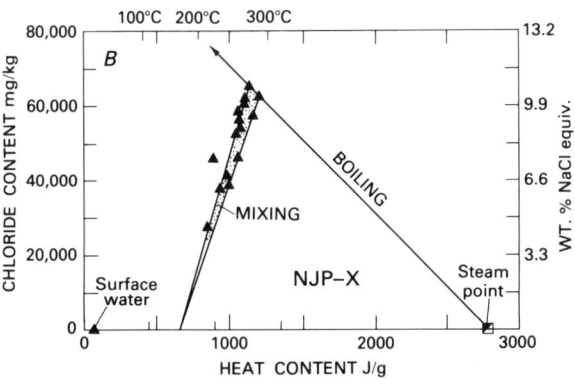

Figure 7.12. a). Plot of homogenization temperature (Th) vs. freezing temperature (Tf) for 221 primary inclusions in a 5-cm band of zoned sphalerite from Creede, Colorado, after Reodder (1977). The numbered areas include all data points from each of the 20 zones sampled, numbered in sequence from zone 1 (earliest) to zone 20 (latest). The number of inclusions in each of the areas outlined are as follows: 1(2), 2(18), 3(1), 4(21), 5(27), 6(9), 7(4), 8(9), 9(8), 10(4), 11(15), 12(2), 13(7), 14(12), 15(14), 16(11), 17(4), 18(32), 19(13), 20(8). b). Plot of heat content vs. chloride content for primary inclusions (solid triangles) replotted from Figure 7.12a. Data points for surface water and steam are also shown. Tick marks for upper and lower temperature scale are offset because of the effect of chloride content.

adequate for fluid-inclusion studies is available, or where the fluid-inclusion evidence for boiling is ambiguous.

The virtual lack of calcite in the Creede district can be explained by the low-calcium content of host rocks resulting from the earlier potassium metasomatism event, and to the low-CO_2 content of the fluids during most of the depositional history. Wetlaufer (1977) has shown that the fluids responsible for deposition of the early (A-Stage = Stage I) Mn-Fe carbonates in the lower parts of the southern portions of the vein systems were similar in their thermal history and chemistry to the later fluids which deposited the bulk of the base- and precious-metal mineralization except that the carbonate-depositing fluids had a higher CO_2 content. It is not unreasonable to suppose that the early Mn-Fe carbonates in the Creede vein systems play the role of calcite in the Broadlands geothermal system and record a period of boiling of relatively gas-rich fluids early in the history of vein filling.

Mixing--Roedder (1977) showed that there was a systematic relation between temperature and salinity for fluid inclusions from the OH vein. He documented this relationship by a detailed growth-zone by growth-zone study of a single, large sphalerite crystal from a single locality (NJP-X) on the OH vein. Twenty distinct growth zones were defined in this crystal, and homogenization and freezing temperatures were measured on sets of inclusions within each zone. The results of his painstaking study are shown in Figure 7.12a. Similar results were obtained for several different localities on the OH vein (Woods et al., 1982). There are many implications of these systematics, perhaps the most important for our present purposes being the mixing of fluids of different temperatures and salinities. Truesdell and Fournier (1975) have shown that plots of heat content (enthalpy) against chloride content are very useful in evaluating the relationship between fluids of different temperatures and composition in geothermal areas. Both enthalpy and chloride content are additive quantities so that trajectories for boiling and mixing are linear on such plots. Roedder's data are replotted as enthalpy-chloride diagrams in Figure 7.12b. It can be seen from this figure that the systematic relationship between temperature and salinity can be explained by mixing of saline, high-temperature waters (from zones 8 and 9 on Figure 7.12a) with fresh waters heated to a temperature of about 160°C. It is possible that waters from zones 8 and 9 are related to each other in that a small amount of boiling of zone 8 waters would yield the slightly more saline, lower temperature water of zone 9. Enthalpy- or chloride-conservation calculations indicate that if the steam is separated from the fluid, only 6 wt.-% of the fluid needs to be converted to steam to produce water 9 from water 8. On the other hand, in order to produce the lowest-temperature, least-saline fluid (water 15), water 8 would have to be mixed with more that its equivalent weight of fresh water at 160°C.

Roedder's documentation of the abrupt temporal variations in the ore fluid along the OH structure was an important time constraint for a later study on the spatial variations in the ore fluid. Using a distinctive growth zone in sphalerite as a time-line throughout the OH vein, Hayba (1984) documented a progressive decrease in both temperature and salinity from the northern, basal end of the vein to localities 200 meters higher and 1000 meters further south. These temperature and salinity gradients are interpreted as the progressive mixing of deeper, saline hydrothermal fluids with overlying, dilute ground waters that have been preheated to approximately 160°C. Independent, isotopic evidence for the presence of a dilute ground water in the OH vein has been documented by Foley et al. (1982) (discussed earlier). It is interesting to note that the estimated temperature of 160°C for the dilute ground waters is within the 100° to 180°C temperature range estimated for the diluent in most New Zealand geothermal systems (Hedenquist and Reid, 1984).

The fluid-inclusion studies of Robinson (1981; Robinson and Norman 1984) also indicate that the deep hydrothermal solutions mixed with shallow ground water in the southern Amethyst vein. Due to the fact that their fluid-inclusion study was done on quartz, it was impossible for them to distinguish growth zones or make detailed time correlations. Instead, a more general approach was taken and the inclusions from one stage of quartz deposition were measured on samples covering a vertical range of 336 meters. Although at each sampled elevation there is a large range in both the temperature and salinity in that stage, there is a general decrease in both temperature and salinity with increasing elevation, which they attribute to mixing.

Additional evidence for mixing comes from the district-wide mineral zonation of the sulfide-rich OH vein in the north to the barite-rich Bulldog vein in the south. In Figure 7.13 the solubility of barite is contoured on a temperature-salinity diagram and mixing and boiling trajectories relevant to Creede are superimposed. It can be seen that barite solubility changes relatively little at the high temperatures and salinities appropriate for input fluids in the northern OH vein, and drops significantly only at salinities below about 6 wt.-% NaCl equivalent (Plumlee and Hayba, 1985). (Note: in sulfate-rich solutions above pH 5, changes in pH have no effect on barite solubility). The topology of Figure 7.13 suggests that fluid mixing was the depositional mechanism for barite at Creede. Most mixing paths (decreasing temperature and salinity) cross solubility contours while boiling paths are parallel to them. Thus, only after the hot, saline fluids, which rose in the northern parts of the district, were diluted significantly did they deposit the large quantities of barite seen in the southern and upper parts of the district. The silver content of the Creede ores is also higher in the southern and upper parts of the district (Barton et al., 1977), suggesting that mixing was an important mechanism of silver deposition at Creede.

<u>Relative importance of boiling and mixing</u>--It was pointed out at the beginning of this section that both boiling and mixing are important processes leading to the deposition of base- and precious-metal ores in epithermal systems. Drummond and Ohmoto (1985), Henley (1985, this volume) and Reed and Spycher (1985, this volume) have all emphasized the efficiency of boiling in the precipitation of both ore and gangue minerals in epithermal environments. Barton et al. (1977) specifically related the precipitation of the Creede ores to cooling and pH changes occasioned by boiling. The evidence from the recent studies of Creede cited above, however, would imply that mixing, not boiling, was the immediate cause of most, if not

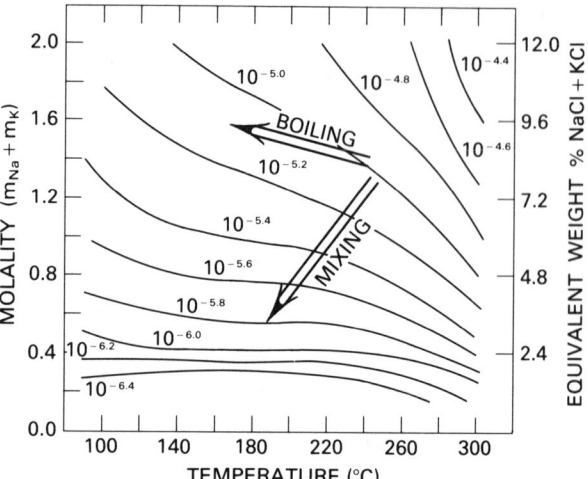

Figure 7.13. Diagram showing the solubility of barite contoured on a temperature vs. salinity plot. Barite solubility was calculated at pH = 5.5, $m_S = 0.02$, $S_{ox}/S_{red} = 10$, $m_{Na}/m_K = 9$. Arrows mark boiling and mixing trajectories.

all ore deposition at Creede. This evidence is summarized below:

1. Fluid-inclusion evidence for boiling has been found only on the northern half of the OH vein, and only in quartz deposited in the latest, silver-poor, stage of mineral deposition on the OH.
2. Fluid-inclusion evidence for substantial amounts of mixing, increasing upward and to the south, have been found along the OH and southern Amethyst vein systems.
3. The deposition of large amounts of barite in the southern parts of the district and its absence in the northern parts is consistent with a mixing model, not with a boiling model of barite deposition.
4. On the OH vein, the maximum temperature measured on B-Stage material (the principal silver-bearing stage at Creede) is 241°C, at least 30° less than the maximum temperature measured from D-Stage material. To produce boiling at that temperature would require that the water table be lower by about 200 meters during B-Stage than it was during D-Stage.
5. In a system dominated by lateral flow (i.e., one that allows for only minor changes in pressure), it is virtually impossible for a fluid to boil after its temperature has been lowered by mixing.

The widespread and intense sericitic alteration that caps the Creede orebodies, the widespread occurrence of vein adularia in the system, and the deposition of massive amounts of Mn-Fe carbonates in the deep parts of the southern portions of the vein systems all argue that very substantial amounts of boiling took place during vein filling. None of these features can, however, be correlated with the major periods of base- and precious-metal deposition. We conclude, therefore, that although boiling can be well documented at Creede by several criteria, there is no evidence that boiling (at least in the ore zone) was related to base- or precious-metal deposition. The evidence for mixing of the deep, saline, ore fluid with overlying fresh ground water during sulfide deposition, on the other hand, is overwhelming. As pointed out in the first section of this chapter, Creede is not unique among the Adularia-Sericite-type deposits in exhibiting a lack of evidence for boiling tied to metal deposition. Only at Colqui is the evidence compelling that precious-metal deposition resulted from boiling of the ore fluid (Kamilli and Ohmoto, 1977).

Summary of Creede Mineralization

On both geologic and geochemical grounds, it has been proposed that the Creede ores were deposited along the top of a deeply circulating hydrothermal system at the interface of that system with the overlying ground water. The concept is illustrated in Figure 7.11. The heat source driving the hydrothermal circulation is speculated to have been an intrusion underlying the district at a depth of 3 to 5 kilometers. This intrusion may have been related to the quartz-latite, ring-fracture volcanism of the Creede caldera cycle, or may have been a quartz porphyry related to the later bimodal basalt-rhyolite volcanism. The ore fluids were dominated by meteoric waters, whose isotopic composition and salinity evolved by processes of evaporation and diagenesis in the playa lake in the moat of the Creede caldera. The episodic introduction of magmatic fluids into the circulating system is suggested by the isotopic composition of rhodochrosite and siderite in the veins. The bulk of the lead in the Creede ores, and therefore, presumably, most of the other metals, appears to have been leached from Precambrian basement rocks at depth. The sulfur in the Creede ores may have come from several sources and its isotopic composition documents a complex mixing and exchange history not now satisfactorily understood. Most of the sulfate in the Creede system appears to have undergone considerable isotopic evolution in the playa lake sediments, but the sulfide sulfur appears to have been buffered by, and perhaps derived from, a large reservoir of magmatic sulfur at depth. Precipitation of the ores along the top of the system appears to have resulted from the dual processes of boiling and mixing, but mixing appears to have been quantitatively the more important mechanism of sulfide deposition. The intense, mixed-layer illite/smectite alteration cap is interpreted to have been generated by condensation along the top of the system of acid volatiles distilled off the deeply circulating ore fluids.

The model includes substantial amounts of fluid-rock interaction and chemical and isotopic exchange in the deeper, higher temperature parts of the system. In the ore zone, however, isotopic exchange between oxidized and reduced aqueous sulfur species was minimal and the aqueous sulfate apparently did not exchange oxygen with the ore fluid nor was there significant oxygen isotopic exchange between the ore fluid and the unaltered wallrocks. The pH, the activity of S_2 gas and redox state of the ore fluids were buffered in the ore zone by reaction with a vein-filling assemblage comprising: Fe-chlorite + hematite + pyrite + K-feldspar + sericite (or mixed-layer illite-smectite clays). Throughout most of the history of vein filling, the redox state of the ore fluid was that of the triple point: Fe-chlorite + pyrite + hematite. During the early part of sulfide deposition, however, numerous excursions to substantially lower oxidation (and sulfidation) states occurred. The excursions have been interpreted to have resulted from episodic introduction of magmatic emanations into the circulating system or to reaction with ferrous silicates episodically exposed to the circulating fluids deep in the system through tectonic adjustments. Alternatively, the relatively oxidized state of the ore fluid during most of the ore deposition cycle could be due to mixing of a deep, reduced fluid with surrounding and overlying oxidized groundwater prior to entering the ore zone, the excursions resulting from lesser amounts of mixing. Present evidence does not allow us to choose between these two alternatives.

THE ACID-SULFATE ENVIRONMENT: SUMMITVILLE AS AN EXAMPLE

The recent thesis by Stoffregen (1985) combined with the earlier work by Steven and Ratté (1960a) and several other studies (Patton, 1917; Mehnert et al., 1973; Lipman, 1975; Perkins and Nieman, 1983) make the Summitville mining district, Colorado, the best documented Acid-Sulfate-type deposit and the logical choice for illustrating the characteristics of Acid-Sulfate-type epithermal systems. Many of the interpretations on Acid-Sulfate-type deposits have been made in light of the important insights gained from the studies done at Goldfield, Nevada (Ashley, 1974; Ransome, 1909). Even so, it should be noted that the observational base for this type of deposit is still much smaller than that for the Adularia-Sericite deposits. Since our experience in the Summitville district is limited, most of the geologic and mineralogic characteristics discussed below, except where noted, are based on the work done by Steven and Ratté (1960a) and by Stoffregen (1985).

As was the case in using Creede as the exemplar for Adularia-Sericite-type deposits, Summitville has some characteristics which are not representative of all of the Acid-Sulfate-type deposits. In particular, Julcani, Lake City II, and Red Mountain are silver-rich, contain relatively more base metals (particularly lead and zinc) and appear to have formed at greater paleodepths. Although these and other differences exist, the similarities in mineralogy, alteration, tectonic setting, and timing of ore deposition relative to host emplacement among Acid-Sulfate-type deposits are striking, and indicate mineralization in a distinct geothermal environment. Ashley (1982) has summarized the observations on a number of deposits of this type into a particularly useful occurrence model which provides further documentation on the characterisitics of Acid-Sulfate-type deposits.

Geologic and Mineralogic Characteristics

<u>Volcanic history</u>--The Summitville mining

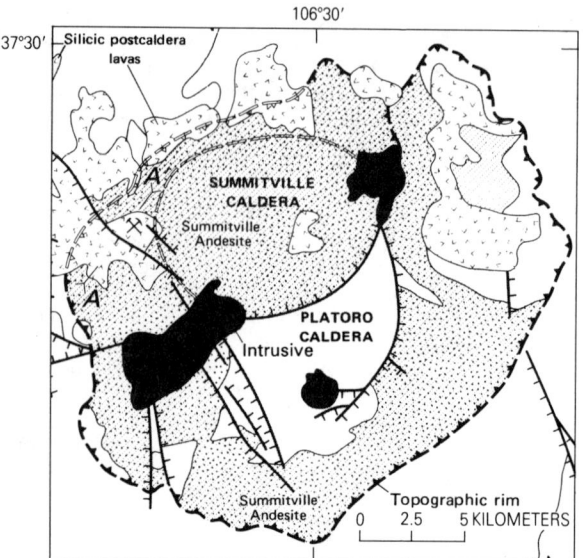

Figure 7.14. Generalized geology of the Platoro and Summitville calderas, modified from Steven and Lipman (1976). Location of Summitville mining district is shown by pick-and-hammer. Control is moderate to good where boundaries are shown by solid symbols. A-A' marks location of cross-section shown in Figure 7.15.

district is located at an elevation of about 12,000 feet on the northwest edge of the Platoro caldera and the younger nested Summitville caldera in the eastern San Juan Caldera complex, Colorado (Lipman, 1975) (Fig. 7.1). The quartz latite porphyry at South Mountain, which hosts the deposit, is a lava dome emplaced 22.8 m.y ago (Mehnert et al., 1973) along the western margin of the Summitville caldera at its intersection with the Pass Creek-Elwood Creek-Platoro fault zone, a major structure cutting across the center of the Platoro-Summitville calderas (Fig. 7.14). Drilling has confirmed Steven and Ratté's (1960a) suggestion that the dome is funnel-shaped in cross section with a narrow intrusive pipe at depth which flares out near the surface (Fig. 7.15). The quartz latite is characterized by 2 to 8 cm K-feldspar phenocrysts in an aphanitic, greenish groundmass; plagioclase phenocrysts are common but are finer grained than the K-feldspar. Also typical are locally resorbed quartz eyes (0.2 to 2 cm), euhedral biotite books (1 to 2 cm), and apatite up to 0.5 cm long (Stoffregen, 1985). The quartz latite at South Mountain is bordered on the north, east, and south by the approximately 29 m.y. old Summitville Andesite which filled the Platoro caldera after collapse and on the west by the rhyodacite of Park Creek (about 28 to 26 m.y.), which comprises lava-dome complexes erupted around the northwest margin of the Summitville caldera after its final collapse (Lipman, 1975). The rhyolite of Cropsy Mountain, a coarsely porphyritic lava flow, locally overlies the quartz latite south of the district. In

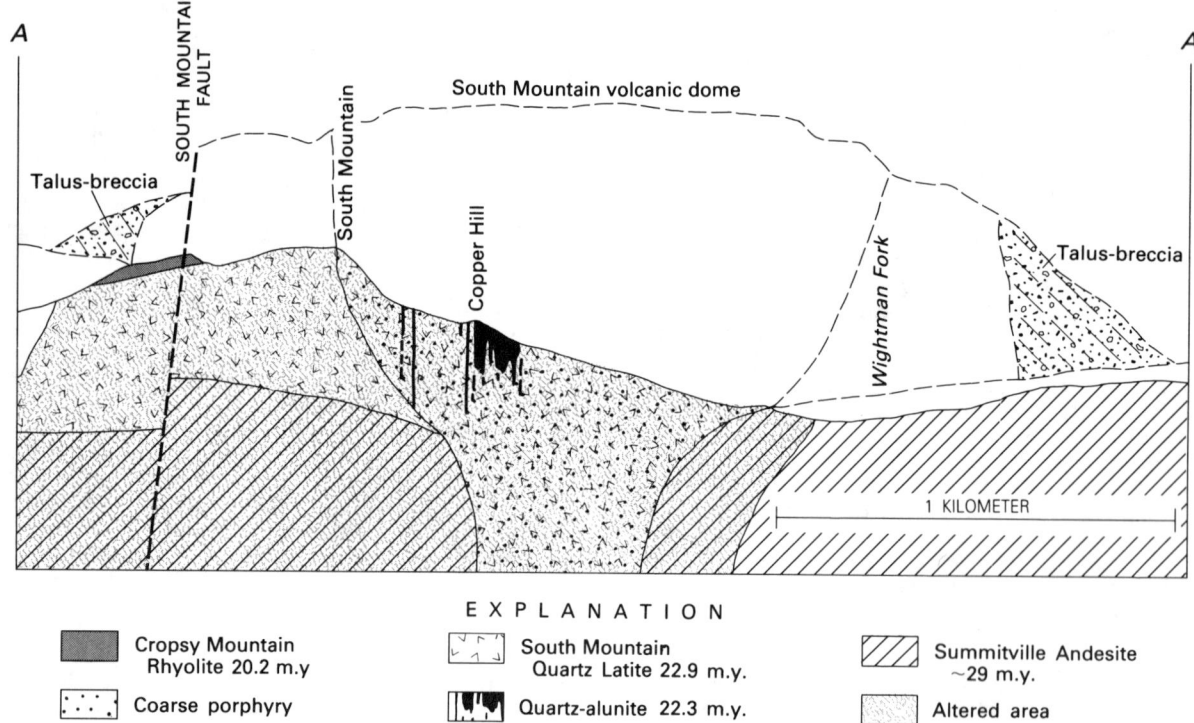

Figure 7.15. Geologic cross-section of the restored South Mountain volcanic dome, modified from Steven and Ratté (1960a). Fault is shown with heavy line; contacts shown with thin line; both are dashed where approximate.

contrast to the older units, the rhyolite of Cropsy Mountain is everywhere unaltered. It has been dated at 20.2 ±0.8 m.y. (Mehnert et al., 1973) and at 18.5 ±1.2 m.y. (Perkins and Nieman, 1983). The age of mineralization is therefore bracketed between 22.8 and 20.2 m.y. in agreement with Mehnert et al.'s (1973) date of 22.3 m.y. on hydrothermal alunite from Summitville.

Ore deposits--The ore bodies are localized along the southwestern margin of the coarsely porphyritic core of the dome, just north of the northwest-trending fault zone at South Mountain (Steven and Ratté, 1960a). The ore occurs in a series of irregular pipes and vein-like masses of quartz and alunite that formed largely by replacement of the original quartz latite. Significant mineralization is confined to a vertical interval of approximately 400 meters, and the surface outcrop of the mineralized zone can be circumscribed by an elipse with axes of 1.5 and 1.0 kilometers (Ratté and Steven, 1960a). Production through 1947 was 113,000 oz of Au, 240,00 oz of Ag, and 433,000 lbs of Cu (Vanderwilt, 1947). Productions since 1947 has been insignificant.

Wallrock alteration--The alteration in the upper part of the deposit shows a well-defined pattern (Fig. 7.16) from an intense zone of acid leaching, referred to as "vuggy silica alteration", surrounded by quartz-alunite alteration grading outward into quartz-kaolinite. Further out, there is generally an abrupt change into an illitic* zone which grades out to a montmorillonite zone. The width of each of the zones is highly variable.

---

*Stoffregen (1985) has used the term illite to refer to a fine-grained (<2 micron) phyllosilicate with a 10-angstrom basal spacing, which does not expand on glycolation, and contains less than 5% smectite layers.

The vuggy silica alteration is interpreted to be the result of the acid dissolution of all the primary rock-forming minerals except quartz. Most of the ore occurs in this very permeable zone, which is characterized by large voids due to the removal of the K-feldspar phenocrysts. At the surface the vuggy silica is quite extensive, but below a depth of about 1000 feet, it becomes much more restricted (Fig. 7.17). In the upper part of the deposit, the surrounding quartz-alunite zone is up to 50 feet wide. Alunite occurs as pseudomorphous replacements of K-feldspar phenocrysts and as matted aggregates replacing the groundmass. With decreasing elevation (between 11,500 and 11,000 feet), alunite becomes insignificant and quartz-kaolinite rather than quartz-alunite surrounds the vuggy silica "vein". The texture of this deeper kaolinite indicates that it has completely replaced pre-existing alunite. Except for this deep zone, the quartz-kaolinite zone is generally thinner (and is locally absent) than the quartz-alunite zone which it surrounds (Stoffregen, 1985).

Outside of the quartz-kaolinite zone, the texture of the alteration shows a marked change from hard and brittle rock, due to a silicified matrix, to a soft, incompetent rock, due to a matrix of clay minerals.

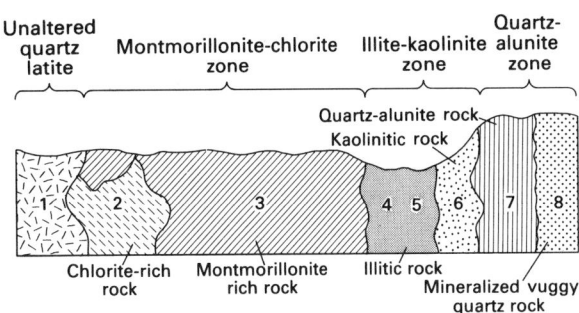

Figure 7.16. Diagram showing hydrothermal alteration pattern in the Summitville district adapted from Steven and Ratté (1960a).

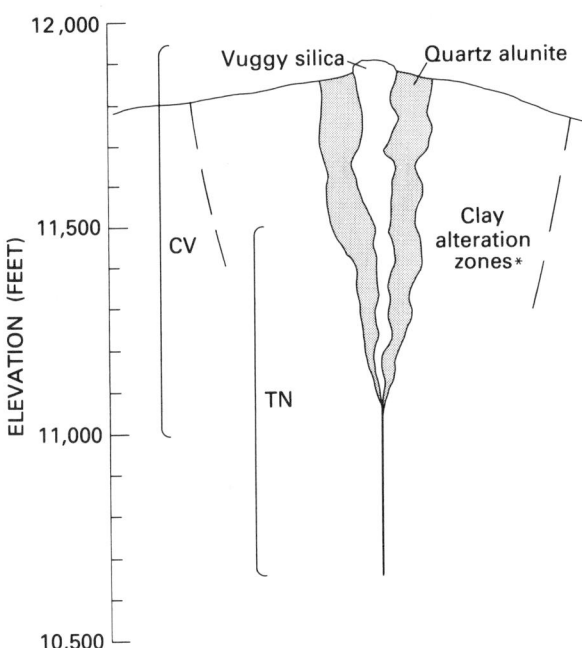

Figure 7.17. Schematic cross section of the alteration patterns and mineral zonation of the Summitville deposit, modified from Stoffregen (1985). The clay alteration zones refer to zones 3-6 in Figure 7.16. CV - covellite, luzonite, enargite, pyrite, marcasite, chalcopyrite, trace sphalerite, sulfur, and gold assemblage; TN - chalcopyrite, tennantite, pyrite, plus minor sphalerite and trace galena assemblage.

The clay envelope around the vuggy silica and quartz-alunite-kaolinite alterations is generally at least 100 feet wide. According to Stoffregen (1985), illite predominates near the quartz-kaolinite zone, but montmorillonite becomes predominant further away from the vein. Kaolinite is present throughout the clay zone, but it decreases in both abundance and crystallinity away from the quartz-kaolinite zone. The illite becomes coarser grained with depth. Mixed-layered illite/smectite clays (having greater than 5% smectite) are generally not present, occurring only locally in the central portion of the deposit where they appear to be related to a post-mineral intrusion.

Mineralogy--Stoffregen (1985) describes the mineralogy of Summitville in terms of three main stages: main, late, and supergene. Mineralization from these three stages was later than much of the acid-sulfate alteration, as evidenced by the presence of ore minerals in cross-cutting fractures and voids in vuggy silica. In addition, a very minor, early stage of mineralization is also present as 10 to 40 micron-sized inclusions in pyrite grains of either pyrrhotite or chalcopyrite + bornite + digenite. The significance of this minor, early stage is not clear.

The character of the main-stage mineralization changes with depth in the deposit (Fig. 7.17). In the deeper parts of the deposit a chalcopyrite-tennantite assemblage predominates. Tennantite usually contains only minor antimony and trace amounts of silver, although rarely it may contain up to 2.0 wt.-% silver. Pyrite plus minor, low-iron (usually $\leq$ 1.0 wt.-%) sphalerite and a trace of galena are part of this deep assemblage. In the upper part of the deposit, which contains the economically significant precious-metal mineralization, the main-stage mineral assemblage consists of covellite, luzonite, and enargite with pyrite, marcasite, locally sulfur, trace amounts of sphalerite, and gold. Chalcopyrite is also present, but it decreases in abundance and is more extensively rimmed by covellite with increasing elevation. Gold is found in the native state. Silver occurs primarily in argentiferous covellite but lesser amounts are also found in either argentite or acanthite (not yet determined), matildite, stromeyerite, and electrum.

Late-stage mineralization at Summitville is characterized by a barite, jarosite, goethite, and gold assemblage found in the uppermost levels of the deposit. Sulfides associated with this assemblage are extremely rare. A yellow phase included in a fluid inclusion in barite (Stoffregen, 1985) suggests the possibility that native sulfur may also be part of this assemblage. This late-stage assemblage, although volumetrically minor, locally contains high grades of gold.

Supergene oxidation extends to a depth of 100 to 200 feet below the surface. Copper is essentially completely removed from the oxidized zone, immediately below which digenite and lesser chalcocite coat and replace other sulfides.

Geochemical Environment

Defining the geochemical environment of the Summitville deposits is limited by the current lack of fluid-inclusion and isotope data. Parameters such as temperature of mineral deposition, chemical composition of the fluids, paleodepth, and origins of the fluids and dissolved constituents are far less well constrained than at Creede. However, Stoffregen (1985) has done an excellent job of limiting the conditions of alteration and mineralization using equilibrium thermodynamics. We will take the same approach, but we will also present some alternative interpretations.

Temperature - salinity ranges -- According to Stoffregen (1985), the lack of fluid-inclusion data is due to a paucity of primary fluid inclusions. He found only three samples with measurable primary inclusions in the small euhedral quartz crystals intergrown with sulfides, lining voids in the vuggy silica and none were found in sphalerite or alunite. His preliminary data, based on 19 measurements, show homogenization temperatures in these samples ranged from $230°$ to $320°C$, with most of the values between $230°$ and $270°C$. Two salinity measurements were 4 and 6 wt.-% NaCl equivalent. Perkins and Nieman (1983) also report temperatures for Summitville of $250°$ to $280°C$, presumably from fluid-inclusion measurements.

In contrast to the lack of good primary inclusions, quartz phenocrysts contain abundant secondary inclusions. The lack of secondary inclusions in quartz phenocrysts outside of the vuggy silica and quartz-alunite alteration zones (Stoffregen, 1985) is consistent with Bruha and Noble's (1983) suggestion that these inclusions represent the fluids responsible for the intense quartz + alunite + pyrite alteration. Bruha and Noble (1983) measured homogenization temperatures of $231°$ to $276°C$, and salinities of 7 to 21 wt.-% NaCl equivalent (averaging 10 wt.-%) in these secondary inclusions from one sample. They also report as many as 5 (unidentified) daughter (or trapped) minerals present in some inclusions. Limited data collected by G. H. Symmes in our laboratory are consistent with those of Bruha and Noble. Based on paragenetic relations and on differences in salinity between these secondary inclusions and the primary inclusions in the vuggy silica zone, Stoffregen (1985) suggests that the high-salinity fluids found in the fractures in quartz phenocrysts represent the alteration fluids, but not the later ore-forming fluids. These very limited data may indicate an evolution from high-salinity to low-salinity waters in the Summitville hydrothermal system, but many more measurements are needed to substantiate any such conclusion.

The high salinity of the fluids which preceded ore deposition is in agreement with observations made by Reynolds (Fluid, Inc., personal communication; also see Bodnar et al., 1985, this volume). He observed a few isolated, healed microfractures in some early quartz, defined by either vapor-rich H_2O+CO_2 inclusions and/or halite-bearing inclusions with small vapor bubbles. While these inclusions are thought by Reynolds to represent early magmatic fluids, their genesis is still uncertain due to the limited information.

Barite, which is an unreliable host for fluid-inclusion data (Ulrich and Bodnar, 1984), is the only hydrothermal mineral with relatively abundant fluid inclusions. Cunningham (1985) reports preliminary temperatures of roughly $100°C$ for inclusions in this late-stage barite associated with the famous gold "boulder" found in talus slopes of the South Mountain dome in 1975. No salinity data were reported.

Paleodepth--Based on geologic reconstruction, Steven and Ratté (1960a) estimate the paleodepth to the top of the ore between 150 and 400 meters. Using a temperature of $250°C$, a salinity of 10 wt.-%, and the lack of evidence for boiling from the limited data on both secondary and primary fluid inclusions, a minimum depth of deposition of 400 meters is estimated from the tables of Haas (1971).

Sources of constituents--No hydrogen-isotope data are available for any of the Acid-Sulfate-type deposits. Goldfield is the only one with oxygen-isotope data, and therefore it is used to help document the source of fluids in this type of deposit. According to Taylor (1973), the $\delta^{18}O$ data from Goldfield are compatible with ore-bearing fluids of essentially 100% meteoric origin. Taylor does note, however, that the overall ^{18}O depletion at Goldfield is appreciably smaller than at nearby Tonopah (an Adularia-Sericite deposit), and suggests that at Goldfield it is likely that either the alteration occurred at lower temperatures (125-200°C), or from fluids with a higher $\delta^{18}O$ value. While Taylor favored the former, recent fluid-inclusion data from Bruha and Noble (1983) indicate temperatures of about 230° to 270°C at Goldfield, thus suggesting that the fluids may have been richer in ^{18}O implying a significant magmatic component. In the absence of deuterium data, it is impossible to do more than speculate on the source of the Goldfield ore fluids. Well-constrained light-stable-isotope studies are badly needed on deposits of the Acid-Sulfate type.

Whitney (1984a,b) has calculated the sulfur speciation in quenched magmatic gases evolved from magmas of various oxidation states. Relatively oxidized magmas such as those at Julcani (Drexler, 1982), and that giving rise to the Fish Canyon ash-flow tuff in the Central San Juan Mountains (Whitney and Stormer, 1983) produce gases rich in SO_2. SO_2 gas is unstable at temperatures below 400°C in the presence of water (Iwasaki and Ozawa, 1960; Holland 1965,1967; Sakai and Matsubaya, 1977; Burnham, 1979) and disproportionates into sulfuric acid and H_2S gas. In our opinion, the intense acid-sulfate alteration characteristic of this deposit type results from the the attack on the wallrock by the H_2SO_4 produced by the disproportionation of SO_2. Although Whitney's calculations are consistent with a magmatic source for the sulfur, they do not demonstrate that such a source was necessary.

Stoffregen (1985), taking a different approach, has shown that the intense acid-sulfate alteration, which preceded ore deposition, can be produced by a fluid whose chemistry is dominated by magmatic gases. He modeled the interactions between an ascending sulfur rich "magmatic" gas (idealized as $H_2O-CO_2-SO_2-H_2S-HCl$) and liquid water using equilibrium thermodynamics. He describes the model as "the interaction of a large mass of the vapor phase with a relatively small amount of liquid, such that the chemistry of the liquid is completely controlled by that of the vapor." The results of his calculations show that at 250°C and vapor-saturation pressure, a solution equilibrating with a vapor phase consisting of subequal amounts of SO_2 and H_2S (totaling about 2 wt.-%) and minor HCl could produce the vuggy silica and acid-sulfate alteration seen at Summitville.

As discussed in a previous section, both the lead and sulfur-isotopic data for Summitville and the other Acid-Sulfate deposits are similar isotopically to the enclosing volcanic rock indicating that lead and sulfur were either derived from the volcanics or from a related magmatic source.

Chemical parameters--Together with the above fluid-inclusion data, the alteration and ore mineralogy can be used to put limits on some of the chemical

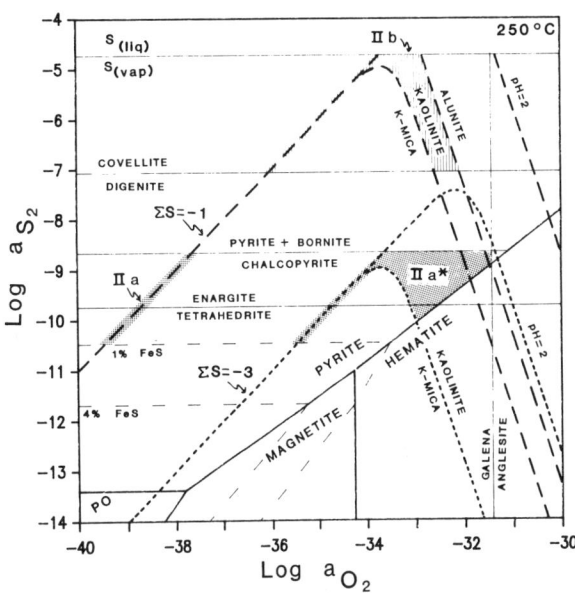

Figure 7.18. Log a_{S_2}-a_{O_2} diagram at 250°C showing the mineral stability fields for the significant minerals in the Summitville ores. Stoffregen's (1985) notation is used for the chalcopyrite-tennantite ore assemblage (IIa), and the covellite ore assemblage (IIb). An alternative to Stoffregen's interpretation for the chalcopyrite-bearing assemblage is shown as "IIa*". The boundaries for this alternative interpretation were calculated at log total sulfur = -3, while Stoffregen used a value of log total sulfur = -1. The heavy dashed lines show the alunite, kaolinite, and Kmica fields and the pH = 2 contour, all calculated at a log total sulfur concentration = -1; the heavy dotted lines show the kaolinite and Kmica fields and the pH = 2 contour, all calculated at a log total sulfur concentration = -3 (the alunite field is not present at this low total sulfur concentration). The contour for 20 mole percent FeS in sphalerite coincides with the pyrrhotite field boundary. Quartz is stable throughout the diagram. The data for the iron-copper sulfides and oxides are summarized in Barton and Skinner (Table 7.2, 1979); the alunite-kaolinite line, interpolated from Hemley et al. (1969); the kaolinite, Kmica, Kfeldspar data are from Henley et al. (1984). Abbreviation: po = pyrrhotite.

parameters of the hydrothermal fluid, such as activities of S_2 and O_2, pH, and total sulfur concentration during the different stages of alteration and mineralization. As shown on Figure 7.18, activity of sulfur is probably the easiest of these parameters to quantify. At Summitville, the quartz-alunite alteration and the later main-stage, upper-level, enargite-luzonite-covellite assemblage (stage IIb) indicate an environment at fairly high sulfur fugacities. The low silver concentrations (usually less than 1.0 wt.-%) which Stoffregen (1985) measured in electrum from the unoxidized zone is consistent with such high sulfur fugacities (Barton and Toulmin, 1964). The deeper assemblage of chalcopyrite and tennantite (stage IIa) specifies an environment at significantly lower sulfur fugacities. The activity of oxygen during the deposition of this deep assemblage is still open to question. On Figure 7.18, field "IIa" represents Stoffregen's (1985) interpretation and field "IIa*" represents an alternative interpretation which will be discussed later. In either case, the sulfur fugacity was significantly lower during the deposition of the deeper chalcopyrite-tennantite assemblage than it was during the upper, covellite assemblage. The maximum iron content of 1.5% FeS in sphalerite limits the lower sulfur fugacity of the deeper assemblage to values greater than about -11 at 250°C.

The calculations by Whitney (1985) show that an SO_2-rich magmatic gas quenched to 250°C would have activities of S_2 and O_2 in the alunite-stable field. Calculations which are consistent with those of Brimhall and Ghiorso (1983) and Stoffregen (1985), reinforce Stoffregen's conclusion that the fluids responsible for the acid-sulfate alteration at Summitville resulted from the condensation of a magmatitic vapor plume into deeply circulating meteoric water.

Delineating the pH, total-sulfur concentration and oxygen activity is considerably more difficult. Figure 7.19 shows a series of activity O_2-pH diagrams at different total-sulfur concentrations relevant to Summitville. Stoffregen's (1985) notation for the different mineral assemblages is used: vuggy silica (Ia), quartz-alunite-pyrite (Ib), low f_{S_2} (chalcopyrite-bearing) ore assemblage (IIa), and high f_{S_2} (covellite dominated) ore assemblage (IIb). As seen on Figure 7.19a, very acidic (pH <3.7) and oxidizing fluids are required to produce the quartz-alunite alteration. Presumably, the extreme base leaching of the vuggy silica alteration was the result of even more acidic solutions. Stoffregen estimates that at a pH less than about 2 appreciable aluminum mobility would inhibit alunite deposition, resulting in a solution that would dissolve most rock-forming minerals except quartz and pyrite, creating the observed vuggy silica assemblage. The alteration sequence of vuggy silica → quartz-alunite → quartz-kaolinite → illite + montmorillonite is clearly the result of decreasing acidity away from the vein. Occasionally the quartz-kaolinite zone is absent and a transition from quartz-alunite to illitic alteration is observed. This may be the result of increasing salinity which causes the kaolinite field to shrink (in the sulfate portion of the diagram) and and finally disappear at a potassium activity of approximately 0.1* at a total sulfur concentration of 0.1 molal.

---

*Stoffregen (1985) calculates the disappearance of the kaolinite field at a potassium activity of 0.005; the difference between our calculations is due to different thermodynamic data (see Henley et al., 1984, p. 81, for discussion of thermodynamic data on the kaolinite-Kmica reaction). Assuming 0.005 as the maximum potassium activity, Stoffregen uses a potassium activity of 0.0001 (log = -4) in his calculations. We feel that is an unreasonably low value based on the few fluid-inclusion salinity measurements which have been done (ranging from about 1 to 3 molal) and based on our calculations for the disappearance of the kaolinite field. We have used a log potassium value of -1.5 based on a 1 molal solution with a Na/K concentration ratio of 10.

Based on the rare occurrence of native sulfur associated with alunite, the total-sulfur concentration during the acid-sulfate alteration was probably between 0.1 and 0.01 molal. Figure 7.19b shows the relative affects on the sulfur and alunite fields caused by reducing the total-sulfur concentration. Total-sulfur concentrations were probably never much greater than 0.1 molal, because at those concentrations the native sulfur field expands greatly and would have been a dominant phase in the assemblage.

Compared to the deeper, stage IIa assemblage, the covellite-dominated stage IIb assemblage is fairly restricted in terms of oxygen activity and pH (Fig. 7.19). The log total-sulfur concentration was probably between -1 and -2, similar to the acid-sulfate alteration, based on the local occurrence of sulfur associated with these ores.

The geochemical environment during deposition of the chalcopyrite-tennantite assemblage, the (deeper) stage IIa ores, is not as well constrained as that of the alteration stages (Ia and Ib) or the higher stage IIa ores. As discussed above, there is no doubt that this assemblage formed at lower activities of S_2 than the covellite-enargite assemblage. However, as seen on Figure 7.19, the chalcopyrite-tennantite assemblage is stable over a wide range of oxygen activities. It is bounded by the chalcopyrite = pyrite + bornite reaction at high S_2 activities and by the 1.5% FeS in sphalerite contour at low S_2 activities. The possible environment for the chalcopyrite-tennantite assemblage is further restricted by Stoffregen's observation that this assemblage (as well as the covellite assemblage) is associated with minor kaolinite. There are two areas on the a_{O_2}-pH diagram where these conditions are met (Figs. 7.19a-c); the field labeled "IIa", at lower oxygen activities, corresponds to Stoffregen's interpretation and field "IIa*", at higher oxygen activities, but lower total sulfur, represents an alternative interpretation. Stoffregen (1985) chose the lower oxygen-activity field because the IIa* field borders on the hematite-stability field and no primary hematite has been identified with the unoxidized ore. Even so, we feel that this upper field represents a likely environment because: (1) it has values of oxygen activity and pH similar to both the preceding acid-sulfate alteration stage and the essentially cogenetic, but higher, stage IIb ores, and

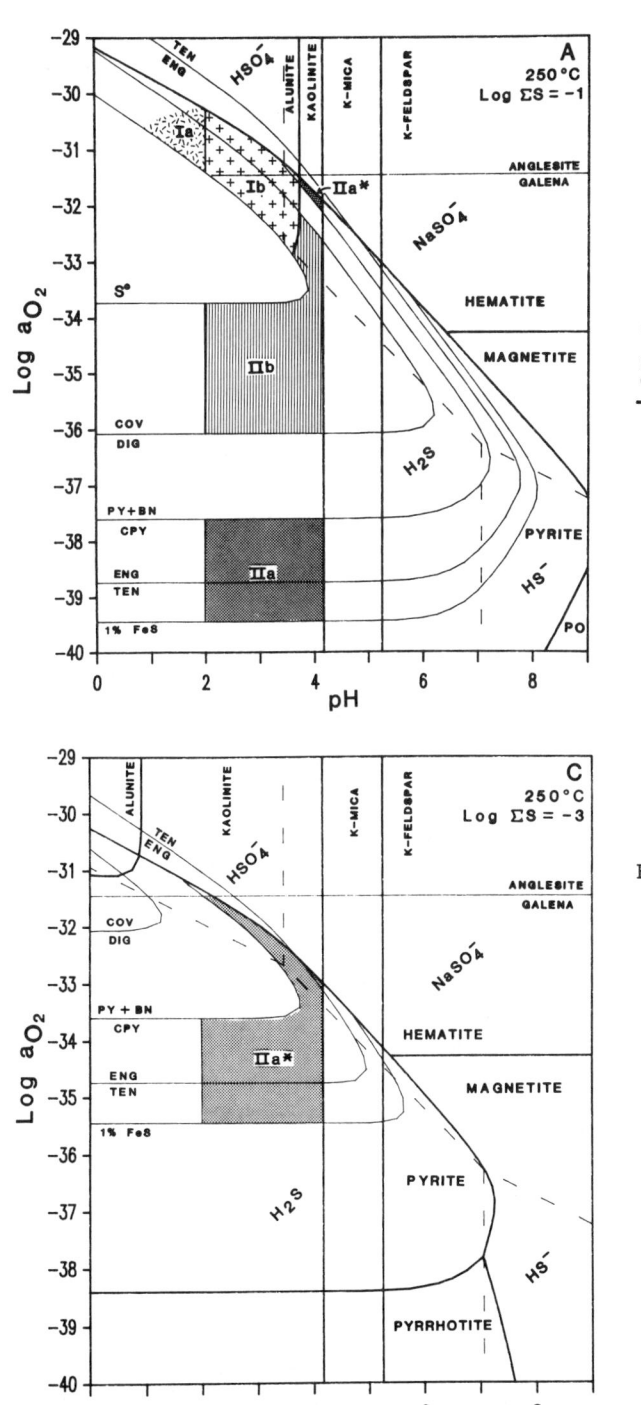

Figure 7.19. A series of log a_{O_2}-pH diagrams constructed for 250°C at different total sulfur concentrations relevant to Summitville. The salinity is 1 molal, with Na+/K+ = 10. Stoffregen's (1985) notation for the different mineral assemblages is used: vuggy silica (Ia), quartz-alunite-pyrite (Ib), low f S_2 (chalcopyrite-bearing) ore assemblage (IIa), and high f S_2 (covellite dominated) ore assemblage (IIb). An alternative to Stoffregen's interpretation for the chalcopyrite-bearing assemblage is shown as "IIa*". Log total sulfur for A). = -1; for B). = -2; and for C). = -3. See Figure 18 for sources of data. Abbreviations: ten = tennantite, eng = engarite, cov = covellite, dig = digenite, py = pyrite, bn = bornite, cpy = chalcopyrite, po = pyrrhotite.

(2) it lies in the sulfate-dominant portion of the diagram which is consistent with the introduction of a sulfur-rich magmatic gas as discussed above (Whitney, 1984a,b; Brimhall and Ghiorso, 1983).

One problematic aspect of this interpretation is formation of tennantite associated with kaolinite in the higher oxygen-activity field (IIa*). This can be resolved by a model which calls on lower total-sulfur concentrations deeper in the deposit (discussed below). As can be seen in Figures 7.19b and c, the tennantite field shifts to lower pH's with decreasing total sulfur. Two other factors which favor tennantite associated with kaolinite are: (1) Stoffregen reported minor Sb substitution in tennantite which appears to increase with elevation in the deposit, and (2) the possible trend of decreasing salinity (potassium activity) with time which was discussed earlier.

As seen in Figure 7.19, except for being slightly less acidic, the chemistry of fluids which deposited the covellite assemblage was very similar to those which produced the acid-sulfate alteration. Because the base of the vuggy silica zone roughly corresponds to the base of the covellite assemblage (Fig. 7.17), it is possible that similar processes may have affected them both. During the alteration stage, Stoffregen (1985) interprets the base of the vuggy silica zone as the level of interaction between the magmatic gases and water. If the magmatic gases continued to be condensed at this same level during ore deposition, it is consistent that the higher sulfur assemblage begins here. Below this level (which presumably would vary with time), the tennantite assemblage would form as a consequence of the lower total-sulfur concentrations and activities. The difference between the alteration- and ore-deposition stages may be related to the increase in permeability following the acid-sulfate leaching which allowed for increased fluid flow resulting in a greater dilution of the magmatic gas by meteoric waters.

Summary of Summitville Mineralization

The work by Stoffregen (1985), Steven and Ratté (1960a), Mehnert et al. (1973) and Perkins and Nieman (1983) has shown that the Summitville ores were deposited shortly after the emplacement of the host volcanic dome on the margins of the Summitville and Plataro calderas. This temporal and spatial magmatic association obviously affected the chemistry of the hydrothermal system. Stoffregen (1985), Whitney (1984a,b), and Brimhall and Ghiorso (1983) have shown that the intense acid-sulfate alteration, which preceded ore deposition, can be produced by a fluid whose chemistry is dominated by magmatic gases. Stoffregen suggests that the base of the vuggy silica alteration represents the level of interaction of these magmatitic gases. Ore deposition, which followed, occurred as two different mineral assemblages, a deeper, chalcopyrite assemblage, and an upper, covellite assemblage. The geochemical conditions which produced these ores is not well understood. Perkins and Nieman (1983) and Stoffregen (1985) have suggested that the change from acid-sulfate leaching to ore deposition reflects a transition from a vapor-dominated geothermal system to a liquid-dominated system. Stoffregen envisions the chemistry of the system to then be dominated by meteoric fluids, which are less oxidized and less acidic. He also suggests that the change in the sulfide assemblages with elevation is due to the influx and mixing with acid-sulfate waters into the upper portion of the deposit as has been modeled by Reed and Spycher (1985, this volume). We agree that the change from alteration to ore deposition represents decrease in the magmatic vapor contribution to the system, but suggest that such a vapor may still have played a significant role in chemistry of the solutions depositing the ores. The change from the lower to the upper sulfide assemblage, which roughly corresponds to the base of the vuggy silica, may still represent the level of the interaction of the magmatic gases. With the present information, it is not possible to differentiate which mechanism is more likely. The final stage of mineralization, the barite, jarosite, goethite, gold assemblage represents the waning stages of hydrothermal activity and the collapse of the system.

GEOTHERMAL INTERPRETATION OF VOLCANIC-HOSTED EPITHERMAL DEPOSITS

Much of the conceptual base on which recent models of epithermal mineralization have been built has come from observations on active geothermal systems (cf. White, 1981). Of particular importance have been observations on the anatomy of active systems and on the identification of processes operating within them. In this synthesis we have attempted to evaluate the observational base for epithermal ore deposits in terms of that for active systems. In summary, it is useful to combine observations on both active geothermal systems and well-studied ore deposits in an attempt to interpret the Adularia-Sericite and Acid-Sulfate-type epithermal deposits in a geothermal framework. In the first chapter in this volume, Henley has summarized most of the pertinent observations on active systems; and the processes operating within them have been explored in detail in the first volume of this series (Henley et al., 1984). The recent review by Henley and Ellis (1983) has established a geothermal interpretation for hydrothermal ore deposits in general. The following discussion leans heavily on these sources and on the preceding discussion in this chapter without further documentation.

Adularia-Sericite Deposits

The observational base on Adularia-Sericite-type deposits suggests that these deposits formed in hydrothermal systems similar to those of the Taupo volcanic zone in New Zealand and the Valles and Yellowstone calderas in the United States, as suggested by Henley and Ellis (1983), and discussed by Henley (1985, this volume). The general structure of such geothermal systems is illustrated in Figure 1.1a in Henley's chapter (1985, this volume), which should be compared with the schematic representation of the Creede hydrothermal system as illustrated in Figure 7.11 of this chapter. From such a comparison to modern geothermal systems, the following characteristics of hydrothermal systems responsible for the formation of Adularia-Sericite-type deposits appear to be particularly important:

1. The ore zone occurs at distances of several kilometers from the heat source that drove the hydrothermal circulation. The ores occupy a narrow vertical interval, the top of which lies at distances of 200-600 meters below the paleo-water table. Au-rich deposits may have formed in a somewhat more shallow environment than did the Ag-rich deposits.

2. The ores appear to have been deposited along the top of a deeply circulating, moderately to strongly concentrated brine at, or close to, the interface of that brine with overlying ground waters.
3. In the upper part of the brine plume, near its interface with the overlying ground waters, fluid flow appears to have been dominated by lateral movement, and discharge to the surface appears to have been confined to small areas constituting only a small fraction of the total projected surface area of the system.
4. Either lithologic aquitards or gravity-driven ground-water flow above the circulating brine plume have formed hydrologic caps on the plume forcing lateral flow across its top; overlying structures are barren of mineralization and often show no evidence of vein filling or wallrock alteration.
5. The waters charging the deeply circulating brine were predominantly of meteoric origin, although some magmatic contribution cannot be ruled out.
6. The fluids in the upwelling plume appear to have been near neutral in pH and to have had their redox and sulfidation states buffered by reactions with iron-bearing minerals in the wallrock or in the vein assemblage.
7. Boiling occurs in the upper parts of the brine plume, and mixing with surrounding ground waters occurs along the sides and, particularly, along the top of the plume. The depth at which boiling commences is strongly dependent on the gas content of the waters, and the degree of mixing dependent on the density contrasts between the plume and the surrounding ground waters. Both processes lead to the precipitation of ore and gangue minerals.
8. The acid required to produce the sericitic alteration associated with the deposits was produced by the condensation, in the upper parts of the system, of acid volatiles distilled off the deeply circulating brine.
9. The metals, and perhaps other components of the ores, appear to have been derived by leaching of wall rocks deep in the system.

Acid-Sulfate Deposits

In contrast to the fairly good database for the more numerous Adularia-Sericite-type deposits, the observational base for the Acid-Sulfate-type deposit is substantially more limited. There is also a similar, but less pronounced difference in the amount of data available on the two environments in active geothermal systems. However, we feel that the data bases on both the ore deposits and the active systems are sufficiently consistent that we are comfortable in using the model of geothermal systems in andesitic volcanic terranes presented as Figure 1.1b in the chapter by Henley (1985, this volume) as a basis for interpreting Acid-Sulfate-type epithermal ore deposits.
1. Observations on Acid-Sulfate-type deposits indicate both a temporal and spatial genetic association with the core of a volcanic dome.
2. The ore zone occupies a narrow vertical interval, the top of which appears to have been within 200-500 meters of the paleo-water table. There is some indication that the Au-rich Acid-Sulfate-type deposits may have formed at shallower levels than have the Ag-rich deposits.
3. Although the data are quite limited, both the sulfur- and lead-isotopic data indicate that they were either derived directly from a magmatic source or from the enclosing volcanic rocks.
4. Light-stable isotopic evidence on the origins of the fluids is almost non-existent but can be interpreted to provide evidence for a magmatic contribution to the fluids responsible for wallrock alteration.
5. The sulfuric acid responsible for the intense acid-sulfate wallrock alteration which preceded mineralization appears to have been generated by the disproportionation of SO_2 gas contained in the magmatic vapor plume into sulfuric acid and H_2S gas.
6. A surficial zone of acid-sulfate alteration due to the atmospheric oxidation of H_2S to form sulfuric acid would also be expected in this type of hydrothermal system, but has not yet been documented in any mining district.
7. The relatively high sulfidation state of the fluids responsible for both the acid-sulfate alteration and the covellite-enargite facies of the sulfide mineralization is an expected consequence of the degassing of a relatively oxidized magma.
8. The increased permeability produced by the intense base leaching associated with the acid-sulfate alteration provided channelways for subsequent ore-fluid migration.
9. The alteration and mineral assemblages of the Acid-Sulfate-type deposit, along with the significant copper contents and the association with porphyritic rocks, suggest that this type of epithermal deposit bears some genetic relation to porphyry-copper systems. The nature of such a relationship, if it exists, remains speculative.

MECHANISMS OF ACID-SULFATE ALTERATION

Intense wallrock alteration to an assemblage consisting of quartz + kaolinite + alunite is the most pronounced characteristic of Acid-Sulfate-type deposits. Most of this alteration takes place in the ore zone at depths of several hundred meters below the water table. However, the same assemblage is produced by sulfuric acid attack on wallrocks in two very different surficial environments. It is important to have a set of criteria that will allow us to distinguish between these three different origins of acid-sulfate alteration so that we may identify the environment in which any particular alteration zone formed. Criteria based on pattern recognition and/or textural evidence can provide compelling evidence to make such identification, but in most cases they do not and there has been considerable confusion in the literature based on misidentification of the origin of acid-sulfate alteration. In most cases, correct identification can be made with the application of light-stable isotope analyses, sometimes in conjunction with K-Ar age determinations. In this section, we will discuss the use of isotopic criteria to make such distinctions.

Bethke (1984) discussed the following three reactions, each of which operates in a different environment (as noted below in parentheses) and results in the formation of sulfuric acid leading to the formation of acid-sulfate alteration

$$4SO_2 + 4H_2O = 3H_2SO_4 + H_2S \quad (primary\ hypogene) \quad (6)$$

$$H_2S + 2O_2 = H_2SO_4 \quad (primary\ supergene) \quad (7)$$

$$2FeS_2 + 7H_2O + 15/2O_2 = 2Fe_2O_3 \cdot 3H_2O + 4H_2SO_4 \quad (8)$$
(secondary supergene)

Reaction (6) is, in our opinion, the most important mechanism by which sulfuric acid is formed in Acid-Sulfate-type ore deposits. It describes the disproportionation of magmatically exsolved SO_2 gas into sulfuric acid and H_2S gas deep in the system, as discussed in the preceding section. We have noted this deep environment of acid-sulfate alteration as "primary hypogene" in recognition of the fact that it occurs deep within the hydrothermal system. A number of active geothermal systems associated with andesitic volcanism, particularly in island-arc environments, contain deep, high-temperature acid waters. The acidity and sulfate content of these waters appear to have been generated by such a disproportionation reaction deep in the system. These include the Tahuangtsui thermal area, Tatun volcanic region, Taiwan, (Chen, 1970), the Matsukawa geothermal area, Japan, (Nakamura et al., 1970), and several other geothermal areas in Japan (Kiyosu and Kurasawa, 1983, 1984). At Matsukawa, intense acid-sulfate alteration also occurs at the surface, but this near-surface alteration probably formed from the oxidation of hydrogen sulfide in the steam-heated waters (reaction 7) as discussed below.

Reaction (7) is the simple oxidation of hydrogen sulfide that produces the spectacular solfataric alteration seen at the surface in active geothermal systems. Because this environment, although surficial, is an integral part of the hydrothermal system, and because the alteration results from wall rock interaction with descending acid waters, we have noted it as "primary supergene." Day and Allen (1925) originally proposed this mechanism to explain alteration seen in the Lassen area, and White (1957) discussed it more extensively (see also Henley and Ellis, 1983; Henley, 1985, this volume; and Reed and Spycher, 1985, this volume). Oxidation of the H_2S occurs when the vapor phase generated by boiling of the deep waters contacts the atmosphere just above the water table. The sulfuric acid so generated percolates back into, and acidifies, the steam-heated ground waters which overly the deeply circulating hydrothermal cell. These surficial, acid, steam-heated waters are essential elements of almost all high-temperature geothermal systems, from which both Acid-Sulfate and Adularia-Sericite types of epithermal ore deposits form. The alteration produced by this mechanism is very intense, but is a surficial phenomenon, rarely extending down to depths of even 50 meters below the surface. These surficial alteration zones are probably not preserved in many ore deposits because they are very soft, shallow, and easily eroded.

The last reaction (8) involves the production of sulfuric acid during supergene alteration of sulfide ore. Because the oxidation antedates the hydrothermal systems (sometimes by millions of years) we have labeled this environment "secondary supergene." Such oxidation has produced alunite at Creede, Goldfield and, undoubtedly, in many other districts. Because this type of acid-sulfate alteration has little to do with the formation of the deposit it is important to be able to distinguish it from the former two mechanisms.

Sulfur isotopes provide a means to differentiate among the three environments outlined above. Field (1966) was the first to use sulfur isotopes to distinguish between hypogene and secondary alunite. Jensen et al. (1971) showed that both hypogene and secondary alunite occurred at Goldfield, and that the two had widely different sulfur-isotopic compositions. The alunite judged to be hypogene on field and petrographic grounds was strongly enriched in ^{34}S compared to that formed by secondary processes (Fig. 7.20). The secondary alunite at Goldfield (and many other districts) has the same range of sulfur-isotopic composition as do the primary sulfides because, due to kinetic effects, there is little or no sulfur-isotopic fractionation during low-temperature oxidation of sulfides. In contrast, hypogene alunite formed at temperatures above approximately 250°C is much heavier in sulfur-isotopic composition than coexisting sulfides because of the rapid isotopic exchange and strong fractionation between aqueous sulfide and sulfate species at moderate to high temperatures, particularly in acid environments (Ohmoto and Lasaga,

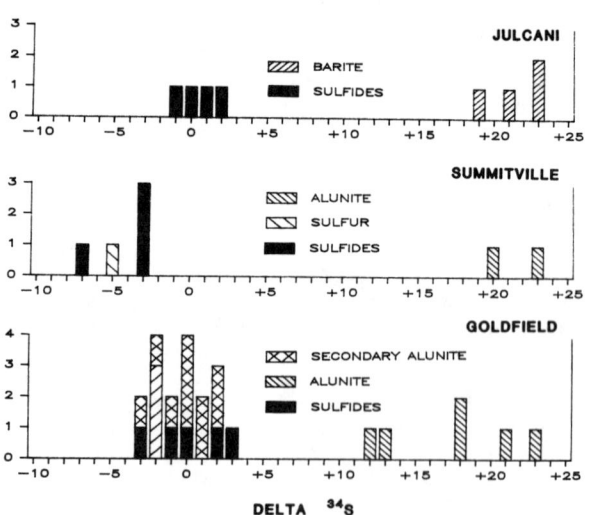

Figure 7.20. $\delta^{34}S$ histograms for Julcani, Summitville, and Goldfield. Data sources: Julcani, Goodell (1977); Summitville, R. O. Rye (U.S.G.S., personal communication, 1985); Goldfield, Jensen et al. (1971).

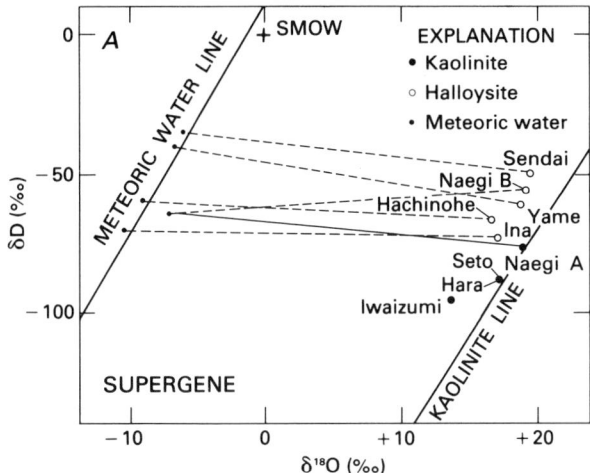

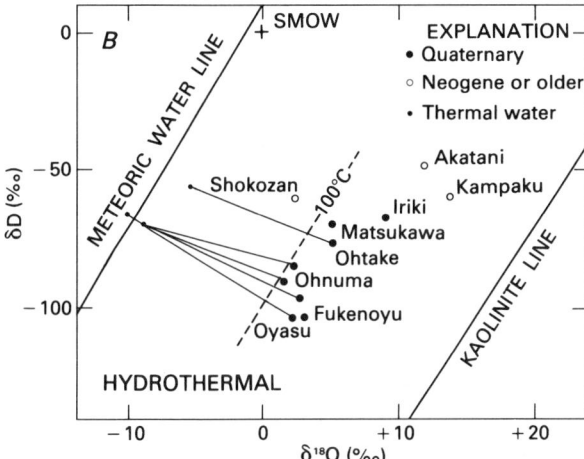

Figure 7.21. a). $\delta D - \delta^{18}O$ diagram for kaolinite and halloysite formed from supergene alteration of some Japanese orebodies, adapted from Marumo et al. (1982). Dashed lines connect mineral analyses with their corresponding surface waters. b). $\delta D - \delta^{18}O$ diagram for hydrothermal kaolinites and for thermal waters near the mineral localities, adapted from Marumo et al. (1982). The dashed line shows the $\delta D - \delta^{18}O$ relation of kaolinite in equilibrium with meteoric water at 100°C.

the watertable over steam-heated waters (reaction 7). When H_2S is oxidized in the steam-heated waters, it undergoes little sulfur-isotopic exchange with the sulfate produced because of slow reaction kinetics at relatively low temperatures (<100°C), as has been demonstrated by Steiner and Rafter (1966) for Wairakei, and by Schoen and Rye (1970) for Yellowstone. Therefore, the alunites formed in both supergene and steam-heated water environments will both have sulfur-isotopic compositions approximately the same as those of their coexisting sulfides.

In order to distinguish the alunites formed from steam-heated waters from the supergene alunite, it is necessary to look at the associated kaolinites. Taylor (1974) showed that hydrogen and oxygen isotopes can be used to differentiate between supergene and primary kaolinite. Figure 7.21a shows a group of kaolinites and halloysites formed by supergene alteration of some Japanese orebodies (Marumo et al., 1982). These data plot very close to the "kaolinite line" (Savin and Epstein, 1970; Lawrence and Taylor, 1971) which represents the locus of hydrogen and oxygen isotopic compositions for kaolinites formed by weathering and assumed to be in equilibrium with meteoric waters along the meteoric water line. Figure 7.21b shows the hydrogen- and oxygen-isotopic compositions of a group of kaolinites from acid-sulfate alteration zones from several geothermal areas (solid circles) and ore deposits (open circles) in Japan. Plots of the oxygen- and hydrogen-isotopic compositions of these kaolinites, formed in the steam-heated water environment, lie considerably removed from the meteoric water line and very close to the dashed line which marks the compositions of kaolinites that would be in equilibrium with meteoric waters at temperatures of 100°C.

Table 7.6 summarizes the isotopic discriminators currently available for determining the origin of acid-sulfate alteration. It indicates that sulfur isotopes can be used to identify alunite formed by the disproportionation of SO_2, and that the hydrogen- and oxygen-isotopic composition of kaolinites can be used to distinguish between primary and secondary supergene origins for shallow acid-sulfate alteration zones. In addition, K/Ar age determinations on alunite can also be used in many districts to distinguish between supergene and primary origins, as has been done at Creede (M. Lanphere, personal communication, 1981) and Round Mountain (Tingley and Berger, 1985).

Recent advances by Pickthorn and O'Neil (1985) and by Bethke, Rye, Wasserman, and Goss (P. Bethke, personal communication, 1985) on the selective analysis of the isotopic composition of oxygen in both the sulfate and hydroxyl sites in alunite is expected to allow the distinction between alunites formed in the steam-heated versus the supergene environment to be made on alunite itself, obviating the need to use associated kaolinite. Such application, however, awaits further definition of the systematics of oxygen- and hydrogen-isotopic fractionation between alunite and water, and the demonstration of its feasibility.

It should be noted that in Acid-Sulfate deposits such as Summitville we may expect to encounter acid-sulfate alteration formed by all three mechanisms discussed above. Although not yet documented for any

1982). Such a relationship has been documented at Summitville and Julcani as well as at Goldfield (Fig. 7.20).

Although sulfur isotopes can be used to distinguish secondary alunite (reaction 8) from hypogene alunite formed by disproportionation of SO_2 (reaction 6), they cannot help in differentiating between supergene alunite formed through the oxidation of primary sulfides by weathering processes (reaction 8) and that formed by the oxidation of H_2S at

Table 7.6--Characteristics of acid-sulfate alteration of different origins (Bethke, 1984)

	Oxidation of H_2S	Disproportionation of SO_2	Supergene oxidation
Reaction (Table 4)	1	2	3
^{34}S alunite	=sulfides	sulfides	=sulfides
^{18}O kaolinite	removed from kaolinite line	removed from kaolinite line	near kaolinite line
K/Ar age alunite	concordant with mineralization	concordant with mineralization	discordant with mineralization

district, we should expect the formation of primary-supergene acid-sulfate alteration (reaction 7) in the shallow parts of the hydrothermal system, probably separated from the deeper hypogene acid-sulfate alteration. In the few districts where sulfur-isotope data on alunite and associated sulfides are available most of the alunite appears to have been formed either by reaction (6) in the deep alteration zone or from supergene oxidation of primary sulfides (reaction 8). However, Henley (1985, this volume) points out that lakes several hundred meters deep filled with acid sulfate-chloride water may form in the craters of stratovolcanoes in andesitic volcanic terranes. Henley cites as an example Lake Ruapehu in New Zealand which is greater than 300 meters deep and contains a sulfate-chloride water of pH = 1.25 at 55°C. The acidity and chemical composition of these lakes is due to the fact that they are fed by volcanic vapors from below as well as by meteoric waters at the surface. In the case of such deep, acid, sulfate-bearing lakes, downward percolation of the lake waters through the underlying volcanic pile would produce intense acid-sulfate alteration at considerable depths beneath the lake (and summit) level. Such downward percolation might be expected to occur as the deep hydrothermal system begins to wane, and it is possible that an earlier deep acid-sulfate alteration zone might be overprinted by a later, descending, acid-sulfate alteration! However, we continue to believe that the bulk of the acid-sulfate alteration in Acid-Sulfate-type ore deposits resulted from the formation of sulfuric acid by the disproportionation of magmatic SO_2 gas.

We should also note that reactions (7) and (8) can also occur in Adularia-Sericite systems. Specifically, because surficial zones of acid-sulfate alteration due to the oxidation of H_2S in the steam-heated water zone are so common in active geothermal systems thought to be analogous to fossil systems represented by Adularia-Sericite-type epithermal ore deposits that we should expect to see them represented in Adularia-Sericite-type epithermal districts. However, to our knowledge no such occurrences have yet been documented in any such district; perhaps because they have been removed by erosion, perhaps because they have not been specifically sought. However, secondary alunite formed as a result of the supergene oxidation of primary sulfides has been documented by both K/Ar age determinations and sulfur-isotope analyses at both Round Mountain and Creede.

ACKNOWLEDGMENTS

We are grateful for the numerous thought-provoking discussions on epithermal and active geothermal systems with Paul Barton and Dick Henley. We also benefitted from talks with Roger Stoffregen, whose comprehensive study at Summitville, Colorado was essential for using Summitville as the archetype for Acid-Sulfate epithermal systems. Work done by Jim Goss on the original compilation of geological data for the sixteen epithermal districts studied in detail is gratefully acknowledged. In addition, we take the full weighty responsibility for our statements and interpretations, and we relieve our colleagues of any such burden.

REFERENCES

Ashley, R. P., 1974, Goldfield mining district; in Guidebook to the Geology of Four Tertiary Volcanic Centers in Central Nevada: Nevada Bureau of Mines and Geology Report 19, p. 49-66.

Ashley, R. P., 1982, Occurrence model for enargite-gold deposits: U.S. Geological Survey Open-File Report 82-795, p. 144-147.

Barnes, H. L., 1979, Solubilities of ore minerals; in Barnes, H. L. (ed.), Geochemistry of Hydrothermal Ore Deposits, 2nd Edition: John Wiley and Sons, New York, p. 404-460.

Barton, P. B., Jr., Bethke, P. M., and Roedder, E., 1977, Environment of ore deposition in the Creede mining district, San Juan Mountains, Colorado: Part III. Progress toward interpretation of the chemistry of the ore-forming fluid for the OH vein: Economic Geology, v. 72, p. 1-24.

Barton, P. B., Jr., and Skinner, B. S., 1979, Sulfide mineral stabilities; in Barnes, H. L. (ed.), Geochemistry of Hydrothermal Ore Deposits, 2nd Edition: John Wiley and Sons, New York, p. 278-403.

Barton, P. B., Jr., and Toulmin, P., III, 1964, The electrum-tarnish method for the determination of the fugacity of sulfur in laboratory sulfide systems: Geochimica et Cosmochimica Acta, v. 28, p. 619-640.

Battory, B. L., 1981, Analysis of the lacustrine sediments of the Creede Formation, Mineral County, Colorado: Unpublished M.S. Thesis, New Mexico Institute of Mining and Technology, New Mexico.

Berger, B. R., and Eimon, P. I., 1983, Conceptual models of epithermal precious metal deposits; in Shanks, W. C. (ed.), Cameron Volume on Unconventional Mineral Deposits: Society of Mining Engineers, New York, p. 191-205.

Berger, B. R., and Silberman, M. L., 1985, Trace-element patterns in hot-spring-type precious-metal deposits; in Berger, B. R., and Bethke, P. M., (eds.), Geology and Geochemistry of Epithermal Systems: Society of Economic Geologists, Reviews in Economic Geology, v. 2.

Bethke, P. M., 1984, Controls on base- and precious-metal mineralization in deeper epithermal environments: U.S. Geological Survey, Open-File Report 84-890, 40 p.

Bethke, P. M., Barton, P. B. Jr., Lanphere, M. A., and Steven, T. A., 1976, Environment of ore deposition in the Creede mining district, San Juan Mountains, Colorado: II. Age of mineralization: Economic Geology, v. 71, p. 1006-1011.

Bethke, P. M., Barton, P. B., Jr., and Rye, R. O., 1973, Hydrogen, oxygen and sulfur isotopic compositions of ore fluids in the Creede district, Mineral County, Colorado (abs.): Economic Geology, v. 68, p. 1205.

Bethke, P. M., and Rye, R. O., 1979, Environment of ore deposition in the Creede mining district, San Juan Mountains, Colorado: Part IV. Source of fluids from oxygen, hydrogen and carbon isotope studies: Economic Geology, v. 74, p. 1832-1851.

Bethke, P. M., Rye, R. O., and Barton, P. B., Jr., 1985, Pre-ore potassium metasomatism, Creede mining district: 114th AIME Annual Meeting, February 24-28, New York, p. 44.

Bodnar, R. J., Reynolds, T. J., and Kuehn, C. A., 1985, Fluid-inclusion systematics in epithermal systems; in Berger, B. R., and Bethke, P. M. (eds.), Geology and Geochemistry of Epithermal Systems: Society of Economic Geologists, Reviews in Economic Geology, v. 2.

Bonham, H. F., Jr., 1986, Models for volcanic-hosted epithermal precious-metal deposits--A review: International Volcanological Congress, February 1986, p. 1-5.

Bonham, H. F., Jr., and Giles, D. L., 1983, Epithermal gold/silver deposits: The geothermal connection; in The Role of Heat in the Development of Energy and Mineral Resources in the Northern Basin and Range Province: Geothermal Resources Council, Special Report 13, 384 p.

Brimhall, G. H., Jr., and Ghiorso, M. S., 1983, Origin and ore-forming consequences of the advanced-argillic alteration process in hypogene environments by magmatic gas contamination of meteoric fluids: Economic Geology v. 78, p. 73-90.

Browne, P. R. L., and Ellis, A. J., 1970, The Ohaki-Broadlands hydrothermal area, New Zealand: Mineralogy and related geochemistry: American Journal of Science, v. 269, p. 97-131.

Bruha, D. I., and Noble, D. C., 1983, Hypogene quartz-alunite +/- pyrite alteration formed by moderately saline, ascendant hydrothermal solutions (abs.): Geological Society of America, Abstracts with Programs, v. 15, no. 5, p. 325.

Buchanan, L. J., 1979, The Las Torres mine, Guanajuato, Mexico: Ore controls of a fossil geothermal system: Unpublished Ph.D. thesis, Colorado School of Mines, 138 p.

Buchanan, L. J., 1981, Precious metal deposits associated with volcanic environments in the southwest; in Dickinson, W. R., and Payne, W. D. (eds.), Relations of Tectonics to Ore Deposits in the South Cordillera, Arizona Geological Society Digest, v. XIV, p. 237-261.

Burnham, C. W., 1979, Magmas and hydrothermal fluids; in Barnes, H. L., ed., Geochemistry of Hydrothermal Ore Deposits, 2nd Edition: John Wiley and Sons, New York, p. 71-136.

Cannaday, F. X., 1955, The OH vein and its relation to the Amethyst fault: Unpublished M.S. thesis, Colorado School of Mines, 57 p.

Casadevall, T., and Ohmoto, H., 1977, Sunnyside mine, Eureka mining district, San Juan County: Geochemistry of gold and base-metal ore deposition in a volcanic environment: Economic Geology, v. 72, p. 1285-1320.

Chaffee, M. A., 1967, A study of the geology and hydrothermal alteration North of the Creede mining district, Mineral, Hinsdale and Saguache Counties, Colorado: Unpublished Ph.D. thesis, University of Arizona (Tucson), 194 p.

Chen, C. H., 1970, Geology and geothermal power potential of the Tatum volcanic region: Geothermics Special Issue 2, p. 1134-1143.

Cumming, G. L., Kesler, S. E., and Krstic, D., 1979, Isotope composition of lead in Mexican mineral deposits: Economic Geology, v. 74, p. 1395-1407.

Cunningham, C. G., 1985, Characteristics of boiling-water-table and carbon dioxide models for epithermal gold deposition; in Tooker, E. W. (ed.), Geologic Characteristics of Sediment- and Volcanic-Hosted Disseminated Gold Deposits-- Search for an Occurrence Model: U.S. Geological Survey, Bulletin 1646, p. 43-46.

Czamanske, G. K., 1974, The FeS content of sphalerite along the chalcopyrite-pyrite-bornite sulfur fugacity buffer: Economic Geology, v. 69, p. 1328-1334.

Day, A. L., and Allen, E. T., 1925, The volcanic activity and hot springs of Lassen Peak: Carnegie Institute of Washington, Pub. No. 360, 190 p.

Doe, B. R., Steven, T. A., Delevaux, M. H., Stacey,

J. S., Lipman, P. W., and Fisher, F. S., 1979, Genesis of ore deposits in the San Juan volcanic field, southwestern Colorado--lead isotope evidence: Economic Geology, v. 74, p. 1-26.

Drexler, J. W., 1982, Magmatic conditions from vitric units of the Julcani district, Peru: Unpublished Ph.D. thesis, Michigan Technological University, Houghton, Michigan.

Drier, J. E., 1976, The geochemical environment of ore deposition in the Pachuca-Real Del Monte District, Hidalgo, Mexico: Unpublished Ph.D. thesis, University of Arizona, 115 p.

Drummond, S. E., and Ohmoto, H., 1985, Chemical evolution and mineral deposition in boiling hydrothermal systems: Economic Geology, v. 80, p. 126-147.

Ellis, A. J., and Mahon, W. A. J., 1977, Chemistry and Geothermal Systems: Academic Press, New York, 392 p.

Emmons, W. H., and Larsen, E. S., 1923, Geology and ore deposits of Creede, Colorado: U.S. Geological Survey, Bulletin 718, 198 p.

Fahley, M. P., 1981, Fluid-inclusion study of the Tonopah district, Nevada: Unpublished M.S. thesis, Colorado School of Mines, 106 p.

Field, C. W., 1966, Sulfur-isotopic method for discriminating between sulfates of hypogene and supergene origin: Economic Geology, v. 61, p. 1428-1435.

Foley, N. K., Bethke, P. M., and Rye, R. O., 1982, A reinterpretation of D_{H2O} values of inclusion fluids in quartz from shallow ore bodies (abs.): Geological Society of America, Abstracts with Programs, v. 14, no. 7, p. 489.

Fournier, R. O., and Truesdell, A. H., 1973, An empirical Na-K-Ca geothermometer for natural waters: Geochimica et Cosmochimica Acta, v. 37, p. 1255-1276.

Giles, D. L., and Nelson, C. E., 1982, Principal features of epithermal lode gold deposits of the Circum-Pacific Rim: Circum-Pacific Energy Minerals Conference, Honolulu, Hawaii, August 22-28, 1982.

Gilzean, M. N., 1984, The nature of the deep hydrothermal system, Red Mountain district, Silverton, Colorado: Unpublished M.S. thesis, University of California (Berkeley), 105 p.

Giudice, P. M., 1980, Mineralization at the convergence of the Amethyst and OH fault systems, Creede district, Mineral County, Colorado: Unpublished M.S. thesis, University of Arizona, 95 p.

Goodell, P. C., 1970, Zoning and paragenesis in the Julcani district, Peru: Unpublsihed Ph.D. thesis, Harvard University, 118 p.

Haas, J. L., Jr., 1971, The effect of salinity on the maximum thermal gradient of a hydrothermal system at hydrostatic pressure: Economic Geology, v. 66, p. 940-946.

Hanaoka, N., 1980, Numerical model experiment of hydrothermal system--topographic effects: Bulletin Geological Survey Japan, v. 31 (7), p. 321-332.

Hayba, D. O., 1983, A compilation of fluid-inclusion and stable-isotope data on selected precious- and base-metal epithermal deposits: U.S. Geological Survey, Open-File Report 83-450, 24 p.

Hayba, D. O., 1984, Documentation of thermal and salinity gradients and interpretation of the hydrologic conditions in the OH vein, Creede, Colorado (abs.): Geological Society of America, Abstracts with Programs, v. 16, no. 6, p. 534.

Heald, P., Hayba, D. O., and Foley, N. K., 1986, Comparative anatomy of volcanic-hosted epithermal deposits: Acid-sulfate and Adularia-sericite types: Economic Geology (in review).

Heald-Wetlaufer, P., and Plumlee, G. S., 1984, Significance of mineral variations in time and space along the Bulldog Mountain vein system with respect to the district-wide hydrology, Creede district, Colorado (abs.): Geological Society of America, Abstracts with Programs, v. 16, no. 6, p. 535.

Heald-Wetlaufer, P., Hayba, D. O., Foley, N. K., and Goss, J. A., 1983, Comparative anatomy of epithermal precious- and base-metal districts in volcanic terranes (abs.): Joint Annual meeting of GAC/MAC/GCU, Victoria, British Columbia, May 11-13, 1983, p. 1.

Healy, J., 1975, Geothermal fields in zones of recent volcanism; in Proceedings Second United Nations Symposium on the Development and Use of Geothermal Resources, U. S. Government Printing Office, Washington, p. 415-422.

Healey, J., and Hochstein, M. P., 1973, Horizontal flow in hydrothermal systems: New Zealand Journal of Hydrology, v. 12, no. 2, p. 71-82.

Hedenquist, J. W., 1983, Waiotapu, New Zealand: The geochemical evolution and mineralization of an active hydrothermal system: Unpublished Ph.D. thesis, University of Auckland, 215 p.

Hedenquist, J. W., and Henley, R. W., 1985, The importance of CO_2 on freezing-point measurements of fluid inclusions: Evidence from active geothermal systems and implications for epithermal ore deposition: Economic Geology, v. 80, p. 1379-1406.

Hedenquist, J. W., and Reid, F. W., 1984, Epithermal gold, concepts for exploration: The Earth Resources Foundation, University of Sydney, Short Course Notes, 2-5 Oct. 1984, 222 p.

Helgeson, H. C., Delany, J. M., Nesbitt, H. W., and Bird, D. K., 1978, Summary and critique of the thermodynamic properties of rock-forming minerals: American Journal of Science, v. 278-A, 229 p.

Hemingway, B. S., Robie, R. A., Kittrick, J. A., Grew, E. S., Nelen, J. A., and London, D., 1984, The heat capacities of osumilite from 298.15 to 1000 K, the thermodynamic properties of two natural chlorites to 500 K, and the thermodynamic properties of petalite to 1800 K: American Mineralogist, v. 69, p. 701-710.

Hemley, J. J., Hostetler, P. B., Gude, A. J., and Mountjoy, W. T., 1969, Some stability relations of alunite: Economic Geology, v. 64, p. 599-612.

Henley, R. D., 1985, The Geothermal Framework of Epithermal Deposits; in Berger, B. R., and Bethke, P. M. (eds.), Geology and Geochemistry of Epithermal Systems: Society of Economic

Geologists, Reviews in Economic Geology, v. 2.

Henley, R. D., and Ellis, A. J., 1983, Geothermal systems ancient and modern: A geochemical review: Earth-Science Reviews, v. 19, p. 1-50.

Henley, R. W., Truesdell, A. H., and Barton, P. B., Jr., 1984, Fluid-Mineral Equilibria in Hydrothermal Systems: Society of Economic Geologists, Reviews in Economic Geology, v. 1, 267 p.

Holland, H. D., 1965, Some applications of thermochemical data to problems of ore deposits, II. Mineral assemblages and the composition of ore-forming fluids: Economic Geology, v. 60, p. 1101-1166.

Holland, H. D., 1967, Gangue minerals in hydrothermal deposits; in Barnes, H. L. (ed.), Geochemistry of Hydrothermal Ore Deposits: Holt, Rinehart, and Winston, Inc., New York, p. 382-436.

Hon, K., Ludwig, K. R., Simmons, K. R., Slack, J. F., and Grauch, R. I., 1985, U-Pb isochron age and Pb isotope systematics of the Golden Fleece vein--Implication for the relationship of mineralization to the Lake City caldera, Western San Juan Mountains, Colorado: Economic Geology, v. 80, p. 410-417.

Horton, D. G., 1983, Argillic alteration associated with the Amethyst vein system, Creede mining district, Colorado: Unpublished Ph.D. thesis, University of Illinois (Urbana-Champaign), 337 p.

Hull, D. A., 1970, Geology of the Puzzle vein, Creede mining district, Colorado: Unpublished Ph.D. thesis, University of Nevada (Reno), 151 p.

Iwasaki, I., and Ozawa, T., 1960, Genesis of sulfate in acid hot springs: Chemical Society of Japan Bulletin, v. 33, p. 1018-1019.

Jensen, M. L., Ashley, R. P., and Albers, J. P., 1971, Primary and secondary sulfates at Goldfield, Nevada: Economic Geology, v. 66, p. 618-626.

Kamilli, R. J., and Ohmoto, H., 1977, Paragenesis, zoning, fluid-inclusion, and isotopic study of the Finlandia vein, Colqui district, Central Peru: Economic Geology, v. 72, p. 950-982.

Kiyosu, T. and Kurahashi, M., 1983, Origin of sulfur species in acid sulfate-chloride thermal waters, northeastern Japan: Geochimica et Cosmochimica Acta, v. 47, p. 1237-1245.

Kiyosu, Y., and Kurahashi, M., 1984, Isotopic geochemistry of acid thermal waters and volcanic gases from Zao volcano in Japan: Journal of Volcanology and Geothermal Research, v. 21, p. 313-331.

Lawrence, J. R. and Taylor, H. P., 1971, Deuterium and oxygen-correlation: Clay minerals and hydroxides in Quaternary soils compared to meteoric waters: Geochimica et Cosmochimica Acta, v. 35, p. 993-1003.

Lipman, P. W., 1975, Evolution of the Platoro caldera complex and related volcanic rocks, southeastern San Juan Mountains, Colorado: U.S. Geological Survey, Professional Paper 852, 128 p.

Lipman, P. W., Doe, B. R., Hedge, C. E., and Steven, T. A., 1978, Petrologic evolution of the San Juan volcanic field, southwestern Colorado: lead and strontium isotope evidence: Geological Society of America Bulletin, v. 8a, p. 59-82.

Lipman, P. W., Fisher, F. S., Mehnert, H. H., Naeser, C. W., Luedke, R. G., and Steven, T. A., 1976, Multiple ages of mid-Tertiary mineralization and alteration in the western San Juan Mountains, Colorado: Economic Geology, v. 71, p. 571-588.

Lipman, P. W., Steven, T. A., Luedke, R. G. and Burbank, W. S., 1973, Revised history of the San Juan, Uncompahgre, Silverton, and Lake City calderas in the western San Juan Mountains, Colorado: U.S. Geological Survey, Journal of Research, v. 1, p. 627-642.

Lipman, P. W., Steven, T. A., and Mehnert, H. H., 1970, Volcanic history of the of the San Juan Mountains, Colorado, as indicated by potassium-argon dating: Geological Society of America Bulletin, v. 81, p. 2329-2352.

Marumo, K., Matsuhisa, Y., and Nagasawa, K., 1982, Hydrogen- and oxygen-isotopic compositions of kaolin minerals in Japan; in Van Olphan, H. and Veniale, F. (eds.), Developments in Sedimentology 35, International Clay Conference 1981, Elsevier, p. 315-320.

McCrink, M. T., 1982, Geology of the Creede Formation, San Juan Mountains, Creede, Colorado: Unpublished M.S. thesis, New Mexico Institute of Mining and Technology, Socorro, 130 p.

McKee, E. H., 1979, Ash-flow sheets and calderas: Their genetic relationship to ore deposits in Nevada: Geological Society of America Special Paper 180, p. 205-211.

Mehnert, H. H., Lipman P. W., and Steven, T. A., 1973, Age of mineralization at Summitville, Colorado, as indicated by K-Ar dating of alunite: Economic Geology, v. 68, p. 399-412.

Misantoni, D. M., 1985, Mineralization along the Midwest fault system, Creede district, Mineral County, Colorado: Unpublished M.S. thesis, Colorado State University, 123 p.

Montoya, J. W., and Hemley, J. J., 1975, Activity relations and stabilities in alkali feldspar and mica alteration reactions: Economic Geology, v. 70, p. 577-583.

Nakamura, H., Sumi, K., Katageri, K., and Iwate, T., 1970, The geological environment of Matsukawa geothermal area, Japan: Geothermics Special Issue 2, p. 221-231.

Nash, J. T., 1975, Fluid-inclusion studies of vein, pipe, and replacement deposits, northwestern San Juan Mountains, Colorado: Economic Geology, v. 70, p. 1448-1462.

Noble, D. C., and Silberman, M. L., 1984, Evolucion volcanica e hidrotermal y cronologia de K-Ar del distrito minero de Julcani, Peru: Sociedad Geologica del Peru, Volumen Jubilar LX Anniversario, p. 1-35.

Ohmoto, H., 1972, Systematics of sulfur and carbon isotopes in hydrothermal ore deposits: Economic Geology, v. 67, p. 551-578.

Ohmoto, H., and Lasaga, A. C., 1982, Kinetics of reactions between aqueous sulfates and sulfides in hydrothermal systems: Geochimica et Cosmochimica Acta, v. 46, p. 1727-1748.

O'Neil, J. R., Silberman, M. L., Fabbi, B. P., and Chesterman, C. W., 1973, Stable-isotope and chemical relations during mineralization in the

Bodie mining district, Mono County, California: Economic Geology, v. 68, p. 765-784.

Patton, H. B., 1917, Geology and ore deposits of the Plataro-Summitville mining district, Colorado: Colorado Geological Survey Bulletin 13, 122 p.

Perkins, R. M., and Nieman, G. W., 1983, Epithermal gold mineralization in the South Mountain volcanic dome, Summitville, Colorado; in The Genesis of Rocky Mountain Ore Deposits: Changes with Time and Tectonics: Denver, Colorado, Denver Region Exploration Geologists Society, p. 165-171.

Petersen, U., Noble, D. C., Arenas, M. J., and Goodell, P. C., 1977, Geology of the Julcani mining district, Peru: Economic Geology, v. 72, p. 931-949.

Pickthorn, W. J., and O'Neil, J. R., 1985, ^{18}O relations in alunite minerals: potential single-mineral thermometer (abs.): Geological Society of America, Abstracts with Programs, v. 17, no. 7, p. 689.

Plumlee, G. S., and Hayba, D. O., 1985, Solubility-temperature-salinity diagrams as a means for interpreting fluid-inclusion/mineral-zoning data from the Creede district, Colorado (abs.): Geological Society of America, Abstracts with Programs, v. 17, no. 7, p. 691.

Ransome, F. L., 1909, The geology and ore deposits of Goldfield, Nevada: U.S. Geological Survey, Professional Paper 66, 258 p.

Ratté, J. C., and Steven, T. A., 1967, Ash-flows and related volcanic rocks associated with the Creede Caldera, San Juan Mountains, Colorado: U.S. Geological Survey, Professional Paper 524-H, 58 p.

Reed, M. H., and Spycher, N. F., 1985, Boiling, Cooling, and Oxidation in Epithermal Systems: A Numerical Modeling Approach; in Berger, B. R., and Bethke, P. M. (eds.), Geology and Geochemistry of Epithermal Systems: Society of Economic Geologists, Reviews in Economic Geology, v. 2.

Rice, J. A., 1984, Controls on silver mineralization in the Creede Formation, Creede, Colorado: Unpublished M.S. thesis, Colorado State University, 135 p.

Robinson, R. W., 1981, Ore mineralogy and fluid-inclusion study of the southern Amethyst vein system, Creede mining district, Colorado: Unpublished M.S. thesis, New Mexico Institute of Mining and Technology, 85 p.

Robinson, R. W., and Norman, D. I., 1984, Mineralogy and fluid-inclusion study of the southern Amethyst vein system, Creede mining district, Colorado: Economic Geology, v. 79, p. 439-447.

Roedder, E., 1960, Primary fluid inclusions in sphalerite crystals from the OH vein, Creede, Colorado (abs.): Geological Society of America Bulletin, v. 71, p. 1958.

Roedder, E., 1963, Studies of fluid inclusions II.: Freezing data and their interpretation: Economic Geology, v. 58, p. 167-211.

Roedder, E., 1965, Evidence from fluid inclusions as to the nature of the ore-forming fluids; in Symposium on Problems of Postmagmatic Ore Deposition, Prague, 1963: Prague, Czechoslovakia Geological Survey, v. 2, p. 375-384.

Roedder, E., 1967, Metastable superheated ice in liquid-water inclusions under high negative pressure: Science, v. 155, no. 3768, p. 1413-1417.

Roedder, E., 1970, Application of an improved crushing microscope stage to studies of the gases in fluid inclusions: Schweizer. Mineralog. U. Petrog. Mitt., v. 50, pt. 1, p. 41-58.

Roedder, E., 1977, Changes in ore fluid with time, from fluid inclusion studies at Creede, Colorado; in Proceedings International Association Genesis Ore Deposits, 4th Symposium, Varna, Bulgaria, 1974: Sofia, Bulgaria, IAGOD, v. 2, p. 179-185.

Roedder, E., 1984, Fluid inclusions: Reviews in Mineralogy, v. 12, Mineralogical Society of America, 644 p.

Roedder, E., and Bodnar, R. J., 1980, Geologic pressure determinations from fluid-inclusion studies: Ann. Rev. Earth Planet. Sci., v. 8, p. 263-301.

Roedder, E., Ringiam, B., and Hall, W. E., 1963, Studies of fluid inclusions III: Extraction and quantitative analyses of inclusions in the milligram range: Economic Geology, v. 58, p. 353-374.

Rytuba, J. J., 1981, Relation of calderas to ore deposits in the western United States; in Dickinson, W. R., and Payne, W. D. (eds.), Relations of Tectonics to Ore Deposits in the South Cordillera, Arizona Geological Society Digest, v. XIV, p. 227-236.

Sakai, H., and Matsubaya, O., 1977, Stable isotope studies of Japanese geothermal systems: Geothermics, v. 5, p. 97-124.

Savin, S. M., and Epstein, S., 1970, The oxygen and hydrogen isotope geochemistry of clay minerals: Geochimica et Cosmochimica Acta, v. 34, p. 25-42.

Schoen, R., and Rye, R. O., 1970, Sulfur isotope distribution in solfataras, Yellowstone National Park: Science, v. 170, p. 1082-1084.

Scott, S. D., and Barnes, H. L., 1971, Sphalerite geothermometry and geobarometry: Economic Geology, v. 66, p. 653-669.

Silberman, M. L., and Ashley, R. P., 1970, Age of ore deposition at Goldfield, Nevada from potassium-argon dating of alunite: Economic Geology, v. 65, p. 352-354.

Silberman, M. L., and Berger, B. R., 1985, Relationship of trace-element patterns to alteration and morphology in epithermal precious-metal deposits; in Berger, B. R., and Bethke, P. M. (eds.), Geology and Geochemistry of Epithermal Systems: Society of Economic Geologists, Reviews in Economic Geology, v. 2.

Sillitoe, R. H., and Bonham, H. F., Jr., 1984, Volcanic landforms and ore deposits: Economic Geology, v. 79, p. 1286-1298.

Slack, J. F., 1980, Multistage vein ores of the Lake City district, western San Juan mountains, Colorado: Economic Geology, v. 75, p. 963-991.

Steiner, A., and Rafter, T. A., 1966, Sulfur isotopes in

pyrite, pyrrhotite, alunite, and anhydrite from steam wells in the Taupo volcanic zone, New Zealand: Economic Geology, v. 61, p. 1115-1129.

Steven, T. A., 1967, Geologic map of the Bristol Mead Quadrangle, Mineral and Hinsdale Counties, Colorado: U.S. Geological Survey, Quadrangle Map GQ-631, scale 1:62,500.

Steven, T. A., and Eaton, G. P., 1975, Environments of ore deposition in the Creede mining district, San Juan Mountains, Colorado: I. Geologic, hydrologic and geophysical setting: Economic Geology, v. 70, p. 1023-1037.

Steven, T. A., and Lipman, P. W., 1973, Geologic map of the Spar City Quadrangle, Mineral County, Colorado: U.S. Geological Survey, Quadrangle Map GQ-1052, scale 1:62,500.

Steven, T. A., and Lipman, P. W., 1976, Calderas of the San Juan volcanic field, southwestern Colorado: U.S. Geological Survey, Professional Paper 958, 35 p.

Steven, T. A., and Ratté, J. C., 1960a, Geology and ore deposits of the Summitville district, San Juan Mountains, Colorado: U.S. Geological Survey, Professional Paper 343, 70 p.

Steven, T. A., and Ratté, J. C., 1960b, Relation of mineralization to caldera subsidence in the Creede district, San Juan Mountains, Colorado; in Short Papers in the Geological Sciences: U.S. Geological Survey, Professional Paper 400-B, p. B14-B17.

Steven, T. A., and Ratté, J. C., 1965, Geology and structural control of ore deposition in the Creede district, San Juan Mountains, Colorado: U.S. Geological Survey, Professional Paper 487, 90 p.

Steven, T. A., and Ratté, J. C., 1973, Geologic map of the Creede Quadrangle, Mineral County, Colorado: U.S. Geological Survey, Quadrangle Map GQ-1053, scale 1:62,500.

Steven, T. A., and Van Lonen, R. E., 1970, Clinoptilolite-bearing tuff beds in the Creede Formation, San Juan Mountains, Colorado: U.S. Geological Survey, Professional Paper 750-C, p. C98-C103.

Stoffregen, R., 1985, Genesis of acid-sulfate alteration and Au-Cu-Ag mineralization at Summitville, Colorado: Unpublished Ph.D. thesis, University of California (Berkeley), 204 p.

Taylor, H. P., 1973, O^{18}/O^{16} evidence for meteoric-hydrothermal alteration and ore deposition in the Tonopah, Comstock lode and Goldfield mining districts, Nevada: Economic Geology, v. 68, p. 747-764.

Taylor, H. P., 1974, The application of oxygen and hydrogen isotope studies to problems of hydrothermal alteration and ore deposition: Economic Geology, v. 69, p. 843-883.

Tingley, J. V., and Berger, B. R., 1985, Lode gold deposits of Round Mountain, Nevada: Nevada Bureau of Mines and Geology, Bulletin 100, 62 p.

Truesdell, A. H., and Fournier, R. O., 1975, Calculation of deep temperatures in geotheraml systems from the chemistry of boiling spring waters of mixed origin; in Proceedings, Second United Nations Symposium on the Development and Use of Geothermal Energy, U. S. Government Printing Office, Washington, D. C., p. 837-844.

Ulrich, M. R., and Bodnar, J. R.,1984, Systematics of stretching of fluid inclusions in barite at 1 atm confining pressure (abs.): Geological Society of America, Abstracts with Programs, v. 16, no. 6, p. 680.

Vanderwilt, J. W., 1947, Mineral resources of Colorado: Colorado State Mineral Resources Board, Colorado, 547 p.

Vergo, N., 1984, Wallrock alteration at the Bulldog Mountain mine, Creede mining district, Colorado: Unpublished M.S. thesis, University of Illinois, 88 p.

Wallace, A. B., 1979, Possible signatures of buried porphyry-copper deposits in middle to late Tertiary volcanic rocks of western Nevada; in Proceedings of the Fifth symposium of the International Association on the Genesis of Ore Deposits, v. 2: Nevada Bureau of Mines Report 33, p. 69-76.

Wason, D. J., 1983, The Bachelor Mountain silver deposit, Creede mining district, Colorado: Unpublsihed M.S. thesis, State University of New York (Stonybrook), 94 p.

Wetlaufer, P. H., 1977, Geochemistry and mineralogy of the carbonates of the Creede mining district, Colorado: U.S. Geological Survey, Open-File Report 77-706, 134 p.

Wetlaufer, P. H., 1978, Chemical similarities of hydrothermal fluids from diverse sources, Creede Ag-Pb-Zn-Cu district, San Juan Mountains, Colorado (abs.): Geological Society of America, Abstracts with Programs, v. 10, no. 7, p. 515.

White, D. E., 1957, Thermal waters of volcanic origin: Geological Society of America Bulletin, v. 68, p. 1637-1658.

White, D. E., 1981, Active geothermal systems and hydrothermal ore deposits: Economic Geology, 75th Anniversary Volume, p. 392-423.

Whitney, J. A., 1984a, Fugacities of sulfurous gases in pyrrhotite-bearing silicic magmas: American Mineralogist, v. 69, p. 69-78.

Whitney, J. A., 1984b, Volatiles in magmatic systems; in Fluid-Mineral Equilibria in Hydrothermal Systems: Society of Economic Geologists, Reviews in Economic Geology, v. 1, p. 155-176.

Whitney, J. A., 1985, Composition and activity of sulfurous species in quenched magmatic gases associated with pyrrhotite-bearing silicic systems: Geological Society of America, Abstracts with Programs, v. 17, no. 7, p. 749.

Whitney, J. A., and Stormer, J. C., Jr., 1983, Igneous sulfides in the Fish Canyon Tuff and the role of sulfur in calc-alkaline magmas: Geology, v. 11, p. 99-102.

Woods, T. L., Bethke, P. M., and Roedder, E., 1982, Fluid-inclusion data at Creede, Colorado in relation to mineral paragenesis: U.S. Geological Survey, Open-File Report 82-313, 61 p.

Zartman, R. E., 1974, Lead isotopic provinces in the cordillera of the western United States and their genetic significance: Economic Geology, v. 69, p. 792-805.

Chapter 8
GEOLOGIC CHARACTERISTICS OF SEDIMENT-HOSTED, DISSEMINATED PRECIOUS-METAL DEPOSITS IN THE WESTERN UNITED STATES

William C. Bagby and Byron R. Berger

INTRODUCTION

Sediment-hosted precious-metal deposits are typically formed in carbonaceous, silty dolomites and limestones or calcareous siltstones and claystones. Gold mineralization is disseminated in the host sedimentary rocks and is exceedingly fine grained, usually less than one micron in size in unoxidized ore. Primary alteration types include silicification, decalcification, argillization, and carbonization. Supergene alteration is dominated by oxidation resulting in the formation of numerous oxides and sulfates and the release of gold from its association with sulfides. Commonly associated trace elements are arsenic, barium, mercury, antimony, and thallium. Deposits of this type are commonly referred to as either Carlin-type deposits, after the large bulk-minable, disseminated-gold deposit in northern Nevada, or as fine-grained or "invisible-gold" deposits. We refer to deposits of this type as sediment-hosted, disseminated precious-metal deposits.

This chapter presents a classification scheme and reviews the geologic characteristics of sediment-hosted, precious-metal deposits. The influences of geology on both mining and the development of genetic and exploration models are discussed. Although deposits of this type occur throughout the western United States, the largest concentration of deposits and also the best understood are in Nevada. We have chosen, therefore, to use selected individual deposits from Nevada as type examples to support the classification scheme and to provide the student with an understanding of the similarities and differences that occur in these deposits. This chapter is thus designed to develop and nurture the knowledge of the comparative geology of sediment-hosted, disseminated precious-metal deposits. This is accomplished by reviewing and comparing regional-, district-, and deposit-scale geologic characteristics.

CLASSIFICATION

Deposits of this type display a variety of geologic and geochemical characteristics that may be used to divide the complete set of deposits into subsets (Tables 8.1 and 8.2). The Carlin deposit has lent its name to this deposit type since its discovery in the early 1960's. Early investigators realized that the Carlin deposit was possibly typical of a class of deposits that had been mined in the past, but for which no classification scheme seemed particularly appropriate. These older known deposits are Getchell, Nevada (Joralemon, 1951), Mercur, Utah (Butler et al., 1920), and Gold Acres, Nevada (Gilluly and Masursky, 1965). As more of these deposits were discovered and developed in the 1970's, it became apparent that Carlin was possibly an end member of the sediment-hosted, disseminated precious-metal deposit type. Thus, "Carlin-type deposit" is a misnomer for some sediment-hosted, disseminated deposits. For example, Carlin is hosted by silty dolomites whereas other deposits are hosted in shales and siltstones. In addition, the gold ore at Carlin occurs in pod-like zones in favorable host lithologies adjacent to vertical faults and it is commonly difficult to visually distinguish ore from unaltered host rock. On the other hand, the ore at some other deposits is more easily distinguished from unaltered rock because of direct association with either intense silicification in the form of "jasperoids" (e.g., Pinson) or with a noticeable increase in silica veining (e.g., Preble). Although "Carlin-type deposits" are considered to have extremely high gold to silver ratios, certain deposits included in this type actually have high silver values, with gold as the major metal (e.g., Dee); others have high silver values and lack gold (e.g., Taylor), but retain similar associated trace elements (Sb, As, Hg) and alteration types.

On the basis of the above characteristics, we have defined two deposit-type subsets: the jasperoidal- and Carlin-type subsets (see Tables 8.1 and 8.2) of which there are gold-rich and silver-rich end members. Jasperoidal-type deposits are those wherein the majority of the gold and/or silver is hosted in jasperoid* or in quartz veins and related silicified wall rocks. On the other hand, Carlin-type deposits are those wherein the gold and/or silver appears to be evenly distributed in the host rocks which do not always appear to be silicifed. Ore zones in Carlin-type deposits are commonly pod-like and extend up to tens of meters away from faults whereas ore zones in the jasperoidal-type deposits are most commonly limited to narrow, shear zones. There is complete gradation between these two subset types and a deposit classified as jasperoidal may have exploration potential for Carlin-type extensions and vice versa. This classification serves as a means of examining and understanding the differences that occur between and within the individual sediment-hosted, disseminated precious-metal deposits. In addition, the separation is useful at the deposit scale for understanding genesis and is a useful concept in regional and district exploration programs.

Table 8.1--Geological aspects of selected sediment-hosted, disseminated precious-metal deposits in the Western United States.

Deposit	Host rocks			Igneous rocks	
	Formation	Age	Lithology	Age	Composition and occurrence
Alligator Ridge	Pilot Shale	Devonian–Mississippian	Thin bedded, calcareous, carbonaceous siltstones and claystones	Tertiary	Siliceous, pumiceous tuff and younger basaltic andesite lava flows
Carlin	Roberts Mountains Formation	Silurian to Early Devonian	Laminated, silty to sandy, carbonaceous, dolomitic limestone	Cretaceous 130 m.y.	Altered quartz diorite and diorite dikes
Cortez	Roberts Mountains Formation	Silurian to Early Devonian	Laminated, silty, argillaceous, carbonaceous, pyrite-bearing limestone with dolomite	Oligocene 33-35 m.y. Jurassic 150 m.y.	Altered dikes and sills of quartz latite; Mill Canyon granodiorite stock
Dee	Devonian Limestone and Vinini Formation	Devonian Ordovician	Massive, fossil-rich limestone	Mesozoic(?) Tertiary(?)	Altered dikes in the Dee mine; intermediate composition
Getchell	Preble Formation Comus Formation	Cambrian Ordovician	Phyllitic shale with interbedded limestone	Cretaceous	Osgood Mountains pluton; Granodioritic with associated intermediate porphyritic dikes. All are altered and in part mineralized.
Gold Acres	Roberts Mountains Formation Valmy Formation Wenban Formation	Silurian to Early Devonian Ordovician Devonian	Carbonates, argillites, and siltstones: All mineralized rocks occur as fault blocks low in the upper plate of the Roberts Mountains thrust.	Cretaceous Tertiary(?)	Altered and mineralized dikes of intermediate composition; Quartz latite sills that are altered.
Horse Canyon	Vinini Formation Wenban Formation	Ordovician Devonian	Siltstones and chert; silty carbonaceous limestone	Tertiary	Altered dikes and sills of quartz latite
Jerritt Canyon	Hanson Creek	Ordovician	Carbonaceous, shaly limestones with chert dolomite, and bioclastic limestones	Mid-Tertiary(?)	Small dikes and plugs of diorite and a small rhyodacite flow 2.4 km SW and 3 km NE of mineralization, respectively

	Structure		
Faults	Folds	Mineralization age	Reference
N-NE, N-NW, and E-W trending normal faults. The NE trending Vantage fault cuts Tertiary tuff. All faults post antiform.	N-NE striking asymetrical antiform plunging to SW. Extensive fracturing and brecciation along crest. Age unknown.	No direct date of mineralization. Tertiary tuff may be altered by gold system placing an lower age constraint.	Klessig, 1984. Ilchick, 1984. Ainsworth and Brimhall, 1983.
Devonian-Mississippian thrust fault (Roberts Mountains thrust), high angle faults trending E, N, and NE and NW. Faults are pre-mineral with some post-mineral movement.	NW directed folds (antiforms and synforms). Major NW-trending antiform in the district. Age of folding is Mesozoic.	No direct date of mineralization. Post-mineral rhyolite lavas and domes dated at 14 m.y. provide upper age constraint and altered Cretaceous dikes (130 m.y.) provide lower age constraint.	Adkins and Pota, 1984. Hausen, 1967. Radtke, 1981.
N, NW, and EW trending, high angle, normal faults. Roberts Mountains thrust surrounds district.	Drag folds associated with faults sympathetic to Roberts Mountains thrust. NW directed regional folds.	Altered and mineralized 33-35 m.y. old quartz latite dikes place a lower age constraint.	Wells, Elliott, and Obradovich, 1971. Wells, Stoiser, and Elliott, 1969. William C. Bagby, U.S. Geological Survey, unpub. field notes.
N, NW, NE, and EW trending high angle faults of Mesozoic(?) and Tertiary age. Roberts Mountains thrust (Dev.-Miss.).	NW directed folds of presumed Mesozoic age.	Unknown, altered dikes remain undated	Wallace and Bergwall, 1984. William C. Bagby, U.S. Geological Survey, unpub. field notes.
N. trending Getchell fault zone includes several strands. Inception of fault Late Cretaceous.	Fold axis plunges 45° NE. On southern limb sediments strike N and dip SE.	Sericite in mineralized granodiorite dated by K-Ar between 87-92 m.y. This is inferred age of gold mineralization.	Joralemon, 1951. Silberman, Berger, and Koski, 1974. Berger and Taylor, 1980.
N, NE faults dip steeply west. Roberts Mountains thrust.	NW directed antiforms and synforms of presumed Mesozoic age.	Not directly dated. Altered Tertiary(?) sill places a lower age constraint.	Wrucke, 1984. Wrucke and Armbrustmacher, 1975.
N, NW, EW trending high angle normal faults of Tertiary age. Roberts Mountains thrust.	NW directed folds.	Altered Tertiary (Oligocene?) dikes place a lower age constraint on mineralization	Coppinger and Cartwright, 1983. William C. Bagby, U.S. Geological Survey, unpub. field notes.
Normal faults striking E-W, N-S N. 20-30 E. Roberts Mountains thrust fault. NS faults post-mineral and high angle.	Regional E-W trending folds.	Unknown	Hawkins, 1973; 1982. Stevens and Hawkins, 1984.

Table 8.1--Geological aspects of selected sediment-hosted, disseminated precious-metal deposits in the Western United States--(continued).

Deposit	Host rocks			Igneous rocks	
	Formation	Age	Lithology	Age	Composition and occurrence
Mercur	Lower Great Blue	Mississippian	Massive, bedded limestone; local bioclastic micrites and wackestones with sparse siltstones	Tertiary 31.6 m.y.	Fine-grained, porphyritic rhyolite plug south of deposit. Believed to post-date gold mineralization. Also coarse-grained porphyritic plugs 1.6 km north of deposit.
Northumberland	Roberts Mountains Formation	Silurian	Laminated silty limestones, shales, and siltstone	Jurassic Tertiary(?)	Altered tonalite and granodiorite dikes and pluton Unaltered siliceous tuffs and altered rhyolitic dikes
Pinson	Comus Formation	Cambrian-Ordovician	Thin-bedded limestone and shale	Cretaceous (90 m.y.)	Intermediate composition dikes and Osgood Mountains pluton
Preble	Preble Formation	Cambrian	Phyllitic shale and interbedded limestone	Cretaceous(?)	Altered dike of intermediate composition
Rain	Webb Formation	Mississippian	Siltstone, shales, and fine-grained sandstone	Tertiary (35 m.y.)	Quartz monzonite stock 6 mi south of deposit
Sterling	Wood Canyon and Bonanza King Formations	Cambrian	Silty and sandy dolomite with minor carbonaceous matter	Tertiary	Quartz latite dikes occur near the deposit.
Taylor	Guilmette Formation	Devonian	Limestone and shaly limestone	Tertiary (35 m.y.)	Rhyolitic dikes altered but not mineralized
Tolman	--	Pennsylvanian	Sandy and calcareous siltstones, silty limestone, and claystone	Tertiary	Altered intermediate to mafic dikes and remnants or rhyolitic ash-flow tuffs. Rhyolitic domes present 5 km SW.

---

*The term jasperoid and jasperoidal are used repeatedly in the text as a rock name and an adjective, respectively. We use the term as an integral part of our classification of deposits of this type and have adopted the definition of Lovering (1972), p. 3): "(1) jasperoids are composed predominantly of silica, which in most places is in the form of aphanitic to fine-grained quartz, and (2) jasperoids form by replacement of the enclosing rock."

REGIONAL GEOLOGIC CHARACTERISTICS OF DEPOSITS IN MINERAL TRENDS AND ISOLATED DEPOSITS

Locations of major known sediment-hosted, disseminated precious-metal deposits in the western United States are given in Figure 8.1. Except for Mercur, Utah, the most thoroughly studied sediment-hosted precious-metal deposits are in Nevada; therefore, we have chosen several of these deposits for discussion in this paper. Some of the Nevada deposits occur along recognized mineral belts that contain

Structure		Mineralization age	Reference
Faults	Folds		
(1) Normal and strike slip faults associated with folding. (2) Normal faults due to Basin and Range extension, (3) Normal faults of minor displacement possibly associated with intrusions.	Northwest trending Late Cretaceous to late Oligocene Ophir anticline and Pole Canyon syncline. Thrust faulting was associated with folding.	Unknown. Likely Tertiary as normal faults are mineralized. Porphyritic plugs are unmineralized.	Tafuri, 1976. Kornze and others, 1984. Edwin W. Tooker, U.S. Geological Survey, oral commun., 1984.
Thrust faults (Dev.-Miss.), NE, N, to NW trending high angle Tertiary and Late Cretaceous faults.	Doming of Paleozoic sediments near Mesozoic pluton and folding of sediments near thrust faults	Altered dikes emplaced between 32-26 m.y. which places lower age constraint provided alteration due to gold hydrothermal system.	Ott, 1983. McKee, 1974. William C. Bagby, U.S. Geological Survey, unpub. field notes.
NE trending high angle fault of Cretaceous(?) age	NE plunging antiform.	Not dated but presumed to be the same as Getchell (87-92 m.y.).	Kretschmer, 1984. Berger, 1980 William C. Bagby, U.S. Geological Survey, unpub. field notes.
N and NE to EW trending high angle faults. Cretaceous(?).	NE directed antiforms and synforms.	Not dated but presumed to be the same as Getchell (87-92 m.y.).	Kretschmer, 1984. Berger, 1980 William C. Bagby, U.S. Geological Survey, unpub. field notes.
N and WNW trending high angle faults.	N-NW plunging antiform.	Unknown.	Knutsen and West, 1984. William C. Bagby, U.S. Geological Survey, unpub. field notes.
Thrust fault contact between Wood Canyon and Bonanza King Formations.	--	Unknown.	William C. Bagby, U.S. Geological Survey, unpub. field notes.
N, NW, NE trending high angle faults of Tertiary age.	Possible S plunging antiform.	The age constraints are provided by post mineralization, mid-Tertiary dikes, and early Tertiary, pre-mineralization faults.	Drewes, 1967. Lovering and Heyl, 1974. Havenstrite, 1983.
Low angle and high angle normal faults. No preferred direction.	Large and small scale folds trending N and NE.	Unknown.	Brady, 1984.

several deposits of this type whereas others are apparently isolated, individual deposits that have not yet been demonstrated to occur with other similar deposits (Fig. 8.2). The geology of these belts and of the areas around isolated deposits provides an understanding of the regional geologic characteristics of these deposits.

The most famous gold belt in Nevada is the Carlin trend. This mineral belt was initially referred to as the Lynn-Railroad belt and was described by Roberts (1966). This belt includes the Rain, Gold Quarry, Maggie Creek, Carlin, Bullion Monarch, Blue Star, Gold Strike, Bootstrap, and Dee sediment-hosted, disseminated precious-metal deposits. Some investigators extend the Carlin belt to the southeast to include the Alligator Ridge deposit (Fig. 8.2). However, detailed studies of the Alligator Ridge area and its relation to extensions of structures within the Carlin trend are needed to determine whether or not there is an association. The Horse Canyon, Cortez, Gold Acres, and Tonkin Springs deposits occur along the Battle Mountain-Eureka belt (Roberts, 1966) which is now more commonly referred to as the Cortez trend. The Getchell, Pinson, and Preble deposits

Table 8.2--Characteristics and form of ore in selected sediment-hosted, disseminated precious-metal deposits in the Western United States.

Deposit	Alteration		Ore bodies
	Hypogene	Supergene	Form
Alligator Ridge	Decarbonization, decalcification, silicification, carbon remobilization, acid oxidation(?)	Oxidation	Pods localized near high angle faults and extending into sediments
Carlin	Decarbonization, silicification, calcification, carbon remobilization, acid leaching with oxidation(?)	Oxidation, calcite remobilization, clay formation	Pods localized along high angle faults and extending into sediments
Cortez	Silicification, acid leaching and oxidation(?), decalcification, and dedolomitization	Deep oxidation resulting in some redistribution of gold	Elongated zones paralleling faults and dikes and notably localized in breccia zones associated with folds
Dee	Silicification and argillization	Oxidation	Zones localized along faults
Getchell	Decarbonatization with silicification and argillization; early calc-silicate skarnification	Minor oxidation	Sheet-like zones localized along strands of Getchell fault and pods in fold hinges
Gold Acres	Silicification, early contact metamorphism	Deep oxidation	Tabular, dipping to SW parallel to thrust faults
Horse Canyon	Decalcification, silicification, carbon mobilization	Shallow oxidation	Localized zones along NNE fractures and in permeable hosts near fractures

	Ore bodies		
Mineralogy	Gold or silver site	Veins	Reference
Oxidized: metallic gold, specular hematite, jarosite, stibiconite, goethite, quartz, barite, calcite, gypsum, alunite, and kaolinite. Unoxidized: stibnite, pyrite, orpiment, realgar, and calcite.	Oxidized: metallic gold; 85% micron-size 15% coarse, visible.	Quartz veinlets cutting jasperoid. Alunite-quartz veinlets cutting jasperoid. Pyrobitumen veins.	Klessig, 1984. Ilchik, 1984. Ainsworth and Brimhall, 1983.
Oxidized: metallic gold, goethite, illite, kaolinite, barite, anhydrite, alunite, dolomite, calcite, quaartz, schuetteite, cinnabar, arsenolite, scorodite, stibiconite, avicennite, and various lead, zinc, and copper oxides. Unoxidized: quartz, calcite, dolomite, illite, pyrite, realgar, orpiment, stibnite, cinnabar, base-metal sulfides, and rare Tl-As-Sb-S minerals.	Oxidized: metallic gold. Unoxidized: gold with mercury, arsenic, tin, and thallium form thin films on pyrite. Gold is locally associated with carbonaceous matter.	Quartz-calcite-orpiment, barite-galena, quartz-pyrobitumen, quartz-pyrite, and calcite veinlets. Calcite zoned away from deposit in general with quartz veins occurring within main ore deposit area.	Hausen, 1967. Radtke, 1981. Radtke and Scheiner, 1970. Hausen, 1981.
Oxidized: quartz, clays, iron oxides, metallic gold, and calicite. Unoxidized: quartz, illite, dolomite, calcite, pyrite.	Oxidized: clusters of particles between silt grains, metallic gold grains in quartz veinlets, grains in limonite pseudomorphs after pyrite. Unoxidized: associated with As and pyrite.	Quartz-pyrite veinlets and post-mineralization calcite veinlets. Calcite veinlets zoned away from ore zone may have formed contemporaneous with decalcification within the ore zone.	Wells, Stoiser, and Elliott, 1969. Wells and Mullens, 1973.
Oxidized: quartz, clays, minor calcite, metallic gold. Unoxidized: quartz, pyrite, stibnite.	Oxidized: metallic gold. Unoxidized: unknown. May be with sulfides in quartz.	Quartz-stibnite and quartz-pyrite veins. Marked increase in gold content with silica veining and silicification.	Wallace and Bergwall, 1984. William C. Bagby, U.S. Geological Survey, unpub. field notes.
Oxidized: metallic gold(?) Unoxidized: pyrite, pyrrhotite, arsenopyrite, marcasite, stibnite, orpiment, realgar, ilsemannite, cinnabar, magnetite, and metallic silver.	Oxidized: no published data. Unoxidized: metallic gold encapsulated by quartz. Gold also associated with pyrite, arsenopyrite, carbonaceous matter, and magnetite. Silver associated with sulfides.	Calcite veins in limestone and quartz veins in phyllitic shale. Stockwork quartz veins cutting igneous rocks are in turn cut by calcite-dolomite veins.	Joraleman, 1951. Berger, 1975.
Oxidized: quartz, kaolinite, iron oxides, sericite, jarosite, gypsum, hexahydrite, dolomite, and calcite. Unoxidized: no published data available.	Oxidized: gold associated with iron oxide-clay fracture coatings.	Unreported.	Wrucke, 1984. Wrucke and Armbrustmacher, 1975.
Oxidized: iron oxides, clay, quartz, metallic gold. Unoxidized: quartz, clay, cinnabar, pyrite, and arsenopyrite.	Gold is recovered from silicified rock and carbonaceous rock. Actual site unknown.	Quartz and quartz-magnesite veins in silicified zones.	Coppinger and Cartwright, 1983. William C. Bagby, U.S. Geological Survey, unpub. field notes.

Table 8.2--Characteristics and form of ore in selected sediment-hosted, disseminated precious-metal deposits in the Western United States--(continued)

Deposit	Alteration		Ore bodies
	Hypogene	Supergene	Form
Jerritt Canyon	Decalcification, silicification, remobilization of organic matter	Oxidization and leaching	Conformable to bedding as pods near faults
Mercur	Decalcification and silicification; minor kaolinite and sericite addition	Oxidization of sulfides to limonite; no effect on silicification or on clays.	Strongly conforms to bedding according to variation in fracture density, chemistry, and proximity of faults.
Northumberland	Early calc-silicate skarn formation, silicification, and intense argillization	Oxidation and clay formation	Tabular zones along tonalite sill-sediment contact, diffuse in breccia zones, and stratiform bodies in sediments.
Pinson	Early calc-silicate skarnification, silicification, and minor argillization	Deep oxidation along fault zones	"A" ore zone is massive jasperoid localized in fault. "B" ore zone in breccia zone of fold hinge.
Preble	Silicification, argillization, dolomitization	Oxidation to about 200 ft	Ore body localized along shear zone
Rain	Silicification, argillization, baritization, leaching, and oxidation(?)	Oxidization	Localized within a high angle fault zone and penetrates into wall rocks
Sterling	--	--	
Taylor	Silicification, late argillization	Significant oxidation resulting in silver enrichment blanket	Tabular silicified bodies on top of Guilmette Formation; localized by faults
Tolman	Silicification followed by late calcification	Oxidation	Concentrated below low angle normal fault

define the Getchell trend along the eastern flank of the Osgood Mountains. Roberts (1966) included the Getchell deposit in the Battle Mountain-Eureka trend. However, the location of Getchell is controlled by north-south structures along the eastern flanks of the Osgood Mountains and it is equivocal whether or not the Battle Mountain-Eureka belt is a factor in the location of either Getchell, Pinson, or Preble.

An important aspect of the regional geologic setting of these deposits is their spatial relationship to Precambrian crystalline basement and accreted terranes. Either the 0.7080 or the 0.7060 strontium isotope isopleth is generally accepted as representing the western extent of Precambrian continental crust in Nevada (Kistler, 1983; Farmer and DePaolo, 1983). We have chosen the 0.7080 isopleth to represent the western extent of Precambrian crystalline basement (Fig 8.2). The dashed line (refer to figure) representing the present-day easternmost extent of allochthonous western assemblage siliceous rocks

	Ore bodies		
Mineralogy	Gold or silver site	Veins	Reference
Unoxidized: pyrite, realgar, orpiment, arsenopyrite, cinnabar, and organic matter. Stibnite, barite, and quartz occur near ore bodies.	Gold 1-4 micron; mineral association inknown.	Quartz, stibnite, and barite veins occur in jasperoid near ore bodies. Calcite and arsenic minerals occur as veins in unoxidized ore.	Hawkins, 1973, 1982.
Unoxidized: pyrite, realgar, orpiment, cinnabar, barite, fluorite. Organic carbon as both kerogen and hydrocarbon fractions. Oxidized: limonite, hematite, scorodite, gypsum, melanterite, rosenite.	Unoxidized: with pyrite/marcasite, in hydrocarbons, in kerogen, and metallic gold. Oxidized: with melanterite and rosenite.	Quartz crystals in vugs in jasperoid, calcite-realgar veins, and orpiment-pyrite-marcasite-organic matter veins. Barite-halloysite veins cut jasperoid.	Tafuri, 1976. Kornze and others, 1984. Jewell, 1985.
Oxidized: micron metallic gold, clay, iron oxides. Unoxidized: carbonaceous matter, silica, pyrite, freibergite, sphalerite, chalcopyrite, and minor molybdenite.	Quartz-silver veins of early mineralization.	Quartz veinlets, cutting jasperoid. Calcite veins cut acid-leached rock and barite veins spatially associated with jasperoid.	Ott, 1983. William C. Bagby, U.S. Geological Survey, unpub. field notes.
Oxidized: silica, goethite, lepidocrocite, hematite, sparse remnant pyrite and marcasite, and metallic gold. Kaolinite and sericite present but minor. Unoxidized: no reduced ore reported.	Micron metallic gold occurs with As-rich pyrite.	Quartz and calcite veins cut jasperoid.	Powers, 1978. Kretschmer, 1984.
Oxidized: iron oxides, minor pyrite, and metallic gold. Unoxidized: pyrite, carbonaceous matter, and minor arsenopyrite, marcasite, chalcopyrite, and sphalerite.	Gold encapsulated in quartz and associated with pyrite. Metallic, micron-size gold in oxidized ore.	Quartz, dolomite, jasperoid breccia, and calcite veins.	Kretschmer, 1984. William C. Bagby, U.S. Geological Survey, unpub. field notes.
Oxidized: barite, silica, jarosite, alunite, iron oxides, kaolinite, illite, metallic gold(?). Unoxidized: none reported.	Not reported.	Quartz-alunite, quartz, barite, and calcite veins.	Knutsen and West, 1984.
--	--	No data available.	
Oxidized: iron oxides, with remnant stibnite, sphalerite, tetrahedrite, chalcopyrite, galena, pyrargyrite. Unoxidized: none reported.	Metallic silver.	Quartz and late stage calcite veinlets.	Lovering and Heyl, 1974. Havenstrite, 1983.
Unoxidized: pyrite, tetrahedrite, cinnabar, barite, calcite, quartz, gold. Oxidized: limonite, gold, same gangue mineralogy as unoxidized ore.	Gold with organic matter and pyrite. Not associated with tetrahedrite.	Calcite with tetrahedrite. Barite with cinnabar. Silica micro-veinlets.	Brady, 1984.

shows the approximate amount of overlap of continental crust by allochthonous terranes, beginning with the middle Paleozoic Antler Orogeny. Allochthonous, early-Paleozoic, western-assemblage rocks were emplaced along the Roberts Mountains thrust in the middle Paleozoic during the Antler Orogeny (Roberts et al., 1958). This terrane was the earliest allochthonous terrane accreted to the western margin of the United States. It was followed in turn by the Golconda and Sonomia terranes emplaced during the late Paleozoic and early Mesozoic, respectively (Coney et al., 1980). Most of the deposits considered in this chapter occur within accreted terranes. The exceptions to this are the Alligator Ridge and Taylor deposits which occur east of any accreted terranes. Deposits within the Carlin and Cortez trends lie within allochthonous rocks that in turn overlie Precambrian crystalline basement. Getchell, Pinson, and Preble in the Getchell trend lie within accreted terrane west of the 0.7080 strontium-isotope isopleth.

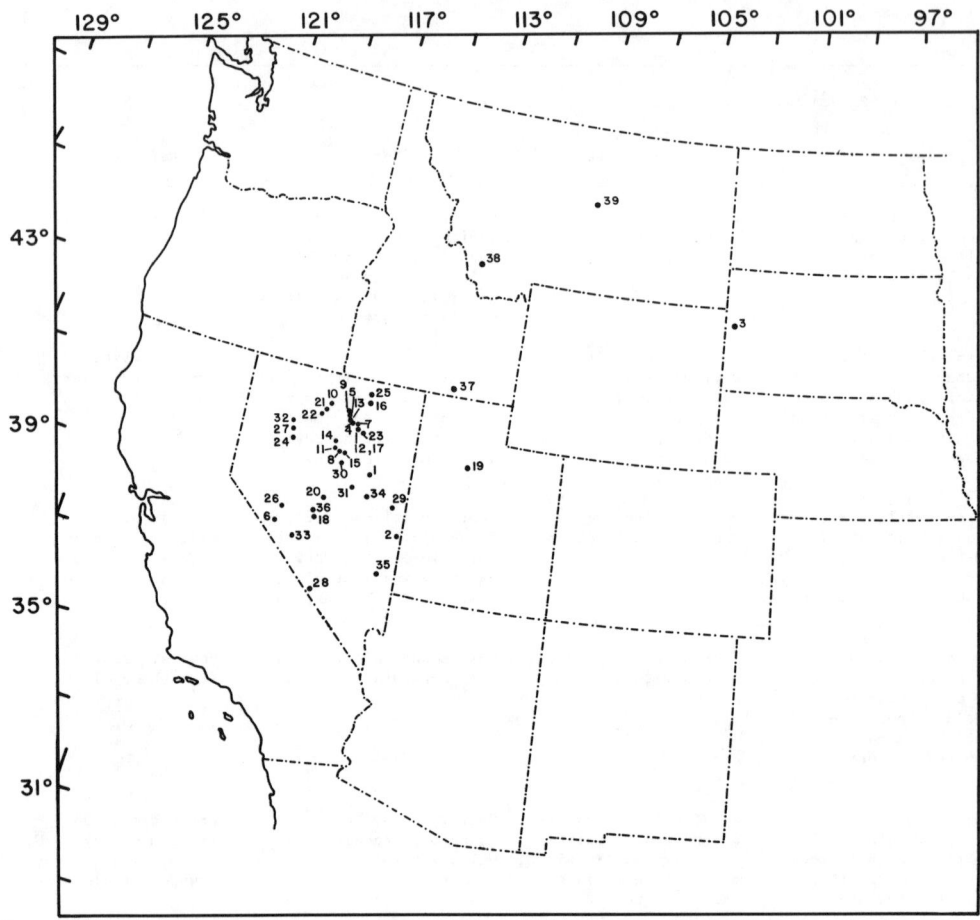

Figure 8.1. Locations of major sediment-hosted, disseminated precious-metal deposits in the western United States. Deposits 1 through 31 are keyed to those in Table 8.5. The other deposits are: (32) Florida Canyon, (33) Weepah, (34) Hamilton, (35) DeLamar, (36) Shale Pit, (37) Tallman, (38) Ermont, and (39) Kendall.

The Getchell Trend

The Getchell trend is localized along the eastern margin of the Osgood Mountains (Fig. 8.3). At Getchell, in the northern part of the trend, mineralization is controlled by splays of the Getchell fault (Berger and Taylor, 1980). It is equivocal whether or not this fault extends the complete distance of the trend. Midway along the trend, at Pinson, the fault is smaller and consists of a single major strand. Continuation of this structure south to Preble is conjecture since the complete length is neither exposed nor has there been detailed geologic mapping. There is at Preble, however, a major north-striking fault that controls mineralization, and that may be a southern extension of the Getchell fault system.

Several formations occur along the trend and serve as host rocks for gold, tungsten, and barite mineralization. These formations include the Cambrian Preble Formation, the Ordovician Comus Formation, and the Osgood Mountains granodiorite. The Preble Formation consists of a sequence of phyllitic shales and thin-bedded to massive limestones (Hotz and Willden, 1964). The lower part of the unit is in depositional contact with the lower Cambrian(?) Osgood Mountains Quartzite. This lower part of the Preble Formation is predominantly phyllitic shale that is variably calcareous and siliceous. About midway up in the Preble Formation the predominant lithology is limestone. The limestone includes thin-bedded limestone separated by shaly partings and thick, massive turbiditic limestone. The upper part of the formation is again predominantly phyllitic shale. The Preble Formation serves as a host rock for gold mineralization at the Getchell, Pinson, and Preble gold deposits and for tungsten skarn deposits around the margin of the pluton.

The Ordovician Comus Formation crops out from the middle part of the trend north past the Getchell aplite, alaskite, and other derivative rocks (Hotz and Willden, 1964). Granodiorite and dacite dikes associated with the pluton are altered and mineralized at Getchell, Pinson, and, possibly, Preble.

The structural history of the Osgood Mountains area is complex. The rocks have experienced both the Antler and Sonoma orogenies. In addition, Erickson and Marsh (1974) identified two episodes of pre-Mesozoic deformation associated with neither the Antler nor the Sonoma orogeny: a Late Cambrian or Early Ordovician (pre-Antler) event and a Late Pennsylvanian or Early Permian (pre-Sonoma) event. The influence of these deformation events on the localization of the later sediment-hosted gold ore is yet unclear, although the isoclinal folds attibuted by Erickson and Marsh (1974) to the pre-Antler event are important ore controls at Getchell.

The southern, western, and northern (Dry Hills) margins of the Osgood Mountains contain Tertiary volcanic rocks which include basalts and basaltic

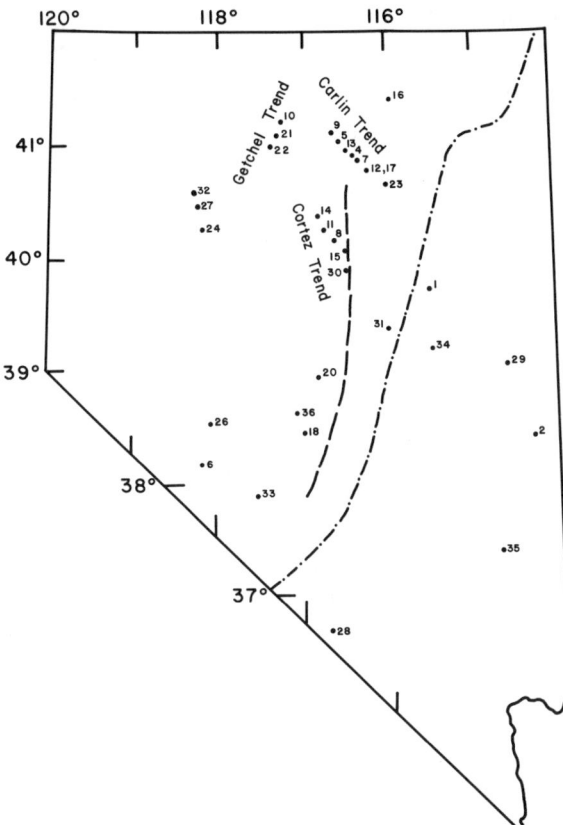

Figure 8.2. Locations of major sediment-hosted, disseminated precious-metal deposits in Nevada. Deposits are numbered as in Figure 8.1. The dash-dot line represents the easternmost extent of allochthonous lower Paleozoic siliceous assemblage rocks. The dashed line is the 0.7080 strontium isotope isopleth from Farmer and DePaolo (1983) and is interpreted as representing the western limit of Precambrian crystalline basement.

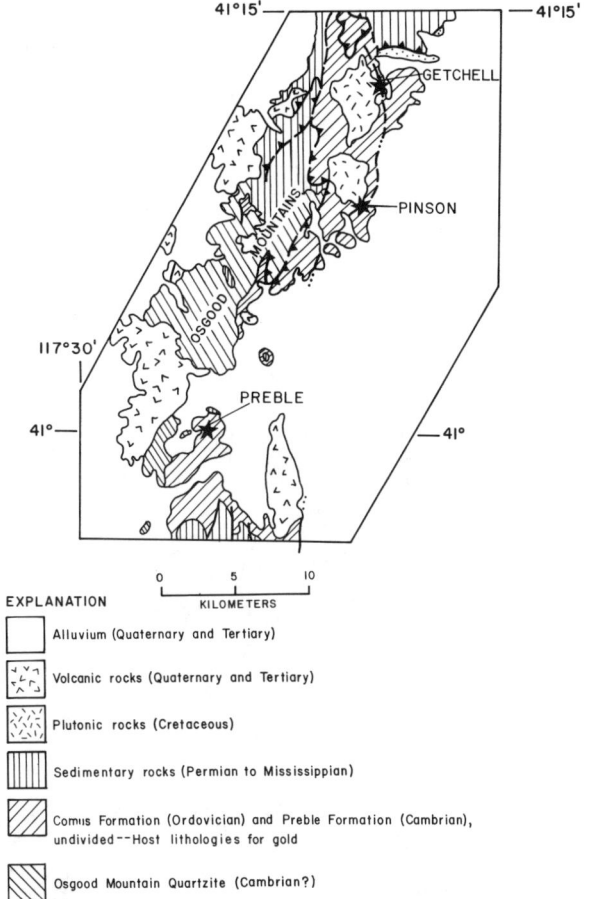

Figure 8.3. Regional geology of the Getchell trend. Getchell, Pinson, and Preble are the major sediment-hosted, disseminated precious-metal deposits that define the trend. Geology is modified from Stewart and Carlson (1984).

mine. It consists of interbedded dolomite, limestone, and shale with subordinate chert, siltstone, and tuffaceous sedimentary rocks (Hotz and Willden, 1964). Greenstones occur in the section near the Getchell mine. This unit is reportedly in fault contact with the Preble Formation wherever the contact is exposed (Hotz and Willden, 1964). The unit is altered and mineralized at the Pinson and Getchell deposits. The rhythmic nature of interbedded shales and carbonate rocks is similar to particular parts of the Preble Formation and when altered, they are difficult to differentiate particularly in the vicinity of the Getchell mine.

The Cretaceous Osgood Mountain stock intrudes the Paleozoic sedimentary rocks. Contact-metasomatized rocks are host to tungsten deposits. The pluton is predominantly granodiorite but includes andesites and represent the youngest igneous activity in the Osgood Mountains. Quaternary basaltic volcanism also occurred in the Edna Mountains a few miles southeast of the Getchell trend.

The Carlin Trend

As mentioned previously, the Carlin trend is identified by an alignment of gold deposits from Rain in the southeast to Dee in the northwest (Figs. 8.2 and 8.4). The southeastern part of the trend does not contain as many identified deposits as the northwestern part.

The Carlin trend strikes northwest and cuts across the northerly trending regional geologic fabric defined by east-directed allochthonous terranes (Fig. 8.2). The belt occurs in the Roberts Mountains allochthon which in turn, at least over most of the belt, overlies Precambrian crystalline basement as inferred from the 0.7080 strontium isotope isopleth. Of importance along this trend are uplifted windows of autochthonous rocks. These windows were the sites of early discoveries of sediment-hosted, disseminated-gold mineralization resulting in the exploration criterion that such windows and the proximity of the Roberts Mountains thrust were both necessary for formation of these deposits. However, many deposits have been discovered away from these windows indicating that their presence is not necessary (e.g., Mercur, Alligator Ridge, Rain).

Rocks of the Roberts Mountains allochthon are from the western, or siliceous, assemblage of lower Paleozoic rocks in northern Nevada (Roberts et al., 1958). This assemblage along the Carlin trend is composed predominantly of interbedded cherts, shales, and siltstones. Carbonate rocks are present but minor. These rocks are in thrust contact with eastern, or carbonate, assemblage lower Paleozoic rocks. The autochthonous rocks are predominantly silty limestones and dolomites with minor shales and siltstones. Rocks in both assemblages serve as hosts for mineralization along the Carlin trend. The major host formations are silty dolomites of the Devonian and Silurian Roberts Mountains Formation, massive fossiliferous limestone of an unnamed Devonian limestone unit (Evans, 1980), and cherts and shales of the Ordovician Vinini Formation. These formations serve as hosts to ore in the northern part of the trend from

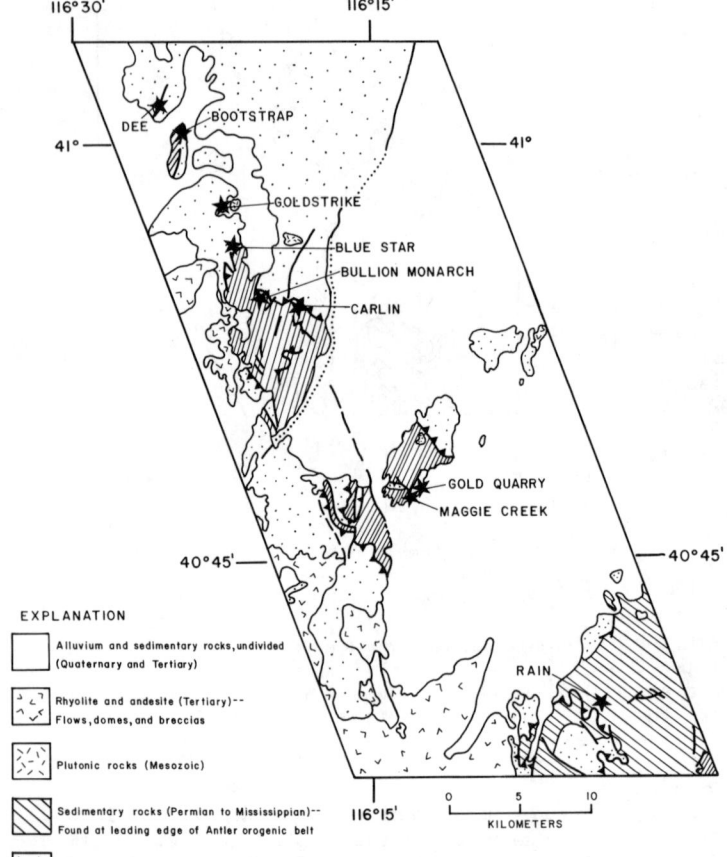

Figure 8.4. Regional geology of the Carlin trend. The major known sediment-hosted, disseminated precious-metal deposits in the trend are Rain, Maggie Creek, Gold Quarry, Carlin, Bullion Monarch, Blue Star, Gold Strike, Bootstrap, and Dee. Geology is modified from Stewart and Carlson (1984).

Maggie Creek-Gold Quarry to Dee to the northwest.

In the southern part of the trend, the Mississippian Webb Formation serves as a host formation to precious metals at the Rain deposit and nearby occurrences. This formation post dates the Antler Orogeny and was formed from debris shed off of the leading edge of the Antler highlands into a foreland basin (Roberts et al., 1958). Shales and siltstones are the predominant lithology in this formation. The upper part of the Devonian Devils Gate Limestone may also serve as a host formation to mineralization in the southern part of the trend. This formation may be correlative with the unnamed Devonian limestone that serves as a host formation in the northern part of the trend (e.g., Bootstrap and Dee).

The Carlin trend is not only identified by an alignment of windows in the allochthon and by ore deposits, but also by Cretaceous and Tertiary plutons. In addition, the trend has positive aeromagnetic anomalies, which, together with exposed plutons, suggest the presence of extensive, buried intrusions, particularly in the Carlin district. These data suggest that there is a fundamental, underlying structural control for the existence of this trend. As mentioned above, the trend strikes northwest and crosscuts the northerly trending fabric defined by Paleozoic tectonism. This suggests that the Carlin trend is younger than accretion of the Roberts Mountains allochthon. Raul Madrid (U.S. Geological Survey, oral communication, 1985) notes that folds within the trend are also northwesterly directed. He suggests that the structures that define the trend were developed during the Mesozoic when subduction was active off the western continental margin and plutons were emplaced east of that subducting margin. (Notably, the northwest-trending Miocene Oregon-Nevada lineament (Stewart et al., 1975) strikes more northerly than the Carlin trend and crosscuts it.) Mineralized Cretaceous igneous rocks in the Carlin to Dee part of the trend suggest that mineralization postdates pluton emplacement and that hydrothermal fluids used the same regional structures that served as controls for Mesozoic pluton and dike emplacement.

Other mineralization in the Carlin trend includes copper mineralization associated with some of the Mesozoic intrusions. Tungsten skarn mineralization is reported from Gold Strike (Morrow and Bettles, 1982). Barite occurs in cherts of the Vinini Formation north of the Dee deposit.

The Cortez Trend

This trend is defined by the alignment of the Gold Acres, Cortez, Horse Canyon, and Tonkin Springs deposits in north-central Nevada (Figs. 8.2 and 8.5). The trend was identified by Roberts (1966) as part of

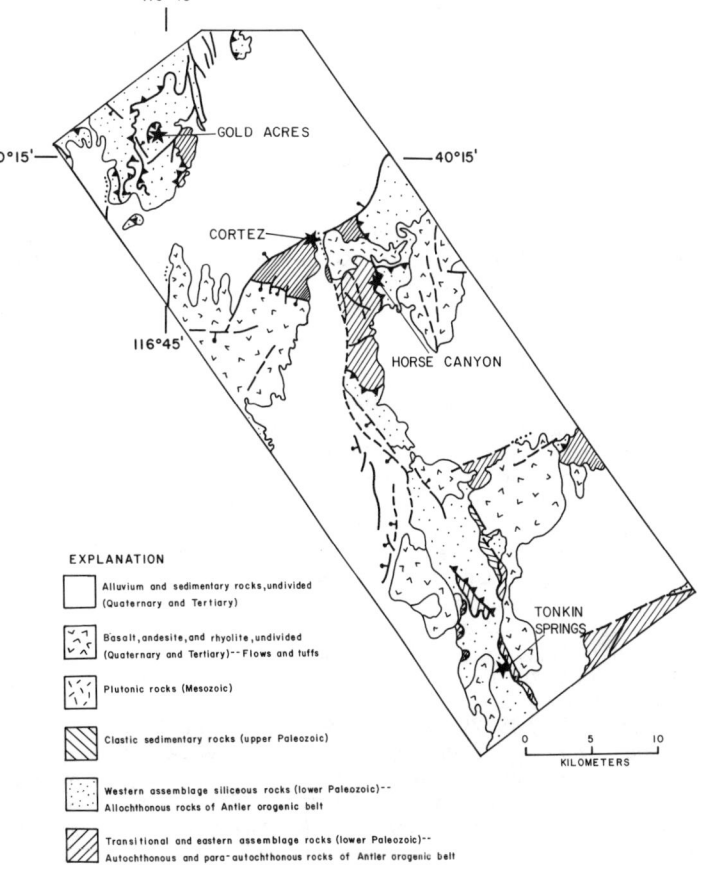

Figure 8.5. Regional geology of the Cortez trend. Gold Acres, Cortez, Horse Canyon, and Tonkin Springs are the major known sediment-hosted, disseminated precious-metal deposits that occur along the trend. Geology is modified from Stewart and Carlson (1984).

the Battle Mountain-Eureka belt, but is now commonly referred to as the Cortez trend after the largest known sediment-hosted, disseminated-gold deposit within the belt. More recently, Rytuba et al. (1984) have suggested that Gold Acres, Cortez, Horse Canyon, and Tonkin Springs are possibly related to caldera development associated with the formation of the Caetano Tuff.

As with the Carlin trend, the Cortez trend contains an alignment of mineral deposits, windows in the Roberts Mountains allochthon, and Mesozoic and Tertiary intrusive rocks. Regional host formations include laminated siltstones in the Devonian and Silurian Roberts Mountains Formation, the Devonian Wenban Limestone, cherts and shales of the Ordovician Vinini Formation, and Mesozoic and Tertiary igneous rocks. These lithologies are little different from those in the Carlin trend. Gold Acres, Cortez, and Horse Canyon all occur within windows of the Roberts Mountains allochthon. Tonkin Springs occurs within upper plate rocks, although an unmineralized Devonian limestone, possibly lower plate, is present. Altered and mineralized igneous rocks within the trend include Mesozoic and Tertiary intrusions. James Rytuba (U.S. Geological Survey, oral communication, 1985) suggests that altered rhyolite dikes at Gold Acres, Cortez, and Horse Canyon are genetically related to the Oligocene Caetano Tuff based on their trace-element chemistry. Silberman and McKee (1971) report a K/Ar age of 94.3 ± 1.9 m.y. on sericite from an altered and mineralized felsite sill in the Gold Acres mine. This age implies a Cretaceous age for the mineralization. The resolution of the actual age of the Gold Acres deposit must await further geochronological studies.

Other mineralization in the Cortez trend includes the Cortez silver deposit in the Cambrian Hamburg Dolomite, the Buckhorn gold deposit in Pliocene volcanic rocks, and bedded barite in Ordovician sedimentary rocks. The Cortez silver deposit is a replacement deposit of mantos and fissure veins in the Hamburg Dolomite (Gilluly and Masursky, 1965; Gilluly and Gates, 1965). The Buckhorn deposit is an epithermal hot-spring related vein deposit.

Isolated Deposits

Several of the sediment-hosted, disseminated precious-metal deposits in Nevada occur as deposits isolated from other known deposits of this type. However, many of these isolated deposits do occur in recognized mineral belts that are defined by regional structures and the alignment of several different types of mineral deposits. Examples of isolated deposits are Taylor, Northumberland, and Alligator Ridge.

The Taylor deposit occurs in the eastern part of the state (Fig. 8.2) and is located in the east-west oriented Hamilton-Ely belt of Roberts (1966). This belt is defined by an alignment of Tertiary intrusions, a positive aeromagnetic anomaly, and ore deposits (Stewart et al., 1977). The ore deposits in this belt include several different types; however, the Taylor mine is the only known major sediment-hosted, disseminated precious-metal deposit. There are several prospects near Taylor that may turn out to be deposits of this type.

The regional geology of the Taylor deposit is characterized by Paleozoic sedimentary rocks intruded by Tertiary rhyolitic rocks. The sedimentary rocks are predominantly limestone, dolomite, and shale. The Devonian Guilmette Formation serves as the host rock for the silver deposits at Taylor. This is a widespread formation and also hosts silver ore at Hamilton, 80 km west of Taylor. The Guilmette Formation is transitional upward with the Devonian to Mississippian Pilot Shale. Although the Pilot Shale is unmineralized in the Taylor area, it serves as the major host for gold ore at Alligator Ridge (see below). The overlying Mississippian Joana Limestone hosts minor silver ore in the Taylor district and hosts major sulfide replacement ore at Ward, 20 km west of Taylor. The Joana Limestone is overlain by the Mississippian Chainman Shale and the Mississippian to Permian Ely Limestone. These units serve as host to copper ore at the Ruth porphyry copper deposit and the Ely Limestone also hosts lead, silver, and zinc ore at Ward.

Northumberland (Fig. 8.2) is located along the eastern margin of the Northumberland caldera. The deposit is located 40 km northeast of the volcanic-hosted Round Mountain gold deposit (Tingley and Berger, 1985) and lies 60 km northeast of the Manhattan silver-gold district (Roberts, 1966; Shawe and Stewart, 1976). Although other sediment-hosted deposits occur south of Northumberland (e.g., the Shale Pit near Round Mountain and the White Caps mine in the Manhattan district), the location of the Northumberland deposit seems to be closely related to caldera development and as such, is a truly isolated deposit of this type.

The regional geology of the Northumberland area is characterized by lower Paleozoic rocks intruded by Mesozoic granodiorite followed by formation of the middle Tertiary Northumberland caldera (McKee, 1974). The Paleozoic rocks consist of both autochthonous and allochthonous rocks of the Antler Orogeny (Kleinhampl and Ziony, 1984). The allochthonous rocks consist of chert, shale, argillite, and some interbedded greenstones, including pillow basalts. These are considered the Ordovician Vinini Formation by McKee (1974). Autochthonous rocks consist of limestone, shale, and argillite mapped as the Ordovician Pogonip Group and the Silurian Masket Shale. Gold and silver ores occur in the Mesozoic intrusions and bedded barite occurs in the allochthonous rocks. Silver-rich quartz veins occur in Cretaceous tonalite bordering the Northumberland deposit. The silver mineralization is apparently older than the sediment-hosted gold.

Alligator Ridge (Fig. 8.2) is located northwest of Ely, Nevada, and appears unrelated to any known mineral belts. It occurs about 20 km south of the Bald Mountain mining district. This district contains limestone replacement ore bodies, quartz vein and contact-metamorphic deposits, and placer deposits (Hose et al., 1976).

The regional stratigraphy in the Alligator Ridge area includes a Devonian through Mississippian section overlain by Tertiary volcanic rocks. The Devonian part of the section consists of the Devils Gate Limestone and the lower part of the Devonian to Mississippian Pilot Shale. These formations are overlain by the Mississippian Joana Limestone, Chainman Shale, and Diamond Peak Formation. This is the same stratigraphic section that occurs in the Taylor district (see above). Alligator Ridge is one of the few deposits of this type that actually has some geologic characteristics that are suggestive of shallow emplacement of the ore (e.g., possible hydrothermal breccias).

GEOLOGIC CHARACTERISTICS OF THREE END-MEMBER, SEDIMENT-HOSTED, DISSEMINATED PRECIOUS-METAL DEPOSITS

The Carlin, Preble, and Taylor deposits are representative of the end-member categories that we recognize in this deposit type. Carlin serves to illustrate the category that is characterized by fine-grained, disseminated-gold ore that is difficult to visually separate from unaltered rock. Preble illustrates the category that is characterized by gold ore associated with intense silicification (jasperoids) and quartz veining. Both of these deposits are gold-rich end-members. Taylor illustrates the silver-rich end-member and is also characterized by intense silicification in the ore zone. There is obviously a gradation between the chemical and geological characteristics of these end-members. We feel, however, that they serve as easily recognized examples of the diversity in this deposit type. The geologic characteristics of these three deposits are reviewed here to provide a basis for comparison of the deposits listed in Tables 8.1 and 8.2 and a summary diagram of the characteristics that define the end-members is given in Table 8.3.

Carlin

The Carlin deposit (Figs. 8.2, 8.4, and 8.6) is probably the most extensively studied deposit of this type. The recognition of this deposit as possibly a new type of gold deposit (Hausen, 1967) influenced exploration models and resulted in government, academic, and industry research into the genesis of the deposit. The following deposit characteristics are summarized from Hausen (1967), Hausen and Kerr (1968), Radtke et al. (1980), Hausen (1981), Hausen et al. (1983), Radtke (1981), and Adkins and Rota (1984).

Host lithology--Most ore at Carlin occurs in the upper part of the Devonian and Silurian Roberts Mountains Formation. The most favorable host rocks in this section are laminated, argillaceous, arenaceous dolomites and calcareous mudstones. Arenaceous peloid wackestone also occurs in this section, but is apparently less favorable as a host for ore.

Igneous rocks--Dikes along faults are exposed in the Carlin pit. Although these dikes are hydrothermally altered, their presumed original composition was felsic, including possibly granodiorite and quartz diorite. Small stocks are exposed north of Carlin that vary in composition from granodiorite to diorite. The stock at the Gold Strike mine has a K-Ar date from biotite of 121 m.y. (Hausen, 1967). Biotite from a dike in the southwestern part of the Carlin Pit yielded a date of 131 m.y. (Morton et al., 1977). This latter date should be used with caution as several unpublished dates from altered rocks in the Carlin mine yield K-Ar ages that vary from late Tertiary to Mesozoic (M. L. Silberman, U.S.G.S., personal communication, 1985).

Structure--The Roberts Mountains thrust fault is exposed in the Carlin pit (Fig. 8.6). This fault brings western siliceous assemblage rocks over eastern carbonate facies rocks, both of early Paleozoic age. The thrust fault is in turn disrupted by sets of normal faults. These faults strike N. 40-45 E., N. 25-30 E., and N. 25 W., in order from oldest to youngest. Both upper and lower plate rocks are folded with fold axes striking N. 45 W.

Alteration and oxidation--There has been both hypogene and supergene alteration of the rocks at the Carlin deposit. Supergene alteration (oxidation and acid-leaching) was so intense in some areas that it is difficult to interpret earlier hypogene alteration. Alteration associated with primary mineralization can only be understood by examining unoxidized, ore-

Table 8.3--Comparison table summarizing geologic characteristics of jasperoidal and pod-like, sediment-hosted, disseminated precious-metal deposits

Jasperoidal, quartz-veinlet type	Disseminated, pod-like type
Quartz veins common	Quartz veins are uncommon
Main ore type is in silicified rock	Main ore type is not silicified
Ore primarily restricted to fault zones	Pod-like ore bodies extend away from faults
Several different silicification stages	Jasperoid may be present
Both gold- and silver-rich varieties	Gold-rich variety most common
Siliceous rocks common	Calcareous rocks common

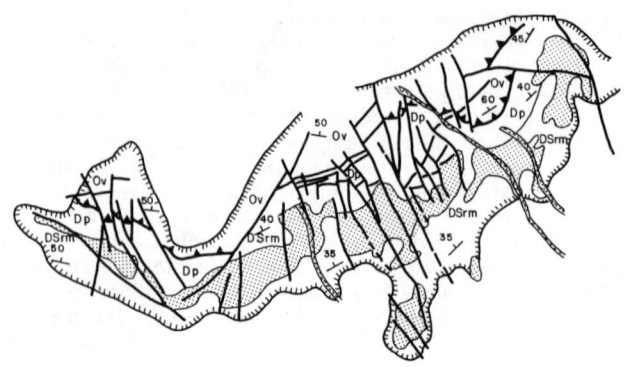

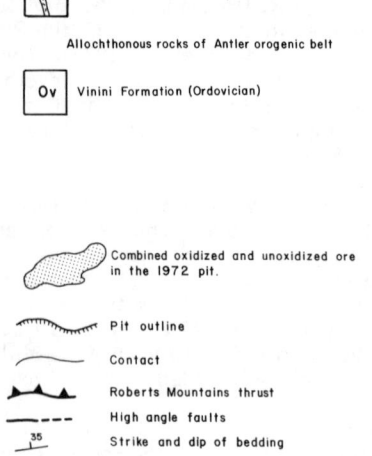

Figure 8.6. Plan map of the early (1972) pit at the Carlin deposit. The stippled pattern represents combined oxidized and unoxidized ore zones within the deposit. Geology is modified from Adkins and Rota (1984).

bearing, and barren rocks. Hypogene alteration consists of several types: (1) decalcification; (2) argillization; (3) silicification; and (4) calcification. Decalcification was early and resulted in the removal of carbonate minerals from the silty dolomites and limestones. Calcite was preferentially leached over dolomite resulting in increased porosity and permeability. Argillization accompanied decalcification resulting in formation of illite from detrital feldspar with minor amounts of montmorillonite and kaolinite. Silicification was a bulk replacement process and resulted in jasperoid formation. Late calcification resulted in the formation of calcite veins.

Hypogene alteration zoning of the overall deposit is clearly defined by calcite and silica distribution. Calcite is sparse in the main ore zone but is abundant above and away from the ore horizon. Silica occurs beneath the ore horizon and is localized along faults.

Supergene alteration includes oxidation of sulfide minerals and acid-leaching of the rocks. This resulted in the formation of iron, arsenic, and antimony oxides and sulfates and in the redistribution of calcite. Supergene oxidation is deepest along major faults and formation contacts.

Ore types--Different ore types at Carlin have been described by Hausen (1967) and Radtke (1981); the following review is based on their information. The ore zones at Carlin are localized near faults (Fig. 8.6). The ore is commonly difficult, if not impossible, to visually separate from waste. Ore control is strictly dependent upon gold assay, and several types of ore are recognized based on metallurgical behavior. Radtke (1981) recognized five primary unoxidized ore types: (1) normal; (2) siliceous; (3) carbonaceous; (4) pyritic; and (5) arsenical.

Normal ore is composed of dolomite, illite, and quartz. It contains minor amounts of kaolinite, sericite, and some remnant calcite. Pyrite is the most abundant sulfide and occurs as remnant diagenetic subhedral cubes and two types of introduced pyrite: dispersed cubes scattered in the rock matrix and clusters of small ($<10\mu$) framboidal pyrite. The origin of the framboidal pyrite is controversial and is interpreted by Hausen (1981) as having formed from bacteria. Gold, in normal ore, coats sulfides and a small amount occurs as metallic gold associated with quartz.

Siliceous ore is less than 10% of the deposit and

is characterized by silica replacement of carbonate. This ore contains 80-95% quartz, 5-10% clay (illite, sericite, and kaolinite), 1-5% dolomite, and small amounts of organic carbon. Gold in this type of ore occurs as metallic gold particles (<10μ) encapsulated in hydrothermal quartz.

Carbonaceous ores are those that contain enough organic carbon to inhibit the complexing of gold with cyanide. Organic carbon in this ore type varies from 1-6% and, except for this high carbon content, carbonaceous ore resembles normal ore. Gold is apparently associated with sulfides, and Radtke (1981) did not observe any metallic gold in this ore type.

Pyritic ore contains 3-10% pyrite compared to the 0.5-3% pyrite content of other unoxidized ore types. Pyritic ore commonly occurs as bands cutting carbonaceous ore. The pyrite occurs as subhedral to euhedral grains up to 0.4 mm in diameter with lesser amounts of framboidal pyrite. The matrix is composed of detrital quartz grains, fine-grained dolomite, and hydrothermal quartz. Detrital dolomite is corroded and little remnant calcite is present. Most of the gold in this ore type occurs as coatings on framboidal and cubic pyrite.

Arsenical ore is characterized by 0.5-10% arsenic, mostly in the form of realgar with some orpiment and a few sulfosalts. The ore is composed of silt-sized quartz grains in a matrix of dolomite grains, clay, and carbonaceous material. Fossil fragments are silicified. Arsenic minerals occur as crosscutting veins and fillings of small (50-300μ) vugs. Gold occurs as films on both framboidal and cubic pyrite.

In addition to pyrite, other minor sulfides and sulfosalts occur at Carlin including realgar, orpiment, stibnite, cinnabar, tetrahedrite, tennantite, sphalerite, galena, molybdenite, chalcopyrite, chalcocite, covellite, lorandite ($TlAsS_2$), christite ($TlHgAsS_3$), weissbergite ($TlSbS_2$), ellisite ($Tl_3 AsS_3$), and carlinite (Tl_2S). Native arsenic has also been reported at Carlin.

Oxidized ore is composed of varying amounts of quartz, clay (illite, with lesser amounts of kaolinite, sericite, and montmorillonite), and dolomite. Oxidized ore contains only about 0.03-0.35% organic carbon and is thus tan or bleached. Limonite staining is common. The oxidized ore is most likely formed by supergene oxidation of the primary sulfide-bearing ores. Radtke (1981) and Hausen (1967), however, both suggest that late-stage oxidation and acid leaching resulting from boiling of a geothermal system at and near the surface was responsible for oxidation and acid leaching at Carlin. Although boiling of hydrothermal waters can occur at depths that are typical of the porphyry environment (Cunningham, 1978), Radtke (1981), and Hausen (1967) appeal to a model that requires the deposit to have formed in either the near-surface or the surface (hot-spring) environment. An alternative hypothesis derived from current research by C. A. Kuehn at Carlin (refer to Bodnar et al., this volume) is that the formation of gold ore at Carlin occurred in a reducing environment at depth (at least one kilometer) and that there were no hypogene oxidizing fluids (with respect to pyrite). In light of this interpretation, all oxidized ore at Carlin formed from supergene alteration of the primary ores at low temperature.

Trace elements--Harris and Radtke (1976) investigated the geochemistry of unoxidized ore at Carlin. Their intent was to test the correlation of gold with mercury, arsenic, and antimony that had already been established at other deposits (e.g., Getchell (Joralemon, 1951); Gold Acres (Wrucke and Armbrustmacher, 1975)). In addition, they statistically analyzed the distribution of gold, barium, copper, molybdenum, lead, zinc, boron, tellurium, selenium, and tungsten in the west, main, and east ore zones. They found high correlations between gold, arsenic, mercury, and antimony. Gold is negatively correlated with barium, and barium has relatively high correlations with the base metals. Gold and tellurium have a significant positive correlation that is strongest in the east ore zone.

Organic geochemistry--Much of the ore in the main and east ore zones is carbonaceous. Hausen and Kerr (1968) noted that concentrates that are rich in carbon commonly are enriched in gold. However, Hausen (1967) notes that the presence of carbonaceous material is no _a priori_ evidence of ore. A similar gold-carbon correlation was noted by Radtke and Scheiner (1970). The difficulty has been in defining the type of gold-carbon association and whether or not the carbonaceous material served as a transporting or precipitating agent for gold. Radtke and Scheiner (1970) theorized several organic compounds that could serve to chelate with gold. Unfortunately, until the chemistry of the hydrothermal solutions responsible for gold transport and deposition are clearly understood, the theoretical gold-carbon chelates can not be suitably tested.

Vein types--The deposit is known for its lack of veins. However, different types of veins do in fact occur, but their spatial and genetic association with gold is not completely understood. Vein types include barite, calcite, quartz, and quartz-pyrobitumen veins. Hausen (1967, p. 73) shows photomicrographs of gold associated with quartz veins. This indicates that at least some of the quartz veining occurred during gold deposition. Unfortunately, due to the fine-grained nature of the gold, it has been difficult for subsequent workers to identify the gold-quartz vein association. For example, quartz-pyrobitumen veins may have formed prior to gold deposition even though many of them now occur in ore. Barite veins are interpreted by Hausen (1967) to be earlier than the gold stage. The statistical analysis of barium, base metals, and gold by Harris and Radtke (1976) indicate a negative correlation between gold and barium that possibly supports the geologic interpretation of the two having been deposited at different times. Calcite veins, possibly formed during gold deposition, occur in zones peripheral to silicified zones and later calcite crosscuts acid-leached, oxidized rocks.

Paragenesis--Development of a detailed paragenetic sequence in this ore deposit is extremely difficult due to the general lack of crosscutting veins that can unequivocally be related to gold or other aspects of hydrothermal activity. Two paragenetic sequences have, however, been published. Radtke (1981) interpreted four stages: (1) early and main that included the introduction of quartz, pyrite, potassium-bearing clays, hydrocarbons, minor barite, and early jasperoid; (2) late main with introduction of As, Sb, Hg, Tl, S, Au, and later base metal minerals; (3) an acid-leaching and oxidation stage with the introduction of major barite, jasperoid, anhydrite, and calcite veins; and, (4) late oxidation caused by weathering. On the other hand, Hausen (1967) placed barite and the base metals very early (Cretaceous) in the paragenesis and gold, realgar, stibnite, native arsenic, jordanite, tennantite, cinnabar, and jarosite midway (Tertiary) in the paragenesis with later, probably Pliocene, supergene oxidation. The major differences between these two paragenetic interpretations are the placement of the base-metal and barite veins and the length of hydrothermal activity. We feel that Cretaceous mineralization is predominantly base metal and not associated with the gold ore at Carlin. In addition, although the oxidized ore at Carlin probably did form by supergene oxidation, the timing of this is probably late Miocene to early Pliocene, a known period of deep weathering throughout Nevada.

Fluid inclusions--Radtke et al. (1980) reported fluid-inclusion analyses by John Slack (U.S. Geological Survey, referenced in Radtke, 1981) on quartz, barite, realgar, and sphalerite from the Carlin deposit. The inclusion data were interpreted in terms of Radtke's paragenesis and indicated main stage (4 samples) temperatures of 200°C increasing to about 300°C during the acid-leaching stage (13 samples). Carl A. Kuehn (U.S. Geological Survey, unpublished data, 1985) is currently studying fluid inclusions at Carlin and has noticed some important compositional characteristics of inclusions in quartz (Kuehn and Bodnar, 1984).

Stable isotopes--Radtke et al. (1980) published stable-isotope analyses for the deposit. They analyzed oxygen, hydrogen, carbon, and sulfur isotopes in veins, rocks, minerals, and fluid inclusions. Their results indicate that hydrothermal solutions were highly exchanged meteoric waters with δD between -140 and -160 per mil. The $\delta^{34}S$ values of 4.2-16.1 per mil were interpreted as indicating diagenetic pyrite as the source of sulfur in the deposit.

Age of mineralization--The age of mineralization at Carlin has not yet been directly dated. An igneous dike that is altered and mineralized within the deposit was dated at 131 ± 4 m.y. (biotite) by Morton et al. (1977). The Gold Strike pluton north of Carlin is altered, contains gold, and has been dated at both 78.4 ± 3.9 m.y. (biotite, Morton et al., 1977) and 121 ± 5 m.y. (biotite, Hausen, 1967). Rhyolitic lavas and domes west of the Carlin pit overlie jasperoid and are not altered or mineralized. These lavas are dated at 14.2 ± 0.3 m.y. (sanidine, McKee et al., 1971). Thus, the age of the deposit could be as old as 130 m.y. but no younger than 14 m.y.

Taylor

The Taylor deposit in eastern Nevada (Figs. 8.2 and 8.7) is the silver-rich end-member in our classification of sediment-hosted disseminated precious-metal deposits. In addition, Taylor has characteristics that are more similar to the jasperoid-quartz vein type (Preble) than the disseminated type (Carlin). The Taylor deposit has not been extensively

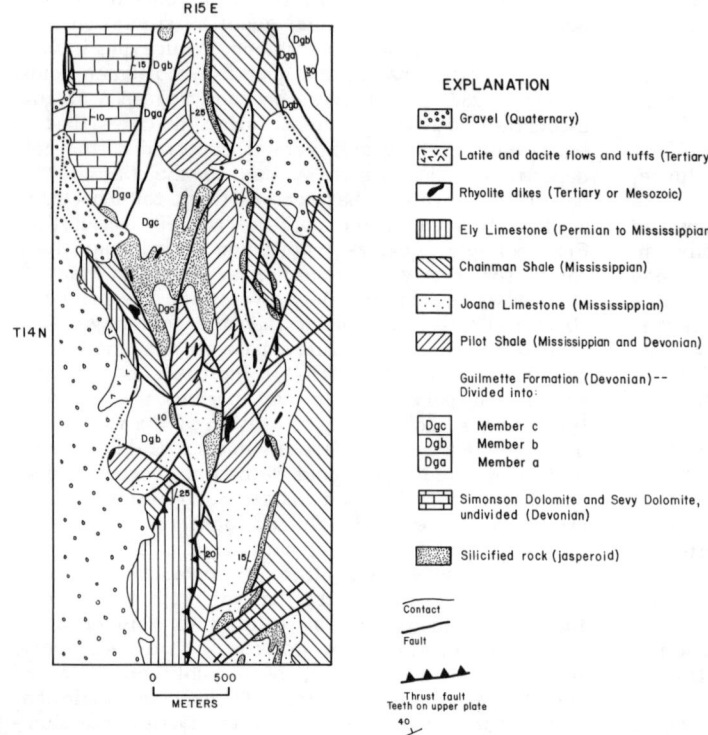

Figure 8.7. Geology of the Taylor mining district. Jasperoids are spatially associated with faults. The large jasperoid near the center of the Figure is the present site of the Taylor deposit. Geology simplified from Drewes (1962, 1967).

studied and, thus, published information is scarce. The following summary about the Taylor deposit is derived from Havenstrite (1983), Hose et al. (1976), Lovering and Heyl (1974), Drewes (1962, 1967), Jeffrey M. Edwards, (Newmont Exploration Limited, unpublished data, 1985), and personal observations (William C. Bagby, U.S. Geological Survey, unpublished field notes, 1985).

Host lithology--The Middle and Upper Devonian Guilmette Formation is the host of silver ore at Taylor. This formation is composed of four members as mapped by Drewes (1967). Member a is a thick-bedded limestone that is locally replaced by a massive coarse-grained dolomite. Member b is shaly limestone and coarse-grained dolomite that intertongues with, and grades laterally into, limestone. Member c includes limestone and shaly limestone with a sandstone marker bed near its base. This member is the major host of ore. These three members together comprise the lower part of the formation. Member d is reef limestone, shaly limestone, sandstone, and conglomerate that is only locally preserved.

Igneous rocks--Rhyolite dikes intrude Paleozoic and Mesozoic rocks in the Taylor district. These dikes strike north and northeast and are hydrothermally altered to clay. Notably, several of the dikes contain xenoliths of jasperoid, the predominant ore type at Taylor (Havenstrite, 1983). The dikes have been mapped as Mesozoic or Tertiary (Drewes, 1962). Havenstrite (1983) reports a date of 35 m.y. for the dikes, but does not indicate how the date was determined.

Structure--Drewes (1962, 1967) mapped normal faults trending north, northwest, and northeast in the Taylor district. These faults cut the rhyolite dikes and are thus considered as late Tertiary to Holocene faults associated with Basin and Range tectonism. Havenstrite (1983) interprets folding in the Taylor district as a large overturned antiform. The axis of the antiform is remarkably coincident with one of Drewes's (1962) normal faults that Lovering and Heyl (1974) refer to as the Taylor fault. Havenstrite (1983) suggests that the ore bodies at Taylor formed in crackle breccias that developed in the hinge and along the flanks of the antiform. Lovering and Heyl (1974) and Drewes (1967) suggest instead that the ore bodies were formed in and near fault breccia zones and that the normal faults controlled hydrothermal fluid flow in the district.

Alteration and oxidation--Silicification and argillization are the only two alteration types that are reasonably well documented in the Taylor district. Thin-bedded and shaly limestones, particulary beneath thick shale horizons, are commonly silicified near faults (Drewes, 1962, 1967). The silicification is a replacement process resulting in the formation of jasperoid. Ore bodies in the Taylor district are limited to these jasperoid zones. Argillization in the district includes the alteration of feldspar in the rhyolite dikes to kaolinite (Drewes, 1962, 1967). The dikes are notably unmineralized. Although sulfides still occur at the surface (Lovering and Heyl, 1974), the Taylor deposit is well oxidized (Havenstrite, 1983). This oxidation remobilized silver, thus resulting in supergene enrichment of the ore. Acids formed during supergene alteration of sulfides imparted a leached, bleached, and spongy appearance to the surface jasperoids.

Ore types--Silver-bearing jasperoid is the predominant ore type at Taylor. Jasperoids formed from limestone along and away from faults as though the faults served as conduits for hydrothermal solutions (Lovering and Heyl, 1974). The ore bodies occur in the upper part (member c) of the Guilmette Formation and are tabular bodies generally about 50 feet thick (Havenstrite, 1983). Sulfides associated with the silicification are stibnite, sphalerite, tetrahedrite, chalcopyrite, pyrite, galena, and pyrargyrite (Lovering and Heyl, 1974). Gangue minerals include calcite, dolomite, sericite, tourmaline, monazite, barite, fluorite, and apatite (Lovering and Heyl, 1974). Limonite is a common consituent of ore. Matrix quartz in highly mineralized jasperoid contains minute inclusions of organic carbon, stibnite, limonite, cerargyrite, and antimony oxides (Lovering and Heyl, 1974). Silver occurs as pyrargyrite (?), cerargyrite, argentite, and metallic silver finely disseminated throughout the jasperoid.

Trace elements--Lovering and Heyl (1974) investigated the geochemistry of jasperoids in the Taylor district and compared them to unmineralized jasperoids from elsewhere in Nevada. They noted that Ag, Au, Cu, Hg, Pb, Sb, Te, and Zn are all anomalous in Taylor jasperoids compared to unmineralized jasperoid (Lovering, 1972).

Organic geochemistry--The only mention of organic carbon in any of the published literature on Taylor is by Lovering and Heyl (1974), who note the presence of minute inclusions of carbon in jasperoid quartz. Havenstrite (1983) appeals to the organic-rich Chainman Shale as a source rock for both silver and silica introduced into the Guilmette Formation.

Vein types--Veins of calcite, quartz, and barite occur in the Taylor district. Barite veins are zoned peripheral to the Taylor deposit (J. M. Edwards, Newmont Exploration Limited, oral communication, 1985) whereas both calcite and quartz veins cross cut jasperoid at the deposit. It is equivocal whether or not any of these veins are part of the same mineralizing event that deposited silver.

Paragenesis--Drewes (1967), Lovering and Heyl (1974), and Havenstrite (1983) all agree that silver and silica were deposited simultaneously from hydrothermal solutions flowing through fault-breccia zones. Havenstrite (1983) indicates that rhyolite dikes and sills must be younger than the ore depositing episode since xenoliths of jasperoid occur in the dikes. Havenstrite (1983) cites this as evidence that magmatic fluids were not associated with the hydrothermal system as proposed by Drewes (1962, 1967) and Lovering and Heyl (1974). However, Lovering and Heyl (1974) make a strong argument that the dikes and the hydrothermal fluids used the same structures as pathways to the upper crust. It is entirely possible that the dikes are simply late intrusive stages of a much larger magmatic-hydrothermal system that had been depositing silver for considerable time prior to dike emplacement.

Fluid inclusions--Jeffrey M. Edwards (Newmont Exploration Limited, oral communication, 1985) has performed preliminary heating studies on quartz, fluorite, and late-stage sphalerite from the Taylor district. He interprets boiling fluids during deposition of the quartz and fluorite at average temperatures of 200°C and 220°C, respectively. Inclusions in late-stage sphalerite (filling vugs and coating jasperoid fragments in breccias) that contain three phases at room temperature and homogenize at about 300°C are interpreted as CO_2-H_2O inclusions.

Age of mineralization--The ore deposits in the Taylor district have not been directly dated. If the rhyolite dikes were emplaced about 35 m.y. ago, are unmineralized, and contain fragments of jasperoid, then they place an upper limit on the age of mineralization. A lower age limit is provided by the age of the host rocks which is middle and late Devonian. The north and northwest-trending ore-bearing faults in the district cut Tertiary volcanic rocks north of the district (Lovering and Heyl, 1974), indicating a Tertiary age for the faulting. This places a tighter constraint on the mineralization age by limiting it to the Tertiary.

Preble

The Preble deposit is the southernmost and smallest known deposit of this type in the Getchell trend. It serves to illustrate the end-member category of gold-rich deposits that contain intense silicification and relatively common quartz veining in the ore zone (Figs. 8.2 and 8.8). Mineralized rock can almost always be visually differentiated from unmineralized rock due to alteration and quartz veining. The following reveiw is summarized from Kretschmer (1984) and William C. Bagby and Raul J. Madrid (U.S. Geological Survey, unpublished data, 1985).

Host lithology--The Cambrian Preble Formation is the host rock for the deposit. The formation consists of phyllitic shale with interbedded, finely laminated and massive limestones. The phyllitic shales host most of the ore. These rocks, where fresh and unaltered, are composed of variable amounts of muscovite, quartz, chlorite, carbonate, and smectite. Metamorphic, folded quartz veins occur throughout the phyllitic shale sequence. Quartz overgrowths on these veins contain gold and are evidence of later silica addition associated with gold deposition.

Igneous rocks--An altered felsic dike, striking northerly and dipping to the east cuts through the deposit. The dike occupies a fault zone that had recurrent movement post-emplacement of the dike. The genetic relationship of the dike to the gold is unknown.

Structure--The sequence is folded and faulted and crenulation cleavage is well developed in the phyllitic shale. At least two periods of deformation have affected the rocks. Early, either pre- or syn-metamorphic veins are folded. These veins consist of calcite where they cut limestone and quartz+calcite where they cut phyllitic shale. North- and east-trending high-angle faults offset the sequence. The fault that contains the felsic dike is one of the main controls of the geometry of the ore body. The north-striking faults have dropped the Preble Formation down to the east.

Alteration and oxidation--Alteration of the phyllitic shale involves an addition of silica and a change in color. This color change is characterized by bleaching of the original phyllite and the greenish sheen becomes buff. Dolomitization of the limestones in the sequence is common near high-angle faults. Jasperoid veins cut the limestone and silica replacement of limestone forms massive jasperoid. Carbonaceous matter has been locally remobilized in fault zones.

Ore types--Ore at the Preble deposit is associated with some form of silicification, either as replacement of limestone or phyllitic shale (jasperoid) or as quartz veining. The quartz veining is best observed in thin section and under the scanning electron microscope as quartz overgrowths on metamorphic quartz veins and as patchy replacement of host rock. The main ore zone at Preble is localized

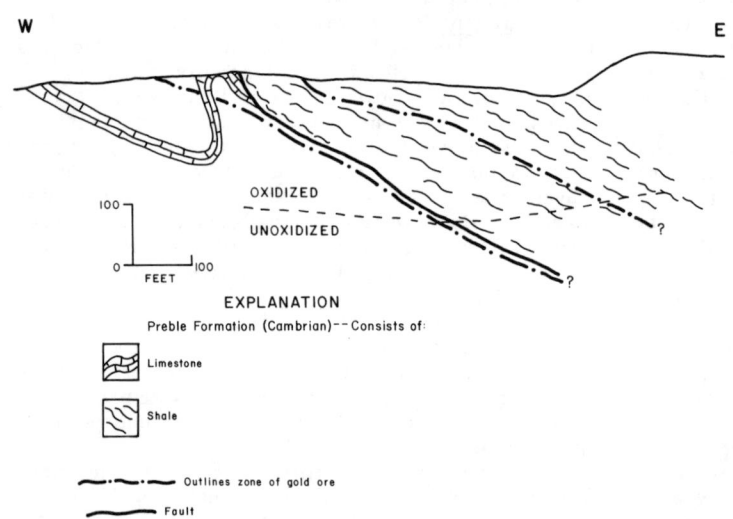

Figure 8.8. Schematic cross section of the main ore zone of the Preble deposit. The ore horizon is restricted to a shear zone associated with an east-dipping normal fault. The diagram is modified from Kretschmer (1984).

within a north-striking fault zone. Mineralization is fairly uniform within the zone and is predominantly limited to the fault and hanging-wall host rocks. Both oxidized and unoxidized ore occur. Gold occurs with sulfides and iron oxides after pyrite. The average size of the gold grains is about 2 microns (Kretschmer, 1984). Pyrite, both cubic and framboidal varieties, is the most abundant sulfide with arsenopyrite, marcasite, chalcopyrite, and sphalerite as minor accessory sulfides (Kretschmer, 1984).

Trace-element geochemistry--Crone (1982) and Crone et al. (1984) investigated the geochemistry of iron oxide-rich fracture coatings over the deposit. Their results indicate a strong geochemical correlation of gold with mercury, arsenic, and antimony. Kretschmer (1984) reports a correlation of thallium, barium, and fluorine with gold.

Organic geochemistry--Unoxidized ore occurs in black, carbonaceous, phyllitic shale. The carbonaceous matter has been heated to high temperatures and is almost graphite. There is no positive correlation between abundance of organic carbon and gold (W. C. Bagby, U.S.G.S., unpublished data, 1985).

Vein types--The Preble deposit is a stockwork veinlet, jasperoidal deposit. There are several generations of veining at Preble, some of which are definitely associated with the period of gold deposition. Vein types include metamorphic quartz, quartz-carbonate, and calcite veins, and later quartz, quartz-calcite, jasperoid, dolomite, and calcite veins. These veins are all recognizable in the field with the possible exception of differentiating between metamorphic and gold-related quartz veins. The jasperoid veins cutting limestones are presumably the veins sampled by Crone (1982) to identify the gold, arsenic, mercury, antimony association.

Fluid inclusions--Fluid-inclusion studies are now in progress on Preble samples (W. C. Bagby, unpublished data, 1985). Metamorphic quartz-carbonate veins contain numerous trails of secondary inclusions. These include both two phase (liquid and vapor) and three phase (liquid CO_2, liquid H_2O, and vapor) inclusions at room temperature that are generally less than 10 microns across.

Isotopes--No stable-isotope data are available for the deposit.

Age of mineralization--M. L. Silberman (U.S. Geological Survey) collected altered felsic dike material from the mineralized zone. Sericite from the dike yielded an $^{40}Ar/^{39}Ar$ age of 100.42 ±1.6 m.y. (L. W. Snee, U.S. Geological Survey, written communication, 1980). The interpretation is that Preble is a similar age to Getchell (Berger, 1980).

GENERAL ASPECTS OF TRACE ELEMENT AND STABLE-ISOTOPE GEOCHEMISTRY

In an attempt to compare trace-element concentrations between deposits, we chose 0.30 ppm Au as a cut-off for selecting samples from the published literature. In addition, we chose to compare only Ba, Hg, Sb, As, Ag, and Au since these elements are generally reported in the literature for deposits of this type. In this way, we hoped to be able to compare the trace-element content of rocks ranging from low-grade ore (0.30 ppm) to higher grades. The published literature commonly has only statistical summaries of data without tabulations of all data. Where this was the case (Carlin, Getchell, and Gold Acres) we used the ranges reported, even though in all three cases, the ranges included gold values less than 0.30 ppm gold. Trace-element geochemistry is given in Table 8.4 and summarized in Figure 8.9.

The highest gold values reported are those from Carlin and Cortez. The Carlin samples represent both oxidized and unoxidized ore from within the early, shallow pit (Radtke et al., 1972) whereas the Cortez samples are all surface samples collected prior to mining (Erickson et al., 1966). Although Radtke et al. (1972) show that oxidized ore has a range of gold

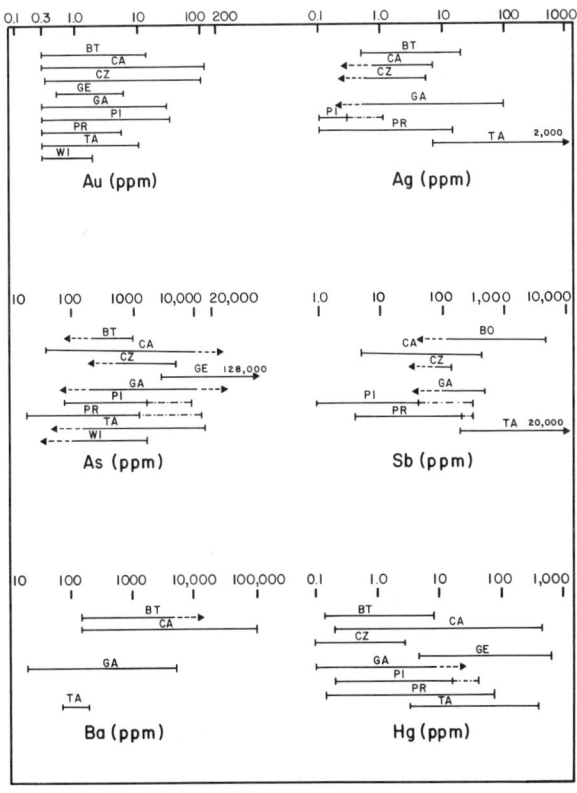

Figure 8.9. Summary of trace element variations in sediment-hosted, disseminated precious-metal deposits. The trace element data are restricted to samples containing 0.30 ppm Au unless otherwise noted in the text or in Table 8.4. The dash-dot extensions to the ranges for Pinson and Preble indicate higher anomalies in iron-oxide fracture coatings than those in rocks (Crone, 1982; Crone et al., 1984). The dashed line extensions with arrows indicate that some values were either greater than or less than the detection limits of the instrument.

Table 8.4--Summary of trace-element geochemistry for selected sediment-hosted, disseminated precious-metal deposits

Deposit (Reference)	Rock type	Au (#)	Ag (#)	As (#)	Sb (#)	Hg (#)	Ba (#)
Bootstrap (Evans, 1974)	Altered dike	0.30-2.0 (2)	0.5-10.0 (2)	200-500 <(2)	100-500 <(2)	1.1-3.0 (2)	200-300 (2)
	Siltstone (carbonate assemblage)	0.4 (1)	2.0 (1)	<200 (1)	200 (1)	0.26 (1)	150 (1)
	Gouge, siliceous breccia (siliceous assemblage)	1-14 (3)	0.7-3.0 (3)	<200-1,000 (3)	100-1,000 (3)	0.14-8.0 (3)	700->5,000 (3)
	Siliceous breccia (carbonate assemblage)	0.3 (2)	0.5-20 (2)	<200 (2)	<100-5,000 (2)	0.8-1.5 (2)	300-700 (2)
Carlin (Radtke and others, 1972)	Limestone Ore (unoxidized)	0.02-88 (120)	<0.7-7 (75)	50->10,000 (60)	5-450 (60)	0.4-453 (120)	150-70,000 (120)
	Oxidized Ore	0.04-125 (250)	<0.7-2 (85)	40->3,000 ((50)	5-450 (50)	0.2-130 (250)	200-100,000 (250)
Cortez (Erickson, et al., 1966)	Limestone (carbonate assemblage)	0.34-116 (11)	tr.-5.5 (11)	<500-5,000 (11)	<100-150 (11)	.02-2.7 (11)	NA
Getchell (Berger, 1975)	Carbonaceous ore	0.99-3.91 (4)	NA	2,800-48,000 (4)	NA	45.8-641.8 (4)	NA
	Siliceous ore	0.51-4.49 (2)	NA	57,000-128,000 (2)	NA	248.9-400.1 (2)	NA
	Argillic ore	3.09-6.00 (2)	NA	64,000-78,000 (2)	NA	89.3-93.9 (2)	NA
Gold Acres (Wrucke and Armbrustmacher, 1975)	Limestone and dolomite	<0.02-30 (137)	<0.5-100 (137)	<200->10,000 (137)	<100-500 (137)	<0.1->10 (137)	20-5,000 (137)

values from 0.04-125 ppm compared with 0.02-88 ppm in unoxidized ore, the medians are 11 and 10 ppm, respectively. With respect to differences in trace-element concentrations between unoxidized and oxidized ore at Carlin, Ag, As, and Ba medians are the same, whereas the Hg median is lower, and the Sb median is higher in the oxidized rocks (Radtke et al., 1972). Silver values are generally less than about 10 ppm in gold-rich members of the deposit type, but do reach as high as 100 ppm (Gold Acres). The silver-rich end member, Taylor, has high gold values of about 10 ppm and high silver values of about 2,000 ppm. In general, for these deposits, As varies from 100 to 1,000 ppm, Sb from 5 to about 200 ppm, Hg from 0.2 to about 30 ppm, and Ba from 30 to about 1,000 ppm. Although there are exceptions to these ranges, there is considerable overlap between the different deposits. The amount of overlap in the concentrations of these elements indicates that the chemical and physical characteristics of the hydrothermal systems were

Tl (#)	Cu (#)	Pb (#)	Zn (#)	Notes
NA	20-150 (2)	15-20 (2)	300-1,500 (2)	LLD: Au (0.05), Ag (0.5), As (200), Sb (100), Hg (0.05), Ba (20.0), Cu (5.0), Pb (10.0), Zn (200.0). Eleven samples analyzed; 3 had 0.05-0.15 ppm Au, two had less than detection limit, and 4 had no Au detected.
NA	50 (1)	10 (1)	<200 (1)	LLD: Au (0.05), Ag (0.5), As (200), Sb (100), Hg (0.05), Ba (20.0), Cu (5.0), Pb (10.0), Zn (200.0). Eleven samples; four had 0.05-0.1 ppm Au; six had no Au detected.
NA	50-100 (3)	10-5- (3)	300-1,000 (3)	LLD: Au (0.05), Ag (0.5), As (200), Sb (100), Hg (0.05), Ba (20.0), Cu (5.0) Pb (10.0), Zn (200.0). Three samples total.
NA	20-100 (2)	<10-20 (2)	<200 (2)	LLD: Au (0.05), Ag (0.5), As (200), Sb (100), Hg (0.05), Ba (20.0), Cu (5.0), Pb (10.0), Zn (200.0). Two samples analyzed. All "jasperoids" analyzed had Au at 0.1 ppm.
<50-50 (4)	7-200 (120)	<7-1,500 (115)	10-850 (50)	LLD: Au (0.02), Ag (0.7), As (10), Sb (0.5), Hg (0.01), Ba (10), Tl (50), Cu (1), Pb (7), Zn (5). Sample suite included 120 unoxidized and 250 oxidized samples. LLD: Au (0.02), Ag (0.7), As
<50-150 (22)	7-200 (250)	<7-200 (243)	10-620 (50)	(10), Sb (0.5), Hg (0.01), Ba (10), Tl (50), Cu (1), Pb (7), Zn (5). Most from upper beds of Roberts Mountains Formation. Samples from pit exposures and drill cuttings.
NA	NA	NA	NA	LLD: Au (?), Ag (?), As (500), Sb (100), Hg (?). Surface samples from Cortez district. Detection limits not recorded.
NA	NA	NA	NA	LLD: Au (?), As (?), Hg (?). Six analyses of carbonaceous ore reported. Two had 0.21 ppm. No detection limits reported.
NA	NA	NA	NA	LLD: Au (?), As (?), Hg (?). Three analyses of siliceous ore reported. One had 0.21 ppm Au. No detection limits reported.
NA	NA	NA	NA	LLD: Au (?), As (?), Hg (?). Two analyses of argillic ore reported. No detection limits reported.
NA	<5-5,000 (137)	<10-300 (137)	<200->10,000 (137)	LLD: Au (0.02), Ag (0.5), As (200), Sb (100), Hg (.01), Ba (20), Cu (5), Pb (10), Zn (200). Samples collected within Gold Acres pit.

similar, even between gold- and silver-rich systems. However, departure from the norm for several of the deposits indicates that the hydrothermal systems were each unique; possibly reflecting intensity of the system (temperature, hydrologic flow rate, lifetime) and also possibly the chemistry of the source and host rocks. A more comprehensive interpretation is currently not possible due to the incompleteness of the comparative data base.

Stable-isotope studies of this deposit type are even more incomplete than trace-element investigations. Rye et al. (1974) and Radtke et al. (1980) investigated stable-isotope systematics at Cortez and Carlin, respectively. Rye (1985) reviewed the conclusions of the preceding papers and added additional data from other deposits for comparison. Figure 8.10 shows the ranges in $\delta^{34}S$ for sulfides and sulfates from different deposits. For the most part, the barites have sulfur isotopic signatures similar to sedimentary barite. On the other hand, sulfides from these deposits

Table 8.4--Summary of trace-element geochemistry for selected sediment-hosted, disseminated precious-metal deposits--(continued)

Deposit (Reference)	Rock type	Au (#)	Ag (#)	As (#)	Sb (#)	Hg (#)	Ba (#)
Pinson (Powers, 1978)	Jasperoid ("A" ore zone)	0.34-34.29 (60)	NA	245-1,650 (60)	0.2-74.8 (60)	NA	NA
(Crone, 1982)	Rock (lithology unrecorded)	0.30-1.65 (6)	<0.1-0.3 (6)	190-1,000 (6)	6.0-41.0 (6)	2.10-16.00 (6)	NA
	Iron oxide fracture coatings	0.34-5.24 (8)	<0.05-1.51 (8)	69-9,000 (8)	2.0-317 (8)	0.38-42 (8)	NA
Preble	Rock (lithology unrecorded)	0.95-1.7 (4)	0.1-0.4 (4)	240-530 (4)	13-34 (4)	6.4-16 (4)	NA
	Iron oxide fracture coatings	0.31-4.73 (10)	<0.05-0.80 (10)	53-13,140 (10)	21.1-330 (10)	0.44-24 (10)	NA
(William C. Bagby, U.S. Geological Survey, unpublished data.)	Shale (drill chips)	0.31-5.7 (28)	0.08-15 (28)	21-1,200 (28)	4.3-210 (28)	0.15-75 (28)	NA
Taylor (Lovering and Heyl, 1974)	Jasperoid	0.2-10.4 (3)	7-2,000 (3)	<150-15,000 (3)	200-20,000 (3)	3.2-380 (3)	70-200 (3)
Windfall (Grove, 1979)	Hamburg Dolomite	0.3-1.9 (27)	NA	<100-5,300 (27)	<1,000 (27)	NA	NA
	Dunderburg Shale	0.4 (3)	NA	1,100-6,600 (3)	<1,000 (3)	NA	NA
	Andesite	0.4-0.7 (2)	NA	<100-3,300 (2)	<1,000 (2)	NA	NA

NA, not analyzed; ND, not detected; and #, number of samples analyzed; all analyses are in ppm; LLD, lower limit of detection.

have isotopic signatures that overlap with sedimentary and organic sulfur. Detailed interpretations of these variations are premature given the lack of detailed paragenetic sequences for the samples.

Oxygen, hydrogen, and carbon isotopes have been examined at Carlin and Cortez (Rye, in press). Large negative δD values indicate dominance of the fluids by meteoric water. Large $\delta^{18}O$ values indicate substantial exchange between fluids and rock in both deposits. Rye (1985) uses these data to suggest that although Cortez and Carlin may be of different age, the hydrothermal systems were very similar.

SUMMARY OF GEOLOGIC CHARACTERISTICS

Sediment-hosted, disseminated precious-metal deposits display considerable variation in their geologic characteristics (Tables 8.1 and 8.2). We have tried to examine this variation by defining two end-

Tl (#)	Cu (#)	Pb (#)	Zn (#)	Notes
NA	NA	NA	NA	LLD: Au (?), Ag (?), As (?), Sb (?). Sixty-one samples analyzed. One had 144 ppm Au (not included in tabulation). Detection limits not listed.
NA	NA	NA	NA	LLD: Au (0.05), Ag (0.1), As (1.0), Sb (5.0), Hg (0.01). Surface samples, lithology unrecorded. Thirty samples analyzed, 22 had no detectable Au; 2 had 0.05 ppm Au.
NA	30-520 (8)	NA	40-1040 (8)	LLD: Au (0.01), Ag (0.05), As (1.0), Sb (2.0), Hg (0.05), Cu (20), Zn (5). Thirty samples, same site as rocks; 22 samples had 0.01-0.30 ppm Au. Twenty-two soil samples; 4 had 0.05-0.10 ppm Au, all others less than 0.05 ppm Au.
NA	NA	NA	NA	LLD: Au (0.05), Ag (0.1), As (1.0), Sb (5.0), Hg (0.01). Thirty-two samples collected; 5 had 0.05-0.25 ppm Au, 23 had <0.05 ppm Au.
NA	20-1,340 (10)	NA	60-1,800 (10)	LLD: Au (0.01), Ag (0.05), As (1.0), Sb (2.0), Hg (0.05), Ba (0.00), Tl (0.00), Cu (20), Pb (0.00), Zn (5). Thirty-two samples; 22 had 0.01-0.26 ppm Au.
1.1-2.4 (28)	NA	NA	NA	LLD: Au (0.05), Ag (0.01), As (0.1), Sb (0.02), Hg (0.01), Tl (0.01). One hundred five samples; 54 had <0.05 ppm Au, 22 had 0.05-0.29 ppm Au, one had 31 ppm Au. All 20 ft composites.
NA	10-2,000 (3)	10-10,000 (3)	<150-700 (3)	LLD: Au (?), Ag (?), As (?), Sb (?), Hg (?), Ba (?), Cu (?), Pb (?), Zn (?). Three "representative samples" from Taylor district. No detection limits reported.
NA	NA	<10-6,400 (27)	<10-2,100 (27)	LLD: Au (0.1), As (100), Sb (1,000), Pb (10), Zn (10).
NA	NA	200-900 (3)	NA	LLD: Au (0.1), As (100), Sb (1,000), Pb (10), Zn (10). High Au value of 23 ppm not included. A total of 92 samples were analyzed for these elements. Thirty-four samples had 0.1-0.2 ppm Au, 13 had trace, and 18 had nil. Fourteen shale samples analyzed; 5 not analyzed for Au, 3 had 0.01-0.2 ppm Au, 2 had tr. Au, and 1 had nil.
NA	NA	100-200 (2)	NA	LLD: Au (0.1), As (100), Sb (1,000), Pb (10), Zn (10). Three andesite samples analyzed; one had 0.01 ppm Au, other two had >0.3 ppm Au.

member subsets that represent extreme cases of the geologic variation of this deposit type. But in addition, there are a large number of similarities that provide cohesiveness and definition to the deposit type. Together, the variations and similarities provide a basic geologic framework from which it is possible to compare known deposits and to develop exploration and assessment programs. One shortcoming of this approach is that it assumes a similar level of understanding for the deposits used in defining the deposit type. This is simply not the case. However, the current interest in this deposit type is resulting in increased industry, academic, and governmental research that will expand our knowledge of these deposits.

Regional and District Scale

The regional and district geologic characteristics seem to show the least variation. Whether the deposit

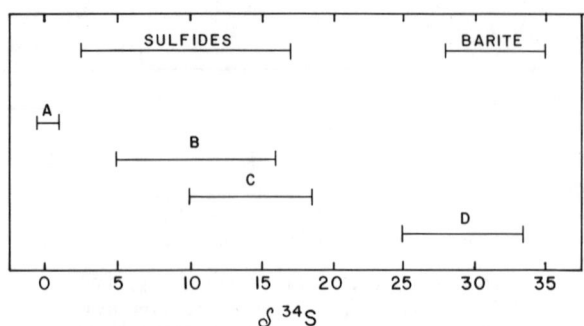

Figure 8.10. Summary of sulfur isotopic variation in sulfides and barite from several sediment-hosted, disseminated precious-metal deposits. The range given for sulfides includes data for pyrite, stibnite, realgar, and orpiment from Carlin, Cortez, Mercur, Jerritt Canyon, White Caps, and Getchell. Barite range includes data for samples from Carlin, Cortez, Getchell, Jerritt Canyon, Mercur, Pinson, Preble, and White Caps. Other ranges are: (A) igneous sulfur; (B) sulfide sulfur from the Roberts Mountains Formation; (C) organic sulfur; and (D) sedimentary sulfate. Data are from Rye (1985), Guenther (1973), and Ohmoto and Rye (1979).

is of the disseminated, pod-like or jasperoidal, quartz-veinlet subset and whether it is gold- or silver-rich is not dependent at our present level of knowledge upon a certain regional geologic environment. Sedimentary rocks of any age that include thinly bedded silty or shaly carbonate rocks or calcareous siltstones or shales provide the ideal host environment. Fluid pathways formed by regional faulting and folding serve to concentrate and direct hydrothermal gold-bearing solutions. A regional heat source is also needed to drive a mineral-depositing hydrothermal system. It may be that regional lithology chemically controls whether or not a deposit is gold- or silver-rich. The data base to support this is incomplete.

Deposit Scale

There are common threads that run through the different deposits.

Host lithology--Thinly bedded silty or shaly, carbonaceous limestones and dolomites, calcareous, carbonaceous shales and siltstones, and bioclastic limestones are the most common types of host lithologies.

Igneous rocks--Intermediate to silicic dikes, plugs, domes, and (or) stocks are present at most deposits; either in the deposit itself or within the district. These rocks range from Cretaceous to middle Tertiary. The genetic association between these rocks and the gold mineralization is unknown.

Structure--High-angle normal and (or) strike-slip faults are present in all deposits. These faults are generally pre-, syn- and post-ore. Thrust faults are present at some deposits, but do not seem to be the major controlling structure for ore deposition. Folds are present, both regionally and within deposits. In fact, many deposits occur near the crest of regional antiforms. The deposit-scale folding in most deposits has not been mapped in detail and, thus, the control of these structures on ore deposition is unknown.

Alteration and oxidation--Hypogene alteration almost invariably includes early decalcification followed by silicification. The disseminated, pod-like deposits contain silicified ore in and near faults, but much of the ore in this subtype is not strongly silicified enough to be termed a jasperoid. The jasperoidal, quartz-veinlet subtype deposits contain mostly silicified ore. Other alteration includes formation of clays and local remobilization of carbonaceous matter. All deposits are oxidized to some extent. Oxidized ore is usually bleached, contains iron oxides and sulfates, and may be acid leached.

Ore types, shapes, and mineralogy--Ore in the jasperoidal, quartz-veinlet subtype is generally restricted to a fault zone with only minor leakage into the surrounding wallrock. On the other hand, disseminated, pod-like deposits characteristically contain ore zones that are podiform and extend away from fault zones. The mineralogy of either subtype always includes pyrite. Other common, but variably present, minerals include cinnabar, stibnite, arsenopyrite, fluorite, barite, calcite, and various thallium and arsenic sulfides and sulfosalts. Oxidized and unoxidized ores are present in both subtypes. However, in some of the smaller jasperoidal, quartz-veinlet deposits, only the oxidized ore is mined.

Site of gold--The gold site has not been identified in all deposits. However, where it is known, the gold particles are always submicroscopic, on the order of 1 to 5 microns. Metallurgical tests indicate that gold is associatied with pyrite, clay, silica, and organic matter. The nature of these associations is not well understood.

Trace-element geochemistry--This deposit type is well known for the trace-element association of Au, As, Sb, Hg, and Tl. These elements are almost universally present in deposits of this type, although concentrations are variable. Other elements that are anomalous in some deposits include W, Te, Se, Cd, and F.

Organic geochemistry--Carbonaceous matter is present in some form in the unoxidized ore of all known deposits. This includes amorphous carbonaceous matter, pyrobitumen, graphite, and kerogen. Hydrocarbons were identified at Mercur (Tafuri, 1976), but are as yet unknown in other deposits. The genetic relationship between gold deposition and carbonaceous matter, if any, is unknown.

Vein types--Veins are most abundant and their relationship to gold seem the clearest in the jasperoidal, quartz-veinlet subset of deposits. However, veins are also present, and possibly very important, in the disseminated, pod-like deposits. Veining typically includes quartz and calcite with variable barite, fluorite, and dolomite veins.

Paragenesis--The general lack of clear crosscutting relations of different veins and their temporal association with gold inhibits the development of detailed paragenetic sequences in these deposits. However, the general paragenesis almost always involves early decalcification and carbon remobilization with later silicification and, in some cases, very late calcite veining. Supergene oxidation and acid-leaching are common, but of variable intensity, at all deposits.

Isotopes--Only the Cortez, Carlin, and Mercur deposits have been examined in detail in terms of stable-isotope systematics. These data suggest the importance of meteoric fluids in the formation of the deposits and sedimentary sulfur as the source of sulfide sulfur. More reconnaissance and detailed stable-isotope studies are necessary to understand the variation in this deposit type.

Age of mineralization--There is evidence that these deposits formed as early as Cretaceous and as late as middle Tertiary. Getchell is the only deposit that has been dated by radiometric techniques resulting in a Cretaceous age for mineralization. Altered and mineralized Tertiary dikes at other deposits indicate a Tertiary age for mineraliztion. It is unlikely that there is a sediment-hosted, disseminateed precious-metal epoch of mineralization. Instead, it is more likely that there is a spread in mineralization ages.

ENVIRONMENT OF FORMATION

Deposits of this type have been classified as epithermal to telethermal (near-surface) deposits since the first descriptions of Getchell (Joralemon, 1951) and Carlin (Hausen, 1967). Joralemon (1951) cited the telescoped nature (meaning the close association of gold, realgar, stibnite, and cinnabar over a short vertical distance) of the mineralization, large vugs and an intensely shattered ore zone as evidence for epithermal deposition at Getchell. Hausen (1967) discussed the difficulty of classifying Carlin due to the lack of vuggy, banded veins that are characteristic of the epithermal environment. He noted, however, that silicified zones at Carlin had the appearance of siliceous sinter and thus chose to classify Carlin as "low-temperature epithermal bordering on telethermal." Both authors referred to Graton's (1933) coinage of telethermal as the extension of epithermal into the near-surface to hot-spring environment. Joralemon (1951) drew correlations between cinnabar-bearing, hot-spring deposits and mercury at Getchell whereas Hausen (1967) drew analogies between trace-element associations (Hg, Sb, Au, and Ag) at Carlin and Steamboat Hot Springs, Nevada.

This deposit type has thus been considered epithermal to possibly telethermal without any rigorous pressure and temperature studies. The epithermal environment was described by Lindgren (1933) as a subset of his genetic classification of hydrothermal ore deposits. He suggested that epithermal deposits are formed by ascending hot waters (hydrothermal) of uncertain origin in a shallow environment that is dominated by rapid changes in both heat and pressure. Lindgren placed temperature constraints of between 50°-$200^{\circ}C$ at moderate pressures to define the epithermal-hydrothermal environment of ore deposition. A modified Lindgren classification interprets "moderate pressures" as between 40-240 bars at depths of 150-915 meters (Ridge, 1968). The telethermal environment as defined by Graton (1933) is the terminous of hydrothermal activity and thus, can be interpreted to represent hot-spring activity. Although Graton (1933) did not assign any depth or temperature constraints to the telethermal environment, presumably it would be the cooler and shallower ends of Lindgren's (1933) epithermal environment.

Research in progress on sediment-hosted, disseminated precious-metal deposits suggests that the environment of deposition for such deposits may in fact actually be deeper than previously assumed, extending depths of formation to deep epithermal and mesothermal zones. The mesothermal environment as defined by Lindgren (1933) is typified by temperatures of 200°-$300^{\circ}C$ at high pressures (interpreted by Ridge (1968) to represent depths of one to three kilometers). Various geological criteria observed in these deposits suggest that the deeper environment may be the case for these deposits. In fact, Lindgren (1933, p. 563) classified Mercur, Utah, and the sediment-hosted, disseminated-gold deposits in the Moccasin district in Montana as mesothermal. It is clear that a complete understanding of the genesis of deposits of this type is still elusive and waits the completion of current research.

EXPLORATION APPLICATION

The differences and similarities in geologic characterisitcs between deposits of this type provide a wealth of recognition criteria for deposits of this type. The regional characterisitics (host lithology, structure, and heat), summarized above, are three criteria that may be used to identify possible areas favorable for exploration. Of course, the alignment of deposits of this type in recognized mineral belts is probably the first and most important criterion used on the regional scale by explorationists.

District- and deposit-scale criteria include structure, alteration, and geochemical characteristics. These criteria are necessarily based upon detailed geologic mapping. For example, alteration types and their spatial relationships to structures and different potential host lithologies must be mapped and reasonably understood prior to geochemical sampling. In addition, vein types and their crosscutting relationships help define a target area. Most of the deposits discussed above have late calcite veining crosscutting oxidized rocks. Jasperoidal breccia and jasperoid veins generally occur near ore, even when they themselves may not carry high gold values. Geochemical surveys, including rock and soil, can be extremely valuable for closing in on a target (Bagby et al., 1984). The ubiquitous gold suite of associated trace elements, arsenic, mercury, and antimony is an important indicator of gold mineralized rock. Extremely high values for these elements are not necessary to define a favorable area. Instead, it may be more significant that the suite of indicator elements is present.

INFLUENCE OF GEOLOGIC CHARACTERISTICS ON MINING

Grade and Tonnage

Sediment-hosted, disseminated precious-metal deposits are variable in grade and tonnage. Table 8.5 lists 31 gold- and silver-rich deposits, 30 for which there are grade and tonnage data. Although the deposits are located throughout the western United States, most are from Nevada. Grade and tonnage values for these deposits are shown on cumulative-frequency graphs in Figure 8.11. Gold grade is plotted against tonnage for 29 of the deposits in Figure 8.12. The cumulative-frequency diagrams show that the median gold grade is 2.5 grams/tonne and the median tonnage is 5.1 metric tonnes. The gold curve also indicates that 90% of the deposits have gold grades less than 7.6 grams/tonne. The largest deposits in terms of gold grade within this suite are Carlin and Jerritt Canyon. The cumulative-frequency tonnage curve indicates that 90% of the deposits have less than 24 million tonnes of ore. The only deposit with greater tonnage is Gold Quarry. A comparison of individual deposits is best displayed by the grade versus tonnage plot in Figure 8.12. The diagonal lines across the figure indicate total contained grams (or ounces) of gold for any given grade and tonnage. This shows that the largest deposits in terms of contained gold are Carlin, Getchell, Jerritt Canyon, and Gold Quarry. All of the other deposits contain less than 50 million grams (about 1.5 million troy ounces) of contained gold per deposit.

One interpretation that can be derived from the grade and tonnage data is that the chances of finding deposits greater than 24 million tonnes are small. Of the sample suite, there are really only 4 outstanding deposits, in terms of contained gold, out of 29. Of those four, two (50%) were discovered during the

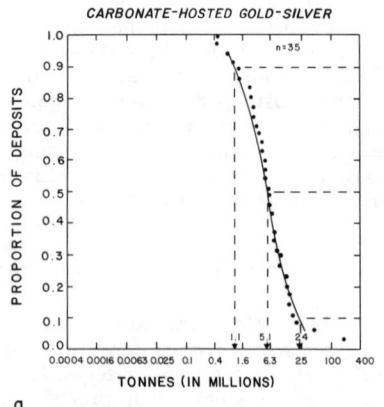

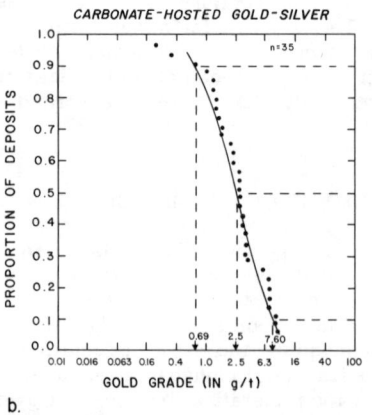

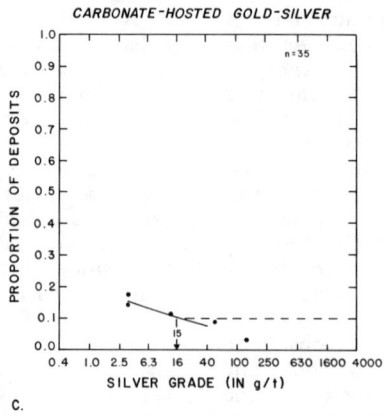

Figure 8.11. Cumulative-frequency diagrams for grade and tonnage of sediment-hosted, disseminated precious-metal deposits. Some data used in these diagrams are confidential and thus, the total number of deposits (n = 35) is greater than those listed in Table 8.5. Figures are from Bagby and Singer (1983).

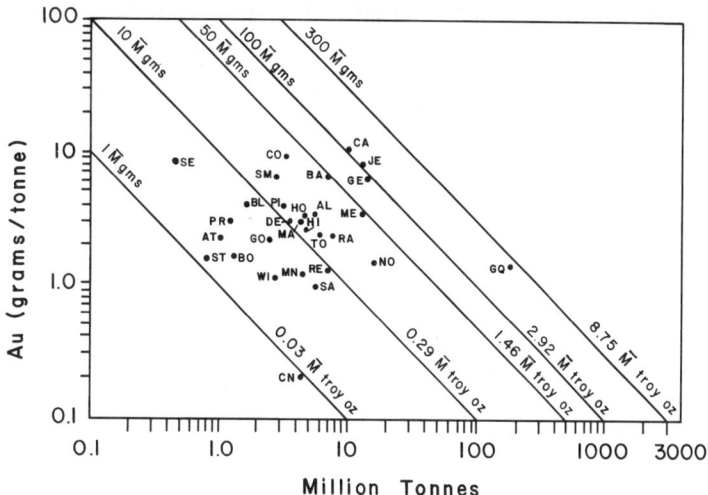

Figure 8.12. Gold grade versus tonnage for gold-rich, sediment-hosted, disseminated precious-metal deposits. The diagonal lines indicate total contained gold for a given grade and tonnage. Deposit abbreviations are those used in Table 8.5. (Gold production for 1984 in the United States was 2.3 million troy ounces (71.5 million grams) as estimated by the U.S. Bureau of Mines, Mineral Commodity Summaries, 1985).

recent increase of exploration activity in the 1970's. Each of the remaining more recently discovered and developed deposits contain less than 50 million grams of gold. The explorationist, however, should not be discouraged. The fact that 50% of these outstanding large deposits were found in recent years is reason enough to hope that more will be discovered, especially in light of our increasing knowledge of the geologic characteristics of this deposit type.

Mineability

Deposits of this type are minable due to the coincidence of several geologic and economic factors. The continued high market value for precious metals, particulary in an otherwise difficult metals economic environment, is the prime reason the exploration for, and development of, low-grade precious-metal deposits continues. The high-market value of precious metals (particularly gold) balanced against mining costs and metal extraction is favorable. Geologic factors that influence minability are oxidation, large tonnages near the surface, and the fine-grained, disseminated nature of the gold. Most of these deposits are deeply oxidized. The oxidation has essentially performed the first metallurgical process by freeing the gold from its association with either sulfide or carbonaceous material. This oxidized ore is therefore amenable to heap-leaching, making deposits with grades as low as 1 gram/tonne profitable to mine. Early returns on oxidized ore can also provide capital for development of circuits necessary to recover metals from the unoxidized ore that is encountered in the later stages of mining. Since these deposits are near surface and the gold is characteristically evenly distributed in its host, the deposits are bulk minable. This lowers mining costs which are important factors in deriving a profit from this type of deposit.

The economic balance sheet for these deposits is important to industry and the favorable figures result in increased capital outlays for exploration and development. This deposit type is therefore presently of interest to the mining industry in the United States. Geologic studies of individual deposits and comparisons between and within deposits increase our knowledge of their occurrence and genesis, and this knowledge leads to further discoveries.

Table 8.5--Grade and tonnage data for sediment-hosted, disseminated precious-metal deposits

	Location	Au grade troy oz/gram	Ag grade troy oz/gram	Tonnage (million) short ton/metric ton	Notes
1. Alligator Ridge (AL)	White Pine Co., NV	0.099/3.39	Unknown	5.9/5.4	1983 reserves plus production to date, company handout.
2. Atlanta (AT)	Lincoln Co., NV	0.08/2.74	1.6/54.86	1.1/1.0	Bonham (1983).
3. Bald Mountain (BA)	Lawrence Co., SD	0.199/6.82	0.372/12.75	7.6/6.9	Norton (1983) Bald Mountain district; several mines.
4. Blue Star (BL)	Eureka Co., NV	0.12/4.11	Unknown	1.8/1.6	Bonham (1983).
5. Bootstrap (BO)	Eureka and Elko Co., NV	0.049/1.68	Present	1.4/1.3	McQuiston and Shoemaker (1981) Calculated from 717,000 tons shipped to Carlin at assumed grade .064 oz; 165,000 tons leach at .044 oz; 500,000 tons leach at .028 oz.
6. Candelaria (CN)	Mineral Co., NV	0.006/0.21	4/137.14	5/4.5	Bonham (1983) Bagby (1983, unpublished field notes).
7. Carlin (CA)	Eureka Co., NV	0.32/10.97	Unknown	11/10	Jackson (1983).
8. Cortez (CO)	Lander Co., NV	0.279/9.57	Unknown	3.6/3.3	Bonham (1983) Actually 3,562,100 tons.
9. Dee (DE)	Elko Co., NV	0.09/3.09	Present	4/3.63	Wallace (in press).
10. Getchell (GE)	Humboldt Co., NV	0.194/6.65	Unknown	15.4/13.97	Bonham (1983) Total production, reserves, and possible.
11. Gold Acres (GO)	Lander Co., NV	0.065/2.23	Unknown	2.8/2.54	Bonham (1983).
12. Gold Quarry (GQ)	Eureka Co., NV	0.043/1.47	Unknown	183/166	Skillings (1984).
13. Gold Strike (GS)	Eureka Co., NV	Unknown	Unknown	Unknown	Bonham (1983) Greater than 100,000 oz Au.
14. Hilltop (HI)	Lander Co., NV	0.079/2.71	Unknown	5.1/4.6	Bonham (1983).
15. Horse Canyon (HO)	Eureka Co., NV	0.10/3.43	Unknown	5.0/4.5	Bagby (1983, unpublished field notes).
16. Jerritt Canyon (JE)	Elko Co., NV	0.243/8.33	Unknown	14.06/12.76	Hawkins (1982), Anonymous (1982).
17. Maggie Creek (MA)	Eureka Co., NV	0.092/3.15	Unknown	4.8/4.35	Anonymous (1980).
18. Manhattan (MN)	Nye Co., NV	0.036/1.23	Unknown	5.0/4.54	Bonham (1983).
19. Mercur (ME)	Tooele Co., UT	0.102/3.50	Unknown	14.31/12.98	Anonymous (1981), Larimer (1983).
20. Northumberland (NO)	Nye Co., NV	0.045/1.54	Unknown	17/15.42	Bonham (1983).
21. Pinson (PI)	Humboldt Co., NV	0.12/4.11	Unknown	3.5/3.18	Wallace (in press).
22. Preble (PR)	Humboldt Co., NV	0.09/3.09	Unknown	1.3/1.18	Wallace (in press) Antoniuk and Crombie (1982) give 1.5 million tons at 0.064 oz .

Table 8.5--Grade and tonnage data for sediment-hosted, disseminated precious-metal deposits--(continued)

		Location	Au grade troy oz/gram	Ag grade troy oz/gram	Tonnage (million) short ton/metric ton	Notes
23.	Rain (RA)	Elko Co., NV	0.083/2.85	Unknown	8.3/7.53	Bonham (1983) Skillings (1984).
24.	Relief Canyon (RE)	Pershing Co., NV	0.04/1.37	Unknown	8.0/7.26	Bonham (1983).
25.	Sammy Creek (SM)	Elko Co., NV	0.2/6.86	Unknown	3/2.72	Wallace (in press).
26.	Santa Fe (SA)	Mineral Co., NV	0.03/1.03	0.26/8.91	6.35/5.76	Anonymous (1984).
27.	Standard (ST)	Pershing Co., NV	0.048/1.65	0.116/3.98	0.884/0.80	Bonham (1983).
28.	Sterling (SE)	Nye Co., NV	0.25/8.57	Unknown	0.500/0.45	Bonham (1983).
29.	Taylor (TA)	White Pine Co., NV	Unknown	3.0/102.86	10/9.07	Bonham (1983).
30.	Tonkin Springs (TO)	Eureka Co., NV	0.084/2.88	Unknown	6.5/5.90	Bagby (1983, unpublished field notes).
31.	Windfall (WI)	Eureka Co., NV	0.034/1.17	Unknown	3.05/2.77	Wilson (1976).

REFERENCES

Adkins, A. R., and Rota, J. C., 1984, General geology of the Carlin gold mine; in J. L. Johnson (ed.), Field Trip Guidebook; Exploration for ore deposits of the North American cordillera: Association of Exploration Geochemists, p. 17-23.

Ainsworth, J. C., and Brimhall, G. H., Jr., 1983, Chemical and mineralogical zoning associated with stratagene, fault-controlled gold mineralization at Alligator Ridge, Nevada (abs.): Geological Society of America, Abstracts with Programs, v. 15, no. 6, p. 513.

Anonymous, 1980, Newmont's Carlin begins production from major new Nevada gold deposit: Intermountain Pay Dirt, issue 12, September, p. 20 and p. 22.

Anonymous, 1981, Joint ventures led by Getty to mine gold at historical Mercur Canyon site: Engineering and Mining Journal, September, p. 39.

Anonymous, 1982, Gold: Freeport Annual Report, p. 14-15.

Anonymous, 1984, Lacana mining, Nevada growth: Mining Journal, v. 303, no. 7778, p. 189-190.

Antoniuk, T., and Crombie, D. R., 1982, The Pinson mine: A Carlin-type gold deposit: Canadian Mining Journal, v. 103, no. 4, p. 61-65.

Bagby, W. C., Pickthorn, W. J., and Goldfarb, R. J., 1984, Distribution of selected trace elements in soils overlying the Dee disseminated-gold deposit, Elko County, Nevada: American Institute of Mining Engineers Preprint 84-319, 9 p.

Bagby, W. C., and Singer, D.A., 1983, Sediment-hosted precious-metal deposit grade-tonnage model; in Singer, D. A. and Mosier, D. L. (eds.), Mineral Deposit Grade-Tonnage Models: U.S. Geological Survey, Open-File Report 83-623, model 5.2.

Berger, B. R., 1975, Geology and geochemistry of the Getchell disseminated-gold deposit, Humboldt County, Nevada: Society of Mining Engineers, AIME, Preprint 75-I-305, 26 p.

Berger, B. R., 1980, Geological and geochemical relationships at the Getchell mine and vicinity, Humboldt County, Nevada: Society of Economic Geologists, 1980 Field Trip Guide, p. 111-135.

Berger, B. R. and Taylor, B. E., 1980, Pre-Cenozoic normal faulting in the Osgood Mountains, Humboldt County, Nevada: Geology, v. 8, p. 594-598.

Bodnar, R. J., Reynolds, T. J., and Kuehn, C. A., 1985, Fluid-inclusion systematics in epithermal systems; in Berger, B. R., and Bethke, P. M. (eds.), Geology and Geochemistry of Epithermal Systems: Society of Economic Geologists, Reviews in Economic Geology, v. 2.

Bonham, H. F., 1983, Reserves, host rocks, and ages of bulk-minable, precious-metal deposits in Nevada; in The Nevada Mineral Industry, 1983: Nevada Bureau of Mines and Geology Special Publication MI-1983, p. 15-16.

Brady, B. T., 1984, Gold in the Black Pine mining district, southeast Cassia County, Idaho: U.S. Geological Survey, Bulletin 1382-E, 15 p.

Butler, B. S., Loughlin, G. F., and Heikes, V. C., 1920, The ore deposits of Utah: U.S. Geological Survey, Professional Paper 111, p. 391-395.

Coney, P. J., Jones, D. L., and Monger, J. W. H., 1980,

Cordilleran suspect terranes: Nature, v. 288, p. 329-333.

Coppinger, W. W. and Cartwright, M. R., 1983, Geology of the Horse Canyon disseminated-gold deposit, Eureka County, Nevada (abs.): Geological Society of America, Abstracts with Programs, v. 15, no. 6, p. 547.

Crone, W. J., 1982, The use of iron/manganese oxide-rich fracture coatings in the geochemical exploration for precious metal deposits, A comparison with standard rock: Unpublished M.S. thesis, University of Nevada (Reno), 93 p.

Crone, W. J., Larson, L. T., Carpenter, R. H., Chao, T. T., and Sanzolone, R. F., 1984, A comparison of iron oxide-rich joint coatings and rock chips as geochemical sampling media in exploration for disseminated gold deposits: Journal of Geochemical Exploration, v. 20, p. 161-178.

Cunningham, C. G., 1978, Pressure gradients and boiling as mechanisms for localizing ore in porphyry systems: U.S. Geological Survey, Journal of Research, v. 6, no. 6, p. 745-754.

Drewes, H., 1962, Stratigraphic and structural controls of mineralization in the Taylor mining district near Ely, Nevada: U.S. Geological Survey, Professional Paper 450-B, p. B1-B3.

Drewes, H., 1967, Geology of the Connors Pass quadrangle, Schell Creek Range, east-central Nevada: U.S. Geological Survey, Professional Paper 557, 93 p.

Erickson, R. L. and Marsh, S. P., 1974, Paleozoic tectonics in the Edna Mountain quadrangle, Nevada: U.S. Geological Survey, Journal of Research, v. 2, no. 3, p. 331-337.

Erickson, R. L., Van Sickle, G. H., Nakagawa, H. M., McCarthy, J. H., and Leong, K. W., 1966, Gold geochemical anomaly in the Cortez district, Nevada: U.S. Geological Survey, Circular 534, 9 p.

Evans, J. G., 1974, Bootstrap window, Elko and Eureka counties, Nevada. Geological summary and anlyses of rock samples: U.S. Geological Survey, Open-File Report 74-369, 19 p.

Evans, J. G., 1980, Geology of the Rodeo Creek NE and Welches Canyon quadrangles, Eureka County, Nevada: U.S. Geological Survey, Bulletin 1473, 81 p.

Farmer, G. L., and DePaolo, D. J., 1983, Origin of Mesozoic and Tertiary granite in the western United States and implications for Pre-Mesozoic crustal structure 1. Nd and Sm isotopic studies in the geocline of the northern Great Basin: Journal of Geophysical Research, v. 88, no. B4, p. 3379-3401.

Gilluly, J., and Gates, O., 1965, Tectonic and igneous geology of the northern Shoshone Range, Nevada, with sections on gravity in Crescent Valley, by Donald Plouff and Economic Geology, by Keith B. Ketner: U.S. Geological Survey, Professional Paper 465, 153 p.

Gilluly, J., and Masursky, H., 1965, Geology of the Cortez quadrangle, Nevada, with a section on gravity and aeromagnetic surveys by D. R. Mabey: U.S. Geological Survey, Bulletin 1175, 117 p.

Graton, L. C., 1933, The depth zones in ore deposition: Economic Geology, v. 6, p. 513-555.

Grove, G. R., 1979, A study of the fine-grained disseminated gold ore of the Windfall mine, Eureka County, Nevada: Unpublished M.S. thesis, University of California (Santa Barbara), 97 p.

Guenther, E. M., 1973, The geology of the Mercur gold camp, Utah: Unpublished M.S. thesis, University of Utah, 79 p.

Harris, M., and Radtke, A. S., 1976, Statistical study of selected trace elements with reference to geology and genesis of the Carlin gold deposit, Nevada: U.S. Geological Survey, Professional Paper 960, 21 p.

Hausen, D. M., 1967, Fine gold occurrence at Carlin, Nevada: Unpublished Ph.D. dissertation, Columbia University, 166 p.

Hausen, D. M., 1981, Process mineralogy of auriferous pyritic ores at Carlin, Nevada; in Hausen, D. M. and Park, W. C. (eds.), Process Mineralogy, Extractive Metalurgy, Mineral Exploration, Energy Resources: Metallurgical Society, AIME, p. 271-289.

Hausen, D. M., Ekburg, C., and Kula, F., 1983, Geochemical and XRD-computer logging method for lithologic ore type classification of Carlin-type gold ores; in Hagni, R. D. (ed.), Process Mineralogy II: Applications in Metallurgy, Ceramics, and Geology: Metallurgical Society, AIME, p. 421-450.

Hausen, D. M., and Kerr, P. F., 1968, Fine-gold occurrence at Carlin, Nevada; in Ridge, J. D. (ed.), Ore Deposits of the United States, 1933-1967, Graton-Sales Volume: Society of Mining Engineers, AIME, New York, p. 908-940.

Havenstrite, S. R., 1983, Geology and ore deposits of the Taylor mining district, White Pine County, Nevada; in Papers given at the precious-metals symposium, Sparks, Nevada, November 17-19, 1980: Nevada Bureau of Mines and Geology, Report 36, p. 14-26.

Hawkins, R. B., 1973, The geology and mineralization of the Jerritt Creek area northern Independence Mountains, Nevada: Unpublished M.S. thesis, Idaho State University, 104 p.

Hawkins, R. B., 1982, Discovery of the Bell gold mine--Jerritt Canyon district, Elko County, Nevada: Mining Congress Journal, February, p. 28-32.

Hose, R. K., Blake, M. C., and Smith, R. M., 1976, Geology and mineral resources of White Pine County, Nevada: Nevada Bureau of Mines and Geology, Bulletin 85, 105 p.

Hotz, P. E., and Willden, R., 1964, Geology and mineral deposits of the Osgood Mountains quadrangle, Humboldt County, Nevada: U.S. Geological Survey, Professional Paper 431, 128 p.

Ilchik, R. P., 1984, Hydrothermal maturation of indigenous organic matter at the Alligator Ridge gold deposits, Nevada: Unpublished M.S. thesis, University of California at Berkeley, 68 p.

Jackson, D., 1983, Carlin gold, a Newmont money generator keeps on renewing itself after sparking the rebirth of gold mining in Nevada:

Engineering and Mining Journal, July, p. 38-43.

Jewell, P. W., 1984, Chemical and thermal evolution of hydrothermal fluids, Mercur gold district, Tooele County, Utah: Unpublished M.S. thesis, University of Utah, 77 p.

Joralemon, P., 1951, The occurrence of gold at the Getchell mine, Nevada: Economic Geology, v. 46, no. 3, p. 267-310.

Kistler, R. W., 1983, Isotope geochemistry of plutons in the northern Great Basin, in The Role of Heat in the Development of Energy and Mineral Resources in the Northern Basin and Range Province: Geothermal Resources Council Special Report No. 13, p. 3-8.

Kleinhampl, F. J. and Ziony, J. I., 1984, Mineral resources of northern Nye County, Nevada: Nevada Bureau of Mines and Geology, Bulletin 99B, 243 p.

Klessig, P.J., 1984, History and geology of the Alligator Ridge gold mine, White Pine County, Nevada; in Wilkins, J., Jr., (ed.), Gold and Silver Deposits of the Basin and Range Province, Western U.S.A.: Arizona Geological Society Digest, v. XV, p. 77-88.

Knutsen, G. C., and West, P. W., 1984, Geology of the Rain disseminated gold deposit, Elko County, Nevada; in Wilkins, J., Jr., (ed.), Gold and Silver Deposits of the Basin and Range Province, Western U.S.A.: Arizona Geological Society Digest, v. XV, p. 73-76.

Kornze, L. D., Faddies, T. B., Goodwin, J. C., and Bryant, M. A., 1984, Geology and geostatistics applied to grade control at the Mercur gold mine, Mercur, Utah: AIME Preprint 84-442, 9 p.

Kretschmer, E. L., 1984, Geology of the Pinson and Preble gold deposits, Humboldt County, Nevada; in Wilkins, J., Jr. (ed.), Gold and Silver Deposits of the Basin and Range Province, Western U.S.A.: Arizona Geological Society Digest, v. XV, p. 59-66.

Kuehn, C. A., and Bodnar, R. J., 1984, P-T-X characteristics of fluids associated with the Carlin sediment-hosted gold deposit (abs.): Geological Society of America, Abstracts with Programs, v. 16, no. 6, p. 566.

Larimer, C., 1983, Mercur lives again as Getty gears up production at new Utah mine and mill: Paydirt, August, p. 8B-10B.

Lindgren, W., 1933, Mineral Deposits, Fourth Edition: McGraw Hill, New York, 930 p.

Lovering, T. G., 1972, Jasperoid in the United States-Its characteristics, origin, and economic significance: U.S. Geological Survey, Professional Paper 710, 164 p.

Lovering, T. G., and Heyl, A. V., 1974, Jasperoid as a guide to mineralization in the Taylor mining district and vicinity near Ely, Nevada: Economic Geology, v. 69, no. 1, p. 46-58.

McKee, E. H., 1974, Northumberland caldera and Northumberland tuff; in Guidebook to the Geology of Four Tertiary Volcanic Centers in Central Nevada: Nevada Bureau of Mines and Geology, Report 19, p. 35-41.

McKee, E. H., Silberman, M. L., Marvin, R. E., Obradovich, J. D., 1971, A summary of radiometric ages of Tertiary volcanic rocks in Nevada and eastern California--Part 1, central Nevada: Isochron/West, No. 2, p. 21-42.

McQuiston, F. W., Jr. and Shoemaker, J. C., 1981, Gold and Silver Leaching, Recovery, and Economics: Society of Mining Engineers, AIME.

Morrow, A. B. and Bettles, K., 1982, Geology of the Gold Strike mine, Elko County, Nevada: Text of talk given at the 88th annual Northwest Mining Association Convention, Spokane, Washington, 4 p.

Morton, J. L., Silberman, M. L., Bonham, H. F., Garside, L. J., and Noble, D. C., 1977, K-Ar ages of volcanic rocks, plutonic rocks, and ore deposits in Nevada and eastern California-Determinations run under the USGS-NBMG cooperative program: Isochron/West, No. 20, p. 19-29.

Norton, J. L., 1983, Bald Mountain gold mining region, northern Black Hills, South Dakota: U.S. Geological Survey, Open-File Report 83-791, 18 p.

Ohmoto, H., and Rye, R. O., 1979, Carbon and sulfur isotopes; in Barnes, H. L., (ed.), Geochemistry of Hydrothermal Systems: Second Edition, John Wiley and Sons, New York, p. 509-567.

Ott, L. E., 1983, Geology and ore localization at the Northumberland gold mine, Nye County, Nevada: Unpublished M.S. thesis, Montana College of Mineral Science and Technology, 52 p.

Powers, S. L., 1978, Jasperoid and disseminated gold at the Ogee-Pinson mine, Humboldt County, Nevada: Unpublished M.S. thesis, University of Nevada (Reno), 112 p.

Radtke, A. S., 1981, Geology of the Carlin gold deposit, Nevada: U.S. Geological Survey, Open-File Report 81-97, 154 p.

Radtke, A. S., Heropoulos, C., Fabbi, B. P., Scheiner, B. J., and Essington, M., 1972, Data on major and minor elements in host rocks and ores, Carlin gold deposit, Nevada: Economic Geology, v. 67, no. 7, p. 975-978.

Radtke, A. S., Rye, R. O., and Dickson, F. W., 1980, Geology and stable-isotope studies of the Carlin gold deposit, Nevada: Economic Geology, v. 75, no. 5, p. 641-672.

Radtke, A. S. and Scheiner, B. J., 1970, Studies of hydrothermal gold deposition (I). Carlin gold deposit, Nevada: The role of carbonaceous materials in gold deposition: Economic Geology, v. 65, no. 2, p. 87-102.

Ridge, J. D., 1968, Changes and developments in concepts of ore genesis-1933 to 1967; in Ridge, J. D. (ed.), Ore Deposits in The United States, 1933-1967, Graton-Sales Volume: Society of Mining Engineers, AIME, p. 1713-1834.

Roberts, R. J., 1966, Metallogenic provinces and mineral belts in Nevada; in Papers presented at the AIME Pacific Southwest Mineral Industry Conference, Sparks, Nevada, May 5-7, 1965: Nevada Bureau of Mines and Geology, Report 13, p. 47-72.

Roberts, R. J., Hotz, P. E., Gilluly, J., and Ferguson, H. G., 1958, Paleozoic rocks of north-central Nevada: American Association of Petroleum

Geologists Bulletin, v. 42, no. 12, p. 2813-2857.

Rye, R. O., 1985, A model for the formation of carbonate-hosted disseminated-gold deposits as indicated by geologic, fluid-inclusion, geochemical, and stable-isotope studies of the Carlin and Cortez deposits, Nevada; in Tooker, E. W. (ed.), Geologic Characteristics of the Sediment- and Volcanic-Hosted Disseminated Gold Deposits--Search for an Occurrence Model: U.S. Geological Survey, Bulletin 1646, p. 35-42.

Rye, R. O., Doe, B. R., and Wells, J. D., 1974, Stable isotope and lead isotope studies of the Cortez, Nevada, gold deposit and surrounding area: U.S. Geological Survey, Journal of Research, v. 2, p. 13-23.

Rytuba, J. J., Madrid, R. J., and McKee, E. H., 1984, Relationship of the Cortez caldera to the Cortez disseminated-gold deposit, Nevada (abs.); in Exploration for ore deposits of the North American cordillera, Symposium of the Association of Exploration Geochemists, Reno, Nevada, March 25-28, 1984, Abstracts with Programs, p. 36.

Shawe, D. R. and Stewart, J. H., 1976, Ore deposits as related to tectonics and magmatism, Nevada and Utah: Transactions of the Society of Mining Engineers, AIME, v. 260, p. 225-260.

Silberman, M. L., Berger, B. R., and Koski, R. A., 1974, K-Ar age relations of granodiorite emplacement and tungsten and gold mineralization near the Getchell mine, Humboldt County, Nevada: Economic Geology v. 69, no. 5, p. 646-656.

Silberman, M. L., and McKee, E. H., 1971, K-Ar ages of granitic plutons in north-central Nevada: Isochron/West, v. 1, p. 15-32.

Skillings, D. N., Jr., 1984, Carlin Gold Mining Company's operations and Gold Quarry project: Skillings Mining Revew, November 24, 1984, p. 4-8.

Stevens, D. L., and Hawkins, R. B., 1984, A comparison of the gold mineralization at Jerritt Canyon, Nevada with other disseminated-gold deposits of the Basin-Range region; in Watson, S. T. (ed.), Transactions of the Third Circum-Pacific Energy and Mineral Resources Conference, p. 339-348.

Stewart, J. H., and Carlson, J. E., 1984, Geologic map of north central Nevada: Nevada Bureau of Mines and Geology, map 50, 1:250,000.

Stewart, J. H., Moore, W. J., and Zietz, Isidore, 1977, East-west patterns of Cenozoic igneous rocks, aeromagnetic anomalies, and mineral deposits, Nevada and Utah: Geological Society of America Bulletin, v. 88, p. 67-77.

Stewart, J. H., Walker, G. W., and Kleinhampl, F. J., 1975, Oregon-Nevada lineament: Geology, v. 3, no. 5, p. 265-268.

Tafuri, W., 1976, Geology and geochemistry of the gold deposits at Mercur, Utah: Unpublished text of talk presented at a symposium by The Geological Society of Nevada and Mackay School of Mines on Geology and Exploration Aspects of Fine-Grained Carlin-Type gold deposits, University of Nevada, Reno, March, 1976.

Tingley, J. V., and Berger, B. R., 1985, Lode gold deposits of Round Mountain, Nevada: Nevada Bureau of Mines and Geology Bulletin 100, 62 p.

Wallace, A. B., Carlin-type disseminated-gold deposits; in Shawe, D. R. (ed.): U.S. Geological Survey, Professional Paper (in preparation).

Wallace, A. B. and Bergwall, F. W., 1984, Geology and gold mineralization at the Dee mine, Elko County, Nevada (abs.): Geological Society of America, Abstracts With Programs, v. 16, no. 6, p. 686.

Wells, J. D., Elliott, J. E., and Obradovich, J. D., 1971, Age of the igneous rocks associated with ore deposits, Cortez-Buckhorn area, Nevada: U.S. Geological Survey, Professional Paper 750-C, p. C127-C135.

Wells, J. D. and Mullens, T. E., 1973, Gold-bearing arsenian pyrite determined by microprobe analysis, Cortez and Carlin gold mines, Nevada: Economic Geology, v. 68, p. 187-201.

Wells, J. D., Stoiser, L. R., and Elliott, J. E., 1969, Geology and geochemistry of the Cortez gold deposit, Nevada: Economic Geology, v. 64, p. 526-537.

Wilson, W. L., 1976, The Eureka Windfall gold mine, (abs.); in Geology and Exploration Aspects of Fine-Grained, Carlin-Type Gold Deposits: A symposium presented by The Geological Society of Nevada and MacKay School of Mines, March 26, 1976, University of Nevada, Reno.

Wrucke, C. T., 1985, Gold Acres, Nevada deposit check list; in Tooker, E. W. (ed.), Geologic Characteristics of the Sediment- and Volcanic-Hosted Disseminated Gold Deposits--Search for an Occurrence Model: U.S. Geological Survey, Bulletin 1646, p. 120-123.

Wrucke, C. T. and Armbrustmacher, T. J., 1975, Geochemical and geologic relations of gold and other elements at the Gold Acres open-pit mine, Lander County, Nevada: U.S. Geological Survey, Professional Paper 860, 27 p.

Chapter 9
RELATIONSHIP OF TRACE-ELEMENT PATTERNS TO ALTERATION AND MORPHOLOGY IN EPITHERMAL PRECIOUS-METAL DEPOSITS

Miles L. Silberman and Byron R. Berger

INTRODUCTION

An epithermal ore deposit is defined as a relatively near-surface deposit formed in a hydrothermal system under low to moderate pressure and a temperature range below about 300°C (Barrett, 1985). This concise definition is a restatement of Lindgren's characteristics of hydrothermal systems of "epithermal" character. A modification of Lindgren's characteristics is tabulated in Table 9.1. These characteristics are both physical and chemical, and we will, in this and the following paper (Berger and Silberman, 1985, this volume), attempt to relate them.

Epithermal lode deposits in the Circum-Pacific region produce approximately 30 million grams of gold annually (Giles and Nelson, 1982) and a larger, but indeterminate, amount of silver. Many epithermal deposits are closely associated with convergent plate boundaries related to present and relatively recent regimes of plate tectonic interaction (Giles and Nelson, 1982; Sawkins, 1984). These mobile regions of the earth's crust are characterized by recent volcanism, high heat flow and tectonic activity, and by the presence of active and recently active geothermal fields, some of which have deposited precious metals and associated metals (Table 9.1) in similar concentrations (but not volumes) to those found in the epithermal ore deposits (Weissberg et al., 1979; Henley, 1985, this volume).

The understanding of processes that occur during the formation of epithermal ore deposits has been advanced in the recent past by the suggestion that these ore deposits are essentially fossil geothermal systems (e.g., White, 1955, 1981; White, 1974; Wetlaufer et al., 1979; Henley and Ellis, 1983; Henley, 1985, this volume). The observed data discussed in the references above demonstrate that epithermal ore deposits and geothermal systems have similar alteration mineralogy, temperatures, fluid compositions and stable-isotope patterns, and geochemical associations. Indeed, features characteristic of geothermal systems, such as siliceous sinter and hydrothermal explosion breccias are found in some epithermal deposits, and can be part of the ore (Barrett, 1985; Silberman et al., 1979; Wallace, 1980, 1984). The analogy between epithermal ore deposits and geothermal systems is supported by the occurrence of ore-grade concentrations of Au and Ag and other associated elements (e.g., As, Sb, Hg, Tl, W) in surface discharge material from several active geothermal systems, such as Steamboat Springs, Nevada, Ohaaki-Broadlands, and Waiotapu, New Zealand, and elsewhere (Weissberg, 1969; Weissberg et al., 1979; Ewers and Keays, 1977; White, 1981), although the amount of this material found to date is small.

Much published information is available on the geology, alteration mineralogy and zoning, fluid composition and isotopic characteristics, geometry of heat sources and flow, and patterns of fluid migration in geothermal systems. The geology and alteration mineralogy and zoning, temperature, isotopic composition and salinity data for ore fluids from many epithermal precious-metal deposits are also well

Table 9.1--Characteristics that classify a hydrothermal system as being epithermal (after Lindgren, 1933)

Depth of formation	Surface to 1000m
Temperature of formation	50° to 300°C
Form of deposits	Thin to large veins, stockworks, disseminations, replacements
Ore textures	Open-space filling, crustification, colloform banding, comb structure, brecciation
Ore elements	Au, Ag, (As, Sb), Hg, [Te, Tl, U], (Pb, Zn, Cu)*
Alteration	Silicification, argillization, sericite, adularia, propylitization
Common features	Fine-grained chalcedonic quartz, quartz pseudomorphs after calcite, brecciation

*[ ] brackets indicate elements seldom present in more than sub-economic concentrations; ( ) parentheses indicate elements often present in economic concentrations but actually less valuable than associated precious metals.

documented (Buchanan, 1981; Heald-Wetlaufer et al., 1983; Hayba et al., 1985, this volume). Detailed data on geochemistry and geochemical zoning in both types of systems are less available, and not nearly as well documented. The geothermal-epithermal analogy is generally accepted, and most models that have been proposed for types of epithermal deposits draw heavily on the literature on geothermal systems. We will, in this and the following paper (Berger and Silberman, 1985, this volume), review geochemical zoning in geothermal systems, and present a number of morphological models and descriptions of varieties of epithermal ore deposits, and summarize our available data on geochemistry, geochemical zoning, and its relationships to alteration mineralogy and physical morphology.

GEOTHERMAL SYSTEMS

Morphology and Characteristics

White et al. (1971) define a geothermal system as a source of heat within the earth's crust, be that from a magmatic intrusion, or regional heat flow, and the rocks and water affected by that heat. The geothermal system includes both the upwelling hot fluids and the marginal convective downflowing, cold recharge waters. Stable-isotopic studies of geothermal systems have demonstrated that, by far, the greatest majority of them contain water predominantly of meteoric origin (White, 1974, 1981). The surface expressions of geothermal systems are hot springs, fumaroles, and other indications of hydrothermal activity such as altered rocks. The surface phenomena are generally a very small fraction of the size of the geothermal system, perhaps on the order of 5 percent (Henley, 1985, this volume).

In order to generate a geothermal system, it is necessary to have a source of heat, a source of water, and a geologic setting that provide original or induced zones of permeability that allow the water to flow and be recharged. The morphology of geothermal systems in volcanic areas, such as are common in the Circum-Pacific region, are described in detail by Henley and Ellis (1983) and Henley (1985, this volume). Examples of these systems include Broadlands and Waiotapu, New Zealand; Steamboat Springs, Nevada; Yellowstone, Montana-Wyoming; and El Tatio, Chile. Henley (1985, this volume) shows typical schematic cross sections of geothermal systems hosted in silicic volcanic terranes, common on continental margin areas, such as Yellowstone and Broadlands and andesitic stratovolcanic terranes, common in island arc areas, such as Matsao, Taiwan.

Alteration Patterns

Alteration patterns in a variety of geothermal systems have been studied in detail. The alteration mineral assemblage varies with temperature, fluid composition, and host-rock primary mineralogy. Most geothermal systems show lateral and vertical changes in alteration mineral assemblages and have considerable variation in their spatial distribution. Since geothermal systems evolve through time, temporal variations in the positions of channels and the water table can cause extremely complex overprinting of alteration assemblages.

The spatial variation of the alteration patterns are illustrated by the results of studies at Steamboat Springs, Nevada (White et al., 1964; Sigvaldeson and White, 1961; 1962). The geothermal system occurs in pre-Tertiary metamorphic and granitic rocks which are overlain by Miocene trachyandesite and andesite flows and pyroclastic rocks, and Late Tertiary and Pleistocene alluvial deposits. These alluvial deposits contain interbedded basaltic andesite flows. A series of rhyolite domes intrude the older volcanic rocks and alluvium. The time-stratigraphic relations are shown in Figure 9.1.

The alteration mineralogy of two drill cores is shown in Figures 9.2 and 9.3, one (GS-5, Fig. 9.2) collared in sinter in a presently active part of the system, and the other (GS-7, Fig. 9.3) collared in an acid-sulfate altered zone in an area above subsurface boiling. Hole GS-5 probably represents a dominantly hypogene alteration assemblage (Sigvaldeson and White, 1962). Sinter occurs from the surface to 84 feet, grading from opaline to chalcedonic with depth. The alluvium which occurs beneath the sinter has been silicified by deposition of hydrothermal quartz with illite, K-feldspar, and iron sulfides. The rocks beneath the alluvium are altered to an assemblage of K-feldspar, illite, albite, chlorite, and residual quartz, which gives way at depth, gradually to a similar assemblage with calcite, greater amounts of chlorite and some sericite. In terms of hydrothermal mineral

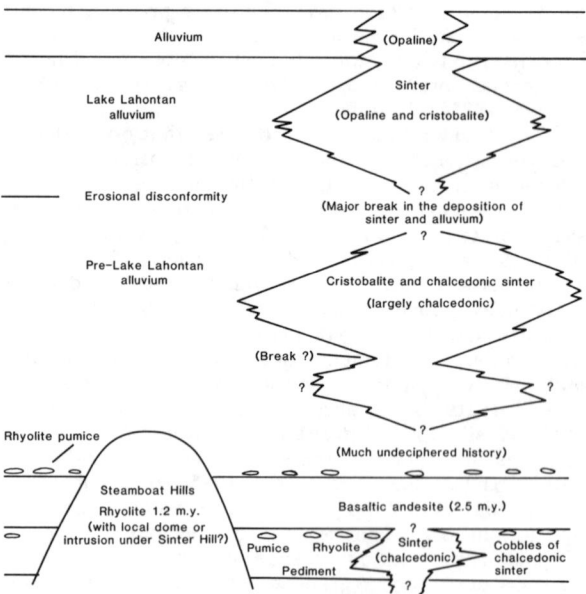

Figure 9.1. Composite time-stratigraphic relations at Steamboat Springs, Nevada, indicating a total lifetime of the geothermal system in excess of 2.5 million years (from Silberman et al., 1979).

assemblages, the zoning would correspond to surface sinter, underlain by silicified rock, overlying an argillic or phyllic assemblage, which slowly gives way at depth to propylitized rock (Rose and Burt, 1979). In detail, the alteration minerals occur in irregular proportions, and there is a gradational change in their relative proportions below the alluvium-volcanic contact (Fig. 9.2). Zones of chalcedony and chalcedony-quartz calcite veining are indicated on Figure 9.2 for GS-5. The content of calcite in the veins increases with depth. The quartz veins in GS-5 range up to 8 feet in thickness (White et al., 1964).

In contrast to GS-5, hole GS-7 (Sigvaldeson and White, 1962) displays an alteration pattern characteristic of acid-sulfate leaching near the surface (Fig. 9.3). The water table is at 114 feet in the hole, and sub-surface boiling has resulted in separation of CO_2 and H_2S gas from the hydrothermal fluid. The H_2S gas condenses in near-surface waters which are oxidizing, and is converted to sulfur which precipitates, and to sulfuric acid. The resultant acidic water percolates downwards, and leaches the granitic rock to a residual, porous mass of opaline silica and residual quartz above the water table. The acid-sulfate waters alter the granitic rock to an assemblage of kaolinite, alunite, and quartz below the water table. This mineralogy gives way at depth to a montmorillonite-dominated assemblage, with kaolinite, illite, residual feldspar, and quartz as the acid is reacted away. With increasing depth chlorite, sericite, and illite are present. The upper alteration zone would be a sulfotaric alteration assemblage, succeeded by advanced-argillic minerals at the water table, giving way gradually to argillic, and then propylitic assemblages. As in GS-5, the alteration minerals occur in variable proportions with depth (Fig. 9.3). At about 358 feet, the assemblage would be considered propylitic. At greater depth, the assemblage again becomes argillic.

In many geothermal systems the drill-core data show that alteration is most intense (and temperatures highest) around fissures where hot water flows. Mineralogy grades to lower temperature assemblages away from these fissures. The fissures themselves tend to contain quartz, K-feldspar, and calcite or wairakite, where boiling has or is occurring (Ellis, 1979).

Permeability is an important controlling factor in alteration. Impermeable rocks are generally less

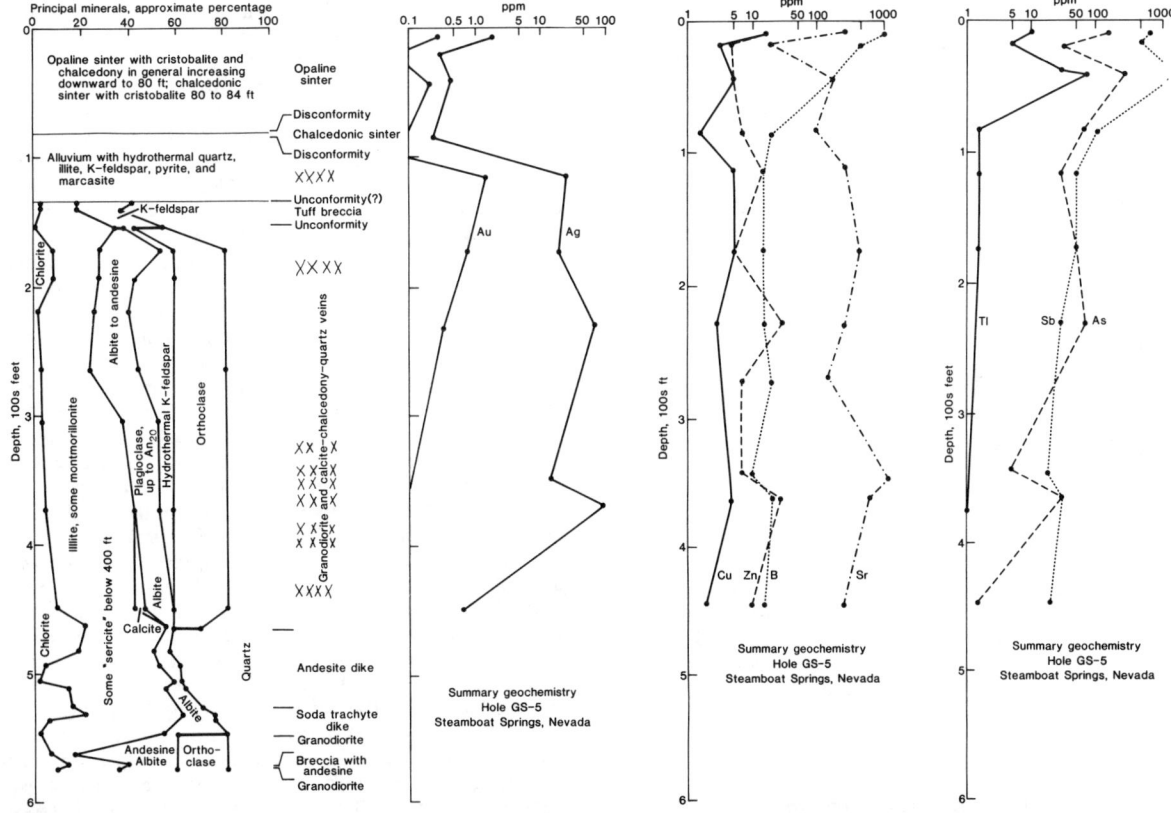

Figure 9.2. Alteration mineralogy and original lithologies in drill hole GS-5, Main Terrace, Steamboat Springs, Nevada. Water table is at or slightly above surface, and springs are still actively flowing. Summary geochemistry for selected elements in sinter and quartz vein from drill core are plotted on the same depth scale (modified from Sigvaldeson and White, 1962).

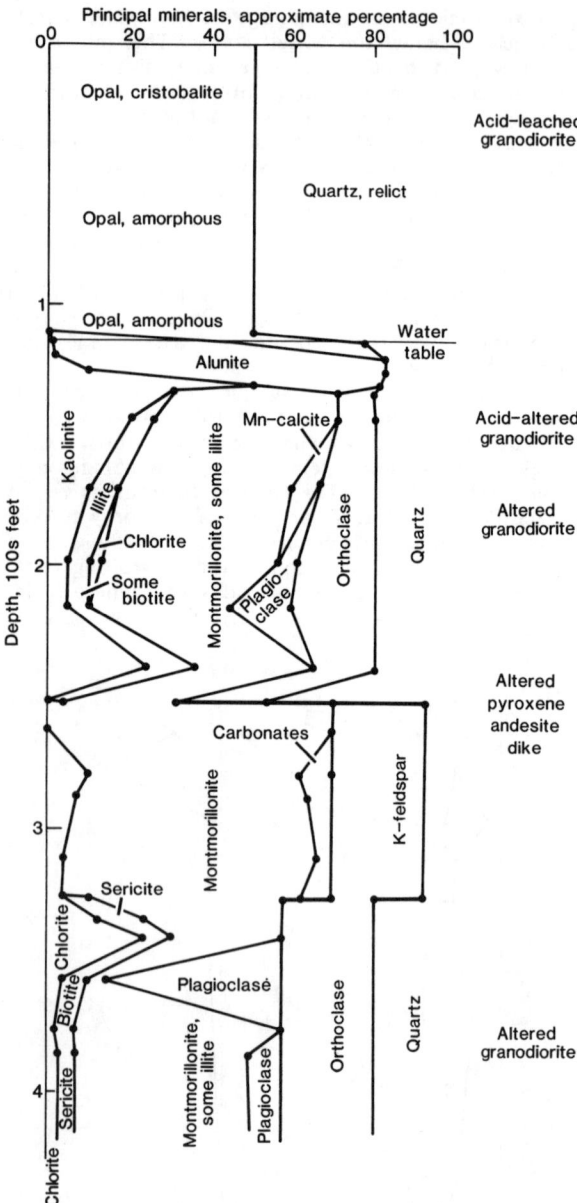

Figure 9.3. Alteration mineralogy and original lithologies in drill hole GS-7, Silica Pit, Steamboat Springs, Nevada. Water table is at -114 feet, and sulfur deposition and solfataric alteration are occurring above the water table (modified from Sigvaldeson and White, 1962).

altered, even in higher temperature zones of geothermal systems than more permeable rocks. In hole GS-5 (Fig. 9.2) at Steamboat, the alluvium is completely altered to an assemblage quartz-Kfeldspar-illite-pyrite, although the temperature in this upper level is only between 130° and 140°C (Sigvaldeson and White, 1962).

Geochemical Zoning

High concentrations of Au and Ag and associated elements are found in surface or near-surface deposits in some geothermal systems (Ewers and Keays, 1977; Weissberg, 1969; Weissberg et al., 1979). Table 9.2 lists chemical data for surface discharge material for several geothermal areas. Au and Ag, along with elements normally associated with them in epithermal deposits (Table 9.1), are strongly concentrated. In some samples, the precious-metal values are in ore grades, particularly the siliceous, antimony-rich precipitates from Steamboat Springs, Nevada, and Broadlands and Waiotapu, New Zealand. However, the actual volume of material enriched to this extent is very small in all of the cases found so far.

Chemical data for drill-core material from Steamboat Springs and Broadlands show that the group of elements As, Sb, Hg, B, Tl, and Au tend to be concentrated in the upper parts of the geothermal systems and decrease with increasing depth. Base metals, Cu, Pb, Zn, and other elements Bi, Se, Te, Co, and to an extent Ag, appear to be precipitated at greater depth in higher temperature zones. However, there are deviations from this simplified scheme, and these variations are significant. Figure 9.2 shows selected chemical data plotted as a function of depth for drill hole GS-5 from Steamboat Springs, Nevada, described earlier. The top four samples are sinter. The remaining, deeper ones are chalcedonic-quartz or chalcedonic quartz-calcite veins cutting the altered rocks below the sinter. The data show strong enrichment of Sb, As, and Tl in the near-surface part of the system. Nearly all of the Tl is in the sinter, with samples below it having about crustal average content for unaltered granitic rock. Sb and As both decrease with depth, with Sb concentrated largely in the upper 100 feet. B is strongly enriched near the surface; below the sinter it is about at crustal average levels for granitic rock. Hg was not analyzed in these samples, but Table 9.2, and data discussed by White (1981) indicate strong enrichment of Hg in the near surface. No cinnabar is found below 50 feet from the present topographic surface. Cu and Zn are present at low concentration levels in the veins and sinter, generally below crustal averages for granitic rock. Pb was not detected in most samples. Zn shows an irregular increase with depth, Cu irregularly decreases with depth. Sr increases irregularly with depth below the sinter, and is probably representative of increasing calcite content of veins with depth (Sigvaldeson and White, 1962; White et al., 1964). Au is detectable in some samples of the sinter, but is highest in veins, just beneath it, then decreases to below detection (0.1 ppm) with greater depth. Ag is also present in the sinter, but is highest in veins at moderate depths, between 84 and 393 feet, and appears to be decreasing again at deeper levels.

There are samples of material precipitated at depth in Steamboat Springs that have elevated base-metal content. Sample 7, Table 9.2, is from a

Table 9.2.—Geochemistry of surficial hot-spring discharge material.

Sample No.	Field No. or description	Area	Rock Type	Au	Ag	As	Sb	Hg	Tl	B	Ba	Be	Bi	Mn	Cu	Pb	Zn	Co	Se	Te	Ref
1	1	BROADLANDS New Zealand	Sb-rich sulfidic-silica ppt, Ohaki Pool	85	500	400	10%	2000	630	NA	NA	NA	NA	NA	NA	25	70	NA	NA	NA	1
2	2		do	56	245	1300	12.8%	ND	2290	NA	NA	ND	2.5	ND	ND	ND	100	ND	221	<20	1
3	7A		Sinter discharge from well	0.45	13	83	159	ND	3.3	ND	ND	ND	2.7	ND	20	20	20	ND	0.5	0.09	2
4	Orange ppt	WAIOTAPU New Zealand	Orange ppt Champagne Pool	80	175	2%	2%	170	320	NA	NA	NA	NA	NA	NA	15	50	NA	NA	NA	3
5	Orange ppt		do	41	26	1.4%	2.6%	NA	520	NA	NA	NA	2.0	NA	ND	ND	100	<25	35	<20	2
6	Sinter		Sinter (Bulk) Champagne Pool	8.4	5.0	NA	NA	NA	NA	NA	NA	NA	NA	NA	NA	NA	NA	NA	NA	NA	3
7	W941C	STEAMBOAT SPRINGS, Nevada	Meta-stibnite & opal discharge from 200-m well	60	400	600	2000	80	2000	2000	NA	NA	NA	NA	2000	400	2000	NA	NA	NA	4
8	W50		Siliceous mud, deep levels	15	50	700	1.5%	100	700	500	NA	NA	NA	NA	20	7	50	NA	NA	NA	4
9	W310D		Sinter with stibnite Surface spring	1.5	1	50	1.0%	30	70	1000	NA	NA	NA	NA	1	NA	0.2	NA	NA	NA	4
10	MTM		Sulfide mud, surface spring	0.1	1.5	35	340	6.5	54	>2000	300	5	N(10)	100	5	N(10)	10	N(5)	NA	NA	5
11	Sinter Hill		Chalcedonic sinter, old terrace	0.1	N(0.5)	N(5)	500	>13	N(.2)	200	300	3	N(10)	50	L(5)	L(10)	L(5)	N(5)	NA	NA	5

[NA = Not analyzed; ND = Not detected; N(5) = Not detected at limit indicated in parentheses; < = Less than amount indicated; > = Greater than amount indicated; values in ppm unless otherwise indicated.]

[1]Weissberg, 1969; [2]Ewers and Keys, 1977; [3]Hedenquist, 1984; [4]White, 1981; [5]this report, Analysts: John Gray, James Domenico.

discharge blast from a 650-ft (200-m) well, consisting of opal and meta-stibnite and has quite significant Cu, Pb, and Zn content. Sample 8, which has moderate amounts of base metals, is believed to be a precipitate from deep levels carried physically to the surface during maximum discharge. Both of these samples are, in fact, also very high in Au, Ag, Sb, As, and Hg--elements normally found near the surface. These two samples from deep within the system are the only material approaching precious-metal ore grade at Steamboat Springs. Unfortunately, the geometry and size of any possible zone of mineralization that they come from cannot be determined from the available data. However, if a deep zone of mineralization exists at Steamboat Springs, it does not appear to conform to the trace-metal zoning indicated by the unmineralized drill core GS-5, but appears to be strongly enriched in both base metals and precious metals and elements considered characteristic of deposits of the latter.

EPITHERMAL ORE DEPOSITS

Morphology and Characteristics

Epithermal ore deposits occur in a wide variety of types and geologic settings within the broad, global framework mentioned earlier. Classification schemes have been based upon plate tectonic concepts (Giles and Nelson, 1982; Sillitoe, 1981; Bonham and Giles, 1981; Sawkins, 1984), association with volcanic landforms (Henley and Ellis, 1983; Sillitoe and Bonham, 1984), physical and mineralogical characteristics (Lindgren, 1933; Giles and Nelson, 1983; Heald-Wetlaufer et al., 1983; Hayba et al., 1985, this volume), and associated magma types and mineralogy (Bonham, 1986). Berger and Eimon (1983) proposed conceptual models of volcanic-hosted epithermal systems based on a combination of physical, mineralogical, and hydrological characteristics. Some students of epithermal deposits fit all varieties of them into a generic model, flexible enough to fit all the variations (Buchanan, 1981; Silberman, 1982). We discuss the morphology of epithermal systems using a hybrid scheme of classification based largely on physical and mineralogical (observational) characteristics, following Berger and Eimon (1983), but modified from recent work of Giles and Nelson (1982); Bethke (1984); Heald-Wetlaufer et al. (1983); Bonham (1986); and Ashley and Berger (1985, in press).

Epithermal ore deposits may occur in a continuum of types ranging from shallow quartz-pyrite stockworks and breccias--the hot-spring environment--to relatively deep veins and fissures, the bonanza environment. It is very important to emphasize the possible continuum of features and characteristics because models proposed for these deposits usually represent end-member cases, and neglect to portray the very important intermediate cases. Table 9.3, modified from the discussion of Giles and Nelson (1982), Bonham (1986), and Ashley and Berger (1985, in press), presents the important characteristics of the deposit types.

In addition to sources of heat and fluids (as in geothermal systems), the precious-metal epithermal deposits need a source of metals, a mechanism of metal transportation to a place of deposition, a mechanism of precipitation, and a sufficient time for the transportation and deposition mechanisms to operate so that economic concentrations of the metals can occur. This question of timing is a critical one, and will be discussed in a later section following the brief description of a few epithermal systems and their geochemical and alteration variations. The nature of the metal sources and mechanisms of transportation and deposition are covered elsewhere in this volume (Henley, 1985; Henley and Brown, 1985; see also Henley and Ellis, 1983; Browne and Ellis, 1970).

Deposit models for these ore systems have been proposed by many authors (cf. Buchanan, 1981; Henley and Ellis, 1983; Hayba et al., 1985, this volume; Bonham, 1986). The early representations tended to be quite general, and were designed to fit all of the observed patterns of mineralization and alteration and metal associations in one context. Lately (Berger and Eimon, 1983; Giles and Nelson, 1982; Bonham, 1986; Ashley and Berger, 1985, in press), the diversity of types of epithermal deposits has been emphasized, and most authors of the recent papers propose multiple models for the general epithermal class of deposits. Schematic drawings of the deposit-type models are generally presented as cross sections through the idealized systems, and show more or less detail on the patterns of alteration and various physical features observed in the deposits. Most of the authors accept the geothermal-epithermal analogy, and connect the various deposits to surficial features of geothermal systems at the top of the cross section. All of these cross sections suffer from their two-dimensional orientation, and do not adequately portray the three dimensional characteristics, and variations that are present in the actual systems (see, for contrast, Vikre, 1985).

Deposit-type models are useful for focusing attention on changes of mineral assemblages and physical characteristics of the deposits with depth, and form a framework within which to discuss geochemical variations. It must be kept in mind, however, that in individual deposits, as in the geothermal systems, genetic processes take place in an extremely complex manner, in which the effects of vertical and lateral variations in temperature due to such effects as positions of the water table, boiling zones, and channels of fluid movement occur. The resultant mineral alteration and metal deposition patterns are made even more complex by changes in these effects with time.

Figures 9.4, 9.5, and 9.6 present rather generalized models from Berger and Eimon (1983) and Ashley and Berger (1985, in press) for a variety of bonanza and hot-spring-type volcanic-hosted precious-metal deposits. Disseminated replacement or "Carlin"-type systems are discussed by Bagby and Berger (1985, this volume).

Alteration Patterns

Study of the alteration mineralogy and zoning in epithermal deposits is progressing rapidly, but the amount of detailed information is not yet as available

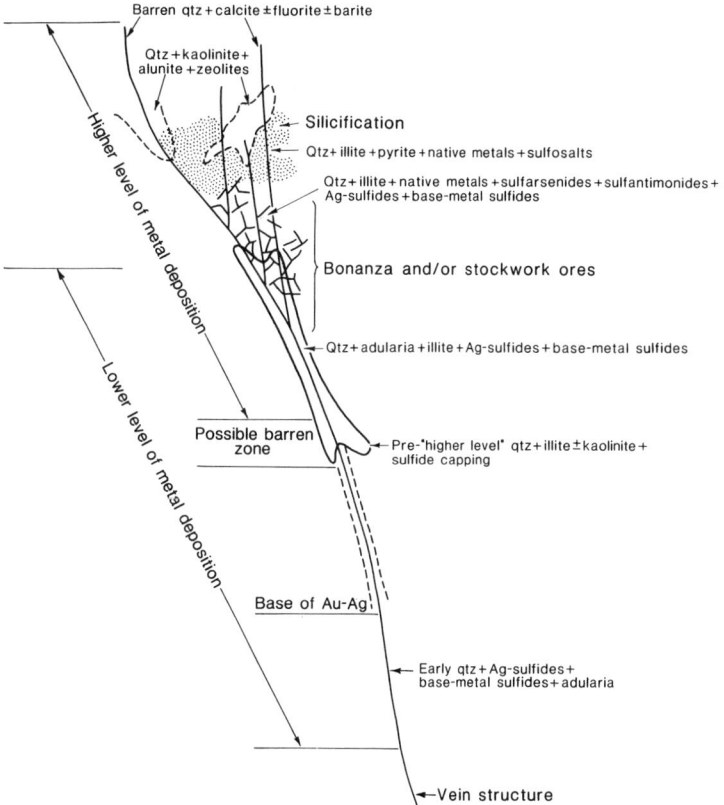

Figure 9.4. Schematic cross section of quartz–adularia or low sulfur bonanza deposit, Bonanza-IA model, showing alteration mineralogy and two zones of mineralization from the "closed cell convection" model of Berger and Eimon (1983).

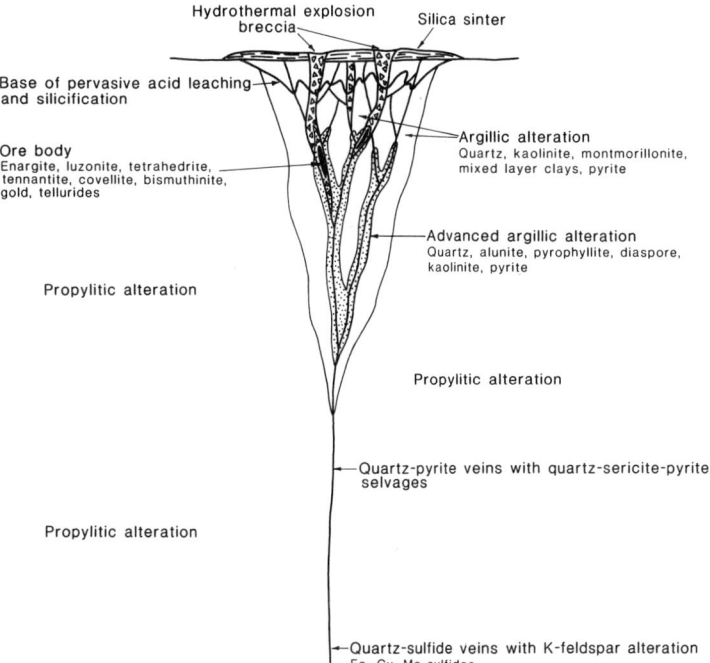

Figure 9.5. Schematic cross section of the quartz–alunite, or high sulfur bonanza deposit, Bonanza-IB model, showing the alteration mineralogy and possible location of ore bodies, and other features from Ashley and Berger (1985, in press).

Table 9.3--Characteristics of epithermal precious-metal systems

Deposit Type	Characteristics	Examples with grade, tonnage, and/or production[1]
Bonanza Vein and Lode	A) **Quartz-adularia or low sulfur type (Fig. 9.4)** Vein and hanging wall stockworks of quartz, calcite, and adularia. Pyrite is major sulfide, with variable amounts of silver sulfides, sulfosalts, native Au, and electrum. Host rocks are usually andesite dacite, rhyolite. Associated elements are As, Sb, variable Hg, variable Cu, Pb, Zn, Mn. Alteration patterns usually silicification, with K-spar proximal, to argillic then propylitic distal, but geometry varies.	A1) **Comstock Lode, NV** (Ag-rich quartz vein and hanging wall stockwork) 8.3×10^6 oz Au 200×10^6 oz Ag from 19 million tons A2) **Pachuca, Mexico** (quartz-vein system, base-metal rich) 6.2×10^6 oz Au 1500×10^6 oz Ag from 100 million tons 3.5% Cu, Pb, Zn A3) **Creede, CO** (Ag-bearing base-metal vein system in southern part of district) 0.15×10^6 oz Au 85.2×10^6 oz Ag with approximately 7.5% Pb, Zn, Cu from 5 million tons
	B) **Quartz-alunite or high sulfur type (Fig. 9.5)** A variant of this type, the Goldfield or enargite-pyrite gold deposits, occur as tabular lodes or pipes of silicified breccia and some veins. Ore is native Au or electrum associated with Cu sulfosalts, and sulfides, pyrite, and complex sulfosalts and tellurides. Host rocks are usually andesite and dacite. Associated elements are Cu, As, Ag, Ba, Bi, Zn, Pb, Te, variable Hg. Alteration is advanced argillic, with alunite in lodes, zoning outwards to argillic halos, then distal propylitization. Both systems structurally focused by major faults, locally controlled by intersections and inflections.	B1) **Goldfield, NV** (quartz-alunite breccia reefs) 4.2×10^6 oz Au 1.7×10^6 oz Ag from 5 1/2 million tons B2) **Summitville, CO** (quartz-alunite pods) 0.24×10^6 oz Au 0.34×10^6 oz Ag 0.427×10^6 lbs Cu

[1]Buchanan, 1981; P. Bethke, USGS, personal communication, 1985, cf. Hayba et al., 1985, this volume.

in the literature as for the porphyry copper and molybdenum deposits. The general patterns of alteration around bonanza- and hot-spring-type systems are shown in the models (Figs. 9.4, 9.5, and 9.6). An outer propylitic zone is nearly ubiquitous, but the nature and mineralogy of the argillic-phyllic zones is highly variable from system to system and depends on temperatures and fluid compositions (particularly pH, and a_{K^+}). In order to document some of the variations that occur, Table 9.4 describes the physical nature of several deposits, along with the mineralogy of the alteration zones, from proximal to the veins or ore zones to distal.

The variability in the alteration zoning in the bonanza and hot-spring systems can result in difficulty when trying to use alteration to determine depth or

Table 9.3--Characterisitcs of epithermal precious-metal systems--(continued)

Deposit Type	Characteristics	Examples with grade, tonnage, and/or production[1]
Hot Springs or silicified quartz stockworks	A) <u>Quartz-adularia or low sulfur type (Fig. 9.6)</u> Silicified host rock, breccias, and stockworks of quartz-sulfide mineralization. Native gold and electrum of micron size associated with pyrite, marcasite, and silver sulfosalts. Sulfide content varies. Host rocks are andesite, dacite, or rhyolite. Host rocks usually affected by silicification with adularia and/or albite, variable calcite, or dolomite. This grades outwards to argillic and/or zeolitic alteration, then to propylitic assemblages. Mineralization due to repeated, episodic, and explosive stockwork veining and brecciation (hydrothermal). Associated elements: Hg, Tl, As, Sb, Ba, (W). Cu, Pb, and Zn usually deep--all highly variable. Sinter frequently present, and may be mineralized. B) <u>Quartz-alunite or high sulfur type</u> These deposits are similar to the A) type except that alteration assemblages are advanced argillic (alunite-bearing) central, zoning out to argillic peripheral zones, and Cu sulfosalts are present in the ore-sulfide assemblages.	A1) <u>Round Mountain, NV</u>[2] 195×10^6 tons at grade of 0.043 opt, Au (reported production to 1982-- 0.85×10^6 oz Au) A2) <u>Borealis, NV</u>[3] 2.3×10^6 tons at grades of 0.10 opt, Au; 0.5 opt, Ag A3) <u>McLaughlin, CA</u>[3] (Hot-spring system, hosted in ophiolitic rocks of Mesozoic age) 22×10^6 tons at a grade of 0.16 opt, Au B1) <u>Paradise Peak NV</u>[3] 10×10^6 tons at a grade of: 0.35 opt, Au; 4.7 opt, Ag
Disseminated replacement	A) Tabular, fault-related stratabound orebodies with micron-sized native Au on pyrite and/or organic carbon particles. Host rocks are thin-bedded calcareous clastic sedimentary rocks--limestones, dolomites, or limy siltstones. Variable silicification and carbon remobilization. Variable amounts of brecciation and veining. Jasperoid development along faults which control fluid access and/or along bedding as replacements, above, below, or within ore. Associated elements: Hg, Tl, As, Sb, Ba (W).	A1) <u>Carlin, NV</u>[4] 11×10^6 tons of 0.29 opt, Au (produced) 6.2×10^6 tons of 0.20 opt Au (reserves) A2) <u>Pinson, NV</u>[4] 4.2×10^6 tons of 0.186 opt, Au (produced) 3.3×10^6 tons of 0.12 opt Au (reserves) A3) <u>Windfall, NV</u>[4] 3×10^6 tons of 0.035 opt Au

[2] Tingley and Berger, 1985.
[3] C. Nelson, consultant, written communication, 1985.
[4] B. Miller, consultant, written communication, 1982; Bagby and Berger, 1985, this volume

proximity to zones of mineralization. Indeed the argillic or phyllic alteration zones are very complex. Changes in geometry and (or) positioning of heat sources with time, sealing of permeable zones by deposition of silica, calcite, etc. will cause overlapping of alteration assemblages. In general, upper level assemblages have illite-montmorillonite, with quartz, which gives way at greater depth to illite-K mica-quartz assemblages. Adularia is usually present in the low-sulfur systems in these zones and is a very common constituent of the silicified zones in the low-sulfur hot-spring systems, where it also tends to occur in the stockwork veins. Solfataric hypogene alteration and supergene alteration can overprint earlier patterns

Table 9.4--Alteration mineral zoning in epithermal precious-metal deposits

Deposit Type	Character of deposit	Alteration mineral zoning
Comstock[1] Lode, NV (Bonanza vein, IA)	Quartz, quartz-pyrite quartz-adularia stockworks in hanging wall of major fault. Minor sulfides (pyrite), sulfosalts, and electrum. Host rock is andesite.	Sericite-quartz adjacent to stockworks, zoning to chlorite-illite albite ± adularia, accompanied occasionally by strong silicification. Outer zone is propylitic, albite-epidote-chlorite-calcite-pyrite-zeolites. Near-surface, proximal advanced argillic assemblages (quartz-alunite) zone outward to kaolinite-pyrophyllite-diaspore-quartz, sericite-quartz, illite-quartz, montmorillonite-quartz, and propylitic.
Aurora, NV[2] (Bonanza vein, IA)	Steeply dipping, massive, fine-grained, multiple-banded quartz-adularia veins. Barren, coarse-grained quartz cuts early fine-grained productive quartz with replaced lamellar calcite. Electrum with pyrite ore minerals. Veins commonly bounded by stockwork zone in brecciated andesite host rock. Veins are along faults.	Deeper levels--Adjacent to veins quartz-albite-adularia-illite grades outwards to quartz-albite-adularia-chlorite±calcite±pyrite. Shallower levels--Wide zone in hanging wall of quartz-illite-montmorillonite± adularia±kaolinite. Footwall assemblage grades from quartz-illite±adularia to quartz-illite-adularia-montmorillonite± chlorite to an outer envelope of albite-quartz-illite-chlorite-montmorillonite-calcite. Propylitic assemblage developed regionally.
Oatman, AZ[3] (Bonanza vein, IA)	Large, tabular, multi-stage quartz-calcite-adularia veins, strongly banded. Very low sulfide content, ore is submicroscopic electrum. Host rocks are andesites and rhyolites.	Silicification immediately above ore shoots-quartz flooding and quartz-calcite stockworking. Illite, minor montmorillonite in hanging wall, less wide in footwall. This zones out to propylitic alteration, consisting of chlorite-pyrite assemblages. Illitic zone specifically characteristic of ore shoots.
Gold-field, NV[4] (Bonanza vein, IB)	Quartz-alunite reefs, consisting of brecciated and fractured, altered dacite, along faults. Ore is Au with sulfides and complex sulfosalts. Occurs generally interstitial to breccia fragments and as coatings. Mineralization occurs only in the "silicified" zones.	Silicified zones consist of quartz-alunite-kaolinite in varying proportions. Pyrophyllite locally present and increases in proportion with depth. This assemblage grades to quartz-kaolinite-illite, then to mont.-kaolinte-quartz, then to a pervasive zone of propylitic alteration, developed regionally. Limonite present in all zones, after pyrite.
Round Mt., NV (Hot-Spring, IIA)	Steeply dipping quartz and quartz-adularia veins; lamellar quartz after calcite. Low-angle quartz and quartz-breccia veins; silicified breccias. Native gold and electrum with pyrite; realgar. Disseminated ore in permeable beds. Host rock is rhyolite ash-flow tuff.	Upper zone of pervasive silicification overlying zone of quartz-illite-chlorite alteration. Veins of adularia crosscut both of above but are most abundant in latter. Deeper zone of quartz-chlorite. Lateral zone of quartz-chlorite-white mica-calicte alteration. Blocks of sinter in central breccia pipe indicate previous presence of sinter overlying system.

Table 9.4--Alteration mineral zoning in epithermal precious-metal deposits--(continued)

Deposit Type	Character of deposit	Alteration mineral zoning
Hasbrouck Mt., NV[5] (Hot-Spring, IIA)	Silicified and adularized breccias and steeply dipping quartz and quartz-adularia veins. Native gold and pyrite, acanthithe in silver-rich veins.	Chalcedonic sinter interbedded with silicified host rocks. Central funnel-shaped zone of quartz-adularia alteration grading outward and downward into quartz-adularia-illite. Deepest alteration is quartz-adularia-albite-illite. Lateral to quartz-adularia-illite is quartz-illite and then quartz-illite-montmorillonite.

[1] Hudson, 1984.
[2] Stone and Osborne, 1984; Osborne and Stone, 1985.
[3] Durning and Buchanan, 1984.
[4] Ashley, 1974.
[5] Graney, 1985

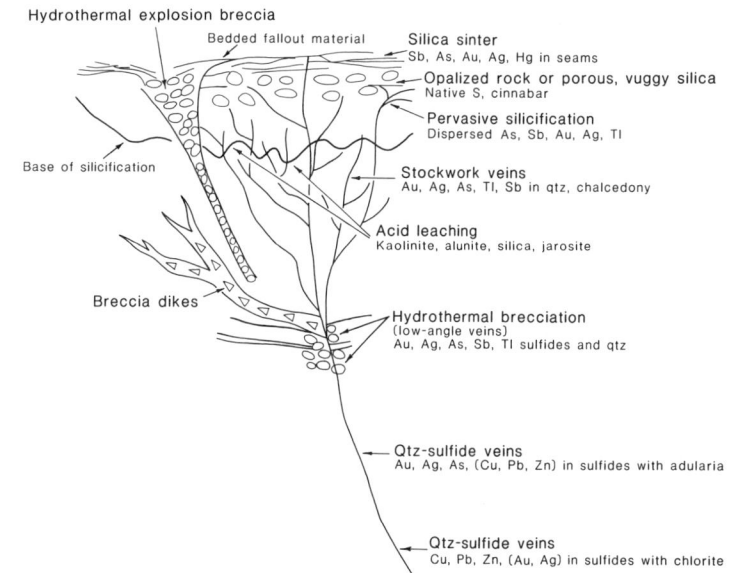

Figure 9.6. Schematic cross section of the quartz-adularia, or low-sulfur hot-springs deposit, or Hot Springs-IIA model showing alteration mineralogy, generalized geochemical associations, and other structural and mineralogical features, from Berger and Eimon (1983). The high-sulfur analog of this type of system has central advanced argillic ± alunite assemblages, surrounded by argillic and propylitic halos, and is termed Hot-spring-IIB model in text.

of alteration, particularly in the upper levels of the systems. One consequence of these complications is that they seriously affect geochemical patterns as the acid-leaching tends to remove many of the pathfinder elements to ore and disperses them laterally, or down to deeper levels, depending on the local hydrology.

The Bonanza-IA model (Fig. 9.4) attempts to depict a two-stage development for zones of mineralization. The competing effects of vertical and lateral changes in temperature, due to the variations mentioned above, and to migration of main-fluid movement channels with time as old channels seal and new fractures open, due to tectonic or even solution processes, can cause additional complications in the development of actual alteration assemblages and their distribution in all of the deposit types.

NATURE OF FLUIDS INVOLVED IN GEOTHERMAL SYSTEMS AND EPITHERMAL ORE DEPOSITS

Stable-isotopic studies, particularly isotopic ratios of oxygen and hydrogen in geothermal systems and epithermal ore deposits have shown that the hydrothermal fluids in both environments are predominantly meteoric water in origin (e.g., Taylor,

1973, 1974; White, 1974; Bethke and Rye, 1979; O'Neil and Silberman, 1974). Fluid-inclusion studies in these ore deposits suggest a temperature range of from 150° to 300°C for deposition. The fluids in most of the systems are dilute, with $NaCl_{eq}$ of 0.5 to 5 wt. percent. The high-sulfur or quartz-alunite systems tend to have somewhat higher salinities, up to about 10 to 12 wt. percent, and temperatures near the upper end of the scale (Hayba, 1983; Bruha and Noble, 1983).

The chemistry of these ore-forming solutions and the nature of the physico-chemical conditions of metal transport and deposition in both epithermal ore deposits and geothermal systems is treated at length elsewhere in this volume (Henley, 1985, this volume; Henley and Brown, 1985, this volume). Hayba et al. (1985, this volume) describe the nature of the geochemical environment associated with the Creede and Summitville epithermal systems in Colorado.

TIMING

Mention has been made of variations of temperature, position of zones of fluid movement, and level of water table and zones of boiling with time. Time is an extremely important factor in the development of epithermal ore deposits, and evidence for the duration of hydrothermal processes and events is contradictory. Geochronological data, largely K-Ar ages on volcanic sequences hosting mineralization and on the products of hydrothermal alteration of wall rocks, and hydrothermal deposition of gangue minerals in veins, breccias, et cetera, summarized by Silberman (1983) and Noble and Silberman (1984) suggest that hydrothermal activity in epithermal ore deposits, geothermal systems, and porphyry copper deposits lasts between about one-half to greater than two and one-half million years with an average on the order of 1.25 m.y. In addition, Silberman (1983) suggested that most epithermal deposits can be related in time to either a stage of local volcanic activity or to a regional pattern of distribution of volcanic rocks, which can be interpreted within the time framework of plate tectonic evolution.

On the other hand, from solution composition theory, Henley (1985, this volume) estimates that quite significant amounts of precious metals can be deposited from even dilute fluids (such as those in the low-sulfur epithermal systems) on time scales of the order of 10^4 years, although this is stated as a minimum figure. These time estimates appear to be in conflict with the evidence from the geochronological studies and from detailed studies at Creede, Colorado, by Barton et al. (1977). However, the geochronological studies completed to date treat a mining district as a whole, and do not focus on putting time limits on the duration of any particular stage or episode of alteration or metal deposition. There is evidence from the physical features of epithermal deposits that mineralizing events are short-lived, but appear to be repetitive. Silberman (1983) cited several lines of evidence for the occurrence of episodic mineralizing events in bonanza vein and hot-springs systems:
1. Stages of brecciation and stockwork veining can be shown to be multiple from cross-cutting relationships, and not all stages are associated with ore deposition.
2. Epithermal quartz veins are frequently repetatively banded, and only some bands carry sulfides and precious metals.
3. In many deposits, alteration assemblages overlap or succeed each other in relatively restricted areas, and only some of those alteration episodes appear to be related to episodes of ore deposition.

Many epithermal ore deposits are spatially associated with small stocks on the order of 2 km ± radius, where the heat of this crystallizing magma is inferred to drive the convecting hydrothermal system (White, 1974, 1981; Norton and Cathles, 1979). Based on heat loss modeling studies, Norton and Cathles (1979) indicate that hydrothermal cells driven by this size of heat source will decay well within about a 100,000-year time span. The geochronological studies at a number of epithermal vein deposits and hot-spring deposits suggest life spans of one-half to one and one-half m.y. (Silberman, 1983; Noble and Silberman, 1984). Given the aforementioned geochronological evidence, additional sources of heat must be supplied to the systems beyond a single small stock or dome to keep the systems active. We suggest that the magmatic systems responsible for driving the hydrothermal cells must be much larger than those surface-exposed stocks of 2 km ± radius, and that the activity must be pulsed and renewed at intervals. Evidence for this pulsing comes from the geochronology of Julcani, Peru (Noble and Silberman, 1984), where eight stages of volcanic and interspersed hydrothermal alteration-mineralization episodes occurred over a total duration of 700,000 years, with individual stages taking between 100,000 and 200,000 years. We conclude that mineralizing episodes may, in fact, be short-lived, but occur within a much longer framework of volcanic evolution and hydrothermal activity.

GEOCHEMICAL ZONING IN EPITHERMAL DEPOSITS

Any epithermal ore deposit viewed at the present resembles a still photograph in an instant of geologic time that catches a very dynamic, evolving, three-dimensional interplay of heat sources, fluid flow, chemical reactions, and structural changes which have combined to produce an economic concentration of metals that will be further modified with additional time by post-hypogene processes. Measurements we make of the geochemistry of an epithermal ore deposit reflect this snapshot view, and the observed complexities of the element distribution patterns caution against the interpretation of the data in a "cook book" fashion. We feel that it is unlikely that a single, all-inclusive element distribution model will be derived that will allow a quick and accurate determination that an economically viable mineral deposit lies within any geochemical or favorable lithological anomaly. At best, we will probably define areas of relative favorability for further investigation.

Geochemical zoning in epithermal systems can be studied on a variety of scales, including regional, district, and orebody or ore-shoot dimensions. Ideally, patterns of geochemical zoning, once they are established, should be usable guides to location of additional orebodies. Most models of epithermal deposits suggest generalized geochemical zoning patterns (Figs. 9.4, 9.5, and 9.6) that presumably may be applied to all epithermal ore deposits. However, we believe that these zoning patterns are specific to individual mineralizing systems, and that a unique, widely applicable model of element zoning to delineate ore in all districts will probably not be achieved. We have found that there is considerable variation in the pathfinder elements associated with precious-metal mineralization, particularly in their lateral and vertical changes in concentration. For example, as shown in the following section, the generalization of decreasing Hg concentration in altered rocks and veins with depth does not always hold.

In any particular deposit or district, the geochemical zoning patterns do tend to have regularity, and once established, can serve as guides to additional reserves or sites of mineralization in that district. That is, they have application to prospecting only after their relationships to known mineralization are established. We will illustrate zoning patterns at a variety of district and orebody scales from actual examples of deposit geochemical studies, in the following section of this report, and in Berger and Silberman (1985, this volume) to follow.

BODIE MINING DISTRICT

The Bodie mining district, Mono County, California, produced approximately 1.5 million ounces of Au from banded quartz-adularia veins that cut Miocene andesites, dacites, and pyroclastic rocks in one of several Tertiary volcanic centers in the region (Kleinhampl et al., 1975). The mining district is an eruptive center whose major structure is an irregular, faulted, north-trending anticline formed by intrusion and doming of the flows and pyroclastic rocks by small dacite intrusions (Fig. 9.7). The intrusive rocks occupy vents from which the volcanic rocks were erupted. Several sets of steeply dipping faults cut all of the rocks exposed, including the intrusions. One prominent set strikes N to NE, and another is normal to this. The major veins and fractures at Bodie also strike N to NE within one of the fault sets. Most of the production came from the northern part of the district, in the vicinity of a small graben filled with tuff breccia that was faulted down into the intrusive dacite of Bodie Bluff during or shortly after its emplacement (Fig. 9.7).

The productive quartz veins cut both the extrusive and intrusive rocks. The veins vary in thickness from about 1 to 90 feet, although most are not more than a few feet thick. The ore minerals are principally native gold and silver. Argentite, pyrite, and sphalerite are present and increase in abundance with depth as the tenor of the gold decreases (Chesterman et al., in press). In the main productive zone, old records quoted by Chesterman et al. (in press) indicate that the ore averaged 1.7 ounces per ton Au, and 3.1 ounces per ton Ag. Adularia is a common constituent of the mineralized quartz veins, with K contents in the range of approximately 3 to 9 percent. The adularia sometimes forms euhedral crystals up to 3 cm long along open fractures (Silberman et al., 1972; Silberman, 1983; O'Neil et al., 1973).

In the southern part of the district, and reportedly at deep levels in the main "bonanza" zone in the southern part of the Bodie Bluff area, a vein system rich in base metals and silver is reported (Chesterman et al., in press). Mineralogy of this vein system is more complex than the simpler quartz-adularia veins to the north and south and consisted of an assemblage of sulfides and sulfosalts. Dump samples of this material assay up to 2 percent Cu and Pb, 1.4 percent Zn, and 220 ppm Ag (M. L. Silberman, unpublished data, 1980).

Wall-rock alteration assemblages in the northern part of the Bodie Bluff area are being determined by petrographic and x-ray diffraction analyses by Peter A. Herrera of The Colorado School of Mines. Figure 9.8 is a schematic cross section along line XX' of Figure 9.7, summarizing preliminary results of these determinations from surface outcrops, underground samples, and drill core supplied by the Homestake Mining Co. An upper zone of quartz-adularia alteration of the intrusive and extrusive rocks gives way at greater depth to an assemblage of adularia-illite-quartz. This deeper alteration assemblage is characteristic of wall-rock mineralogy in the main bonanza production zone. At even greater depths, and peripheral to the zone of economic mineralization, propylitic assemblages occur. The graben at Bodie Bluff contains silicified fall-out and fall-back breccias of hydrothermal origin, which contain a variety of clasts of intrusive and extrusive rock, and siliceous sinter. Argillic alteration occurs beneath the silicified breccias (P. A. Herrera, written communication, 1985). Hematite-quartz-adularia matrix hydrothermal breccias with K-silicate altered clasts cut the quartz-adularia altered rocks at the top of Bodie Bluff. The breccias and their host rocks are themselves cut by chalcedonic, banded quartz veins, which are mineralized (Silberman, 1982, 1984; P. A. Herrera, written communication, 1985).

Silberman (1982, 1984) suggested that the Bodie mining district is a mineralized geothermal center with most of its upper near-surface features preserved. The alteration patterns are quite similar to those of the Hot Springs A model, and Herrera's studies, including the important documentation of the presence of sinter in explosion breccias at Bodie Bluff, appear to confirm this suggestion. In all of the previous studies at Bodie, the sinter was not identified. Bodie has features characteristic of both Hot Springs A and Bonanza A deposits. It was mined largely for bonanza-type lode quartz veins, but there are numerous sets of these, and large, structurally focused vein systems such as those at Oatman or the Comstock Lode are not present. A broad, blanket-like zone of quartz-adularia alteration, accompanied by hydrothermal breccias and mineralized stockwork veining are characteristics frequently found in the hot-spring environment.

Detailed stable-isotopic analyses of the mineralized quartz veins and altered wall rocks from

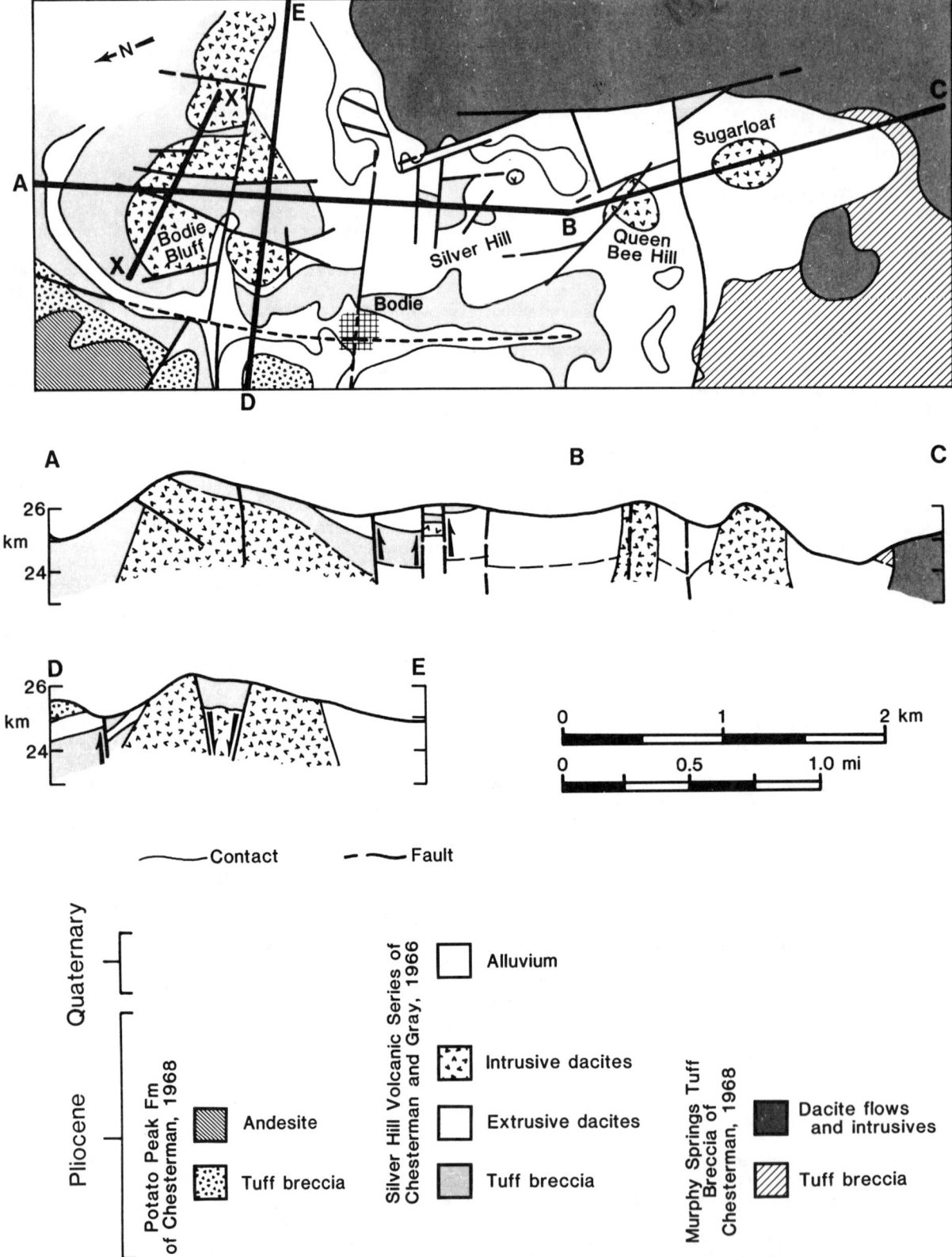

Figure 9.7. Generalized geology and cross sections of the Bodie mining district, Mono County, California (modified from Silberman et al., 1972).

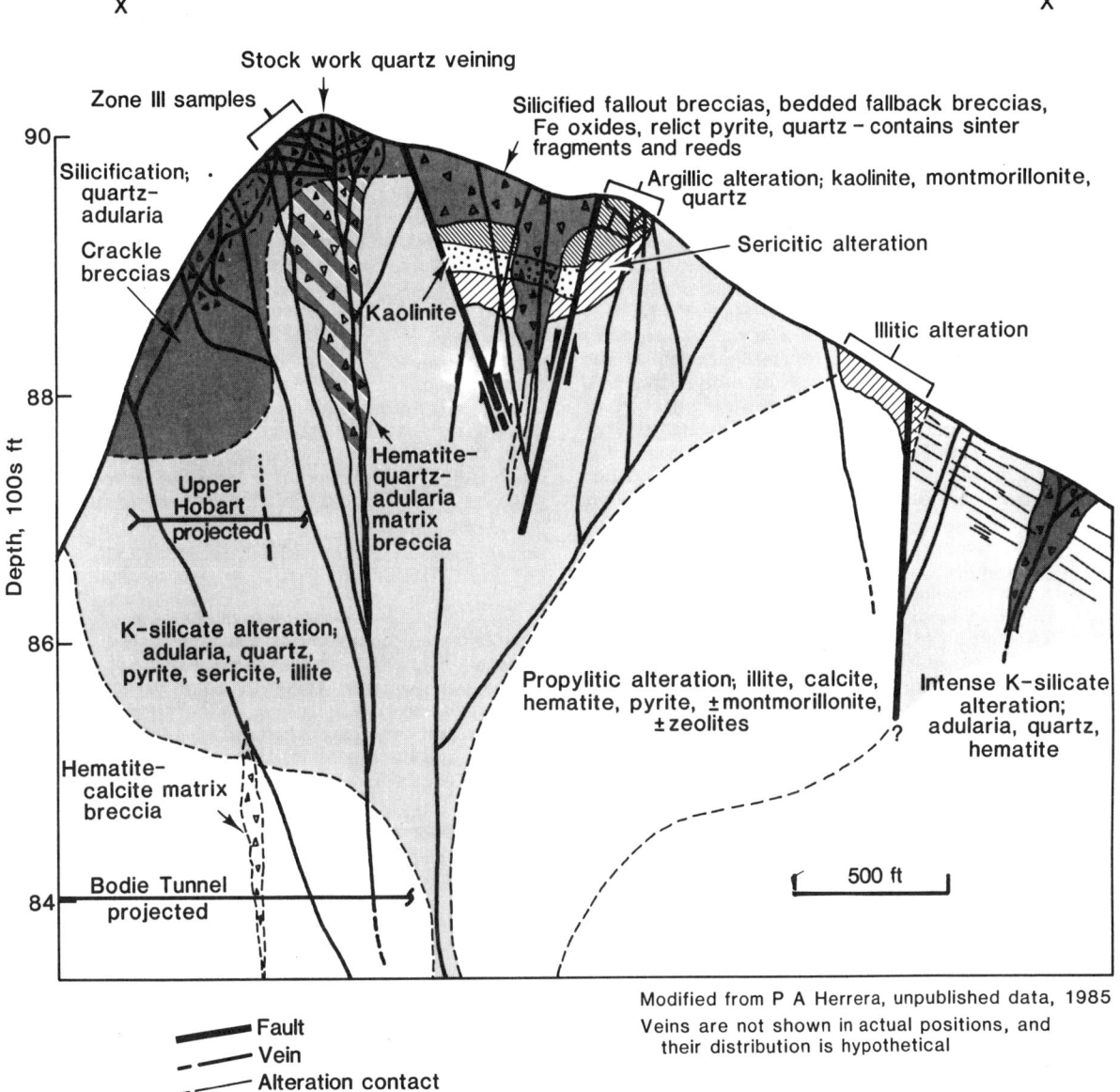

Figure 9.8. Schematic alteration cross section of the northern part of Bodie Bluff (section XX') showing alteration mineralogy and relationships between important structural and hydrothermal features. Samples from Zone III, the Upper Hobart Tunnel, and the Bodie Tunnel form the basis for the geochemical discussion in the text (modified from P. A. Herrera, USGS, unpublished data, 1985).

the K-silicate and propylitic zones and fluid-inclusion temperature and salinity determinations (O'Neil et al., 1973; Nash, 1972) indicate that mineralization and alteration were produced by fluids of meteoric origin that were heated to temperatures of between 215° and 245°C, with salinities of less than 0.5 %-equivalent NaCl. These ore fluids are isotopically and chemically similar to present-day geothermal fluids in the Bodie Hills area (O'Neil et al., 1973).

O'Neil et al. (1973) discussed major and selected minor element chemical relations of alteration in the Bodie Bluff area; however, they did not attempt to delineate vertical or lateral zoning patterns. They documented that K-silicate alteration, which accompanied emplacement of the ore-zone veins, resulted in a net gain in K_2O, a loss in Na_2O, CaO, and MgO, and little change in SiO_2 and Al_2O_3. Sr was depleted and Rb gained during the process. The degree of alteration appeared to O'Neil et al. (1973) to correlate with depletion of ^{18}O, and was explained as progressive replacement and recrystallization of the host rocks by K-feldspar (adularia) of similar chemical and isotopic composition as that of the veins. P. A. Herrera is currently completing a detailed trace-element and major-element chemical study of the Bodie Bluff area that will attempt to relate alteration assemblages and elemental chemical variation in detail.

Trace-element analyses at Bodie were carried out by O'Neil et al. (1973) during the isotopic and major-element variation sampling. The results of some of these analyses were briefly discussed by Silberman (1982, 1984). Data from three sets of samples collected over a vertical range of approximately 600 feet are summarized in Figures 9.9 and 9.10. Quartz veins and altered wall rock were sampled from the quartz-adularia alteration zone at the top of Bodie Bluff (zone III of Fig. 9.8) at elevations of 8,900 to 9,000 feet; in the K-silicate (adularia-illite-quartz) zone of the Upper Hobart Tunnel at an elevation of 8,740 feet; and, in the propylitic alteration zone of the Bodie Tunnel at an elevation of 8,400 feet, as structurally deep within the district as is currently accessible. The Upper Hobart Tunnel is in the productive zone, and one vein (approximately 2 feet wide) in the presently accessible part of the workings was stoped. The Bodie Tunnel was largely a haulage adit, but a moderately large vein (approximately 1 foot wide) was followed with a drift and partly stoped. Quartz veins in the Upper Hobart Tunnel are, with the exception of the one stoped, on the order of a few inches in width. The veins in the Bodie Tunnel are, in general, the same width to even thinner. Quartz veins in zone III are normally on the order of one to three inches in width (occasionally wider) and are vuggy, porcellaneous, and micro-crystalline to chalcedonic in texture. The deeper veins are fine grained, vuggy, and sheeted. All veins have considerable limonite along fractures and vugs.

The data for selected trace elements in vein samples and wall rock from zone III and the Upper Hobart and Bodie tunnels are plotted as bar graphs on a logarithmic scale (Figs. 9.9 and 9.10). For the Upper Hobart and Bodie tunnels, the plots are organized from the center (nearest the center of Bodie Bluff) to the periphery of the mineralized zone, from top to bottom of the figure. The adits are projected onto the line of section, as both are slightly north of it.

Figures 9.11 and 9.12 show trace-element contents of samples collected at selected intervals laterally away from stoped, mineralized quartz veins in the Upper Hobart Tunnel and Bodie Tunnel. The plots are organized with the elemental concentration of the vein in the center and that of the surrounding wall rock on either side. The zoning for Bodie Tunnel (BT) and Upper Hobart (UH) are plotted together in Figure 9.12 so that differences in elemental content as a function of depth (and alteration assemblage) can be compared. Figure 9.11 also shows the $\delta^{18}O$ with distance from the vein in the upper Hobart tunnel. The two sets of plots (Figs. 9.9, 9.10, 9.11, and 9.12) allow a comparison of geochemical zoning to be made on both a district and ore-shoot scale--a scale difference of, in this case, about one order of magnitude.

We show the trace-element data for descriptive purposes of delineating elemental zoning in an epithermal system and make no conclusion here about the physico-chemical causes of this zoning. The actual alteration mineral assemblage of the samples from each suite are shown on Figure 9.8, the schematic cross section of Bodie Bluff.

Large-scale Vertical Zoning at Bodie Bluff-- the Big Picture

Quartz veins--The general pattern of elemental concentration changes are well illustrated by the bar graphs (Figs. 9.9 and 9.10). Gold is present in chalcedonic quartz veins in zone III in significant quantities, usually between 1 and 10 ppm. Gold is highest in the quartz veins of the Upper Hobart Tunnel, and decreases in concentration in veins of the Bodie Tunnel level to generally less than a ppm, although the majority of veins sampled have concentrations above the level of determination. Figure 9.9 shows that the gold content of the veins is higher towards the periphery of the district than in veins towards the center of Bodie Bluff. Silver is highest in the quartz veins of zone III, where it reaches concentrations of between 10 and 100 ppm. It drops slightly in the veins of the Upper Hobart, and is lower still in the Bodie Tunnel, although a few high concentrations in samples are found there. Silver also tends to be higher in the peripheral part of the Bodie Tunnel, relative to that part of the tunnel nearer the center of Bodie Bluff. Arsenic tends to be slightly higher and more consistent in the quartz veins of the Upper Hobart relative to those of zone III, but is highest in the quartz veins of the Bodie Tunnel. Antimony, although more irregular in concentration, essentially follows the same pattern. Mercury is very variable, reaching its highest single sample concentrations in the Bodie Tunnel, but is basically much the same level of concentration in general in all three levels.

The base metals, Cu, Pb, and Zn, are at low levels of concentration in the quartz veins at Bodie, and are at or near levels of crustal average for intermediate rocks (Parker, 1967). Copper is about the same concentration in the quartz veins of zone III and the Upper Hobart, and is highest in the interior part of the Bodie Tunnel. Lead increases from zone III veins to the Upper Hobart, and again, is highest in the interior part of the Bodie Tunnel. Zinc is present in consistently measurable amounts only in the Bodie Tunnel. The pattern of Mn concentration is complex. It is most consistent in the veins of zone III, although individual samples are higher in the Upper Hobart. It reaches its highest levels of concentration in the peripheral zone of the Bodie Tunnel, whereas it tends to be lower in general in the interior part of the Bodie Tunnel than in the two shallower zones. Barium is about the same in veins of zone III and the Upper Hobart, and is highest in the Bodie Tunnel veins. Strontium is present consistently only in the veins of the Bodie Tunnel. It is generally below detectability in the Upper Hobart, and present at low levels in zone III veins.

In summary, in the quartz veins of Bodie Bluff, Au is highest at intermediate depth levels, although significant concentrations occur in the shallow level, zone III. Silver decreases with depth, and is on average lower in the productive zone veins than nearer the paleo-surface. Arsenic and Sb increase with depth in the system; Hg shows little variation in concentration. Copper, Pb, and Zn increase with depth, as do Ba and Sr. Manganese distribution is complex, and highest in the deep, peripheral zone of the Bodie Bluff system. There also appears to be a strong lateral zoning in the deeper, propylitic part of the system with Cu and Pb highest near the center of the bluff, and Au, Mn, and Ba (the latter weakly) higher in the peripheral part of this alteration zone. Other elements, such as B, Cr, and Ni were not detectable in enough samples for patterns to be established.

It is instructive to compare the variation patterns of the Bodie quartz veins to the silicious deposits of Steamboat Springs, Nevada (Fig. 9.2), including sinter and quartz veins. The Bodie patterns are different. In particular, the elements As, Sb, and Ag show a pattern of vertical zoning different from that in the Steamboat Springs geothermal system, whereas Sr, Cu, and Zn, which increase with depth at Bodie, are similar to the zoning observed at Steamboat Springs. Mercury, which is strongly concentrated near the surface at Steamboat, maintains about the same level of concentration throughout the Bodie Bluff hydrothermal system.

Altered wall rocks--The patterns of vertical variation in elemental concentration of the wall rocks at Bodie are highly irregular and trends are more difficult to identify (Fig. 9.10). The data for zone III are separated into a set of results for samples collected adjacent to quartz veins, or containing quartz stringers (upper bar graphs) and those not adjacent to quartz veins (lower bar graphs). Gold and Ag in zone III rocks are highest in samples adjacent to or containing quartz veins, whereas As and Sb are highest in rocks away from quartz veins. It is rare for Au to be greater than 1 ppm and Ag greater than 10 ppm in wall rocks anywhere in the Bodie Bluff part of the system, although a few higher concentrations do occur. Gold contents are quite irregular, and are highest in the wall rocks of zone III, in the rocks adjacent to or containing quartz veins. Gold is still present in significant levels in the Upper Hobart Tunnel, and drops off considerably at the Bodie Tunnel level, although most samples still have detectable amounts present. Thus, gold appears to irregularly decrease with depth in the system. The highest Ag values are found in wall rocks adjacent to veins in zone III and in the wall rocks of the Upper Hobart Tunnel, and are lower elsewhere. Mercury is irregularly high in both the Upper Hobart and Bodie tunnels, and lower in zone III. Cu is consistently present, although irregular, at about the intermediate composition rock crustal average in zone III and Upper Hobart Tunnel wall rocks, and lower in the Bodie Tunnel. Zinc and Mn appear highest in the Bodie Tunnel wall rocks. The other elements are just too irregular in concentration to specify variation.

Thallium was not analyzed in the O'Neil et al. (1973) sample suite. P. A. Herrera (written communication, 1985) reports Tl at highest concentrations in altered wall rocks, particularly silicified-hematitic breccias, in zone III (with a range of concentration of about 5 to 15 ppm), with concentrations generally decreasing at greater depth in the system.

Detailed Lateral Zoning

Ore-shoot scale variation in element concentrations at distances on the order of inches and feet from mineralized quartz veins was studied by two sampling traverses, approximately perpendicular to two veins--one vein, two feet wide, which dips at about 85° SW, in the Upper Hobart Tunnel, and another, about one foot wide, which dips 60° SW in the Bodie Tunnel. Several other sampling traverses around veins in other accessible underground workings are in the process of being analyzed and compiled (P. A. Herrera, unpublished data, 1985).

Figure 9.11 from data of O'Neil et al. (1973) shows that there is some regularity in the distribution of K, Rb, Sr, and Au and $\delta^{18}O$ relative to the two-foot-wide, mineralized quartz vein in the Upper Hobart Tunnel. A channel sample of the vein contained 4.2 ppm Au. The wall rocks are considerably lower than that at the vein margins, and drop off laterally. The elements Rb and K, which increase during K-silicate alteration, increase as the vein is approached, and Sr decreases. In like manner, $\delta^{18}O$ becomes more negative as the vein is approached representing depletion of the heavy isotope of oxygen. Figure 9.12 shows summary plots of two longer traverses around the same Upper Hobart vein, and a vein in the Bodie Tunnel, that were drifted along and stoped in places. These plots do not show the regularity of the lateral variation shown by the K, Rb, Sr, Au, and $\delta^{18}O$ data. In most of these samples, the gold content was below the .05 ppm detectability limit and was not plotted. Silver falls off rapidly in the wall rocks with the exception of a high-grade sample in the hanging wall of the Bodie Tunnel vein. This sample had quartz veinlets, and is mineralized with 15 ppm Au and elevated contents of Cu, Zn, As, and Hg as well. The distribution patterns of most of the elements plotted are too irregular to serve as guides to the

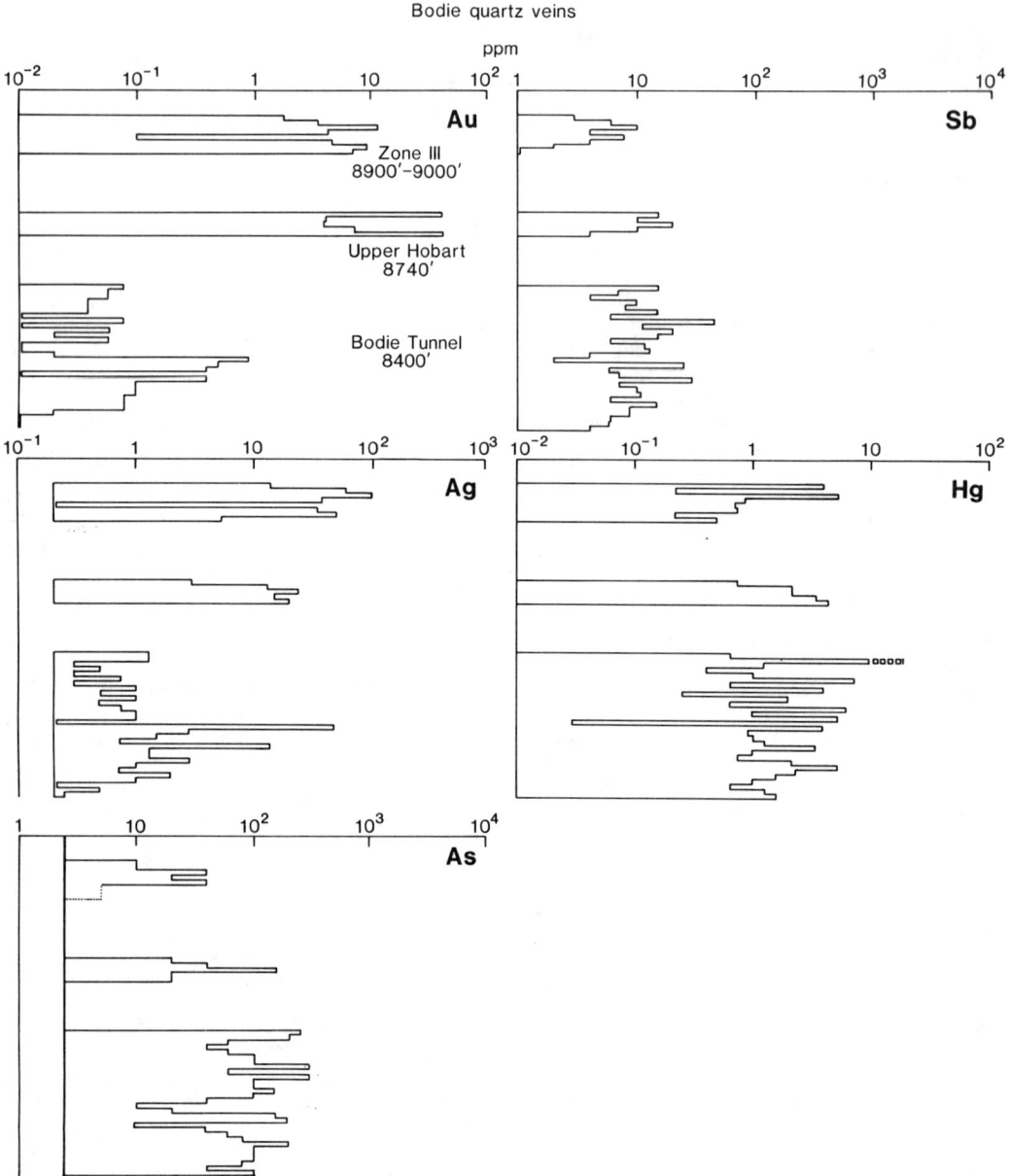

Figure 9.9. Selected trace-element content of quartz veins from three different vertical levels in the northern part of Bodie Bluff.

proximity of mineralized veins with the exception of Hg, and possibly Mn. Manganese forms a halo around the Bodie Tunnel vein that is better developed in the footwall from about 5 to 20 feet from the vein. Mercury forms a pronounced halo at about 10 to 25 feet from both veins (the limit of the traverse).

The profile sampling at the detailed scale (inches to feet) shows that ^{18}O depletion, K and Rb addition, and a decrease in Sr indicate proximity to a mineralized vein, and that the veins appear to have a Hg halo surrounding them. Manganese appears to halo the vein in the propylitized zone, but not the vein in the K-silicate altered productive zone. On the district-wide scale of vertical zoning, it would be very difficult to predict one's proximity to the zone of economic mineralization from the trace-element data--particularly the pathfinder elements that the typical epithermal models suggest can be used for this

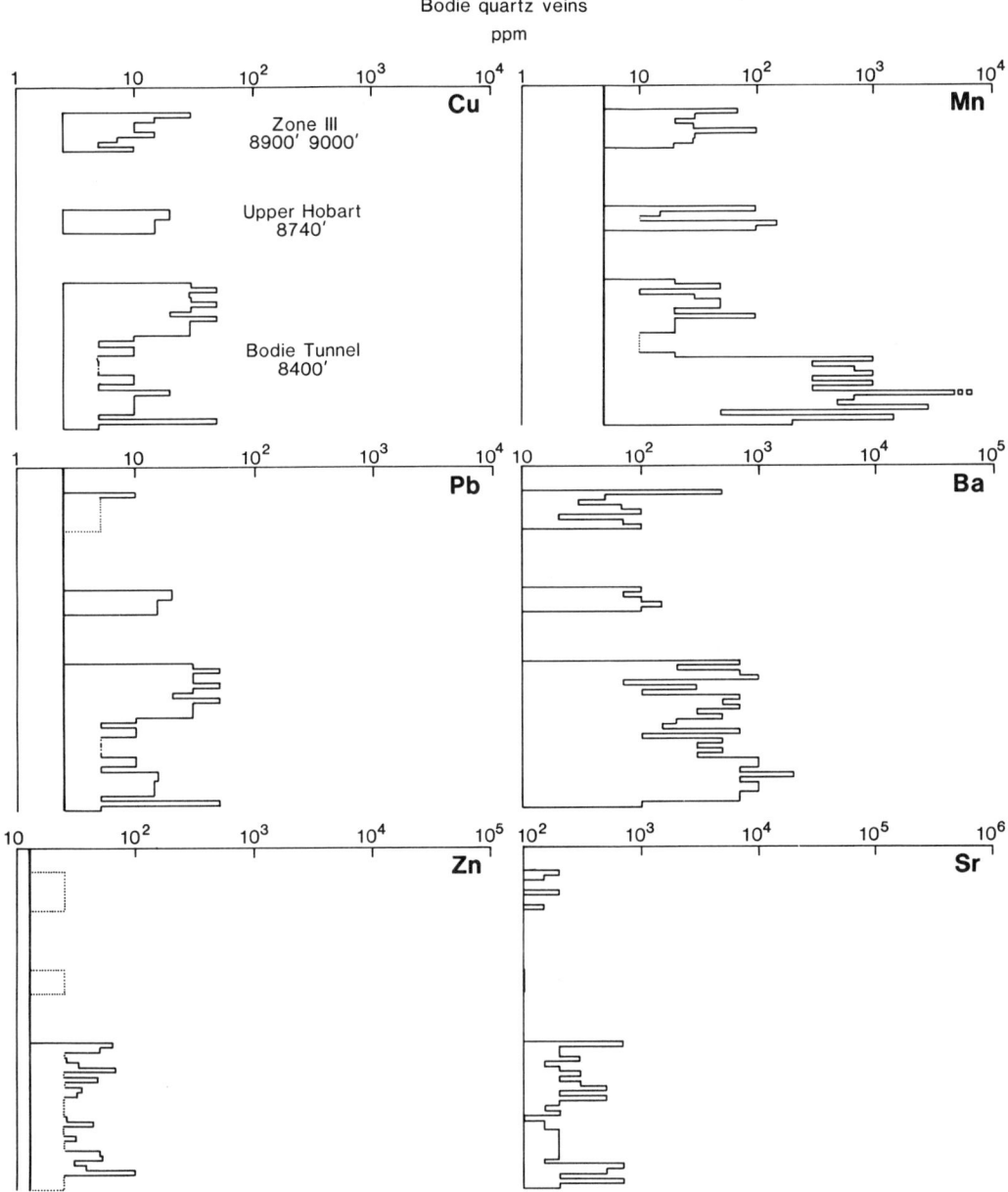

Figure 9.9. (cont'd.)

purpose. In fact, in the quartz veins at Bodie, the pathfinder elements As and Sb tend to increase with depth, and rather than being concentrated at shallow levels above the zone of mineralization, Hg does not vary much with depth. In the wall rocks, the contrasts in these elemental concentrations with depth, if present, are very weak. On the scale of the area of Bodie Bluff (less than 20% of the area of the mining district), it appears that a Mn halo might be the most useful geochemical indicator of a direction towards economic mineralization. To recognize this halo, quartz veins should be sampled and analyzed and not whole-rock samples.

Although the trace-element patterns are quite complex, a pattern of zoning appears to be present at Bodie such that a scheme of qualitative evaluation could be set up. Good signals for the approach to the zone of economic mineralization would be increasing

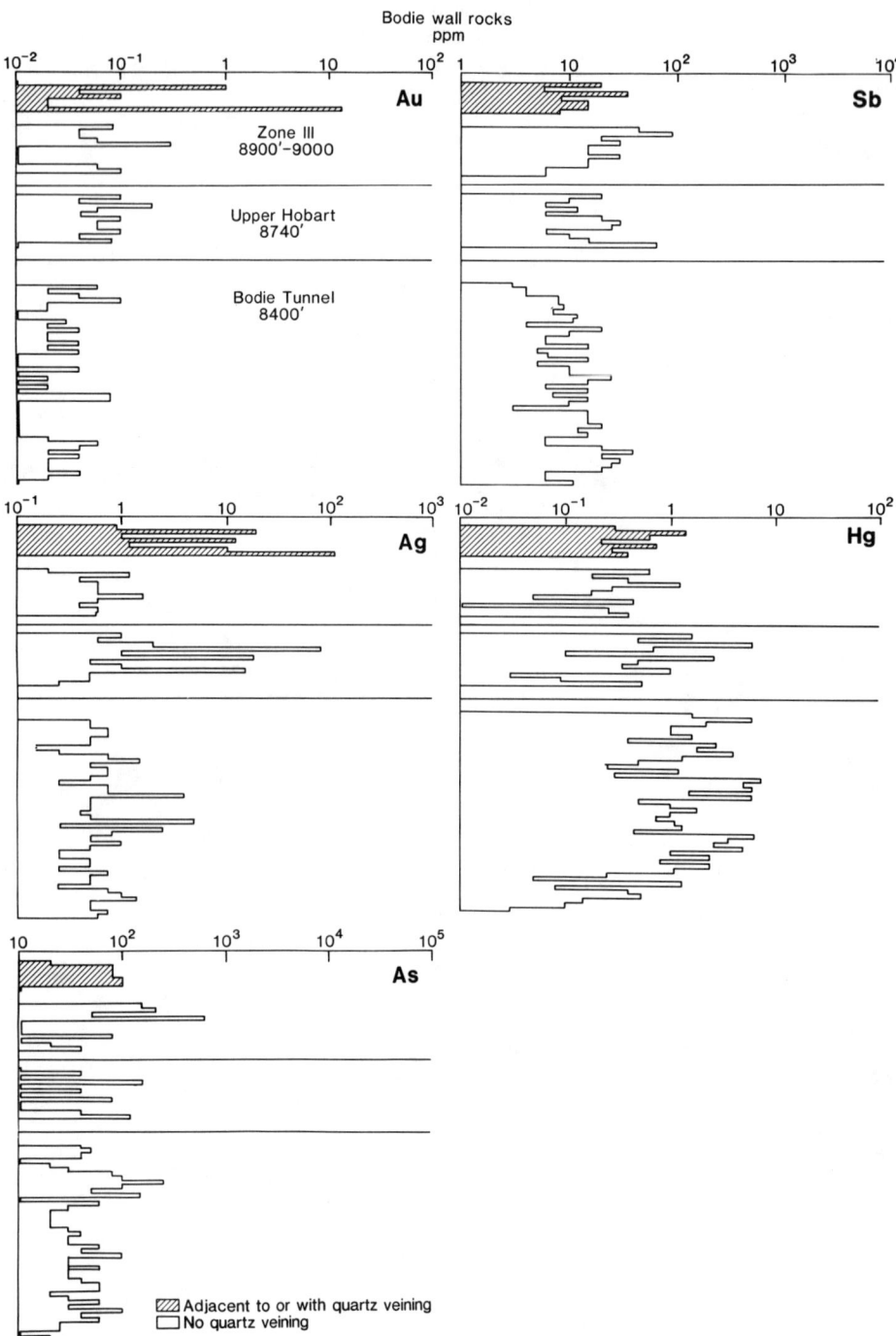

Figure 9.10. Selected trace-element concentrations for altered wall rocks from three different vertical levels in the northern part of Bodie Bluff. For Zone III, samples adjacent to or containing quartz veins are plotted separately from those with no veining. Alteration assemblages of the wall rocks are given on the schematic cross section of northern Bodie Bluff (Fig. 9.8).

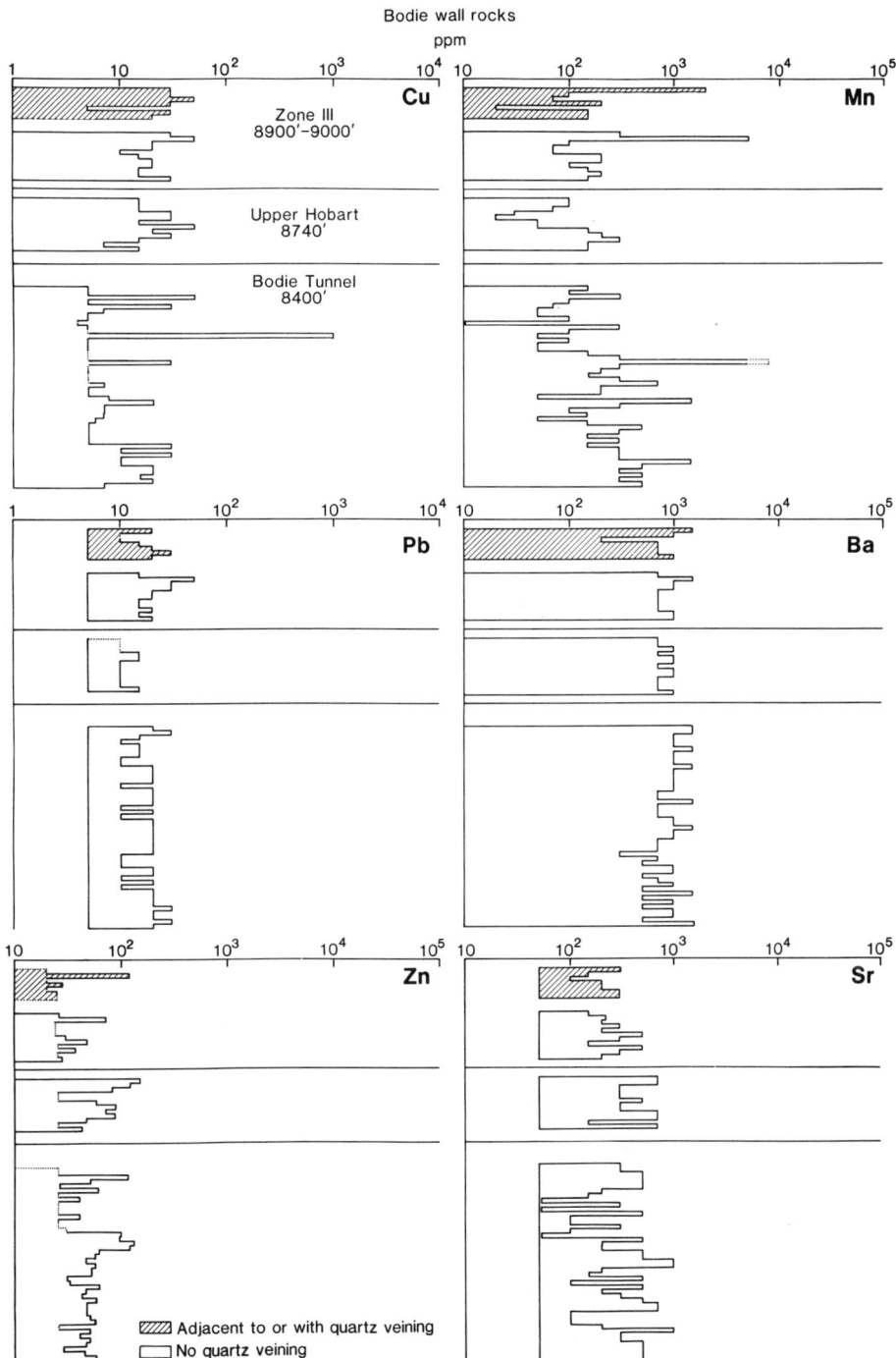

Figure 9.10. (cont'd.)

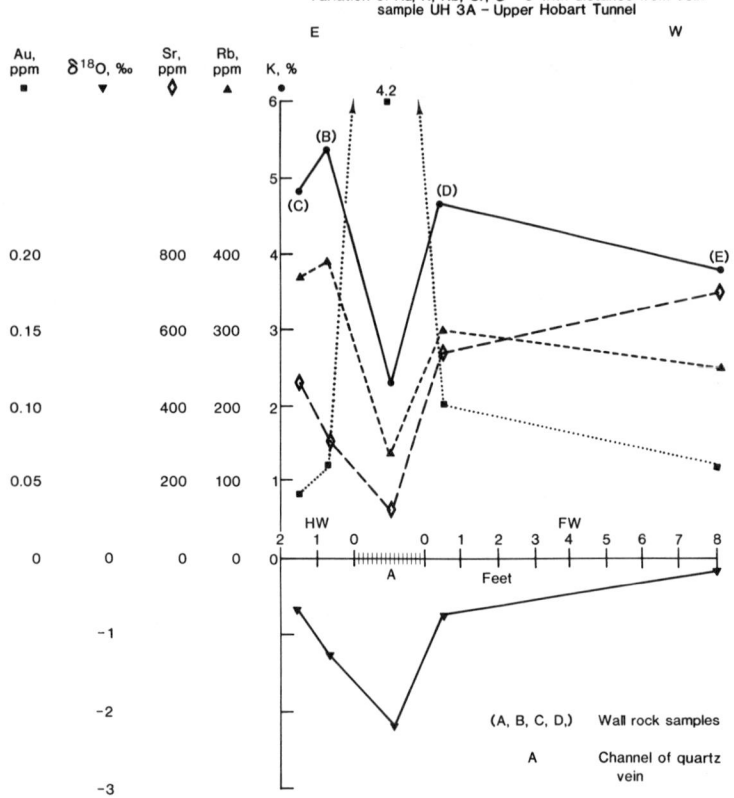

Figure 9.11. Variation of the concentration of Au, K, Rb, Sr, and oxygen isotope ratio with distance from a two-foot-wide quartz vein in the Upper Hobart Tunnel, northern Bodie Bluff.

Au and slight increases in As, Sb, and Hg in quartz veins. This should be associated with a slight decrease in Ag, and slight increases in Pb and Ba. Evidently, when Sr and Zn are consistently present, and Pb, Ba, and Mn are at maximum values, the ore zone is above the area being sampled (Figs. 9.9 and 9.10). These exploration geochemical guides, however, may be useful only because they have already been empirically related to the actual zone of productive mineralization. One could argue the point that the best indicator of economic Au mineralization at Bodie is Au in the quartz veins!

Geochemical zoning patterns at Bodie do not conform, in total, to those predicted by the epithermal models or the geothermal analogy, but they do have some regularity, and could with future refinement be tested as predictive tools.

The alteration mineral assemblages and physical features of the rocks appear to be much better guides to vertical and lateral position in the district. Upward widening of the zones of hematite-quartz-adularia matrix breccias are useful for delineating near-surface supra-vein positions. These breccias are much like similar features in zones of silicification above ore shoots at Oatman, Arizona (Durning and Buchanan, 1984). At Bodie, although they certainly occur above the general bonanza vein zone, we have not yet been able to associate one of these breccias with a specific, subsurface vein.

PARAMOUNT MINING DISTRICT-- VERTICAL ZONING

The gold prospect in the Paramount mining district, located about 5 miles NNE of Bodie, is hosted in rhyolitic pyroclastic and epiclastic rocks and dacite lava flows. The area produced Hg (Kleinhampl et al., 1975). At the surface, extensive zones of silicification, brecciation, and argillic alteration are exposed. Siliceous sinter occurs as large disaggregated blocks and is exposed near the old Paramount Hg workings. The Homestake Mining Company and Houston Oil and Minerals have drilled the prospect for the possibility of a disseminated hot-spring-type volcanic-hosted gold deposit. The Homestake Mining Company allowed us to sample and analyze a 425-foot reverse circulation rotary drill hole that encountered a zone of Au mineralization about 50 feet thick averaging 1.4 ppm Au. The hole was collared in alluvium near the blocks of sinter, and passed into volcanic breccias and pyroclastic rocks to a depth of 230 feet. Below 230 feet it encountered dacite and andesite flows, and remained in the rocks to total depth. The alluvium and volcanic rocks are altered to assemblages characteristic of moderate to intense silicification, argillic alteration, and chloritization (Homestake Mining Co., written communication, 1985). We have not yet done x-ray petrographic mineral assemblage determinations of the drill cuttings and have only generalized data for alteration assemblage determination. The drill-hole logs indicate that several zones of chalcedonic quartz veining are present, and an interface between a zone of oxides (hematite and jarosite) and a zone of sulfides plus oxides (pyrite plus hematite and limonite) occurs at about 170 feet depth. Sulfides, principally pyrite, vary from about 10% immediately beneath the oxide-sulfide interface to about 2%, generally decreasing with depth (Fig. 9.13).

Figure 9.14 summarizes available chemical data for 5-foot intervals of the hole as bar graphs.

The zone of gold mineralization occurs between 110 and 160 feet depth, where the Au content is consistently above 1 ppm, with a maximum value of 1.9 ppm, immediately above the oxide-sulfide interface. Part of the gold-mineralized zone contains 5 to 10 percent chalcedonic quartz veins, but the rest of the zone has no chalcedonic veins listed on the drill log. Alteration in the zone of gold mineralization is also variable, consisting of alternating zones of moderate to intense silicification and argillization. Deeper in the hole (215-240 feet), a 25-foot zone of Au mineralization varies from about 0.5 to 1 ppm, and is coincident with the occurrence of 1 to 5 percent chalcedonic quartz veining in silicified rock. A few other intervals of 5 or 10 feet with greater than 0.5 ppm Au intervals occur lower in the hole, but these intervals appear unrelated to either sulfide content or quartz veining. Although irregular, higher concentrations of Au occur beneath the "mineralized" zone than above the zone near the surface, where it increases irregularly towards the zone.

Silver is generally low near the surface in the drill hole, although some high intervals are found, and shows a very irregular pattern throughout the rest of the hole. It appears to be slightly lower in the gold-mineralized zone (at about 2 ppm) than the 50-foot interval in the hanging wall and the 75-foot interval in the footwall, where the Ag contents vary from 3 to 5 ppm. There is a greater consistency in the Ag concentration in the 75 feet or so below the oxide-sulfide interface where about 10 percent sulfides are present. Near the bottom of the hole, a few intervals have as high as 5 to 20 ppm Ag and are unrelated to any obvious veining or alteration type.

The mercury concentration is high in most of the hole, as it is elsewhere in the Paramount district (Kleinhampl et al., 1975; M. L. Silberman, unpublished data, 1985). The Hg content in the drill hole is above 10 ppm near the surface decreasing to about 50-foot depth, where it increases again to greater than 10 ppm. The higher values (>10 ppm) appear to occur in the 50-foot interval above the gold-mineralized zone. It is still at or near 10 ppm for about 100 feet below the mineralized zone and then irregularly decreases to total depth. Within the Au zone per se, the Hg content is lower (4 to 8 ppm) than in the hanging wall of the zone, and slightly lower than the proximal part of the footwall. The Hg distribution suggests a Hg halo that is better defined in the hanging wall.

Thallium concentrations are generally high, two to ten times the crustal average for felsic rocks (Parker, 1967). Thallium is lowest in concentration and irregular in distribution near surface, and increases at a depth of 50 feet and remains mostly above 10 ppm for another 50 feet. It decreases again in the Au-mineralized zone, then increases strongly just below the zone where the concentration remains mostly above 10 ppm for approximately 100 feet. This footwall zone of high Tl concentration corresponds approximately to sulfide contents above 10 percent. The thallium increases again near the bottom of the hole, but with the exception of one 5-foot interval, is below 10 ppm. Thallium appears to halo the Au zone, with a sharper peak to the halo in the hanging wall.

Arsenic and antimony both have complex patterns of vertical distribution and are consistently above 200 and 100 ppm, respectively. Arsenic tends to decrease from near the surface to the Au-mineralized zone, where it varies between 500 and 1,000 ppm. It falls sharply at the oxide-sulfide interface and increases again in the zone of 10 percent sulfide beneath the interface. Below the high-sulfide zone it decreases, and increases again near the bottom. Antimony is distributed in a similar pattern, but shows a more general decrease with depth, interrupted by the mineralized zone where a slight positive anomaly occurs in part of the zone. Low concentrations in the footwall As and Sb appear more related to the oxide-sulfide interface than to the gold mineralized zone.

The boron concentration is 100 ppm just below the surface, and decreases irregularly with depth. The gold-mineralized zone interrupts the decrease and has a poorly defined halo of relatively high B surrounding it. Barium is slightly lower in the Au-mineralized zone (700 to 1,500 ppm) than above and below it. The highest concentrations (2,000 to 3,000 ppm) are found in the 35-foot interval just above the Au-mineralized zone. Strontium also appears to be at a minimum concentration in the Au-mineralized zone, and increases in both hanging and footwalls with the highest concentrations occurring in the footwall, increasing irregularly to about 100 feet above TD, then decreasing again. The Sr pattern is reminiscent of that found for lateral variation of Sr around the vein in the Upper Hobart Tunnel at Bodie (Fig. 9.11).

Molybdenum is highest in the Au-mineralized zone reaching levels of 30 to 50 ppm. It is generally at lower concentrations both above and below the zone. Outside of the Au interval the highest Mo contents are where the sulfide content is highest. Molybdenum is relatively high at Paramount, whereas at Bodie and many of the other epithermal systems from which we have data, it is present at very low levels, although at Round Mountain, Aurora, and Divide, Nevada, Mo is present in relatively significant concentrations.

Manganese forms a well-defined halo in the hanging wall of the gold-mineralized zone where it reaches values up to 2,000 ppm in the 60-foot interval above the zone. It is at a minimum within the zone and increases for a 60- to 70-foot interval below it with concentrations between 1,000 and 5,000 ppm. Manganese drops off below this level, although an occasional high interval is encountered.

Copper and lead concentrations have relatively similar distribution patterns. They increase irregularly from the surface to about 100-foot depth, are relatively low in the Au-mineralized zone, and increase below this, particularly in the interval with the highest sulfide content. Both Cu and Pb form an

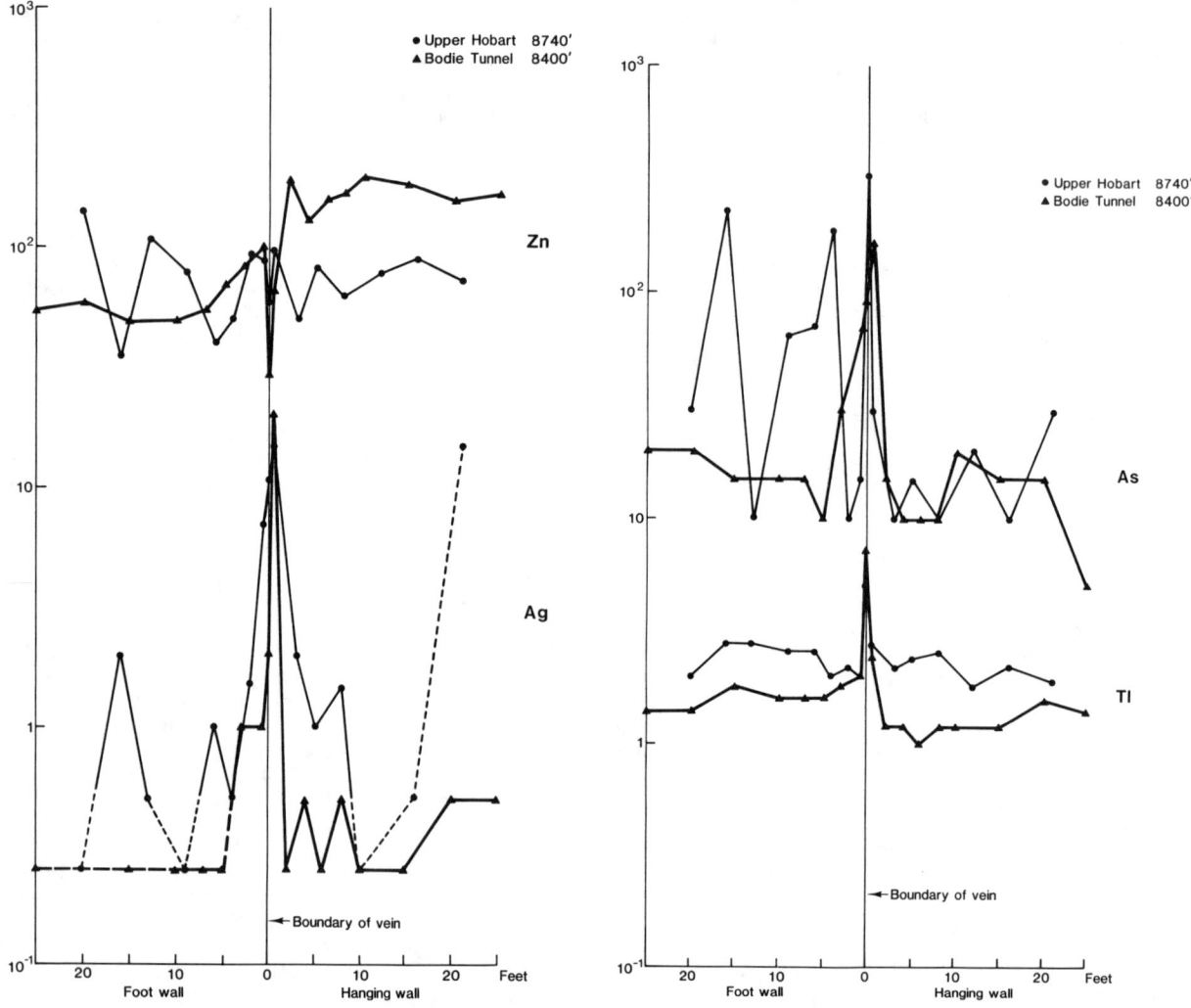

Figure 9.12. Variation of the concentration of selected trace elements with distance from a mineralized quartz vein in the Upper Hobart Tunnel, and in the Bodie Tunnel, northern Bodie Bluff. Hg at both levels, and Mn in the Bodie Tunnel appear to halo the vein.

ostensible halo around the Au-mineralized zone, but largely because low values are found within this zone. Zinc irregularly decreases from near surface to total depth, except for irregular highs, which break the pattern in the Au-mineralized zone. Co is near the detection limit at the surface, first is detectable at a depth of 60 feet, and then decreases towards the Au-mineralized zone, where it is below detection. Below the zone, Co is at relatively high values where the sulfide content is high, and then decreases with depth. The near total absence of Co in the Au-mineralized zone makes it appear to form a halo around the zone, but the Co distribution is too irregular to be useful as a pathfinder to gold mineralization. Nickel has its highest concentration within and just below the zone of highest sulfide content and shows little relationship to the Au-mineralized zone. Chromium also shows little relationship to the Au-mineralized zone, and has its highest concentrations midway down the hole, in a large interval (150 feet) that includes both the Au and high-sulfide zones. Vanadium decreases irregularly from near the surface to about 100 feet above total depth, where it increases slightly again.

In summary, the distribution of some elements (Au, As, and Sb) appears to be related to the oxide-sulfide interface, while other distributions (Ag, Cu, Pb, Co, and Ni) appear to be related to the total sulfide content, and others (Hg, Tl, Mn, B, Sr, and Ba) appear to form halos of varying degrees of definition around the Au-mineralized zone. Vanadium and zinc show little definable relationship to either feature, the sulfide or gold-mineralized zones.

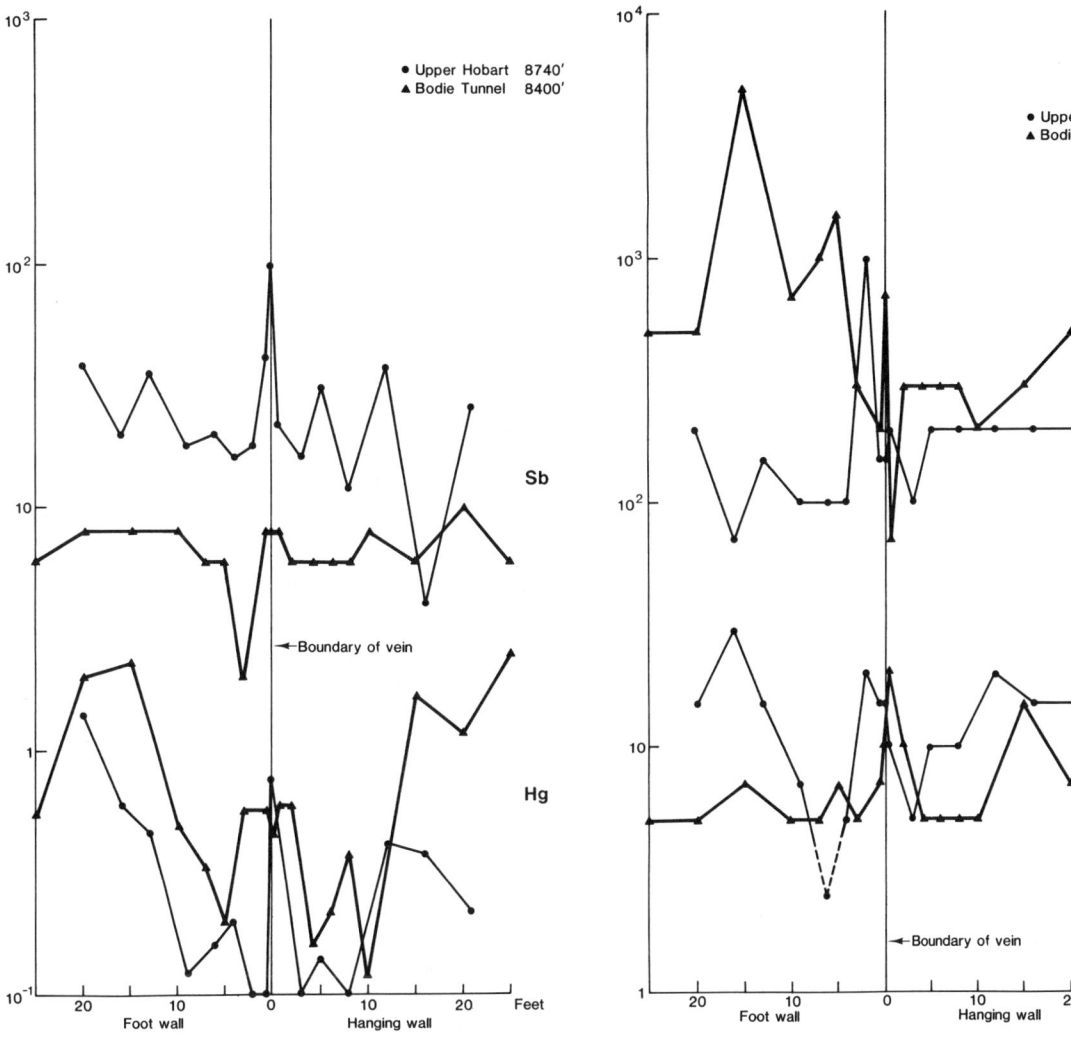

Figure 9.12. (cont'd.)

From the perspective of an explorationist, proximity to the gold-mineralized zone is best indicated by Hg, Tl, and Mn and to a lesser extent by Ba and B. In this hydrothermal system, proximal increases in Mn, Tl, and Hg, and perhaps Ba, below a B-enriched zone might serve as exploration geochemical guides to gold mineralization. We feel that other halos, such as those for Pb, Cu, Co, and Sr are too poorly defined to be valuable in exploration. We are perhaps most intrigued by the interruption of the trace-element distribution patterns by the partly chalcedonic-veined, Au-mineralized zone. The interpretation of the trace-element patterns is complicated by many factors--the oxide-sulfide interface, the high-sulfide zone, and the presence of quartz veining. At Paramount, we have not as yet related the element distribution patterns to alteration mineralogy, and our understanding and interpretation of the patterns will perhaps change when this part of our research is completed.

SUMMARY

The study of geochemical zoning patterns in epithermal ore deposits is important to evaluating the use of geochemical pathfinders for locating potential ore-grade mineralization. To accomplish this objective, it is important to be able to relate the physical and geochemical morphologies of the hydrothermal systems. Conceptual models of epithermal ore-deposit types are of necessity generalizations of the geometry,

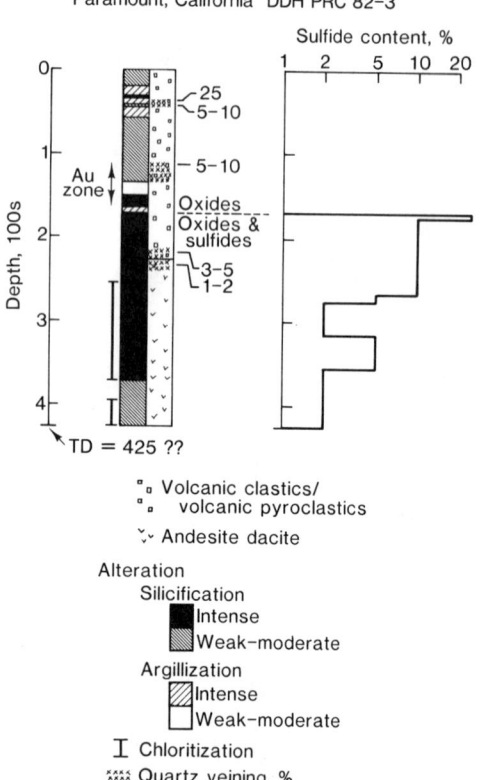

Figure 9.13. Generalized lithology, alteration mineralogy, sulfide content, and percentage of chalcedonic quartz veining in drill hole PRC 82-3, from the Paramount prospect, Bodie Hills, Mono County, California (modified from information supplied by Homestake Mining Co.).

alteration, structural controls, ore and gangue mineralogy, and trace-element geochemistry of these deposit types. Variations in the patterns due to duration, structural complexity, effects of host-rock properties such as permeability and chemical reactivity, the overprinting of features by changes in the position of active vents, elevation changes in the water table, and changes in the levels of boiling make the actual relationships in deposits very complex. How complex the actual patterns can be is illustrated by chemical variations, lateral and vertical at Bodie, California, and vertical at Paramount, California. In addition, the geochemical patterns in actual deposits can be quite different from that reported in active geothermal systems. From the geochemical results at Steamboat Springs, Nevada, and Bodie and Paramount, California, we conclude that the presence of mineralization in these epithermal systems causes deviations from an all-inclusive, standard "geothermal zoning" of elements. The deviations in the trace-element patterns are related to the physico-chemical processes that occur during the deposition of the ores, and if the processes can be demonstrated to be regular, perhaps the trace-element patterns will be useful for predicting the location of ore-grade mineralization in any given epithermal system.

Both the Bodie and Paramount hydrothermal systems illustrate vertical trace-element zoning in only a portion of a mineralizing system, and, at that, on the scale of hundreds of feet. The lateral-zoning example at Bodie illustrates the types of patterns that can be expected around individual ore shoots. In order to derive geochemical patterns as guides to mineralization at the district-wide scale, much more data than were summarized above are necessary.

In the following paper, Berger and Silberman (1985, this volume) summarize the results of detailed chemical analyses of rotary and diamond drill holes from two hot-spring-type gold deposits, Round Mountain and Hasbrouck Mountain, Nevada. In both studies, drill holes which transected the mineralized zones and their weakly mineralized-altered margins were analyzed chemically and the trace-element results were combined with data on alteration mineralogy. The combination of mineralogy and trace-element chemistry produces a better large-scale, three-dimensional representation of the physical and chemical morphologies of hydrothermal systems than we were able to present from the essentially two-dimensional study at Bodie, and the one-dimensional study at Paramount. Round Mountain and Hasbrouck Mountain also show considerable variation from a standard "geothermal system" metal zoning model, but both also demonstrate regularity in their elemental distribution patterns that might be of predictive significance.

ACKNOWLEDGMENTS

We would like to express our appreciation to Peter Herrera of the Colorado School of Mines for providing us his information and insights on the geology of the Bodie Mining District, and to Bob Blakestad of Homestake Mining Company for providing access and information on Bodie and on the Paramount prospect. Analytical data summarized within this report were provided over a 17 year period of time by many chemists of the Branch of Exploration Geochemistry of the U.S.G.S. We are particularly indebted to R. M. O'Leary for providing wet chemistry (acid digestion-AA), and M. S. Erickson for providing Emission Spectrographic data for the many hundreds of samples submitted for analyses. The manuscript has benefitted markedly from critical reviews by Bob Blakestad and Andy Wallace, but we bear sole responsibility for the conclusions reached within.

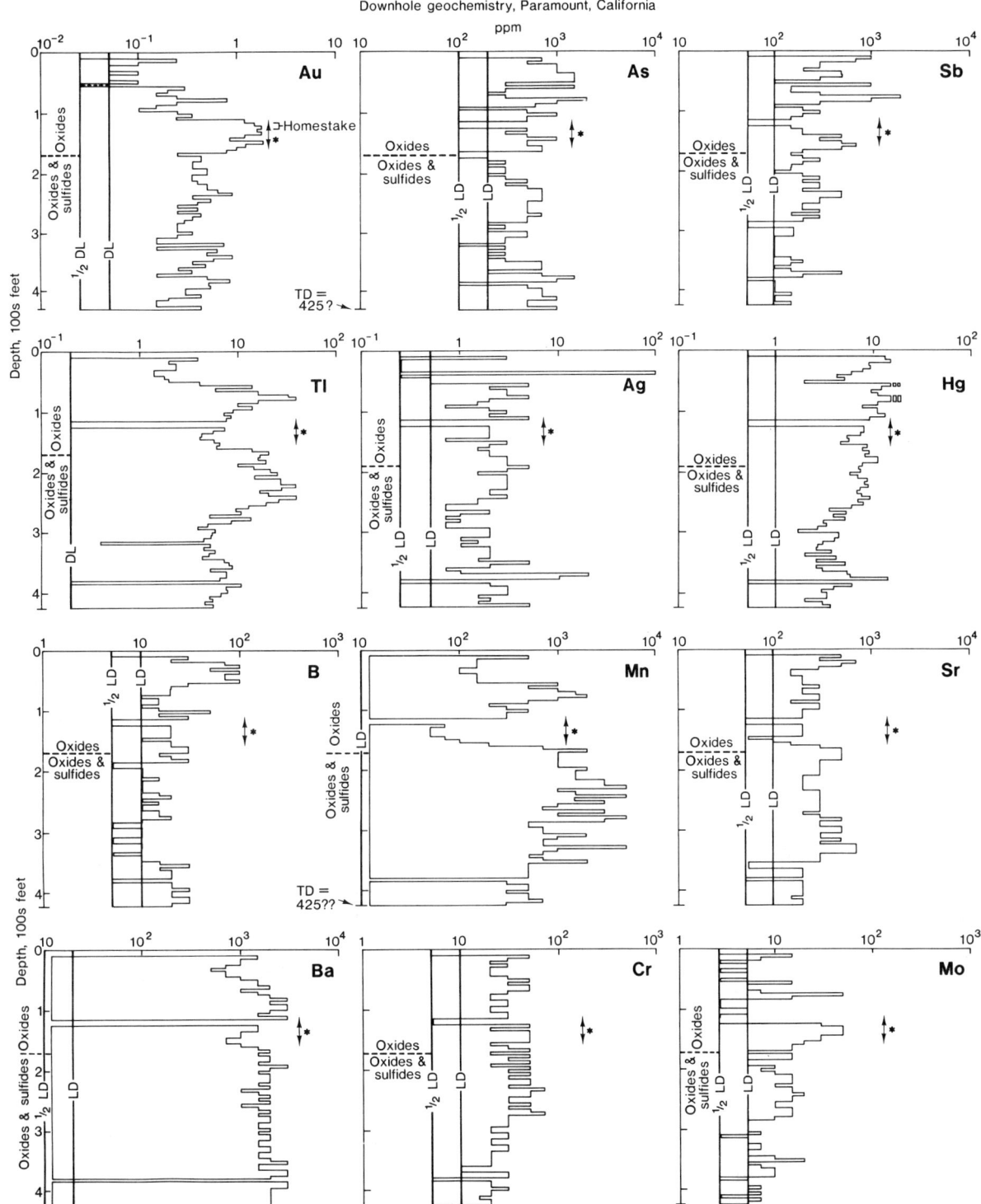

Figure 9.14. Summary geochemistry of selected trace elements of 5-foot intervals of cuttings from drill hole PRC 82-3, from the Paramount prospect, Mono County, California, plotted as a function of depth. DL is detection limit for the analytical method used. 1/2 DL is one half that value.

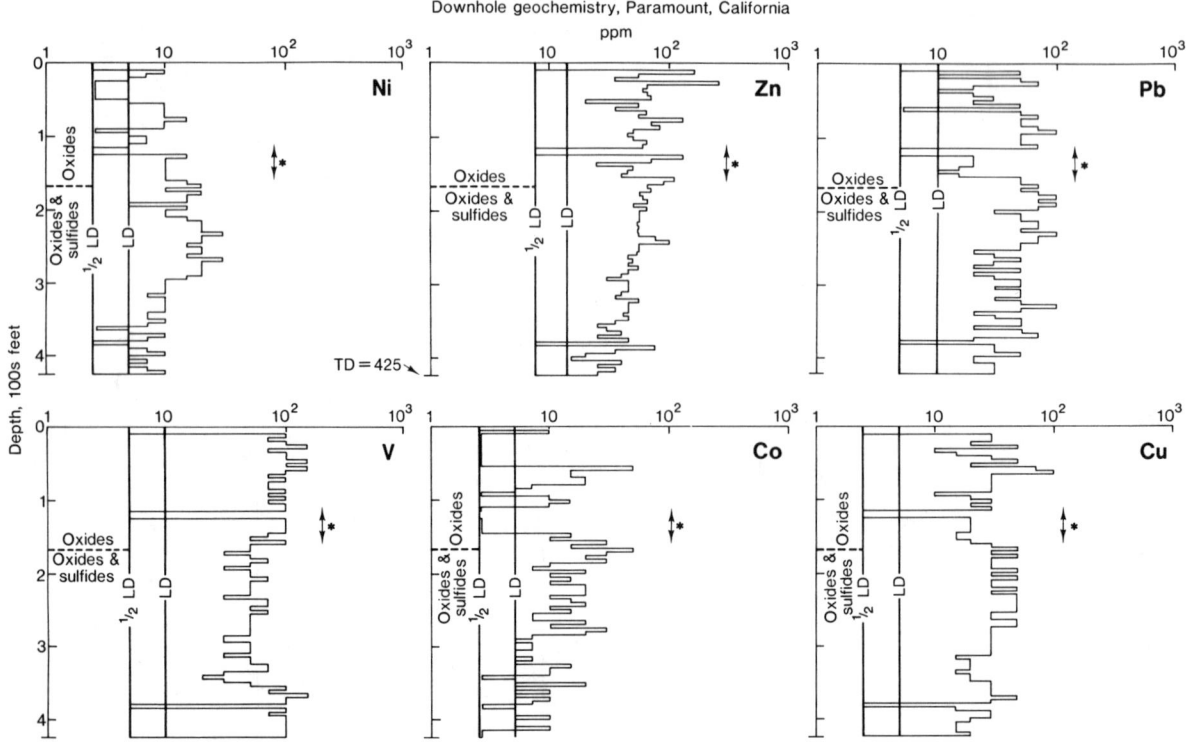

Figure 9.14. (cont'd.)

REFERENCES

Ashley, R. P., 1974, Goldfield mining district: Nevada Bureau of Mines and Geology Report 19, p. 49-66.

Ashley, R. P., and Berger, B. R., 1985, Precious metals in volcanic terranes: U.S. Geological Survey, Circular (in press).

Bagby, W. C., and Berger, B. R., 1985, Comparative anatomy of sediment-hosted epithermal precious-metal deposits; in Berger, B. R., and Bethke, P. M. (eds.), Geology and Geochemistry of Epithermal Systems: Society of Economic Geologists, Reviews in Economic Geology, v. 2.

Barrett, R. A., 1985, The geology, mineralization, and geochemistry of the Milestone hot-spring silver-gold deposit near the Delamar silver mine, Owyhee County, Idaho (abs.): Geological Society of America, Abstracts with Programs, Rocky Mountain Section, p. 207.

Barton, P. B., Jr., Bethke, P. M., and Roedder, E., 1977, Environment of ore deposition in the Creede mining district, San Juan Mountains, Colorado: Part III. Progress toward interpretation of the chemistry of the ore-forming fluid for the OH vein: Economic Geology, v. 72, p. 1-24.

Berger, B. R., and Eimon, P. I., 1983, Conceptual models of epithermal precious-metals deposits; in Shanks, W. C. III (ed.), Cameron Volume on Unconventional Mineral Deposits, Society of Mining Engineers, p. 191-205.

Berger, B. R., and Silberman, M. L., 1985, Relationships of trace-metal patterns to geology in hot-spring-type precious-metal deposits; in Berger, B. R., and Bethke, P. M. (eds.), Geology and Geochemistry of Epithermal Systems: Society of Economic Geologists, Reviews in Economic Geology, v. 2.

Bethke, P. M., 1984, Controls on base- and precious-metal mineralization in deeper epithermal environments: U.S. Geological Survey, Open-File Report 84-890, 14 p.

Bethke, P. M., and Rye, R. O., 1979, Environment of ore deposition in the Creede mining district, San Juan Mountains, Colorado: Part IV. Source of fluids from oxygen, hydrogen, and carbon isotopes: Economic Geology, v. 74, p. 1832-1851.

Bonham, H. F., Jr., 1986, Models for volcanic-hosted epithermal precious metal deposits: A review; International Volcanological Congress, February 1986, p. 1-5.

Bonham, H. F., Jr., and Giles, D. L., 1983, Epithermal gold-silver deposits: Geothermal Resources Council, Special Report 13, p. 257-262.

Browne, P. R. L., and Ellis, A. J., 1970, The Ohaki-Broadlands hydrothermal area, New Zealand, mineralogy and related geochemistry: American Journal of Science, v. 269, no. 2, p. 97-131.

Bruha, D. J., and Noble, D. C., 1983, Hypogene quartz-alunite ± pyrite alteration formed by saline, ascendant hydrothermal solutions (abs.): Geological Society of America, Abstracts with Programs, v. 15, no. 5, p. 325.

Buchanan, L. J., 1981, Precious-metal deposits associated with volcanic environments in the southwest; in Dickinson, W. R., and Payne, W. D. (eds.), Relations of Tectonics to Ore Deposits in the South Cordillera: Arizona Geological Society Digest, v. XIV, p. 237-262.

Chesterman, C. W., Chapman, R. H., and Gray, C. H., Jr., 1985, Geology and ore deposits of the Bodie mining district, Mono County, California: California Division of Mines and Geology Bulletin (in press).

Durning, W. P., and Buchanan, L. J., 1984, The geology and ore deposits of Oatman, AZ; in Wilkins, J., Jr. (ed.), Gold and Silver Deposits of the Basin and Range Province, Western United States: Arizona Geological Society Digest, v. XV, p. 141-158.

Ellis, A. J., 1979, Explored geothermal systems; in Barnes, H. L. (ed.), Geochemistry of Hydrothermal Ore Deposits, Second Edition: John Wiley and Sons, New York, p. 632-683.

Ewers, G. R., and Keays, R. R., 1977, Volatile- and precious-metal zoning in the Broadlands geothermal field, New Zealand: Economic Geology, v. 72, p. 1337-1354.

Giles, D. L., and Nelson, C. E., 1982, Principal features of epithermal lode gold deposits of the Circum-Pacific Rim: Transactions, Third Circum-Pacific Energy and Mineral Resources Conference, p. 273-278.

Graney, J. R., 1985, Controls of alteration and precious-metal mineralization in a fossil hydrothermal system, Hasbrouck, Mountain, Nevada (abs.): Geological Society of America, Abstracts with Programs, v. 16, no. 6, p. 523.

Hayba, D. O., 1983, A compilation of fluid inclusion and stable isotope data on selected precious- and base-metal epithermal deposits: U.S. Geological Survey, Open-File Report 83-450, 24 p.

Hayba, D. O., Bethke, P. M., Heald, P., and Foley, N. K., 1985, Geologic, mineralogic, and geochemical characteristics of volcanic-hosted epithermal precious-metal deposits; in Berger, B. R., and Bethke, P. M. (eds.), Geology and Geochemistry of Epithermal Systems: Society of Economic Geologists, Reviews in Economic Geology, v. 2.

Heald-Wetlaufer, P., Hayba, D. O., Foley, N. K., and Goss, J. A., 1983, Comparative anatomy of epithermal precious- and base-metal districts hosted by volcanic rocks: U.S. Geological Survey, Open-File Report 83-710, 16 p.

Hedenquist, J. W., 1984, Waiotapu geothermal field; in Field Trip and Conference Guide: Geothermal Research Center, D.S.I.R., p. F1-F16.

Henley, R. W., 1985, Geothermal framework for epithermal ore deposits; in Berger, B. R., and Bethke, P. M. (eds.), Geology and Geochemistry of Epithermal Systems: Society of Economic Geologists, Reviews in Economic Geology, v. 2.

Henley, R. W., and Brown, K. L., 1985, A practical guide to the thermodynamics of geothermal fluids and hydrothermal ore deposits; in Berger B. R., and Bethke, P. M. (eds.), Geology and Geochemistry of Epithermal Systems: Society of Economic Geologists, Reviews in Economic Geology, v. 2.

Henley, R. W., and Ellis, A. J., 1983, Geothermal systems, ancient and modern--A geochemical review: Earth Science Reviews, v. 19, p. 1-50.

Honda, S., and Muffler, L. J. P., 1970, Hydrothermal alteration in core from research drill hole Y-1, Upper Geyser Basin, Yellowstone National Park, Wyoming: American Mineralogist, v. 55, p. 1714-1737.

Hudson, D. M., 1984, Geology of the Steamboat Springs-Virginia City region, Nevada; in Johnson, J. L. (ed.), Exploration for Ore Deposits in the North American Cordillera, Field Trip Guidebook: The Association of Exploration Geochemists, p. FT11, 1-10.

Kleinhampl, F. J., Silberman, M. L., Chesterman, C. W., Chapman, R. H., and Gray, C. H., 1975, Aeromagnetic and limited gravity studies and generalized geology of the Bodie Hills region, Nevada and California: U.S. Geological Survey, Bulletin 1324, 38 p.

Lindgren, W., 1933, Mineral Deposits: McGraw Hill Co., New York, 930 p.

Mills, B. A., 1984, Geology of the Round Mountain gold deposit, Nye County, Nevada; in Wilkins, J., Jr. (ed.), Gold and Silver Deposits of the Basin and Range Province, Western USA: Arizona Geological Society Digest, v. XV, p. 89-100.

Nash, J. T., 1972, Fluid inclusion studies of some gold deposits in Nevada: U.S. Geological Survey, Professional Paper 800-C, p. 15-19.

Noble, P. C., and Silberman, M. L., 1984, Evolucion volcanica e hydrotermal y cronologia de K-Ar del Distrito de Julcani, Peru: Sociedad Geologica del Peru, v. 50, p. 1-35.

Norton, Dennis, and Cathles, L. M., 1979, Thermal aspects of ore deposition; in Barnes, H. L. (ed.), Geochemistry of Hydrothermal Ore Deposits, Second Edition, John Wiley and Sons, New York, p. 611-631.

O'Neil, J. R., and Silberman, M. L., 1974, Stable isotope relations in epithermal Au-Ag deposits: Economic Geology, v. 69, p. 902-907.

O'Neil, J. R., Silberman, M. L., Fabbi, B. P., and Chesterman, C. W., 1973, Stable isotope and chemical relations during mineralization in the Bodie mining district, Mono County, California: Economic Geology, v. 68, p. 765-784.

Osborne, M. A., and Stone, J. G., 1985, Alteration and mineralization of a portion of the Aurora mining district, Mineral County, Nevada: Geological Society of America, Abstracts with Programs, Rocky Mountain Section, p. 260.

Rose, A. W., and Burt, D. M., 1979, Hydrothermal alteration; in Barnes, H. L. (ed.), Geochemistry of Hydrothermal Ore Deposits: Second Edition, John Wiley and Sons, New York, p. 173-235.

Sawkins, F. J., 1984, Metal Deposits in Relation to Plate Tectonics: Springer-Verlag, 340 p.

Sigvaldeson, G. E., and White, D. E., 1961, Hydrothermal alteration in two drill holes at Steamboat Springs, Washoe County, Nevada: U.S. Geological Survey, Professional Paper 424-D, p. 116-122.

Sigvaldeson, G. E., and White, D. E., 1962, Hydrothermal alteration in drill holes GS-5 and GS-7, Steamboat Springs, Nevada: U.S. Geological Survey, Professional Paper 450-D, p. 113-117.

Silberman, M. L., 1982, Hot-spring type, large-tonnage, low-grade gold deposits; in Erickson, R. L. (compiler), Characteristics of Mineral Deposit Occurrence: U.S. Geological Survey, Open-File Report 82-795, p. 131-143.

Silberman, M. L., 1983, Geochronology of hydrothermal alteration and mineralization-- Tertiary epithermal precious-metal deposits in the Great Basin: Geothermal Resources Council, Special Report No. 13, p. 287-303.

Silberman, M. L., 1984, Field guide to the Bodie mining district, Mono County, California--with annotated road log: Society of Economic Geologists, 1984, Nevada Field Guide (in press).

Silberman, M. L., Bonham, H. F., Jr., Garside, L. J., and Ashley, R. R. 1979, Timing of hydrothermal alteration-mineralization and igneous activity in the Tonopah mining district and vicinity, Nye and Esmeralda counties, Nevada; in Ridge, J. D. (ed.), Proceedings of the Fifth Quadrennial Symposium, International Association on the Genesis of Ore Deposits: Nevada Bureau of Mines and Geology, Report 33, p. 119-126.

Silberman, M. L., Chesterman, C. W., Kleinhampel, F. J., and Gray, C. H., Jr., 1972, K-Ar ages of volcanic rock and gold-bearing quartz-adularia veins in the Bodie mining district, Mono County, California: Economic Geology, v. 67, p. 597-604.

Silberman, M. L., Stewart, J. H., and McKee, E. H., 1976, Igneous activity, tectonics, and hydrothermal precious-metal mineralization in the Great Basin during Cenozoic time: Transactions, Society of Mining Engineers, AIME, v. 260, p. 253-263.

Silberman, M. L., White, D. E., Keith, T. E. C., and Doctor, R. D., 1979, Duration of hydrothermal activity at Steamboat Springs, Nevada from ages of spatially associated volcanic rocks: U.S. Geological Survey, Professional Paper 458-D, 13 p.

Sillitoe, R. H., 1981, Ore deposits in Cordilleran and island-arc settings; in Dickinson, W. R., and Payne, W. D. (eds.), Relations of Tectonics to Ore Deposits in the South Cordillera: Arizona Geological Society Digest, v. XIV, p. 49-70.

Sillitoe, R. H., and Bonham, H. F., 1984, Volcanic landforms and ore deposits: Economic Geology, v. 79, p. 1286-1298.

Steiner, A., 1968, Clay minerals in hydrothermally altered rocks at Wairakei, New Zealand: Clay and Clay Minerals, v. 16, p. 193-213.

Stewart, J. H., Carlson, J. E., and Johannesen, D. C., 1982, Geologic map of the Walker Lake $1^{\circ} \times 2^{\circ}$ quadrangle, California and Nevada: U.S. Geological Survey, Miscellaneous Field Studies Map MF-1382-A, scale 1:250,000.

Stone, J. G., and Osborne, M. A., 1984, Road Guide-- Aurora mining district: Society of Economic Geologists, 1984 Field Guide (in press).

Taylor, H. P., Jr., 1973, O^{18}/O^{16} evidence for meteoric-hydrothermal alteration and ore deposition in the Tonopah, Comstock Lode, and Goldfield mining districts, Nevada: Economic Geology, v. 60, p. 747-764.

Taylor, H. P., Jr., 1974, The application of oxygen and hydrogen isotope studies to problems of hydrothermal alteration and ore deposition: Economic Geology, v. 69, p. 843-883.

Thompson, G. A., and White, D. E., 1964, Regional geology of the Steamboat Springs area, Washoe County, Nevada: U.S. Geological Survey, Professional Paper 488-A, 52 p.

Tingley, J. V., and Berger, 1985, Lode gold deposits of Round Mountain, Nevada: Nevada Bureau of Mines and Geology Bulletin 100, 62 p.

Vikre, T. G., 1985, Precious metal vein systems in the National district, Humboldt County, Nevada: Economic Geology, v. 80, p. 360-393.

Wallace, A. B., 1980, Geology of the Sulphur district, southwest Humboldt County, Nevada: Society of Economic Geologists, 1980 Field Conference, p. 80-91 (in press).

Wallace, A. B., and Friberg, R. S., 1984, Geology and mineral deposits of the Sulphur mining district, Humboldt County and Pershing County, Nevada; in Johnson, J. L. (ed.), Exploration for Ore Deposits in the North American Cordillera, Field Trip Guidebook: The Association of Exploration Geochemists, p. FT8, 1-10.

Weissberg, B. G., 1969, Gold-silver ore-grade precipitates from New Zealand thermal waters: Economic Geology, v. 64, p. 95-108.

Weissberg, B. G., Browne, P. R. C., and Seward, T. M., 1979, Ore metals in active geothermal systems; in Barnes, H. L. (ed.), Geochemistry of Hydrothermal Ore Deposits, Second Edition: John Wiley and Sons, New York, p. 739-780.

Wetlaufer, P. H., Bethke, P. M., Barton, P. B., Jr., and Rye, R. O., 1979, The Creede Ag-Pb-Zn-Cu-Au district, central San Juan Mountains, Colorado: A fossil geothermal system: Fifth Symposium, International Association on the Genesis of Ore Deposits, Snowbird, Utah, v. II, p. 159-164.

White, D. E., 1955, Thermal springs and epithermal ore deposits; in Bateman, A. M. (ed.): Economic Geology, 50th Anniversary Volume, p. 99-154.

White, D. E., 1974, Diverse origins of hydrothermal ore fluids: Economic Geology, v. 69, p. 954-973.

White, D. E., 1981, Active geothermal systems and epithermal ore deposits: Economic Geology, 75th Anniversary Volume, p. 392-423.

White, D. E., Muffler, L. J. P., and Truesdell, A. H., 1971, Vapor-dominated hydrothermal systems compared with hot-water systems: Economic Geology, v. 66, p. 75-97.

White, D. E., Thompson, G. A., and Sandberg, C. A., 1964, Rock structure and geologic history of Steamboat Springs thermal area, Washoe County, Nevada: U.S. Geological Survey, Professional Paper 458-B, 63 p.

Chapter 10
RELATIONSHIPS OF TRACE-ELEMENT PATTERNS TO GEOLOGY IN HOT-SPRING-TYPE PRECIOUS-METAL DEPOSITS

Byron R. Berger and Miles L. Silberman

INTRODUCTION

Those epithermal precious-metal deposits where ore was precipitated within 100-300 m of the earth's surface such that the direct interaction of hydrothermal fluids with the surface is a major cause of ore-mineral precipitation in the upper part of the system make up the subclass known as hot-spring-type deposits (Berger and Eimon, 1983; Berger, 1985). The deposits were emplaced as small veins, stockworks, and explosive breccias in association with non-marine volcanism, generally calc-alkaline in composition. Henley (1985b, this volume) and Hayba et al. (1985, this volume) prefer to not separate hot-spring deposits as a separate class or subtype of epithermal deposits. However, we have chosen to treat hot-spring related deposits separately because of the importance of hydrothermal eruptions and accompanying brecciation to near-surface ore deposition and exploration recognition criteria (Adams, 1985, this volume).

Active geothermal systems have long been thought to be modern analogs of epithermal systems (cf. White, 1955; Weissberg et al., 1979), but it wasn't until the recent discovery of the McLaughlin gold deposit in California and the publication of data on Round Mountain, Nevada (Berger and Tingley, 1980; Tingley and Berger, 1985) and Hasbrouck Mountain, Nevada (Silberman et al., 1979; Graney, 1984) that there became a widespread recognition among explorationists of the geological and geochemical characteristics and resource importance of fossil hot-spring systems. Subsequently, study in the Bodie, California mining district by P. Herrera and M. L. Silberman (Silberman and Berger, 1985, this volume) has further linked fossil hot-spring systems to the deeper-emplaced bonanza-type epithermal vein deposits by documenting a continuum of features from the surface springs to the deep veins. Recent research at Broadlands, New Zealand by K. Brown (Henley and Brown, 1985, this volume) has shown that the geothermal fluids are near saturation with respect to gold, and electrum is actively being precipitated at the wellhead. This implies that alkaline-chloride thermal waters in general may be saturated with gold, and that the source of the gold is of less importance than the precipitation mechanisms. Although no economic concentrations of gold have yet been discovered at Broadlands, recent exploration in an active geothermal area on Lihir Island, New Guinea has resulted in the discovery of a potentially economic gold deposit (R. Henley, D.S.I.R., personal communication, 1985). Thus the link between modern geothermal systems and epithermal-type vein deposits has been closed.

The trace-element chemistry of modern geothermal systems has been studied by several workers including Weissberg (1969), Weissberg et al. (1979), Ewers and Keays (1977), White (1981), and Henneberger (1983). These studies documented the high concentrations of arsenic, antimony, mercury, thallium, and tungsten in the upper parts of geothermal systems in addition to gold and silver. The trace-element distribution in relationship to alteration in drill hole GS-5 at Steamboat Springs, Nevada was summarized by Silberman and Berger (1985, this volume). Heretofore, similar geochemical studies of hot-spring-type precious-metal deposits have not been published. It is the purpose of this paper to summarize the trace-element patterns for gold, arsenic, antimony, mercury, and thallium in two well-studied hot-spring-type systems, Hasbrouck Mountain and Round Mountain, Nevada, and to relate these patterns to the geology and hydrothermal alteration in each system.

CONTROLS ON TRACE-ELEMENT PATTERNS

The system-wide trace-element patterns observed in ore deposits represent the summation of a multiplicity of processes that are related to the time and space history of the geothermal system, the variations in the fluid chemistry of the system, the chemistry of the host rocks, and the physical nature of the heat source and the hydrothermal system including fracturing, permeability, and brecciation.

Silberman et al. (1979) documented (a) the length of time that hot-spring activity has been taking place at Steamboat Springs, Nevada, and (b) the episodic nature of the geothermal activity (Table 10.1). They found that hot-spring activity has been taking place intermittently for more than 2 million years. Henneberger (1983) determined that there have been four separate, overlapping stages of geothermal activity at Ohakuri, New Zealand (Fig. 10.1). Henneberger showed that the location of the hydrothermal activity migrated with time. Sufficient time elapsed between the stages that erosion took place between stage 1 and stage 2. Using active thermal systems both studies illustrate the time and space complexities resulting in superimposed alterations and trace-element patterns that are probably present in all hydrothermal systems, including those that are mineralized. Similar timing and spatial variations in the physical aspects of hydrothermal systems was documented by Lloyd (1972) at Orakeikorako, New Zealand, where hot-spring activity

Table 10.1--Statigraphic and age relationships of geothermal activity and the emplacement of rhyolite intrusions and andesite flows in the Steamboat Hot Springs geothermal field, Nevada (from Silberman et al., 1979)

Stratigraphic Unit	Radiometric Age
- Recent alluvium; interbedded opaline sinter	
- Lake Lahontan alluvium; interbedded opaline and cristobalite sinter	
-------Erosional disconformity-------	
- Pre-Lake Lahontan alluvium; interbedded cristobalite and chalcedonic sinter	
- Interbedded, older chalcedonic sinter	
- Rhyolite pumice; intrusion of Steamboat Hills Rhyolite	1.2 m.y.
- Basaltic andesite	2.5 m.y.
- Rhyolite pumice; interbedded chalcedonic sinter	
- Alluvium; interbedded chalcedonic sinter	
- Granodiorite (bedrock)	

has resulted in a large number of scattered hydrothermal explosion craters, the foci of which changed with time. Phreatic eruptions occur over areas where the heat flow is sufficiently high to locally exceed the boiling point with depth relation (Elder, 1981). The implication at Orakeikorako is that, as at Ohakuri, the hottest part of the surface hydrothermal activity has migrated with time. Inactive sinter terraces at different elevations and in different areas from the presently active zone of fluid flow and sinter development at Steamboat Springs, Nevada, also document the migration of fluids through time (White et al., 1964; Silberman et al., 1979). Tilted sinter and sinter within some sedimentary deposits and beneath volcanic rocks at Steamboat and younger sinter illustrate the long history of intermittent activity with periods of erosion.

Henley (1985a,b; Henley and Ellis, 1983) reviewed geothermal fluid chemistry and the chemical environments of transport and deposition of precious metals. He suggests that the ore and associated trace-element distributions observed in epithermal deposits result from the salinity and gas content of the hydrothermal fluids. In turn, the total salinity of any given geothermal system is dependent upon the host-rock compositions and tectonic setting (Henley, 1985b, this volume). Basaltic- and silicic-hosted volcanic systems have lower salinities than andesite-hosted systems. Gold-bearing systems have lower salinities and high sulfur contents whereas silver-base metal bearing systems have relatively higher salinities. In the Taupo Volcanic Zone, New Zealand, Henley (1985a) has found a coincidence of high gas flux (CO_2 + H_2S) and gold, and he hypothesizes that the gas is derived from a deep source, thus requiring a structural setting that allows upward flow from deep crustal (or even upper mantle?) regions.

The permeability of the host rocks is one of the most important physical attributes affecting the hydrothermal solutions, and is therefore an important control on hydrothermal alteration and trace-element patterns. Unless hydrothermal systems are small relative to the length and width of fractures, the permeability will be fracture-controlled as was demonstrated by Elder (1981) in the Taupo, New Zealand volcanic zone. Porosity-controlled permeability is of increased importance in small hydrothermal systems and locally in the larger systems. Faults, joints, and fractures are all important in controlling fluid flow, particularly when they constitute extensive, through-going structural systems. Porosity permeability is an intrinsic aspect of each individual host lithology. Siltstones, shales, and welded tuffs have low porosity permeability whereas pumiceous tuffs, tuff breccias, and lapilli tuffs are relatively permeable. Within any given stratigraphic layer the porosity permeability may vary resulting in stratabound zones of relatively higher and

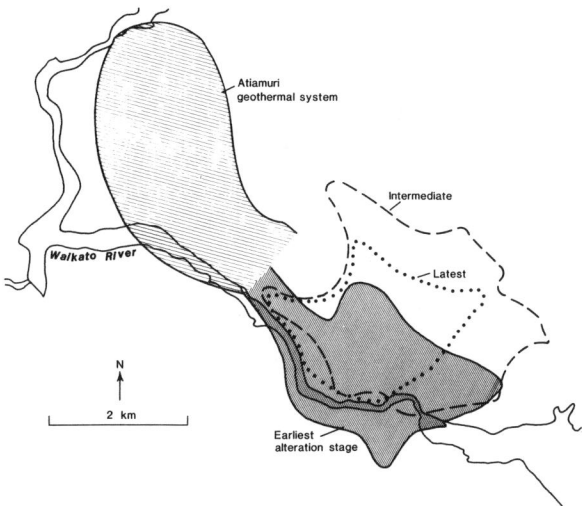

Figure 10.1. The location of distinct stages of hydrothermal activity at Ohakuri, New Zealand, based upon stratigraphic and alteration studies by Henneberger (1983). Early activity was characterized by zones of quartz-adularia, silicification, and zeolitization. Intermediate activity is recognized by the addition of kaolinite alteration probably resulting from a change in the water table. Late-stage activity is characterized by silicification and a diminishing volume of hydrothermal activity.

lower permeability. The physical nature of these zones may also change with time. The deposition of minerals (e.g., quartz, calcite, zeolites, etc.) may seal porous zones, thereby reducing permeability. Subsequent hydrothermal fracturing or tectonic activity may induce fracture-controlled permeability, reopening "locally sealed" parts of the system.

When the system consists of two-phase flow, such as a vapor-liquid mixture, the flow of one phase affects the flow rate of the other phase. Differential flow rates can result in varying vapor-liquid ratios within the system; these ratios are important controls on the partitioning of trace elements between the liquid and the vapor. Therefore, trace-element patterns may be dependent upon the vapor-liquid ratio in the fluid as well as the gas composition of the vapor phase. As the deeper-heated waters near the surface, processes occur which modify the fluid flow (Elder, 1981). Permeability varies laterally due to surface fractures, and topography becomes an important influence. A subsurface boiling water table may result in steaming ground where extreme leaching and acid-sulfate alteration assemblages may be adjacent to alteration produced by alkaline-chloride waters. Patchy areas of hot ground may be surrounded by cold ground due to the downward percolation of cold ground water (Elder, 1981). All of these surface processes result in complex trace-element distribution patterns, particularly when combined with temporal migration of the thermal activity as discussed previously.

Hydrothermal breccias are an essential attribute of the hot-spring subtype of epithermal precious-metal deposits. They are commonly associated with high-grade ores (Berger and Eimon, 1983; Berger, 1985; Sillitoe, 1985; Tingley and Berger, 1985). Hedenquist and Henley (1985) classified hydrothermal eruptions as either shallow or deep. Shallow eruptions occur within a few meters of the surface where steam flow is hampered by mineral deposition or where there are changes in the near-surface hydrology. Deep hydrothermal eruptions occur below the steam condensate zone (Hedenquist and Henley, 1985) due to hydraulic fracturing (Grindley and Browne, 1976). The deeper breccias serve to focus fluid flow, thus explaining the important association of breccias with ore in hot-spring-type deposits. Hedenquist and Henley (1985) hypothesize that hydrothermal eruptions occur where there is localized overpressuring in the upper 300 m of the geothermal system. The local sealing of fluid-flow paths by mineral deposition causes slight increases in fluid pressure with fluids predominantly flowing to other unblocked flow channels. Vapor accumulates in the sealed fractures and gas exsolution causes increased pressure on the seal from the gas cap. Hydraulic fracturing will set off a hydrothermal eruption--the fracturing either due to the gas cap exceeding lithostatic pressure or tectonic activity reopening the sealed pathways.

TRACE-ELEMENT PATTERNS IN STUDIED DEPOSITS

Hasbrouck Mountain, Nevada

Hasbrouck Mountain is located in the Divide mining district about three miles south of the town of Tonopah in west-central Nevada. Gold was first discovered on Gold Hill just east of Hasbrouck Mountain in 1902, and small gold mines in the district produced intermittently until high-grade silver ore was discovered in 1917 below the mined levels on Gold Hill (Bonham and Garside, 1979). The greatest period of production in the district was between 1920 and 1929. The Kernic vein on Hasbrouck Mountain was unusual in the early history of the district in that it was mined largely for its silver content instead of gold. The exploited mineralization on Hasbrouck Mountain was restricted to narrow veins that often consisted of silica-cemented breccias, and there is no record of any significant production. Total gold-silver production in the district was about $3.5 million (Bonham and Garside, 1979). In the 1970's, Hasbrouck Mountain was explored for its low-grade, large-tonnage gold potential, and an ore body of about 3 million tons averaging about 0.05 ounces Au/ton was discovered (A. Wallace, Cordex Exploration, written communication, 1982). Subsequent geologic studies on Hasbrouck Mountain (Silberman et al., 1979; Bonham and Garside, 1979; Graney, 1984) showed the ore system to be related to hot-spring activity as evidenced by the presence of chalcedonic sinter and extensive, multiple episodes of hydrothermal brecciation, including an eruption breccia interbedded with Siebert Formation

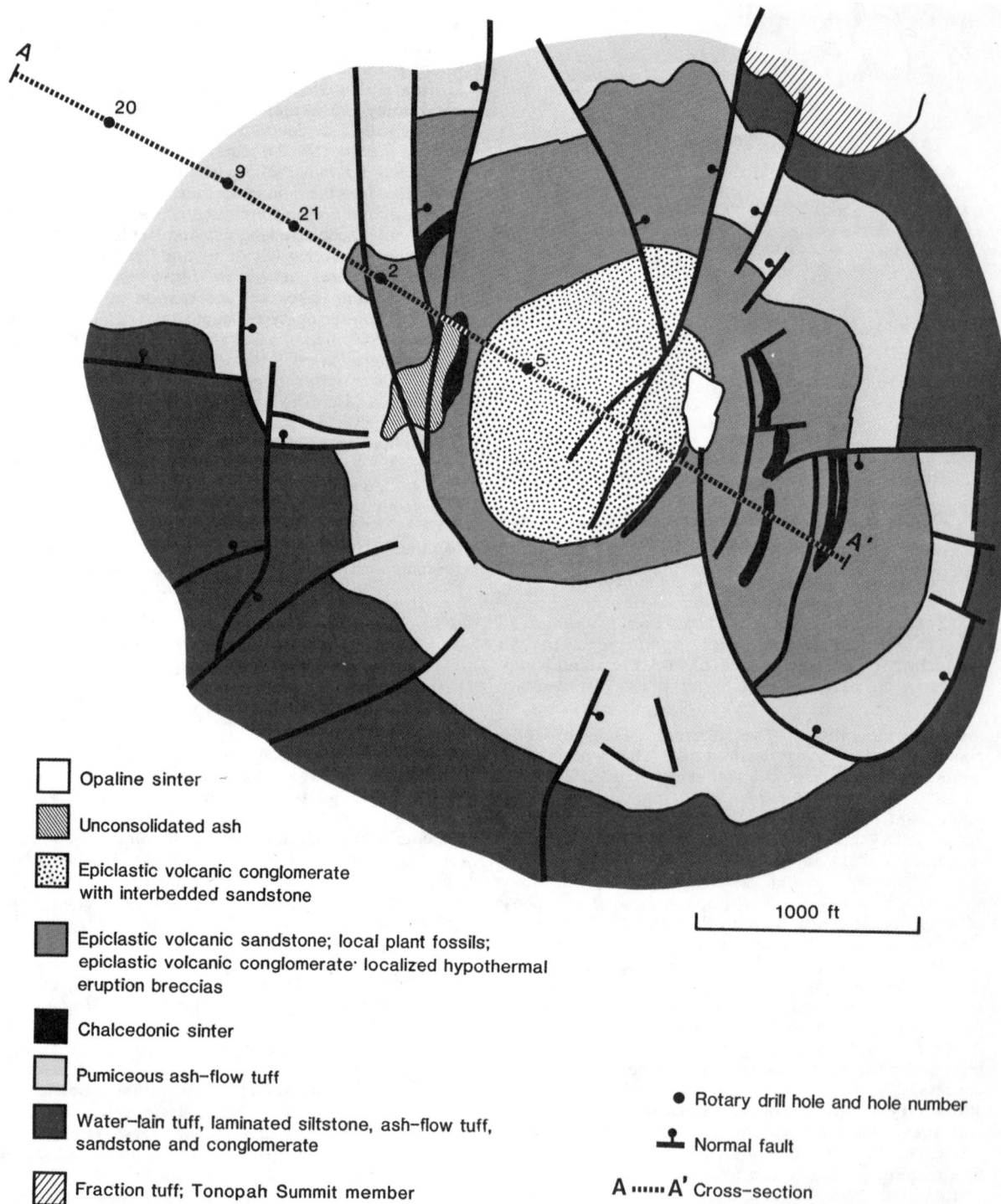

Figure 10.2. The generalized geology of Hasbrouck Mountain, Nevada, based upon the work of Graney (1984). A–A' is the line of the northwest-southeast cross section shown in Figure 10.3, with the dots showing the number and location of rotary drill holes. The top of Hasbrouck Mountain is approximately located southeast of RDH-5 along the cross-section line near the northeast-trending faults.

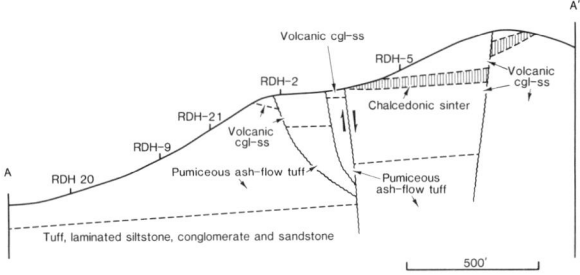

Figure 10.3. A northeast-looking, generalized geologic cross section of Hasbrouck Mountain, Nevada, also showing the locations of rotary drill holes for which alteration and trace element data are given in Figures 10.4 a-e. Refer to Table 10.2 for more detailed stratigraphic information.

sandstones (R. Henley, D.S.I.R., personal communication, 1984).

The areal geology of the Divide district was described by Bonham and Garside (1979) and the geology of Hasbrouck Mountain by Graney (1984). A simplified version of the geologic map of Hasbrouck Mountain is shown in Figure 10.2, and a cross section is shown in Figure 10.3. The oldest stratigraphic units are the Tonopah Summit and King Tonopah members of the Fraction Tuff, a quartz latite ash-flow tuff (Table 10.2). Silberman et al. (1979) report potassium-argon ages of 18.2 to 19.9 m.y. on biotite and alkali-feldspar from the tuff. The Fraction Tuff is overlain by the middle Miocene Siebert Formation, a composite unit of intercalated sedimentary, volcaniclastic, and volcanic rocks which yield K-Ar ages of 16.2 to 15.5 (Silberman et al., 1979). Rhyolite domes and related dikes of approximately the same age intrude the older units. Bonham and Garside (1979) report two episodes of dome emplacement. One is the Oddie rhyolite which is consanguinous with the mineralization in the district; the second being the post-mineralization Brougher domes. A number of northerly trending normal faults transect Hasbrouck Mountain. The stratigraphy as well as the structure constitutes an important control on the trace-element distribution patterns. The lowermost Siebert unit consists of fluvial sandstones and conglomerates with some ash-flow tuff beds. Water-lain tuffs, ash-flow tuff, epiclastic volcanic siltstone, volcanic conglomerate, and sandstone form a sequence that underlies chalcedonic sinter. Volcanic sandstone, siltstone, conglomerate, and ash-flow tuff are interbedded with and overlie the sinter and make up the youngest Siebert Formation exposed on Hasbrouck Mountain. The ash-falls and ash-flows are commonly poorly welded and some contain abundant pumice lapilli and therefore are not as easily fracturable as the welded tuffs and thinly laminated water-lain tuff. Also, the poorly welded material has a relatively higher porosity permeability than the welded and thinly laminated rocks. Hydrothermal activity was taking place locally at Hasbrouck Mountain during the development of the volcanic field hosting the activity. K-Ar ages of hydrothermal adularia and sericite from Hasbrouck are in the range of 16.3 to 15.3 m.y. (Silberman et al., 1979). Local volcanic units, including the rhyolite plugs and other flow units, are in the range of 16.8 to 14.8 m.y.

The detailed and careful study by Graney (1984) outlined two discrete foci of gold mineralization based upon the distribution of gold and the alteration patterns. These two foci appear as "mushroom-shaped" ore zones with concentric alteration envelopes. We interpret these ore zones as the sites of the primary conduits along which the hydrothermal fluids were flowing. Figures 10.4a through e show the hydrothermal alteration in a fence of drill holes that transects the two mineralized conduits. RDH-5 (Fig. 10.4a) and RDH-2 (Fig. 10.4b) penetrate the central portion of each conduit. Quartz-adularia alteration gives way to an assemblage of quartz-adularia-albite at depth. The quartz-adularia zones in the two foci coalesce in the upper part of the system rendering the appearance on the surface of only one relatively large alteration zone. RDH-21 (Fig. 10.4c) was collared in quartz-adularia-altered rock and went into quartz-adularia-illite alteration beneath it. The patterns in drill holes 2, 5, and 21 suggest that in the main fluid conduits two-feldspar alteration marks the axial root of the ore-bearing quartz-adularia zone, and that both the albitic and adularia zones are flanked by illitic alteration. RDH-9 (Fig. 10.4d) and RDH-20 (Fig. 10.4e) are peripheral to the main conduit areas and show a marked increase in illitic alteration and the appearance of montmorillonite as an alteration product. The data presented by Graney (1984) indicate that illitic alteration is also present as a major alteration mineral beneath the adularia- and albite-bearing zones.

Exposures in underground workings of the quartz-adularia zone show it to be extensively brecciated. The breccias may be small or very extensive, and appear to have happened repeatedly in single locations. There are two major classes of brecciation-- uncemented and cemented. The uncemented breccias consist of rock fragments of varying sizes in a finely comminuted matrix of rock flour and rubble. These breccias are interpreted to have resulted from sudden eruptions of locally over-pressured vapor-enriched fluids with the brecciation resulting from the rapid expansion of the gases upon breaching of the localized sealing and consequent boiling. The continuity of finely laminated beds across this type of breccia indicates that there has been little or no vertical transport of rock fragments during the explosive eruptions. The cemented breccias are similar in texture and structure to the uncemented variety except that the rock fragments are supported by a matrix of silica and disseminated pyrite. The ore occurs as quartz-adularia veins and veinlets within and around the breccias as well as within the silica-cemented breccias.

The trace-element patterns observed at Hasbrouck Mountain are complex, and in all likelihood are due to the interaction of several physical and chemical processes taking place episodically within the fluid conduits. As a consequence of this complex genesis, in detail the distributions of individual

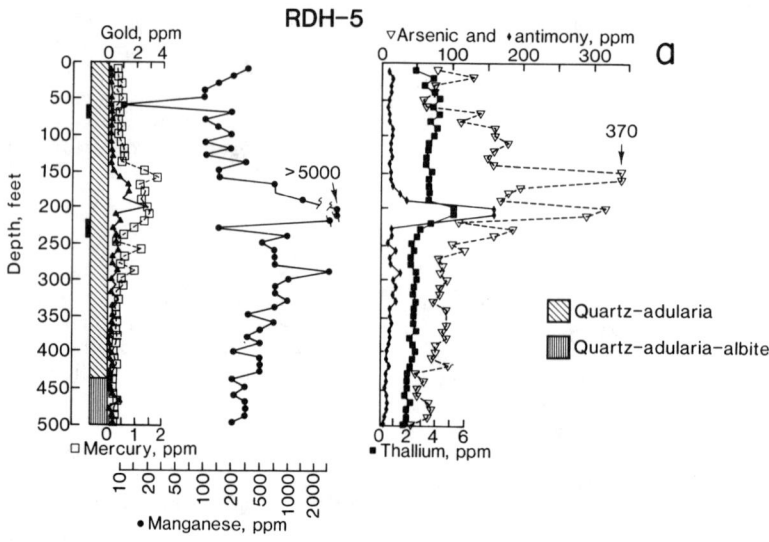
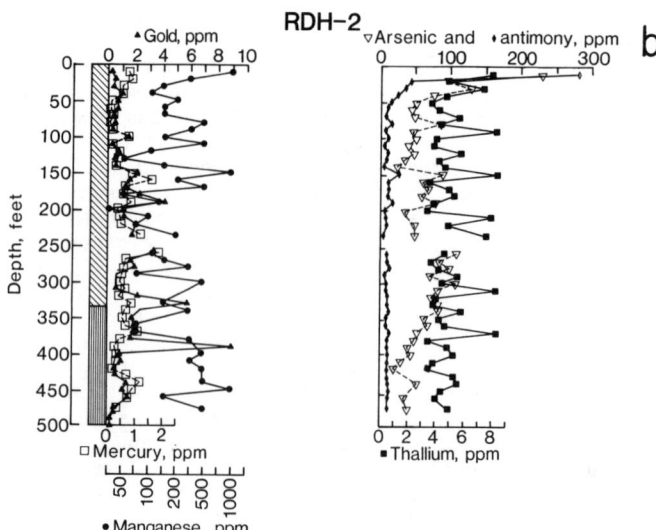
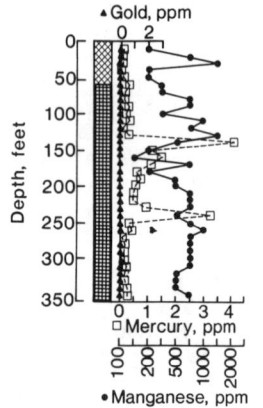
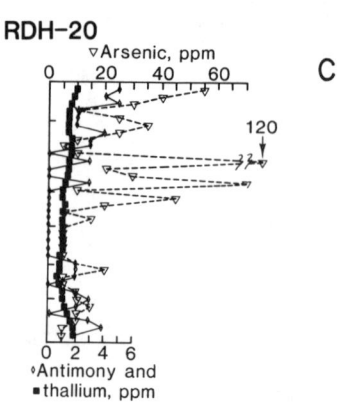

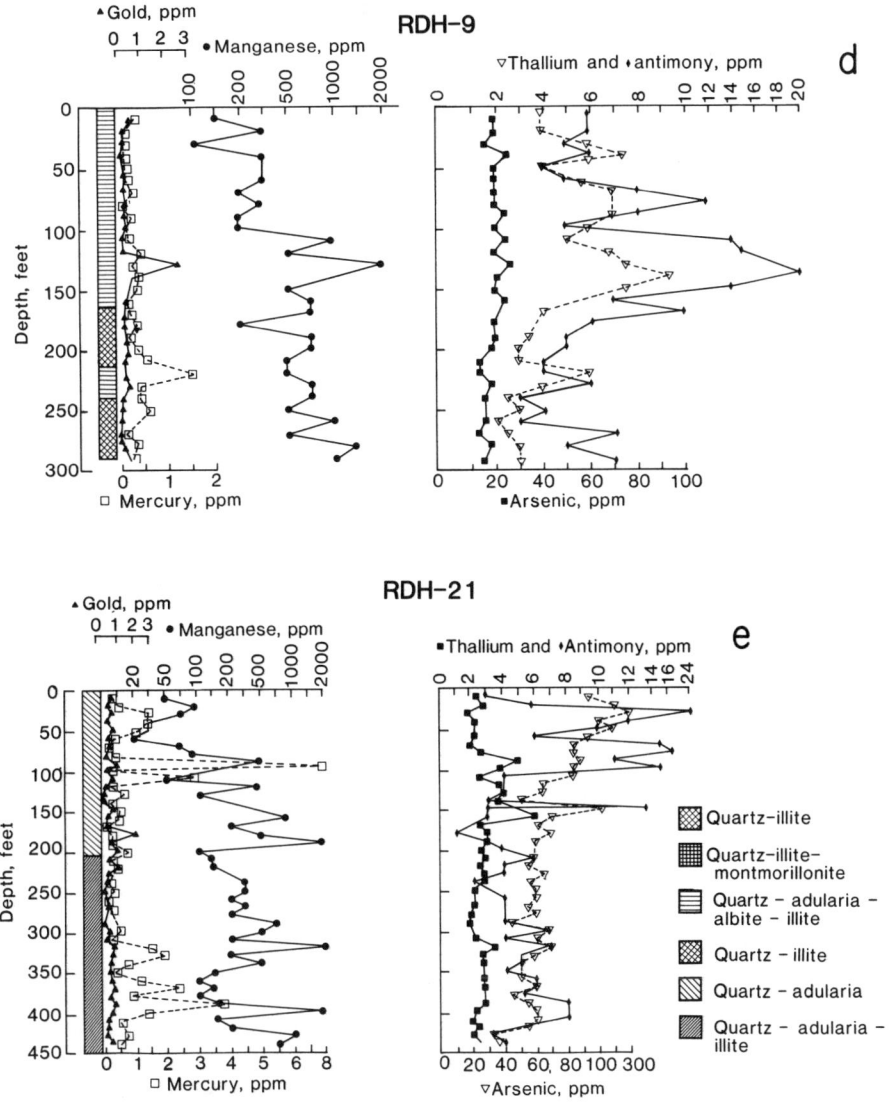

Figure 10.4.a-e. Hydrothermal alteration and data for selected trace elements in rotary drill holes along cross section A-A' (refer Figs. 10.2 and 10.3).

    a). Data from RDH-5 which penetrated the central part of the eastern-most focus of hydrothermal activity. The heavy black line from 50-70-foot depth shows the location of chalcedonic sinter (refer to Fig. 10.3) in the hole. The heavy black line from 220-240-foot depth marks an interval of intense brecciation.

    b). Data from RDH-2 which penetrated the central part of the western focus of hydrothermal activity.

    c). Data from RDH-20 which penetrated the most distal alteration zone away from the main focus of hydrothermal activity.

    d). Data from RDH-9 which penetrated lateral alteration zones adjacent to the main focus of hydrothermal activity.

    e). Data from RDH-21 which penetrated a small lateral wedge of quartz-adularia altered rock carrying ore-grade gold and bottomed in a peripheral alteration zone.

Table 10.2--Stratigraphic relationships at Hasbrouck Mountain, Nevada (after Graney, 1984; Bonham and Garside, 1979; ages from Silberman et al., 1979)

- Tertiary opaline sinter (some plant fragments)

- Unaltered Tertiary air-fall ash

--------------Main Mineralizing Event----------------(15.3-16.3 m.y.)

- Miocene Divide Andesite; quartz latite dike
  (approximate age 14.8-16.9 m.y.)

- Miocene Siebert Formation (15.5-16.2 m.y.)

  •Volcanic conglomerate; some interbedded sandstone and ash-flow tuff

  •Volcanic sandstone and siltstone; some plant fragments

  •Chalcedonic sinter

  •Volcanic conglomerate with some sandstone; local hydrothermal fall-back breccias

  •Pumiceous ash-flow tuff

  •Water-lain tuff; interbedded volcanic siltstone and sandstone

  •Crystal-lithic ash-flow tuff

  •Fluvial sandstone and conglomerate

----------------------Unconformity----------------------------------

- Miocene Fraction Tuff

  •King Tonopah Member; vitric-lithic rhyolite ash-flow tuff
   (approximate age 18.2-19.9 m.y.)

  •Tonopah Summit Member; vitric quartz latite to rhyolite ash-flow tuff

---

elements and the relationships between the various elements are themselves complex and vary depending on the scale at which the system is evaluated. Figures 10.4a-e show the vertical distributions of selected trace elements in a fence of drill holes referred to above. Assuming that the foci of the hydrothermal system are outlined by the distribution of quartz-adularia alteration, the concentrations in drill cuttings of gold, silver, arsenic, antimony, mercury, and thallium are relatively more enriched in the potassically altered central parts of the system (Table 10.3), while zinc, lead, manganese, and boron are enriched in all alteration zones peripheral to the quartz-adularia alteration zones. When the concentrations of elements in the quartz-adularia-illite zones (peripheral to the quartz-adularia) are compared to the more distal quartz-illite-montmorillonite zones, thallium, mercury, and zinc are more enriched in the outer quartz-illite-montmorillonite zones and gold and arsenic are relatively more abundant in the quartz-adularia-illite zones. Therefore, at the scale of the whole system, there appears to be an inner zone of arsenic, antimony, thallium and mercury related to the precious-metal ore and an outer zone of mercury, boron, and manganese that bounds the ore zones. The results are reminiscent of the elemental zoning surrounding the Au mineralized zone at Paramount (Silberman and Berger, 1985, this volume). Although ore occurs only in association with quartz-adularia alteration, gold is above the analytical detection level throughout the altered area.

The trace-element distributions in the drill holes shown in Figures 10.4a-e are complex when examined in detail. RDH-5 (Fig. 10.4a) was collared stratigraphically higher than the chalcedonic sinter (shown as heavy black line from 60-70-foot depth on Figure 10.4a) and penetrated the entire breadth of the

Table 10.3--The concentration of selected trace elements associated with the quartz-adularia zones at Hasbrouck Mountain, Nevada, vis-a-vis concentrations associated with all other alteration zones

Quartz-adularia zone (n = 193 samples)			All other alteration zones (n = 205 samples)		
Element	Minimum (ppm)	Maximum (ppm)	Element	Minimum (ppm)	Maximum (ppm)
Au	0.05	6.2	Au	0.05	2.5
Hg	0.06	7.0	Hg	0.02	4.1
As	3.8	550	As	2.0	200.0
Sb	1.0	280	Sb	1.0	52.0
Tl	1.4	8.4	Tl	0.6	12.0
W	1.0	270	W	3.0	79.0

ore-bearing quartz-adularia alteration zone. The top of the higher-grade gold zone occurs about 80 feet beneath the sinter. Mercury, antimony, and arsenic mimic the gold distribution pattern. Manganese also appears to be higher in the gold zone in this hole. The heavy black line between 220-240-foot depth in RDH-5 (Fig. 10.4a) marks an interval of intense brecciation. Manganese shows a marked decrease within the breccia, thereby effectively forming a halo around the breccia zone. Manganese oxides are commonly observed adjacent to breccias both on the surface and in underground workings. As a general rule gold, arsenic, and antimony show a weak negative correlation with depth in both RDH-5 and -2 (Fig. 10.4b), the central conduits of the hydrothermal system. Mercury and thallium show no particular trend with depth in these holes. Peripheral to the main part of the hydrothermal system (RDH-21, -9, and -20, Figs. 10.4c-e), the gold, arsenic, and antimony continue to decrease with depth as does thallium to a slight degree, but the mercury tends to show an increase with depth.

Round Mountain, Nevada

Round Mountain is located in the Round Mountain mining district about 50 miles north of the town of Tonopah, Nevada, on the western slope of the Toquima Range. Gold was initially discovered at Round Mountain in the 1890's, but production did not take place until 1906, and significant lode mining continued until 1942. Through 1969, about 350,000 ounces of gold and 360,000 ounces of silver were produced (Tingley and Berger, 1985). In the 1960's renewed exploration outlined substantial tonnages of low-grade gold-silver mineralization containing 9.4 million ounces gold and 16.9 million ounces silver, and open-pit mining has been ongoing since 1977. Based upon geologic and geochemical information gathered by Tingley and Berger (1985) at Round Mountain, the ore deposit was interpreted by them to have been formed in the upper portion of a hot-spring system. Their evidence includes a breccia pipe with included blocks of sandstone, conglomerate, and sinter 100-200 feet down in the pipe beneath their original stratigraphic level, a hydrothermal-eruption (?) breccia interbedded with epiclastic rocks and syn-sedimentary quartz-sulfide beds within the sediments.

The regional geology around Round Mountain was mapped by Shawe (1981), and the detailed deposit geology (Fig. 10.5) by Tingley and Berger (1985). A schematic stratigraphic column is shown in Figure 10.5. The oldest rocks in the district are complexly folded Cambrian to Ordovician schist, quartzite, argillite, and limestone intruded by a Cretaceous muscovite-biotite granite. In the Fairview mine (Fig. 10.5), the granite is pervasively altered with milky quartz veins. Late Oligocene to early Miocene rhyolite ash-flow tuffs are the main host rocks to ore. The tuffs yield K-Ar ages of 26.1 to 24.7 m.y. on biotite and sanidine. The lowest unit is a welded ash-flow tuff megabreccia of undetermined thickness. Blocks of Paleozoic sediments and Cretaceous granite occur in the breccia and vary in size from pebbles to blocks several hundred feet across. A complex, thick quartz-sanidine rhyolite ash-flow unit overlies the megabreccia consisting of a lower 400- to 500-foot-thick non-welded base grading upward into an 800-foot-thick welded tuff which is then transitional upward into a 75- to 100-foot-thick non-welded top. The ash-flows are capped by a sequence of epiclastic rocks including water-lain tuff, sandstone, conglomerate, and bedded silica (interpreted by Tingley and Berger (1985) to be subaqueous chalcedonic sinter).

Gold mineralization is controlled primarily by northwest-trending fractures (Fig. 10.6) and occurs as low-angle and vertical vein and sheet zones, breccia-filling and stockworks in the welded tuff, and as microveinlets and disseminations in the lower non-welded tuff immediately above the megabreccia. All of these mineralization types are genetically related and their location and variation in form mainly reflect variations in physical properties of the host rocks. The ore minerals are electrum, free gold, and auriferous pyrite. K-Ar age determinations on adularia and

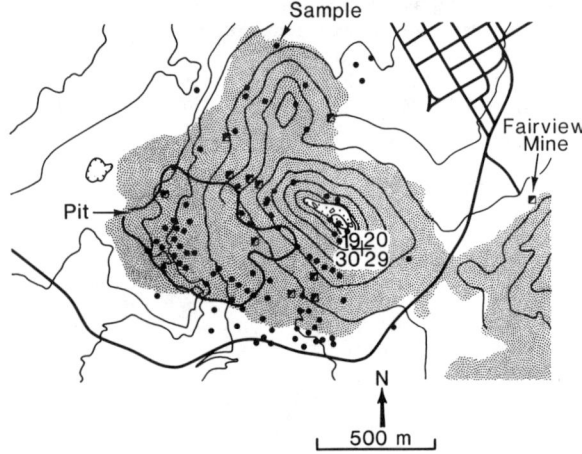

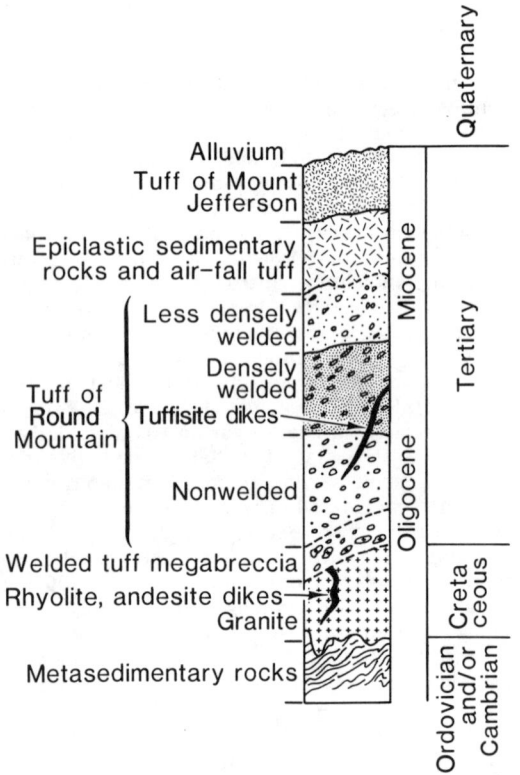

Figure 10.5. A generalized geologic map and schematic stratigraphic column for Round Mountain, Nevada (from Tingley and Berger, 1985). The sample locations for data used to construct Figures 10.7-10.10 are shown as black dots. The open-pit outline shows the mine as it was in late 1979. The area of bedrock outcrop is shown in a stippled pattern.

sericite within veins and adjacent to breccias range from 25.1 to 25.5 m.y. (Tingley and Berger, 1985).

All of the veins contain quartz and adularia and have irregular, but persistent, alteration envelopes of initially sericitic-argillic and then propylitic mineral assemblages. The gold occurs as free gold and electrum on the quartz and adularia crystals and within sulfides in the veins. The phyllic-argillic alteration reflects the presence of plagioclase totally replaced by white mica (a mixture of sericite and illite), biotite completely altered to white mica and chlorite, and pumice replaced by mixtures of white mica and kaolinite. Sanidine phenocrysts are embayed. The propylitic alteration consists of plagioclase partially altered to white mica, biotite variably altered to chlorite-white mica, and the rock matrix altered to a clay-white mica mixture.

The breccia zones and stockworks are common, but only one area in the Sunnyside mine beneath Round Mountain has been mined to any extent. The breccias formed from the repeated fracturing, quartz veining and silica flooding, and rebrecciation of the host tuff. Vuggy quartz often cements the breccia fragments. Gold occurs as free gold and electrum and within sulfides in the vugs and as inclusions in sulfides in the silica matrix of the breccia. A zone of intense phyllic-argillic alteration surrounds the silicified breccias.

A disseminated, blanket-like deposit occurs in the lower non-welded portion of the ore-bearing ash-flow tuff, and the alteration in this zone consists of quartz, white mica, and possibly adularia. The mineralization occurs as microveinlets of quartz, sulfides filling pumice cavities, and disseminated sulfides throughout the tuff. Quartz-chlorite alteration generally underlies the ore-bearing rock.

The distribution of mineralization at Round Mountain is complex and cannot be simplified into the distinct and separate centers of hydrothermal activity in the same manner that Graney (1984) was able to do at Hasbrouck Mountain. Nevertheless, there is a general zone beneath Round Mountain and Stebbins Hill defined by pervasive silicification centered about Round Mountain and even more extensive phyllic-argillic alterations that accompany vein and breccia ore, and a second zone on the southwestern side of Round Mountain where veins transecting propylitically altered rock predominate and the phyllic-argillic alteration is confined to narrow selvages along the veins (Fig. 10.6). The area of pervasive alteration (hachured pattern on Figure 10.6) was interpreted by Tingley and Berger (1985) to be a linear chain of closely spaced fluid channels and the alteration envelopes around the individual channels have coalesced to create the apparent single alteration zone. The channels themselves are marked by the presence of hydrothermal breccias, each breccia representing multiple events of cementation with quartz, adularia, and sulfides and subsequent rupturing. Ore is found both within sulfides and vugs in the breccias and in sulfides along low-angle veins that in general dip towards the breccia centers and appear to be concentric around the breccias. Ore-bearing high-angle veins are found to both crosscut the breccias and be truncated by the breccias. Tingley and Berger interpret the low-angle veins to have been

Figure 10.6. Generalized structure and alteration map of Round Mountain, Nevada (modified from Tingley and Berger, 1985). The pervasive areas of silicification and phyllically altered rock occur where there are discrete foci of hydrothermal brecciation in the subsurface. Quartz-adularia veins occur along all of the northwest-trending fractures and some of the northeast-trending fractures.

derived from the multiple injection of vein constituents along low-angle cooling joints in the welded tuff during the hydrothermal explosions that created the breccia centers. The sequential injection of material has dilated the low-angle joints to produce the thick veins; the vein thicknesses decrease away from the inferred centers of fluid flow and brecciation. The second zone of mineralization probably represents several separated fluid channels and a long, linear hydrothermal vent area known as the Automatic structure.

As a consequence of the complex mineralization history at Round Mountain, the geochemical patterns (Figs. 10.7-10.10) have an even more complicated expression and genesis than those at Hasbrouck Mountain. In general, areas that are enriched in gold and silver (Fig. 10.7) are also anomalous in the pathfinder elements arsenic and antimony (Fig. 10.8), thallium and mercury (Fig. 10.9), and molybdenum and tungsten (Fig. 10.10). The gold deposits are not characteristically enriched in copper, zinc, and lead, although locally these base metals occur in anomalous amounts. Zinc and copper are most commonly at highest concentration along with silver in areas of manganese enrichment. The mineralized area characterized by pervasive alteration is marked by an enrichment of arsenic, antimony, and thallium and a depletion of calcium and manganese. Tungsten, silver, and molybdenum are also enriched in this area with the tungsten restricted to that part of the area with the most intense silicification. Gold is present in amounts exceeding 0.05 parts per million throughout this area of pervasive alteration (Figs. 10.6 and 10.7), but also overlaps in its entirety the adjacent zone on the southwest side of Round Mountain and is thus an indicator element of the whole breadth of the highly mineralized area. There is a small-scale geochemical zoning that is discernible around each of the centers of brecciation in the pervasively altered zone. Manganese oxides occur in abundance along the low-angle veins out away from the breccias. The manganese is seen in some instances to completely ring breccia pipes and pods. As mentioned, the manganese is enriched in silver and the base metals. Gold along the low-angle veins and within sulfides in the breccias averages about 650 fine, whereas crystalline gold in vugs within the breccias is greater than 900 fine.

The second alteration zone on the southwest side of Round Mountain contains anomalous arsenic, antimony, and thallium within the veins, but there is no pervasive saturation of large concentrations of these elements out into the propylitcally altered wallrock. Very high-grade concentrations of free gold

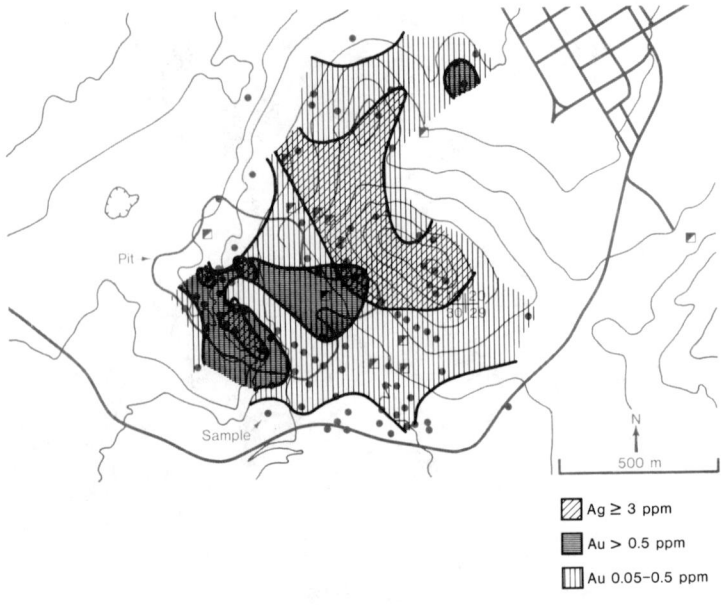

Figure 10.7. The distribution of gold and silver in rock samples from the surface at Round Mountain, Nevada (from Tingley and Berger, 1985).

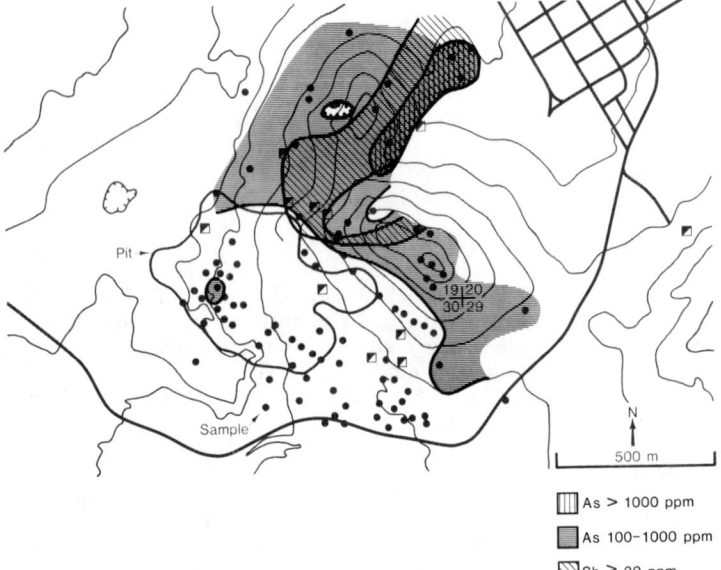

Figure 10.8. The distribution of arsenic and antimony in rock samples from the surface at Round Mountain, Nevada (from Tingley and Berger, 1985).

were stoped at two locations, the northwest (Sphinx Pit) and southeast (No. 2 Glory Hole) ends of the Automatic structure. Near these high-grade zones the Automatic structure is intensely brecciated and argillized and the fractures are commonly coated with fluorite. Away from the Automatic structure, per se, breccias are not extensive suggesting that slightly different physical processes occurred during mineralization in this zone as compared to the pervasively altered zone.

Arsenic, antimony, mercury, and thallium appear to have the highest concentrations in surface samples in the structurally high parts of the hydrothermal system (Figs. 10.8 and 10.9). Antimony is highest in areas of more intense argillic alteration such as on the east side of Stebbins Hill. In order to document the changes in metal concentrations with depth, it was necessary to analyze a series of rotary drill holes across both of the generalized altered areas. The apparent decrease in surface samples in the more volatile elements with depth is also demonstrated from the analyses of drill cuttings. The lowest whole-rock concentrations of arsenic and antimony occur in the blanket-like deposit in the lower non-welded tuff, although selected samples of quartz microveinlets in this deposit type are still highly anomalous in arsenic.

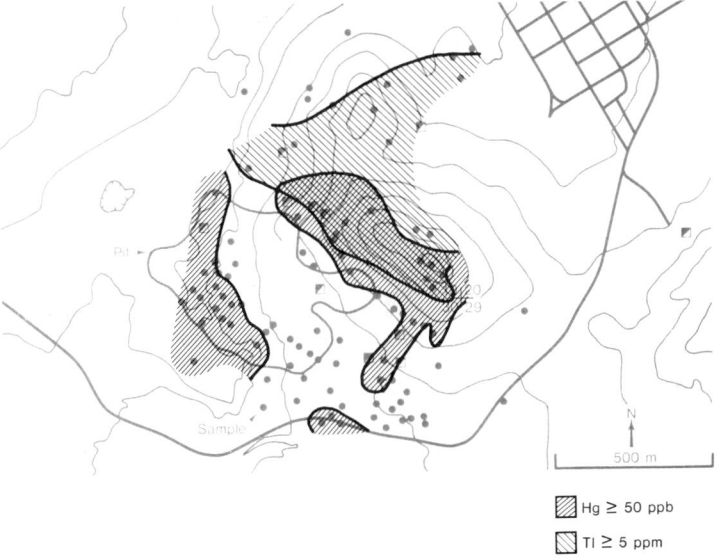

Figure 10.9. The distribution of mercury and thallium in rock samples from the surface at Round Mountain, Nevada (from Tingley and Berger, 1985).

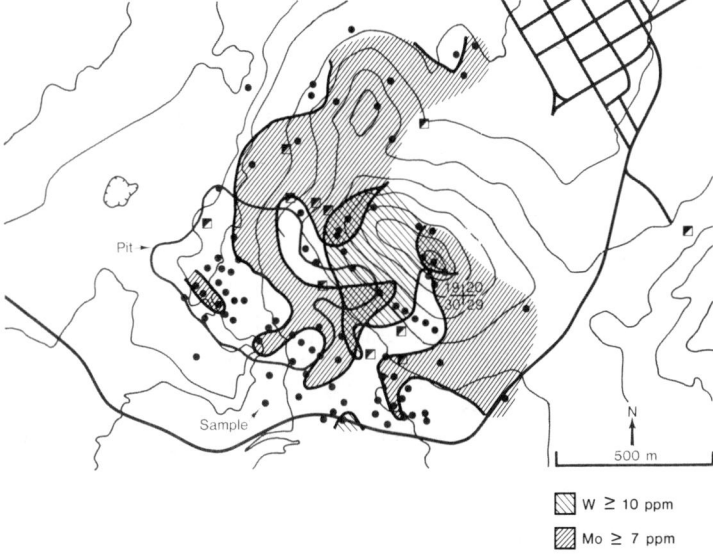

Figure 10.10. The distribution of tungsten and molybdenum in rock samples from the surface at Round Mountain, Nevada (from Tingley and Berger, 1985).

Although thallium decreases with depth in the lower non-welded tuff, the decrease in concentration is less dramatic than with either arsenic or antimony.

DISCUSSION

The major goals in studying the trace-element patterns at Hasbrouck Mountain and Round Mountain are to provide insights into the origin of ore-bearing hot-spring systems and to provide data and interpretations useful to explorationists. Based upon the detailed study of these two deposits, it was found that they have a number of features in common. Both of the deposits studied concentrated similar suites of elements--gold, silver, arsenic, antimony, mercury, thallium, molybdenum, and tungsten. Both deposits have a paucity of base metals associated with the ore. Additionally, both are related to non-marine rhyolitic volcanic activity with evidence of consanguineous high-level intrusive activity, and at both deposits the timing of mineralization overlaps the period of active volcanism. The hydrothermal fluids appear to have been focused into specific channels with repeated episodes of localized self-sealing through mineral deposition followed by brecciation and subsequent ore deposition. It is not known if tectonic activity, hydrothermal overpressuring due to a gas cap as suggested by Hedenquist and Henley (1985), or both, caused the brecciation or rupturing of the seal. The occurrence of breccia veins and dikes ostensibly unrelated to through-going fractures suggests that overpressuring plays some role in hydrothermal eruptions although active tectonism is known to accompany volcanic and intrusive activity.

For the most part, ore-grade concentrations of the precious metals are associated with the precipitation of the gangue minerals quartz and adularia. This assemblage is deposited in the main hydrothermal flues. The pathfinder elements arsenic, antimony, thallium, and mercury also have the highest concentrations within the main fluid channels and are thus effective indicators of the foci of hydrothermal activity. At Hasbrouck Mountain and Round Mountain, mercury is anomalous both within the core potassic alteration zone and in the outer periphery of the hydrothermal system. Manganese shows a similar pattern as mercury by forming a peripheral ring around the main areas of fluid flow. Silberman and Berger (1985, this volume) documented Hg and Mn halos around individual veins or mineralized zones in similar parts of mineralized systems at Paramount and Bodie. Thus, these elements give evidence of being more generally useful for delineating areas of mineralization. However, the Bodie data and differences in the distribution of these elements in the four hydrothermal systems described show that considerable variation occurs.

Within the core potassic zone, the vertical trace-element patterns appear to be dependent upon the relative permeabilities of the rocks. More porous portions of sedimentary beds in the Siebert Formation at Hasbrouck Mountain show increased concentrations of all the pathfinder elements (Figs. 10.3 and 10.4). Areas of brecciation of both Hasbrouck Mountain and Round Mountain also are enriched in these same trace elements.

Lateral flow may become particularly important when higher elevations flank the upper parts of the systems and cold surface waters force the fluid to flow outward (Hanaoka, 1980) or when the fluids flow laterally beneath an impermeable layer. In situations where lateral flow has taken place, a zone of argillic alteration underlies the main fluid channel, thus giving the appearance of a lateral rather than an underlying alteration assemblage. Lateral flow leads to hot-spring vent areas that are up to several kilometers away from the main heat source as at Waiotapu in New Zealand (Hedenquist and Henley, 1985), and R. Henley (D.S.I.R., personal communication, 1983) speculates that the Fairview mine at Round Mountain may be a lateral flow vent area. The alteration at the Fairview is consistent with this interpretation in that intense quartz-white mica alteration underlies the quartz-adularia ore-bearing rock. A similar interpretation may be made of the Hasbrouck Mountain mineralization in that illitic alteration underlies the potassic zone with the "feeder" structures dipping to the east towards Gold Hill where there has been a focus of intrusive activity.

The compilation of geochemical data from Hasbrouck Mountain and Round Mountain shows that the physical and alteration features of the two deposits are of considerable importance to the patterns observed. Deposit-specific structural features and host-rock permeabilities appear to be very important controls on the focusing of hydrothermal fluids into individual flues in each deposit. Therefore, geochemical and alteration zoning will vary laterally and vertically between different deposits and no theoretical trace-element patterns or specific concentrations of selected pathfinder elements should be expected. However, geochemical zoning can be determined for any given system on an empirical basis and may then be used as a guide to mineralization in that district. Elements that serve as direct pathfinders to the precious metals in one deposit may be less useful in another deposit, although most hot-spring-type mineral deposits probably contain the same general suite of elements--gold, silver, arsenic, antimony, thallium, mercury, tungsten, and molybdenum. Occasionally, elements such as selenium and tellurium may be important pathfinder elements (e.g., DeLamar, Idaho), although they were not found to be of importance in the two deposits discussed in this paper.

REFERENCES

Adams, S. S., 1985, Using geological information to develop exploration strategies for epithermal deposits; in Berger, B. R., and Bethke, P. M. (eds.), Geology and Geochemistry of Epithermal Systems: Society of Economic Geologists, Reviews in Economic Geology, v. 2.

Berger, B. R., 1985, Hot-spring type gold deposits; in Tooker, E. W. (ed.), Geologic Characteristics of Sediment and Volcanic-Hosted Disseminated Gold Deposits--Search for an Occurrence Model: U.S. Geological Survey, Bulletin 1646.

Berger, B. R., and Eimon, P. L., 1983, Conceptual models of epithermal precious-metals deposits; in Shanks, W. C., III (ed.), Cameron Volume on Unconventional Mineral Deposits: Society of Mining Engineers, p. 191-205.

Berger, B. R., and Tingley, J. V., 1980, Geology and geochemistry of the Round Mountain gold deposit, Nye County, Nevada (abs.): Precious Metals Symposium, Nevada Bureau of Mines and Geology, Reno, NV, p. 18c.

Bonham, H. F., Jr., and Garside, L. J., 1979, Geology of the Tonopah, Lone Mountain, Klondike, and northern Mud Lake quadrangles, Nevada: Nevada Bureau of Mines and Geology Bulletin 93, 142 p.

Elder, J. W., 1981, Geothermal Systems: Academic Press, London, 508 p.

Ewers, G. R., and Keays, R. R., 1977, Volatile- and precious-metal zoning in the Broadlands geothermal field, New Zealand: Economic Geology, v. 72, p. 1337-1354.

Graney, J. R., 1984, Controls of alteration and precious-metal mineralization in a fossil hydrothermal system, Hasbrouck Mountain, Nevada (abs.): Geological Society of America, Abstracts with Programs, v. 16, no. 6, p. 523.

Grindley, G. W., and Browne, P. R. L., 1976, Structural and hydrological factors controlling the permeability of some hot water geothermal fields: United Nations Symposium Development and Use Geothermal Resources, 2nd, San Francisco, Proceedings, p. 377-386.

Hanaoka, N., 1980, Numerical model experiment of hydrothermal system-topographic effects: Bulletin of the Geological Survey of Japan, v. 31, p. 321-332.

Hayba, D. O., Bethke, P. M., Heald, P., and Foley, N. K., 1985, Geologic, mineralogic, and geochemical characteristics of volcanic-hosted epithermal precious-metal deposits; in Berger, B. R., and Bethke, P. M. (eds.), Geology and Geochemistry of Epithermal Systems: Society of Economic Geologists, Reviews in Economic Geology, v. 2.

Hedenquist, J. W., and Henley, R. W., 1985, Hydrothermal eruptions in the Waiotapu geothermal system, New Zealand: Their origin, associated breccias, and relation to precious-metal mineralization: Economic Geology, v. 80, p. 1640-1668.

Henley, R. W., 1985a, Ore transport and deposition in epithermal environments: Proceedings Symposium on Stable Isotopes and Fluid Processes in Mineralization, University of Brisbane, Geological Society Australia Special Publication, p. 1-43.

Henley, R. W., 1985b, The geothermal framework of epithermal deposits; in Berger, B. R., and Bethke, P. M. (eds.), Geology and Geochemistry of Epithermal Systems, Society of Economic Geologists, Reviews in Economic Geology, v. 2.

Henley, R. W., and Brown, K. L., 1985, A practical guide to the thermodynamics of geothermal fluids and hydrothermal ore deposits; in Berger, B. R., and Bethke, P. M. (eds.), Geology and Geochemistry of Epithermal Systems, Society of Economic Geologists, Reviews in Economic Geology, v. 2.

Henley, R. W., and Ellis, A. J., 1983, Geothermal systems, ancient and modern: Earth Science Reviews, v. 19, p. 1-50.

Henneberger, R. C., 1983, Petrology and evolution of the Ohakuri hydrothermal system, Taupo volcanic zone, New Zealand: Unpublished M.S. thesis, University of Auckland, 141 p.

Lloyd, E. F., 1972, Geology and hot springs of Orakeikorako: New Zealand Geological Survey, Bulletin 37.

Shawe, D. R., 1981, Geologic map of the Round Mountain quadrangle, Nye County, Nevada: U.S. Geological Survey, Open-File Report 81-515.

Silberman, M. L., and Berger, 1985, Relationship of trace-element patterns to alteration and morphology in epithermal precious-metal deposits; in Berger, B. R., and Bethke, P. M. (eds.), Geology and Geochemistry of Epithermal Systems, Society of Economic Geologists, Reviews in Economic Geology, v. 2.

Silberman, M. L., Bonham, H. F., Jr., Garside, L. J., and Ashley, R. R., 1979, Timing of hydrothermal alteration-mineralization and igneous activity in the Tonopah mining district and vicinity, Nye and Esmeralda counties, Nevada; in Ridge, J. D. (ed.), Proceedings of the Fifth Quadrennial Symposium, International Association on the Genesis of Ore Deposits: Nevada Bureau of Mines and Geology, Report 33, p. 119-126.

Silberman, M. L., White, D. E., Keith, T. E. C., and Doctor, R. D., 1979, Duration of hydrothermal activity of Steamboat Springs, Nevada from ages of spatially associated volcanic rocks: U.S. Geological Survey, Professional Paper 458-D, 13 p.

Sillitoe, R. H., 1985, Ore-related breccias in volcanoplutonic arcs: Economic Geology, v. 80, p. 1467-1514.

Tingley, J. V., and Berger, B. R., 1985, Lode gold deposits of Round Mountain, Nevada: Nevada Bureau of Mines and Geology Bulletin 100, 62 p.

Weissberg, B. G., 1969, Gold-silver ore-grade precipitates from New Zealand thermal waters: Economic Geology, v. 64, p. 95-108.

Weissberg, B. G., Browne, P. R. C., and Seward, T. M., 1979, Ore metals in active geothermal systems; in Barnes, H. L. (ed.), Geochemistry of Hydrothermal Ore Deposits, Second Edition: John Wiley and Sons, New York, p. 739-780.

White, D. E., 1955, Thermal springs and epithermal ore deposits; in Bateman, A. M. (ed.): Economic Geology, 50th Anniversary Volume, p. 99-154.

White, D. E., 1981, Active geothermal systems and hydrothermal ore deposits: Economic Geology, 75th anniversary volume, p. 392-423.

White, D. E., Thompson G. A., and Sandberg, C. A., 1964, Rock structure and geologic history of Steamboat Springs thermal area, Washoe County, Nevada: U.S. Geological Survey, Professional Paper 458-B, 63 p.

Chapter 11
BOILING, COOLING, AND OXIDATION IN EPITHERMAL SYSTEMS: A NUMERICAL MODELING APPROACH
Mark H. Reed and Nicolas F. Spycher

INTRODUCTION

Some active geothermal systems are currently depositing gold, silver, and base metals, and most "epithermal" ore deposits formed in once-active geothermal systems (e.g., White, 1981; Henley, 1985, this volume). Boiling of hot (100°-300°C) ground water in such systems is a process of fundamental significance because it fixes temperature gradients (e.g., White et al., 1971; Muffler et al., 1971; Henley and Ellis, 1983) and causes precipitation of sulfide, carbonate, and silicate minerals (e.g., Buchanan, 1981; Berger and Eimon, 1983). The gas phase, including H_2O, CO_2, and H_2S, when condensed and oxidized near the surface, produces acid waters that generate argillic alteration of rocks and which may trigger deposition of precious metals. The geologic and hydrologic framework of a boiling geothermal system is depicted in Figure 11.1, based in part on White et al. (1971), Henley and Ellis (1983), Berger and Eimon (1983), and Steven and Eaton (1975). Figure 11.2 corresponds to Figure 11.1, showing in flow-diagram form the chemical components and processes in the hydrothermal system. These include boiling (A, Figs. 11.1 and 11.2), condensation of the boiled gas in rock (B), oxidation of the gas by the atmosphere (C), condensation followed by oxidation of the gas in cool, fractured ground (D), mixing of acid ground waters with the boiled liquid (E), and mixing of cold ground water with the boiled liquid (F). All of these processes shape the chemistry of geothermal systems and several of them are responsible for ore formation in epithermal systems. We present here some results of detailed calculations of heterogeneous equilibria (Reed, 1982) that apply to the boiling, cooling, fluid-fluid mixing, condensation, and oxidation depicted in Figures 11.1 and 11.2. Our focus here is on epithermal ore formation, so most details of the calculation approach in general and many details of reaction chemistry are reserved for publication elsewhere (Reed and Spycher, in preparation).

The results presented here are an extension and refinement of earlier work (Reed and Spycher, 1983) on boiling of a dilute, gas-rich water based on that from Broadlands Well No. 2 (Mahon and Finlayson, 1972). In a related study using homogeneous equilibrium calculations on numerous geothermal waters, Reed and Spycher (1984) showed how the pH and mineral solubilities in the aqueous phase are influenced by boiling and dilution. More recently, Drummond and Ohmoto (1985) used homogeneous and partial heterogeneous equilibrium calculations to evaluate the effects of boiling on mineral precipitation in epithermal systems. Drummond and Ohmoto (1985) provide a detailed discussion of several important features of boiling systems that we omit here, including: open- vs. closed-system boiling; kinetics of redox reactions; and some effects of major element solution chemistry on the size of pH changes caused by boiling. Our calculations are for a water resembling that from Broadlands Well No. 2 (Mahon and Finlayson, 1972), reconstructed (Reed and Spycher, 1984) from analyses of the liquid and gas at 25°C (Table 11.1). Trace-metal contents for this water are modified from Weissberg et al. (1979). After completing the study presented here, we learned that the long-accepted values for gold concentration in the Broadlands water err in being too low (Henley and Brown, 1985, this volume) thus, with respect to gold, our water does not match the actual Broadlands water as closely as we intended. The results are, nevertheless, valid in their own right, applying to a Broadlands-like water with a low initial gold concentration. The initial water composition given in Table 11.1 is in equilibrium numerically with chlorite, galena, muscovite, and sphalerite at 278°C, resulting in slightly decreased molalities of several components relative to the Broadlands water. A Broadlands-like water was selected for initial study for several reasons: (a) a complete analysis is available, including trace metals and gases, (b) it is from a well-studied hydrothermal system, providing abundant field data for comparison with our results (Browne and Ellis, 1970; Mahon and Finlayson, 1972; Ewers and Keays, 1977; Grant, 1977; Weissberg et al., 1979), and (c) Broadlands waters have large CO_2, H_2S, and CH_4 gas fractions; thus, the chemical effects of degassing should be distinct.

For the purpose of exploring the geochemical processes illustrated in Figure 11.2, we started by boiling 1.015 kg of the initial solution (1 kg of H_2O plus 15 grams of dissolved species). All successive operations on the liquid plus gas system (condensation, mixing, oxidation, etc., Fig. 11.2) discussed and illustrated below were effected without re-normalizing to a unit of gas or unit of liquid. Thus, all extensive quantities presented below are "per 1.015 kilogram of initial unboiled fluid."

BOILING

When a homogenous aqueous phase rises in a hydrothermal system (Fig. 11.1, A), it experiences a decrease in pressure resulting in boiling. Boiling induces a temperature decrease and a pH increase which causes minerals to precipitate. Thus, the initially homogenous aqueous phase separates into several phases including gas, liquid, and minerals. The

Table 11.1--Boiling with fractionation

Temperature (deg.C.)	101	151	202	230	242	250	262	278	278
Pressure (bars)	1.02	4.69	14.7	24.7	30.4	35.0	43.9	67.8	
Gas wt. %	36	28	19	13	11	8.4	3.6	0.0	(ppm)
pH	7.80	7.31	6.99	7.00	6.94	6.85	6.65	6.41	
Component species (total molality)									
H^+	.259E-02	.350E-02	.693E-02	.116E-01	.184E-01	.294E-01	.784E-01	.363E+00	
Cl^-	.476E-01	.420E-01	.374E-01	.363E-01	.345E-01	.332E-01	.316E-01	.308E-01	1076
F^-	.377E-03	.333E-03	.296E-03	.287E-03	.274E-03	.263E-03	.251E-03	.244E-03	4.6
SO_4^{--}	.472E-05	.118E-06	-.529E-05	-.143E-04	-.271E-04	-.483E-04	-.159E-03	-.425E-02	
SO_4^{--}(ion)	.395E-05	.739E-06	.145E-06	.805E-07	.457E-07	.258E-07	.782E-08	.295E-09	
CO_3^{--}	.309E-02	.331E-02	.482E-02	.708E-02	.104E-01	.157E-01	.399E-01	.183E+00	10842
HS^-	.109E-03	.147E-03	.231E-03	.336E-03	.461E-03	.642E-03	.134E-02	.731E-02	239
SiO_2	.611E-02	.539E-02	.480E-02	.657E-02	.741E-02	.799E-02	.890E-02	.869E-02	515
Al^{+++}	.208E-08	.668E-07	.802E-06	.986E-06	.118E-05	.137E-05	.177E-05	.230E-05	0.06
Ca^{++}	.415E-04	.367E-04	.327E-04	.317E-04	.352E-04	.385E-04	.366E-04	.356E-04	1.4
Mg^{++}	.978E-07	.863E-07	.769E-07	.745E-07	.732E-07	.760E-07	.947E-07	.347E-06	0.008
Fe^{++}	.109E-12	.194E-11	.215E-09	.446E-08	.216E-07	.498E-07	.171E-06	.119E-05	0.07
K^+	.574E-02	.507E-02	.452E-02	.438E-02	.417E-02	.401E-02	.382E-02	.372E-02	143
Na^+	.458E-01	.404E-01	.360E-01	.349E-01	.333E-01	.320E-01	.305E-01	.296E-01	672
Zn^{++}	.957E-12	.670E-11	.715E-10	.203E-09	.347E-09	.541E-09	.125E-08	.400E-08	0.0003
Cu^+	.587E-08	.159E-07	.145E-07	.183E-07	.225E-07	.216E-07	.206E-07	.200E-08	0.001
Pb^{++}	.799E-10	.112E-08	.299E-08	.290E-08	.276E-08	.269E-08	.289E-08	.359E-08	0.0007
Ag^+	.117E-08	.256E-08	.228E-08	.221E-08	.211E-08	.202E-08	.193E-08	.187E-08	0.0002
Au^+	.857E-09	.756E-09	.673E-09	.653E-09	.622E-09	.598E-09	.570E-09	.554E-09	0.0001
Hg^{++}	.127E-08	.112E-08	.997E-09	.967E-09	.921E-09	.886E-09	.843E-09	.820E-09	0.0002
Gases (mole %)									
H_2O	99.1	98.8	98.2	97.9	97.2	96.1	91.8	74.9	
CO_2	0.8880	1.17	1.72	1.98	2.65	3.72	7.80	21.4	
H_2S	0.0150	0.0195	0.0282	0.0318	0.0416	0.0566	0.106	0.193	
CH_4	0.0195	0.0245	0.0337	0.0353	0.0495	0.0760	0.204	3.35	
H_2	0.00737	0.0141	0.0319	0.0522	0.0651	0.0757	0.0935	0.150	
Minerals (moles/deg.C.)									
Clinochlore				.332E-15	.103E-09	.212E-09	.836E-09		
Daphnite				.202E-14	.550E-09	.111E-08	.444E-08		
Galena	.492E-11	.308E-10				.249E-10	.312E-10		
Sphalerite	.303E-13	.247E-12	.248E-11	.862E-11	.189E-10	.449E-10	.863E-10		
Quartz			.488E-04	.791E-04	.966E-04	.108E-03	.631E-04		
Microcline	.165E-09	.316E-08	.320E-08	.147E-07	.237E-07	.334E-07			
Calcite				.153E-06	.909E-06				
Chalcopyrite			.183E-10	.105E-08					
Bornite	.334E-14	.314E-12							
Chalcocite	.622E-10								
Acanthite	.244E-10								

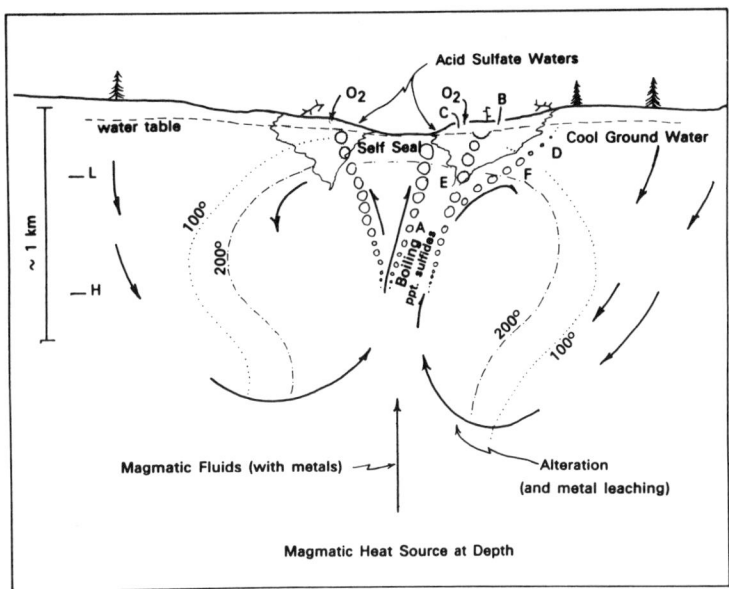

Figure 11.1. Schematic diagram of a boiling hydrothermal system based in part on White et al. (1971), Henley and Ellis (1983), Berger and Eimon (1983), and Steven and Eaton (1975). Ascending hot waters begin to boil (indicated by circles) at the depth labeled H, if pressure is hydrostatic or at some shallower depth between H and L if pressure exceeds hydrostatic but is less than lithostatic. Boiling and various mixtures of boiled hot water and gases with near-surface ground water and atmospheric oxygen produce hot-water compositions and ore-forming environments labeled by letters A through F. These reaction features are identified by letter and described in Figure 11.2 and in the text.

gas, liquid, and some of the minerals are themselves solutions. At equilibrium in any such mixture of specific bulk composition at pressure (P) and temperature (T), overall heterogeneous equilibrium can be calculated (Reed, 1982) yielding: (a) the activities, molalities, and distribution of aqueous species; (b) the masses, identities, and compositions of minerals; and (c) the mass and composition of the gas phase. In the results presented here, such calculations are carried out for stepwise decreases in P and T along a predetermined, nearly isoenthalpic boiling curve (explained below). By this means we simulate the chemical effects of boiling during the ascent of a geothermal water initially at 278°C and 67.8 bars to the surface where T = 100°C and P = 1 bar.

The calculation of the results presented here departs from the approach of Reed (1982) only in treating the gas phase as ideal rather than non-ideal. Eight components, H_2O, CO_2, H_2S, CH_4, HCl, HF, H_2, and S_2, are used to represent the gas phase. Where oxidized gases are important (as in oxidation calculations below), O_2 is included also. Thus, all chemically significant species except NH_3 (Giggenbach, 1980) are included. The inclusion of economically interesting but chemically insignificant gases composed of Hg, As, and Sb is the object of future work.

Solid solutions of chlorite and white mica were treated using ideal multi-site mixing (Kerrick and Darken, 1975). All Henry's Law constants for gas-liquid partitioning are implicitly accomodated in the equilibrium constants for the appropriate mass action equations (equation (12) of Reed, 1982). Dependence of Henry's Law constants on salinity (e.g., Ellis and Golding, 1963; Drummond, 1981) is implicitly accounted for in the activity coefficients for neutral aqueous species (e.g., Helgeson, 1969).

Isoenthalpic boiling is one common and useful assumption for the behavior of geothermal fluids (Elder, 1965; Grant et al., 1984; Henley et al., 1984, p. 9, 143; Drummond and Ohmoto, 1985). We determined a nearly isoenthalpic equilibrium at many temperatures by traversing pressure from high to low at each temperature. By calculating the enthalpy of the gas-liquid mixture along each isothermal pressure traverse, we determine a pressure at which the enthalpy equals that of the pure liquid at the temperature and pressure of incipient gas separation (278°C and 67.8 bars for our water). This fixes a series of P-T points along an isoenthalpic boiling curve and it provides a wealth of information concerning super- and sub-isoenthalpic cooling trajectories. The latter information is presented separately in a following section.

A better way to determine an isoenthalpic boiling curve would be to incorporate an enthalpy equation in the system of simultaneous equations, having specified an initial enthalpy. Then, the isoenthalpic P-T curve could be determined by successsively decrementing P (or T) and solving for T (or P) that satisfies the enthalpy equation. We are currently developing this capability.

Having fixed a P-T boiling curve, the heterogeneous equilibrium calculations are carried out by stepping down the P-T curve in increments of 5° to 16°C. For boiling along such a curve, two cases are treated: (a) boiling with mineral fractionation and (b) boiling at constant bulk composition. A third possibility, boiling with gas fractionation (Rayleigh distillation, e.g., Drummond and Ohmoto, 1985) with or without mineral fractionation, is also of interest, but we have not yet calculated it. Boiling in a system that is closed with respect to gas separation allows continuous chemical communication between the liquid and the gas separating from it as they ascend in a fracture system. This appears to be a reasonable first approximation of real systems.

Boiling with mineral fractionation is executed by

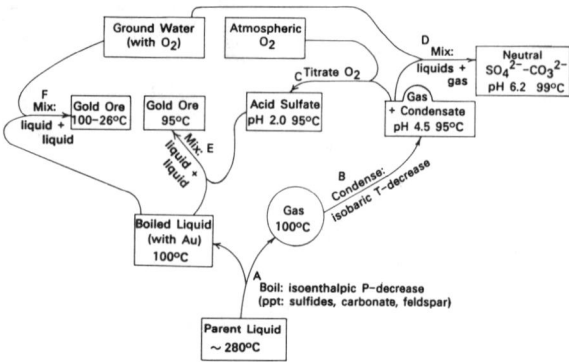

Figure 11.2. Flow diagram showing some of the boiling, mixing, and reaction processes that a deep geothermal water experiences. The processes depicted here, labeled A through F, correspond to those with the same labels in Figure 11.1 and described in the text. The notations along the arrows indicate the type of calculation presented here for the corresponding processes. Boiling (A) is presented in Figures 11.3, 11.4, and 11.7; oxidation of condensate and gas (C) is in Figure 11.10; mixing of gases with ground water (D) is in Figure 11.11; mixing of acid-sulfate water with boiled water (E) is in Figures 11.12 and 11.13; mixing of ground water with boiled water is in Figure 11.14.

BOILING RESULTS

Calculated results of boiling with mineral fractionation are given in Figures 11.3a-i and Table 11.1. Results for closed-system boiling are depicted in Figures 11.4a-d. Cooling-only results are shown in Figures 11.5a-d. These Figures are best read from right to left, corresponding to the ascent of fluids in the geothermal system.

It is convenient in the following discussion to use the term, "early" to refer to chemical events toward the high-temperature end of these plots. Also, reference to "rapidity" of mineral precipitation or molality changes refers to a rate of change with decreasing temperature and corresponds to the slope of curves in the graphs. In all cases, 1.015 kilograms of fluid were used at the start; thus, results expressed in grams or moles refer to quantities of gases or minerals per 1.015 kilogram of initial liquid. Since minerals were fractionated in the calculation shown in Figure 11.3, the mineral Figures 11.3b and 11.3d refer to quantities precipitated <u>per degree</u> of temperature change (per 1.015 kilogram of initial fluid); i.e., they represent a finite-difference approximation of the derivative of mineral abundance with respect to temperature, whereas the mineral figures in 11.4 and 11.5 refer to the absolute mass of minerals present. Quartz disappears at 200°C in Figure 11.3b because it was arbitrarily suppressed at that point to simulate its failure to maintain equilibrium with the aqueous phase at low temperatures (Fournier, 1985, this volume; Reed, 1985).

subtracting from the bulk composition before each P-T decrement an amount of the various chemical components corresponding to those which were fixed in minerals on the just-finished P-T step. This corresponds in nature to an ascending boiling fluid "leaving behind" the minerals precipitated from it. They are thus not available to back-react at lower P and T. In the constant bulk composition case, all mineral precipitates are kept in chemical contact with the solution throughout boiling. They are available to back-react and to constrain the solution chemistry. Another case discussed below is a "cooling only" calculation, for which the pressure is held high enough throughout incremental cooling to prohibit formation of a gas phase. This corresponds to cooling entirely by conduction and provides an instructive comparison to the boiling calculations.

The thermochemical data for these calculations are compiled from diverse sources, including Helgeson (1969); Helgeson and Kirkham (1974a, b; 1976); Walther and Helgeson (1977); Helgeson et al. (1978); Reed and Spycher (1984); Wolery (1979); Barnes (1979); and Schwartzenbach and Widmer (1966). All of the data for complexes of metals with sulfides and bisulfides (e.g., Pb, Zn, Cd, Hg, Au) are from the primary references given by Barnes (1979). For most of these data, we corrected the constants to a standard state of infinite dilution using a modified Debye-Huckle equation (Helgeson et al., 1981), and we extrapolated measured values to cover the temperature range from 300°C to 100°C.

DISCUSSION OF BOILING AND COOLING

The profound effect of boiling on the hydrothermal chemistry can best be appreciated by comparing the boiling calculations to a simple cooling calculation (Fig. 11.5) on the same water composition. In the cooling-only case, pH <u>decreases</u> (Figs. 11.5b and 11.6) as weak acids (H_2CO_3, HCL, HSO_4^-) dissociate with decreasing temperature. In the boiling reactions, pH increases, as discussed below. The influence of this pH effect relative to temperature change, among others, is explored below by comparing the boiling calculation results to the cooling-only results.

Figure 11.3a shows the composition of the gas phase continuously in equilibrium with the aqueous phase throughout boiling. (Note that the corresponding plot for the closed-system boiling calculation closely resembles Figure 11.3a, and is omitted.) Scrutiny of curves for the predominant "dry" gases, CO_2, CH_4, H_2S, and H_2, shows that these gases are rapidly fractionated from the aqueous phase in the first 5°C to 25°C of boiling (278°-250°C). Below 250°C, the abundance of H_2O gas steadily increases while the dry gases change little. This early degassing effect divides the boiling into two stages which makes a distinct imprint on the solid phase assemblage. The early stage of rapid degassing of CO_2 causes an abrupt increase in pH (decrease in a_{H^+}, Figs. 11.3c and 11.4b) as the following reaction is displaced by CO_2 escape from the aqueous phase

$$HCO_3^-_{(aq)} + H^+_{(aq)} \longrightarrow H_2O_{(aq)} + CO_2_{(g)} \quad (1)$$

(Note, here and below, an arrow is used to indicate a reaction forced in the indicated direction by the boiling or cooling process occurring over a P-T interval. An equal sign is used for equilibria at specific P-T points.) Degassing of aqueous H_2S

$$HS^-_{(aq)} + H^+_{(aq)} \longrightarrow H_2S_{(g)} \quad (2)$$

could also contribute to pH increase, but since the quantity of H_2S is quite small (Fig. 11.3a) relative to CO_2, its influence is minor. Thus, it is primarily reaction 1 that results in a change in pH from 6.4 at 278°C to 7.0 at 245°C to 7.8 at 100°C (Fig. 11.3c).

In contrast to the major gases, the absolute amounts of HCl and HF in the gas phase reach maxima in the range of 225°C to 250°C (Fig. 11.3a). Their early rise is a reflection of removal of molecular HF and HCl present in the aqueous phase at high temperatures, e.g.

$$HCl_{(aq)} \longrightarrow HCl_{(g)}$$

With decreasing temperature, the dielectric constant of water increases (Helgeson and Kirkham, 1974a), stabilizing ionic species, leading to dissociation of aqueous HCl which displaces the above reaction back to the left.

Sulfide and Carbonate Mineral Precipitation

Figures 11.3b and 11.4a show that the highest rates of mineral precipitation prevail over the short temperature interval between 278°C and 245°C where the temperature decrease combines with the pH increase to force the following example sulfide and carbonate reactions

$$Zn^{2+}_{(aq)} + HS^-_{(aq)} \longrightarrow ZnS_{(sphalerite)} + H^+_{(aq)} \quad (3)$$

$$Ca^{2+}_{(aq)} + HCO_3^-_{(aq)} \longrightarrow CaCO_3_{(calcite)} + H^+_{(aq)} \quad (4)$$

The greater significance of pH increases to these reactions relative to temperature decrease is apparent from the absence of calcite, sphalerite, chalcopyrite, and galena at high temperature in the cooling calculation (Fig. 11.5a), relative to the boiling calculations (Figs. 11.3b and 11.4a). The more acid conditions in the cooling-only case (Fig. 11.6) displace reactions 3 and 4 to the left, inhibiting mineral precipitation. For sulfides, this occurs despite the increase in metal-ion molalities (Fig. 11.5c) due to dissociation of their chloride and sulfide complexes (Fig. 11.5d).

Over the same short temperature interval (278°-245°C) where most mineral mass precipitates, the loss of the reduced gases, H_2 and CH_4, causes oxidation of aqueous sulfide to sulfate, increasing sulfate molality as shown in Figure 11.3i. Degassing also causes total aqueous carbonate and sulfide (represented as $CO_3^=$ and HS^-, Fig. 11.3c) to decrease while the molality of individual ion $CO_3^=$ increases (Fig. 11.3i), because the increase in pH displaces the following reaction to the right

$$HCO_3^-_{(aq)} \longrightarrow H^+_{(aq)} + CO_3^=_{(aq)} \quad (5)$$

Individual ion HS^- decreases during boiling (Fig. 11.3f) due to loss of sulfide to the gas phase, but the decrease is subdued relative to the change in total molality of sulfide (Fig. 11.3c) because the increase in pH results in dissociation of aqueous H_2S

$$H_2S_{(aq)} \longrightarrow H^+_{(aq)} + HS^-_{(aq)} \quad (6)$$

Transfer of $10^{-2.5}$ moles of H_2S to the gas phase (Fig. 11.3a) entirely dominates over sulfide mineral precipitation (all in the range of 10^{-6} moles or less, Fig. 11.4a) in removal of sulfide from the aqueous phase. This is partly a consequence of low initial base-metal contents in the solution, consistent with its low salinity (and, therefore, low Cl^- content; see further discussion below).

Despite separation of sulfide into the gas phase, the high-sulfide concentration in this water makes sulfide and bisulfide complexes dominant (Figs. 11.3h, 11.4d, and 11.5d) for all ore metals except Zn, for which a bisulfide is secondary to a chloride, and Pb, for which a carbonate is dominant over a short interval. All of the complexes become less stable with decreasing temperature, which tends to release the metals to form sulfide minerals. Breakup of bisulfide complexes due to H_2S escape in boiling further promotes metal precipitation, despite the moderating effect of H_2S dissociation, reaction (6).

Figure 11.3d, which shows the relative rate of precipitation of sulfide minerals with decreasing temperature, indicates that sphalerite and galena precipitate from this water before chalcopyrite. The early precipitation of sphalerite results from the combined effects of increasing pH (reaction 3) and dissociation of $ZnCl^+$ (Fig. 11.3h) with decreasing temperature. Figures 11.3b and 11.3d show that there is a gap in galena precipitation between 245°C and 200°C. This is a consequence of a shift in the dominant Pb complex ligand from HS^- to $CO_3^=$ and back again to HS^- (Fig. 11.3h). The carbonate complex takes over temporarily as $m_{CO_3^=}$ increases (Fig. 11.3f) owing to increasing pH (reaction 5).

The late precipitation of chalcopyrite (after sphalerite and some galena) shown in Figure 11.3d is a consequence of its very stable bisulfide complex (Fig. 11.3h). Henley et al. (1984) points out that Broadlands waters precipitate more chalcopyrite than sphalerite or galena late in the boiling process.

Because the water is sulfide rich, there is sufficient sulfide to stabilize the high-sulfur copper minerals, bornite and chalcopyrite, over chalcopyrite with decreasing temperature (Fig. 11.3d), as discussed

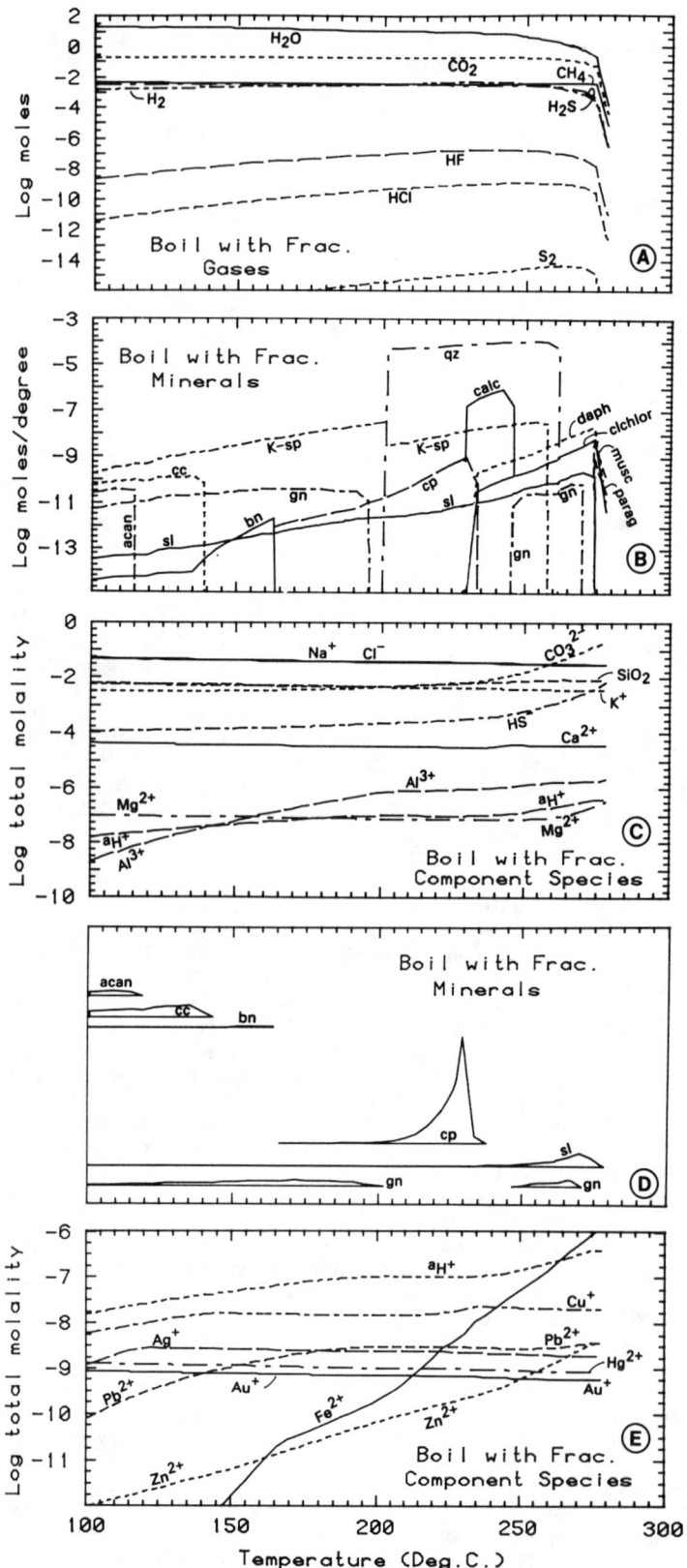

Figure 11.3. Calculated isoenthalpic boiling of 1.015 kg of a Broadlands-like water with fractionation of minerals: a). Moles of species in the gas phase, continuously in equilibrium with minerals and the aqueous phase; b). Rate of precipitation of minerals in moles per degree of temperature change; c). Total molality of a portion of the aqueous component species (not individual ion molality). Activity of hydrogen ion is also shown, for which the scale refers to log activity; d). Paragenetic-style diagram showing relative rate of sulfide mineral precipitation on a weight basis; e). Total molality of ore-metal component species; Fe^{2+} refers to total aqueous Fe.

Figure 11.3 (cont'd.) f). and g). Individual species molalities of component species; h). Individual species molalities of the principal ore-metal complexes; i). Individual species molalities of carbonate, sulfate, and sulfide species. Mineral abbreviations used here and in the following Figures are as follows: acanthite, acan; bornite, bn; calcite, calc; chalcocite, cc; chalcopyrite, cp; chlorite, chl; cinnabar, cinn; clinochlore, clchl; covellite, cv; daphnite (Fe-chlorite), daph; graphite, graph; hematite, hem; kaolinite, kaol; microcline, K-sp; muscovite, musc; pyrite, py; quartz, qz; sphalerite, sl.

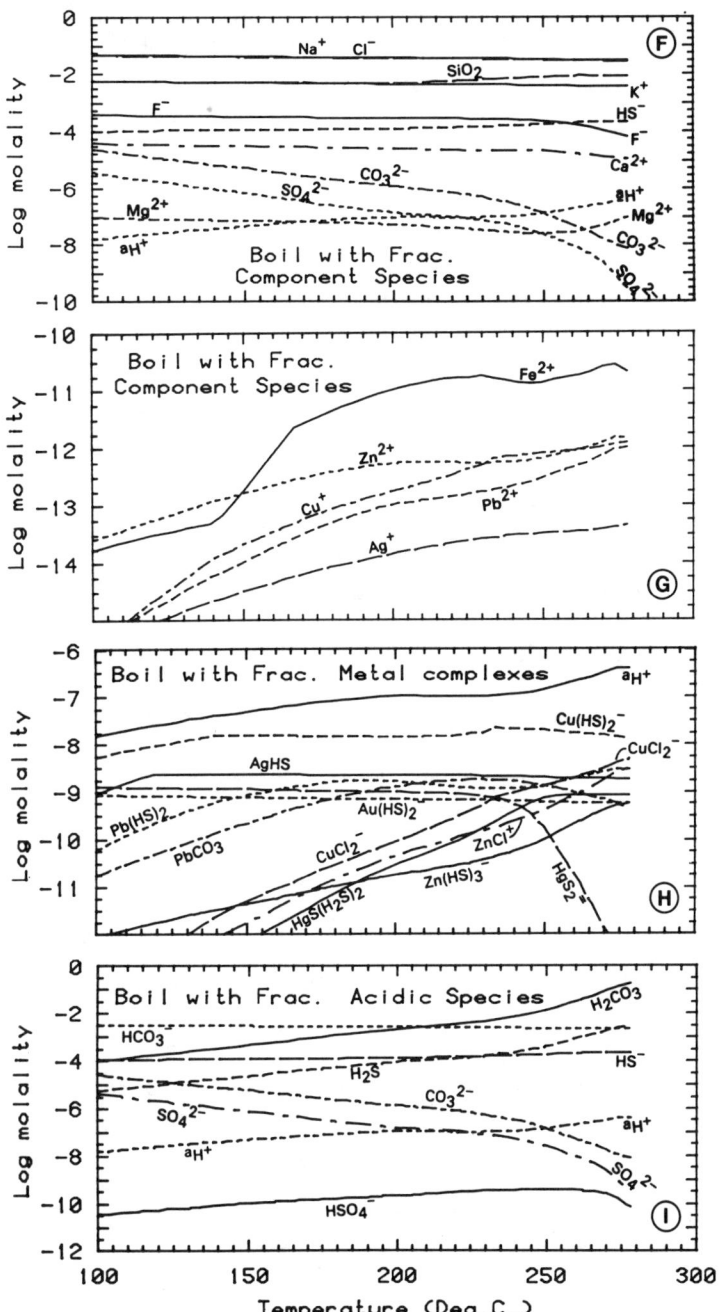

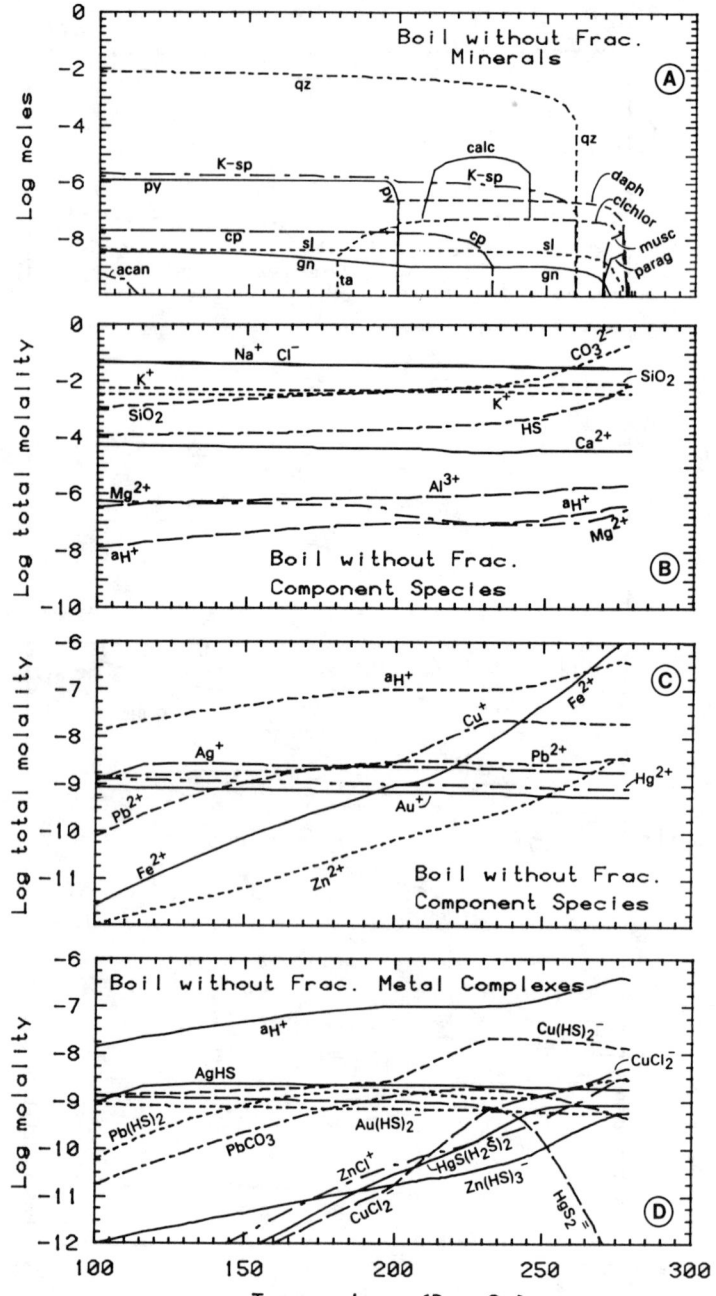

Figure 11.4. Calculated isoenthalpic boiling of a Broadlands-like water without fractionation of minerals. The gas phase composition is essentially the same as in Figure 11.3a. a). Moles of minerals present in equilibrium with the gas and aqueous phases during boiling. See caption to Figure 11.3 for mineral abbreviations. b). Total molality of a portion of the component species. Activity of hydrogen ion is also shown, for which the ordinate refers to log activity. c). Total molality of ore-metal component species. Fe^{2+} refers to total aqueous Fe. d). Individual species molalities of principal complexes of ore metals.

Figure 11.5. Cooling of a Broadlands-like water without boiling and without mineral fractionation. (a) Moles of minerals present in equilibrium with the aqueous phase. See caption for Figure 11.3 for mineral abbreviations. (b) Total molality of a portion of the component species. Activity of hydrogen ion is also shown, for which the ordinate refers to log activity. (c) Total molality of ore-metal component species. (d) Individual species molalities of principal complexes of ore metals.

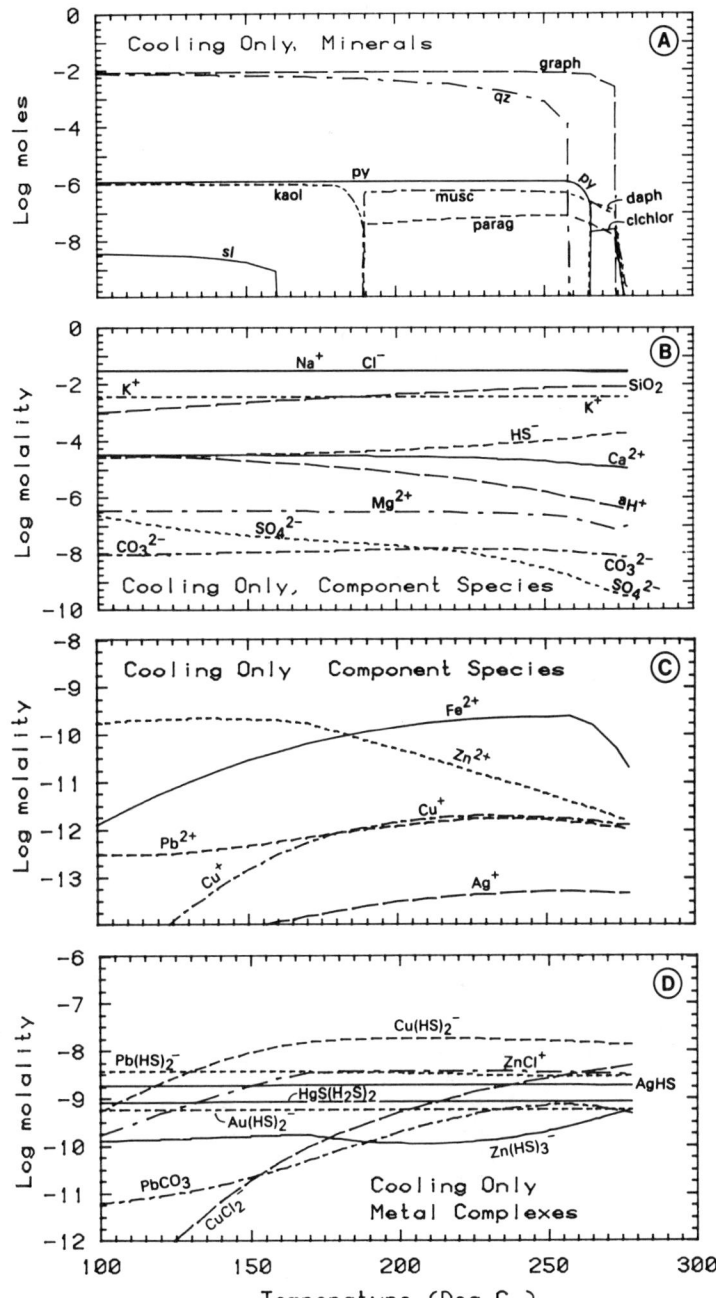

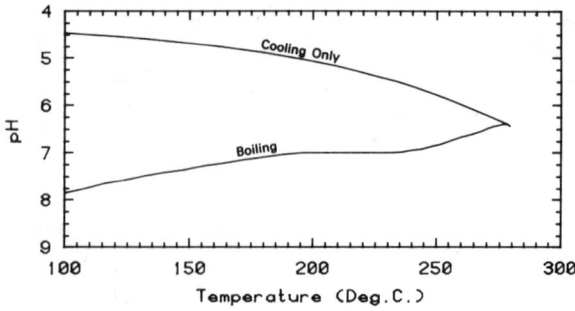

Figure 11.6. Comparison of pH in calculations of boiling (lower curve, both boiling calculations) and cooling only (upper curve). See text for discussion of differences.

by Reed (1985). The high-sulfide concentration also holds Ag^+ in the stable $AgHS$ complex (Fig. 11.3h) throughout cooling from $278°$ to $115°$, where acanthite precipitates (Figs. 11.3b and 11.3d). In contrast to silver, gold and mercury never precipitate because their sulfide and bisulfide complexes (Fig. 11.3h) are quite stable over the entire temperature and pH range of boiling. Gold and mercury behavior are more thoroughly discussed below.

Precipitation of Silicates

In both boiling calculations, solid solutions of chlorite and muscovite precipitate as well as quartz and K-spar (Figs. 11.3b and 11.4a). Solid solution compositions (calculated using ideal and ideal multi-site mixing) are not plotted separately, but compositions can be read from the figures by comparing the moles of solution end members. The calculated chlorite is iron rich, reflecting the high Fe/Mg ratio of the starting solution (Table 11.1), and the rapid dumping of Fe from solution during cooling as $Fe(OH)_4^-$ dissociates.

In contrast to the sulfides and carbonates, the very presence of silicates is not a consequence of the pH increase upon boiling. The effects of temperature decrease control silicate precipitation, as is obvious for quartz but perhaps not so obvious for the others. Muscovite and K-spar precipitate because Al is liberated from $Al(OH)_4^-$ as its stability decreases with decreasing temperature (Reed and Spycher, 1984). This displaces the following example reaction to the right

$$K^+ + Al(OH)_4^- + 3SiO_2 \rightarrow KAlSi_3O_8 + 2H_2O \quad (7)$$
$$(aq) \quad (aq) \quad (aq) \quad (K\text{-spar}) \quad (aq)$$

Chlorite precipitates in response to this Al supply combined with Fe supplied by dissociating $Fe(OH)_4^-$. The validity of these conclusions is indicated by the nearly quantitative removal of aqueous Al and Fe in both the boiling reactions and in the cooling-only reaction (Fig. 11.5a), where pH decreases (Figs. 11.5b and 11.6). The high pH of the boiling waters fixes feldspar rather than mica in accordance with the following reaction

$$6SiO_2 + KAl_3Si_3O_{10}(OH)_2 + 2K^+ = 3KAlSi_3O_8 + 2H^+ \quad (8)$$
$$(aq) \quad (muscovite) \quad (K\text{-spar}) \quad (aq)$$

This reaction also accounts for the replacement of mica by feldspar with decreasing temperature (and increasing pH) shown in Figures 11.3b and 11.4a.

Boiling Without Fractionation and Cooling Only

As discussed above, most of the chemical processes in the closed-system boiling calculation are the same as in the boiling with mineral fractionation. The only important differences are: (a) the low-temperature sulfide assemblage is dominated by pyrite (Fig. 11.4b) which is entirely absent in the fractionation case (Fig. 11.3d) and (b) talc appears as a silicate phase in the closed-system case (Fig. 11.4a). Both of these differences are a consequence of the re-dissolution of the chlorite at low temperature in the closed-system case whereas chlorite is fractionated out of the system (taking Fe and Mg with it) in the fractionation case. Back-reaction of the iron-rich chlorite at $200°C$ (Fig. 11.4a) provides iron for pyrite, magnesium for talc, and extra aluminum for K-spar.

The purpose in executing the cooling-only calculation (Fig. 11.5) was to isolate the effects of temperature change itself from the effects of boiling. The cooling-only calculation results are valuable for this purpose, but are not necessarily valuable as a model of reality. The single most significant difference between the cooling and boiling calculations is that dissociation of weak acids upon cooling causes pH to decrease whereas pH increases in the boiling calculation (Fig. 11.6). This is why very little sulfide precipitates and phyllosilicates dominate instead of feldspar and carbonate (Fig. 11.5a). The decrease in pH with cooling displaces sulfide into H_2S (reaction 6) accounting for the decrease in m_{HS^-} in Figure 11.5b. Despite this decrease, metal-bisulfide complexes (Fig. 11.5d) do not dissociate sufficiently to overcome the effect of the acid conditions in enhancing sulfide mineral solubility.

Another conspicuous difference in the cooling calculation is that graphite precipitates. The initial water is enriched in aqueous methane which escapes into the gas upon boiling. If boiling is not allowed, the hydrocarbon is forced to go into graphite upon cooling. (Graphite is the only reduced carbon phase, except methane, available in the numerical model.)

SUPER- AND SUB-ISOENTHALPIC BOILING

In addition to the nearly isoenthalpic boiling and cooling-only cases discussed above, we explored numerically the closed-system geochemical space corresponding to fluid ascent with partial conductive cooling and fluid ascent with excess enthalpy boiling. These results are shown in Figure 11.7 in diagrams of temperature vs. gas weight percent. The figures are contoured with mineral stability boundaries, pH, and mole percent CO_2 gas in the gas phase. The diagrams show the isoenthalpic boiling path (A-D, Fig. 11.7a)

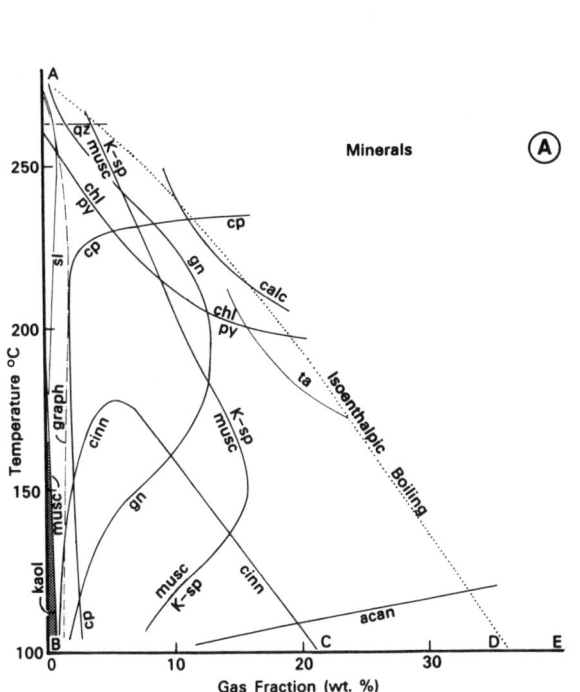
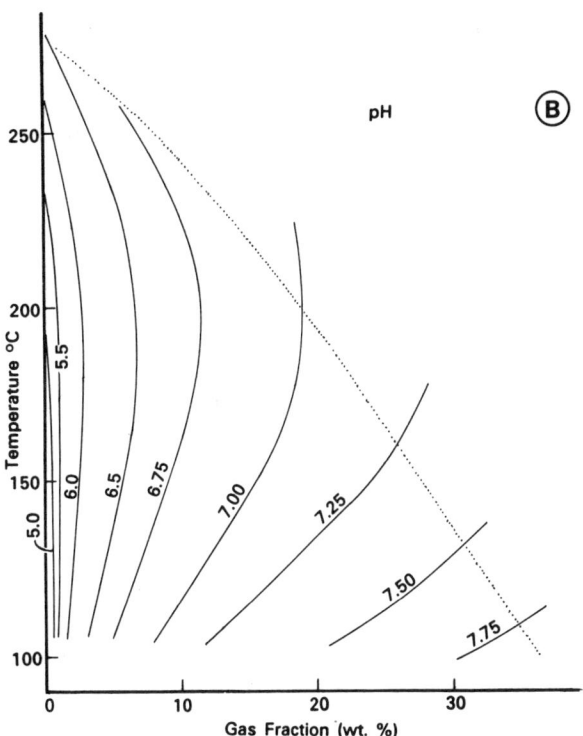
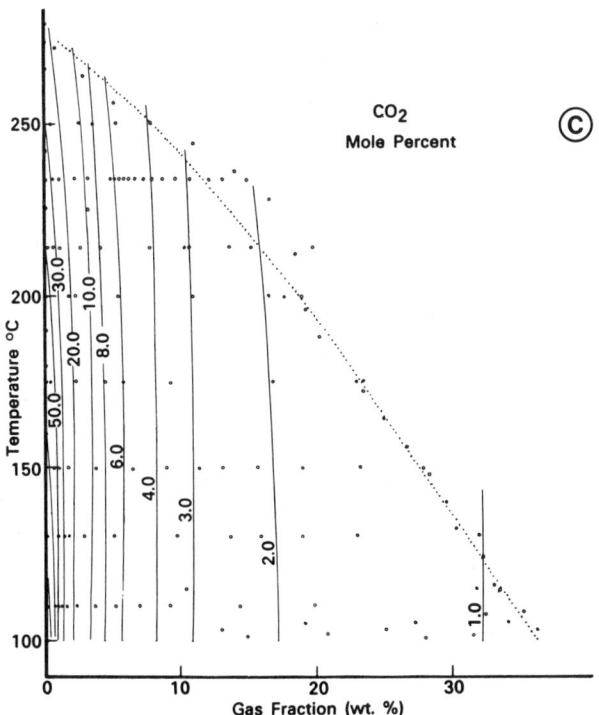

Figure 11.7. Calculated results of closed-system boiling of a Broadlands-like water under a range of conditions corresponding to super-isoenthalpic to conductively cooled. The abcissa expresses the gas fraction (including H_2O and the dry gases) of the total (liquid plus gas) in weight percent. At any given temperature on the ordinate, only one pressure yields a gas fraction corresponding to isoenthalpic boiling conditions, indicated by curve A-D, in part a). Other pressures yield more or less gas, corresponding to super- and sub-isoenthalpic boiling conditions. See text. a). Mineral phase diagram showing fields for the presence or absence of the indicated minerals. The shaded area at lower left is the stability field for kaolinite; to the right of the shaded area, moscovite is stable rather than kaolinite. See Figure 11.3 for mineral abbreviations. b). pH of boiling waters. Dotted line shows isoenthalpic boiling curve, A-D from part a). c). Mole-percent CO_2 in the gas phase in equilibrium with the boiling waters. Contours for 70 and 90% CO_2 are not labeled but are shown to the left of the 50% contour at low temperature and low-gas fraction. The dotted line shows the isoenthalpic boiling curve, A-D from part a). The other small circles show the points for which calculations were done to produce Figure 11.7.

used in the previously discussed calculations depicted in Figure 11.4 and the cooling-only path (A-B, Fig. 11.7a), representing complete conductive cooling as depicted in Figure 11.5.

The isoenthalpic curve (Fig. 11.7) represents the temperature-steam fraction trajectory of a fluid which exchanges no heat with its wall rocks (Elder, 1965; Grant et al., 1984). All heat used to vaporize the water is supplied from the water itself. As the heat is removed, the water cools. An isoenthalpic path would be expected under steady-state flow conditions in a hydrothermal system such that the thermal gradient in the wall rock has been previously fixed by the boiling fluid (Grant et al., 1984).

All trajectories between A-D and A-B (Fig. 11.7) represent fluids that boil during ascent, but which simultaneously lose heat to the wall rock. Such trajectories apply when pressure increases over time due to self-sealing, such that the depth of first boiling ascends (from H toward L, Fig. 11.1). As the depth of boiling becomes shallower (because boiling is prohibited by increased pressure due to self-sealing) ascending hot waters encounter rock that is cooler than boiling along the newly established pressure-depth regime allows; thus, the rocks extract heat from the water. Depending on the time rate of self-sealing (and consequent rate of pressure change with time) the extent of such heat extraction by the rocks could result in cooling trajectories spanning much of the range between A-D and A-B (Fig. 11.7a). Super-isoenthalpic trajectories, lying to the right of A-D (e.g., A-E, Fig. 11.7), represent boiling fluids that extract excess heat from the wall rock as they ascend. This produces steam quantities in excess of that which the ascending water alone could produce. Super-isoenthalpic conditions would prevail, for example, when new heat is introduced at shallow levels by magma intrusion. A more common cause would be breaking of a self-sealed system, resulting in a reversion to hydrostatic pressure from super-hydrostatic conditions. This would cause the depth of first boiling to deepen (downward toward H, and below, Fig. 11.1), exposing hot wall rock to newly lowered pressure conditions. These same thermal effects are discussed by Goguel (1982) and Truesdell (1979) in the context of exploitation of geothermal energy reservoirs.

The common occurrence of self-sealing and re-breaking in active geothermal systems is well established (e.g., White et al., 1971; Muffler et al., 1971; Keith et al., 1978; Facca and Tonani, 1967; Henley and Ellis, 1983) as is the common occurrence of hydrothermal breccias in epithermal ore deposits (e.g., Berger and Eimon, 1983; Nelson and Giles, 1985; Hedenquist and Henley, 1985). Thus, it is more probable that many of the sub- and super-isoenthalpic trajectories represented in Figure 11.7 are visited during the active lifetime of epithermal ore-forming systems. Chemical implications of this are discussed further below.

The results shown in Figure 11.7 are based on calculations of overall heterogeneous equilibrium for each of the individual points shown in Figure 11.7c. Most of these points were calculated using isothermal pressure-drop traverses at various temperatures. Although each such traverse is interesting in its own right, it is most useful to combine all results on a few diagrams, then explore the various ascent trajectories discussed above. The mineral phase boundaries shown in Figure 11.7a are best understood as a consequence of the pH variations shown in Figure 11.7b, which are primarily a consequence of the CO_2 degassing represented in Figure 11.7c. The rapid pH increase with early boiling (corresponding to small gas fractions, Fig. 11.7b), also apparent in Figures 11.3c and 11.4b, are a result of the early rapid escape of most CO_2 into the gas phase as indicated in Figure 11.7c and plotted in moles in Figure 11.3a. Figure 11.7b shows that essentially all boiling paths emanating from point A result in pH increases except conductive cooling paths near path A-B. The monotonic decrease in pH along path A-B is shown in Figure 11.5b.

The pH control on gangue silicate mineralogy, for example, is apparent from the approximate parallelism of the muscovite-K-spar boundary in Figure 11.7a and the pH contours of Figure 11.7b. K-spar forms on the high-pH side of the boundary in accordance with reaction 8. According to the phase diagram, all boiling trajectories that result in more than 10 percent gas at $100°C$ will produce deep muscovite and shallower K-spar. The particular depth where the transition from muscovite to K-spar occurs is fixed by the point of intersection of the ascent trajectory (e.g., A-C) with the phase boundary in Figure 11.7a.

The cinnabar field in Figure 11.7 is particularly instructive because it demonstrates the effects of competing reactions on the solubility of a sulfide mineral. According to Figure 11.7a, isoenthalpic boiling does not precipitate cinnabar from this water, but neither does conductive cooling. Consider an isothermal P-drop traverse (left to right, Fig. 11.7) at a temperature between $100°C$ and $170°C$; cinnabar remains soluble in the absence of boiling (zero gas fraction, Fig. 11.7) because the large H_2S concentration in the aqueous phase displaces the following reaction to the left

$$\underset{\text{(aq)}}{HgS(H_2S)_2} = \underset{\text{(cinnabar)}}{HgS} + \underset{\text{(aq)}}{2H_2S} \quad (9)$$

The protonated mercury complex and H_2S are strongly favored by the low pH (Fig. 11.7). Upon boiling, reaction 9 is displaced to the right as H_2S (gas) escapes, resulting in cinnabar precipitation. However, as boiling proceeds, the consequent increase in pH (Fig. 11.7b) shifts sulfide equilibria such that the mercury-sulfide complex, $HgS_2^=$, increases in concentration, causing cinnabar to redissolve (reaction 10) despite the further loss of H_2S gas

$$\underset{\text{(cinnabar)}}{HgS} + \underset{\text{(aq)}}{HS^-} \longrightarrow \underset{\text{(aq)}}{HgS_2^=} + \underset{\text{(aq)}}{H^+} \quad (10)$$

In the context of possible fluid ascent trajectories, this behavior of cinnabar means that only paths with partial conductive cooling of the fluid will produce cinnabar.

The cinnabar and muscovite-K-spar examples illustrate the profound effect of the boiling heat budget on the expected epithermal vein mineralogy. To the extent that the heat budget of boiling fluid "parcels" varies through time, fluid trajectories on Figure 11.7a vary and vein mineralogy must vary. This is one probable source of mineral banding in epithermal veins. As a system experiences self-sealing and re-breaking, trajectories shift from isoenthalpic to sub-isoenthalpic to super-isoenthalpic on Figure 11.7a. Such a fluid, at its 230°C isotherm (which itself shifts in space) would first precipitate quartz, calcite, K-spar, Fe-chlorite, chalcopyrite, sphalerite, and galena (Fig. 11.4a), then (depending on where the sub-isoenthalpic trajectory falls) perhaps quartz, muscovite, pyrite, and sphalerite (but not galena, chalcopyrite, chlorite, K-spar, and calcite), as indicated by Figure 11.7a at 230° and 5-percent gas fraction. At this stage the much lower pH of the fluid (Fig. 11.7b) might also etch earlier formed minerals and cause muscovite to replace earlier formed K-spar (e.g., Hedenquist and Henley, in preparation, 1985). Upon re-breaking and a swing to super-isoenthalpic conditions, the fluid again would precipitate carbonate, feldspar, etc., at its 230° isotherm.

Instead of following a fluid parcel of given temperature through the history outlined above, we may consider how physical and chemical conditions vary at a point fixed in space (in a vein cavity, for example). A gradual pressure increase due to sealing would correspond to traversing toward high temperature in Figures 11.3 and 11.4 for any given position in a vein. Thus, a point initially at 250°C (Fig. 11.3b) where quartz, K-spar, chlorite, sphalerite, and galena are precipitating could experience a shift to 278°C as pressure increases, resulting in etching of the previously formed minerals, overlapping precipitation of muscovite, and replacement of K-spar by muscovite. Subsequent pressure drop would result in a return to precipitation of the lower temperature assemblage. Other more complicated changes could occur, depending on the rate of pressure change, resulting in sub- and super-isoenthalpic boiling (Fig. 11.7a) at the given point.

BOILING AND GOLD PRECIPITATION

The large sulfide content of the Broadlands-like water used here stabilizes the $Au(HS)_2^-$ complex (Fig. 11.3h) precluding gold precipitation throughout the boiling range from 278°C to 100°C. In the calculation of boiling with mineral fractionation (Fig. 11.3), gold is undersaturated by more than two orders of magnitude at 278°C (log Q/K = -2.44), but this decreases to undersaturation by half an order of magnitude after boiling to 100°C (log Q/K = -0.57). Thus, if the solution had been saturated with gold at the point of incipient boiling (as in Drummond and Ohmoto, 1985, who used a large gold concentration) or if we had used the new value for gold concentration from Henley and Brown (1985, this volume) which is 23 times greater than the 0.1 ppb that we assumed (Table 11.1), gold would have precipitated due to boiling. The critical reaction for gold precipitation in the acid to neutral pH range is the following

$$8Au(HS)_2^- + 6H^+ + 4H_2O \rightarrow 8Au^0 + SO_4^= + 15H_2S \quad (11)$$
$$\text{(aq)} \quad \text{(aq)} \quad \text{(aq)} \quad \text{(gold)} \quad \text{(aq)} \quad \text{(g or aq)}$$

This reaction entails reduction of Au^+ to Au^0 by sulfide which is thereby oxidized to sulfate. Reaction 11 makes it clear that pH increase induced by boiling competes with loss of H_2S gas in determining whether gold precipitates or not (see also, Drummond and Ohmoto, 1985). The thermochemical data used here for gold-bisulfide complexes (Seward, 1973) indicate that as pH increases in the range calculated here (6.4 to 7.8), gold solubility increases (reaction 11) as H_2S dissociates to HS^-. Thus, the calculated approach toward gold saturation upon boiling is principally a consequence of loss of H_2S to the gas phase, displacing reaction 11 toward gold saturation (see also, Hedenquist and Henley, 1985).

The precipitation of sulfide minerals constitutes another sink for sulfide which promotes gold precipitation. However, in the case calculated, the escape of H_2S to the gas phase is far more significant than sulfide mineral formation as a sink for aqueous sulfide (compare Figs. 11.3a and 11.4a), as discussed in a preceding section. In any case, gold did not precipitate because of the excess sulfide content of original water. Instead, it stayed in the aqueous phase and, in the geologic context, it would be transported to the surface hot-spring environment (see below).

In order for gold to precipitate with sulfides in the deep vein environment as it did, for example, at Comstock (Bastin, 1922), it is necessary to create a sulfide-deficient water by removing essentially all of the aqueous sulfide to minerals and gas. This is best accomplished by boiling waters which have a large ratio of metals (Fe, Cu, Pb, Zn) to sulfide so that precipitation of pyrite, chalcopyrite, galena, and sphalerite along with H_2S gas removes all sulfide from the aqueous phase. The low salinity of the Broadlands-like water precludes high-metal contents because chloride is necessary to complex significant concentrations of base metals in the presence of high-sulfide concentrations. Other calculations (Reed, 1985) on more saline waters (e.g., .5 m NaCl) show that chloride complexes of Pb, Zn, and Cu dominate, even in high-sulfide waters. A large metal/sulfide ratio is necessary to assure sulfide depletion with consequent gold precipitation, and large salinities are necessary to large base-metal concentrations. Thus, the boiling of saline waters should produce gold precipitation with sulfides in accordance with the following example reaction in which galena and gold precipitation are coupled

$$8Au(HS)_2^- + 4H_2O + 15PbCl_2$$
$$\text{(aq)} \quad \text{(aq)} \quad \text{(aq)}$$
$$\rightarrow 8Au^0 + SO_4^= + 30 Cl^- + 15PbS + 24H^+ \quad (12)$$
$$\text{(gold)} \quad \text{(aq)} \quad \text{(aq)} \quad \text{(galena)} \quad \text{(aq)}$$

As boiling proceeds, CO_2 escapes, driving up pH (reaction 1); this decrease in H^+ activity drives sulfide precipitation (reaction 3), which in sulfide-deficient waters is coupled to gold precipitation as in reaction (12). Thus, the formation of a hot-spring gold deposit (see below) as opposed to a base-metal-vein gold deposit may result primarily from the distinction between sulfide-deficient and sulfide-excess boiling waters, which itself is tied to the distinction of low-salinity vs. high-salinity waters. Fluid-inclusion data support this distinction.

THE HOT-SPRING ENVIRONMENT

In the hot-spring environment at and just below the surface over a boiling hydrothermal system, the effects of cool temperatures, atmospheric oxidation, and the influx of meteoric ground waters determine patterns of rock alteration, water chemistry, and ore formation. Figures 11.1 and 11.2 show the near-surface geochemical processes that we chose to simulate numerically for the purpose of understanding the origins of hot-spring gold deposits (Berger and Eimon, 1983; Henley and Ellis, 1983). In the process of unravelling the chemistry of gold deposition, we generated numerical models of the formation of acid condensates, acid-sulfate waters, and neutral carbonate-sulfate waters.

All results were produced using the same fundamental approach for multi-component heterogeneous equilibrium calculations (Reed, 1982) that we applied to boiling. The sequence of operations on the gas and liquid phases left over from boiling are as follows (refer to Figs. 11.1 and 11.2 for points designated below by letters):
1. Condense the gas phase in open space by cooling from 100°C to 93°C (B).
2. Oxidize the gas with atmospheric O_2 (C).
3. Condense the gas in cold, oxygenated ground water and react at 99°C (D).
4. Titrate (D) the acid-sulfate water from (C) into the boiled aqueous phase from (A).
5. Titrate (F) the cold, oxygenated ground water into the boiled aqueous phase from (A).

The results of each of these calculations are presented below in the order of the above list.

Condensation of the Boiled Gas

When the gas phase from a boiling hydrothermal system escapes into cool, fractured rock above the boiling water table, the H_2O fraction largely condenses (White, 1957; White et al., 1971) and carries with it small amounts of dissolved CO_2 and H_2S. Subsequent, direct atmospheric oxidation (Fig. 11.8) produces acid-sulfate waters. We calculated the chemical characteristics of the condensate in order to provide a starting composition for oxidation, but also to understand better how the condensate may act alone in altering wall rocks and re-dissolving vein minerals.

We separated the gas phase from the liquid phase left over from the boiling calculation at 101°C (Table 11.1, Fig. 11.3a) and numerically condensed it by cooling the gas alone to 93°C at a pressure of 1.01 bar. Computationally, such a condensation is no different than the boiling calculation. There is a gas phase in equilibrium with a liquid phase. As temperature decreases at constant pressure, gas condenses to liquid and chemical potentials of all species are equalized between the two phases. The resultant liquid composition at 95°C is given in Table 11.2 and pH and gas composition are shown in Figure 11.9. Gas and liquid compositions are also shown at the extreme left-hand ends of Figures 11.10a and 11.10b.

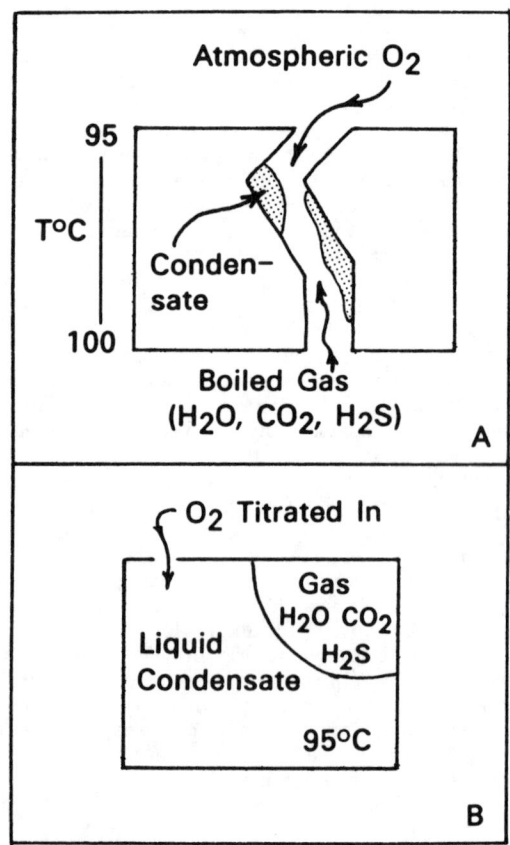

Figure 11.8. a). Schematic diagram showing the boiled gas phase passing through a temperature gradient in fractured rock above the water table. The H_2O condenses, carrying some CO_2 and H_2S with it. Atmospheric oxygen enters from above, oxidizes H_2S to sulfuric acid which dissolves in the liquid condensate. b). Diagram of a semi-closed system approximation to the scheme in part a), as set up for numerical oxidation at 95°C (see text and Fig. 11.10). Oxygen gas is titrated into a two-phase mixture of condensate plus gas, but the original gas phase remains in contact with the liquid throughout the calculation. Thus, the H_2S is gradually consumed by oxidation, rather than being continuously re-supplied from below.

Table 11.2—Water compositions

	Cold ground water		Condensed gas (95°C)		Acid-sulfate water (0.03 moles of added O_2)		Gas + ground water (72 kg of added water)	
pH	7 (25°C)		4.5 (95°C)		2.06 (95°C)		6.3 (99°C)	
	(molality)	(ppm)	(molality)	(ppm)	(molality)	(ppm)	(molality)	(ppm)
Cl^-	.733e-04	2.6	.115e-10	.0000004	.115e-10	.0000004	.743e-04	2.6
F^-	.174e-04	0.33	.590e-08	.0001	.590e-08	.0001	.176e-04	0.3
$CO_3^=$	.202e-02	121	.203e-02	126	.187e-02	116	.115e-02	68
HS^-	---	---	.118e-03	4.0	---	---	---	---
CH_4 aq.	---	---	.280e-05	0.04	---	---	---	---
$SO_4^=$	.323e-04	3.1			.848e-02	842	.748e-04	7.1
O_2 aq.	.250e-03	8			.184e-04	0.6	.493e-04	1.5
SiO_2	.105e-02	63					.106e-02	63
Al^{+++}	.111e-06	0.003					.113e-06	0.003
Ca^{++}	.724e-04	2.9					.733e-04	2.9
Mg^{++}	.535e-04	1.3					.542e-04	1.3
Fe^{++}	.179e-06	0.01					.469e-14	0.0
K^+	.358e-04	1.4					.363e-03	1.4
Na^+	.457e-04	10.5					.463e-03	10.5

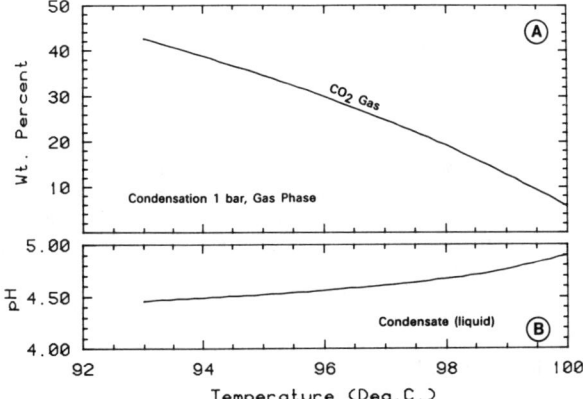

Figure 11.9. Calculated process of condensation of the gas phase boiled off the Broadlands-like water. The CO_2 content of the gas phase in equilibrium with the liquid during condensation from 100°C to 93°C at 1.01 bar is given in a). The liquid contains water with dissolved H_2S, CO_2, etc., yielding the pH shown in b). pH decreases with cooling as more and more CO_2 dissolves in the aqueous phase, despite the fact that the proportion of CO_2 in the gas phase increases a).

During condensation, pH decreases from 4.9 at 100°C to 4.5 at 95°C (Fig. 11.9) while the CO_2 fraction of the gas phase increases from less than 10 percent to more than 40 percent by weight (2.4 to 18 mole percent). The remainder of the gas is H_2O, except for CH_4, H_2, and H_2S, which also increase in proportion, but which constitute less than 1 percent of the gas. The pH of 4.5 at 95°C is entirely a consequence of carbonic acid, the molality of which increases during cooling (as the solubility of CO_2 increases) despite the fact that the proportion of CO_2 in the gas phase also increases. Condensation thus produces a mildly acidic liquid, which at 95°C has a mass of 353 grams per initial 1.015 kg of unboiled water. The associated gas is quite CO_2 rich, as also indicated at low temperature and small gas fraction in Figure 11.7c where CO_2 gas fractions exceed 70 mole percent. Such CO_2-rich gases reported from the boundaries of hot hydrothermal systems (Mahon et al., 1980; Goguel, 1982) are undoubtedly a consequence of condensation of the H_2O gas fraction as calculated here.

The calculated pH of 4.5 for the condensed liquid is certainly acidic enough and its volume great enough to effect alteration of volcanic wall rocks to illite or kaolinite (depending on temperature and K^+ activity), and it may be responsible for some of the illitic wall-rock alteration reported in the upper parts of vein systems, particularly in the hanging wall (Berger and

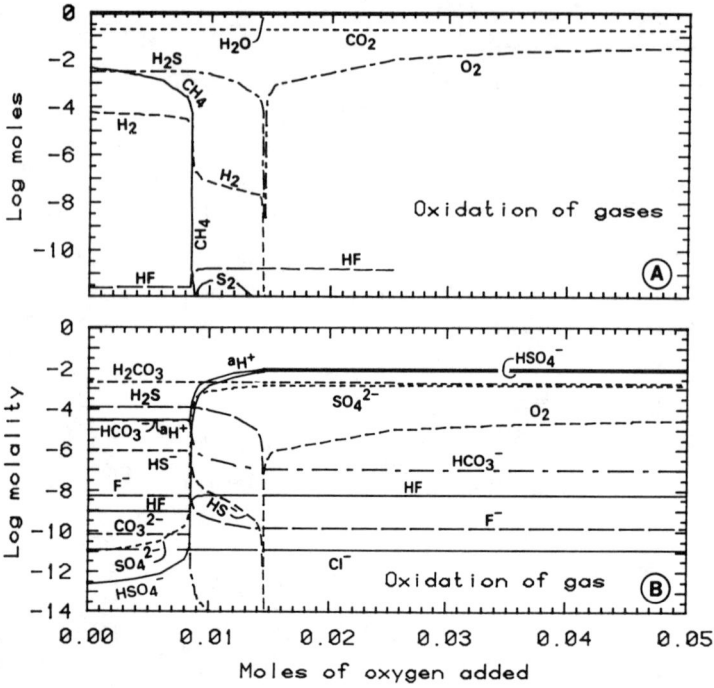

Figure 11.10. Calculated oxidation of 353 grams of the boiled gas phase by atmospheric O_2 to produce an acid-sulfate water. a). The composition of the gas phase in equilibrium with the aqueous phase as O_2 is titrated into the liquid plus gas mixture. CH_4 was allowed to oxidize despite the fact that such oxidation is kinetically retarded. See text. b). Composition of the aqueous phase, showing molalities of individual species. Activity of hydrogen ion is also shown, for which the ordinate refers to log activity.

Eimon, 1983). The condensate is not acidic enough, however, to cause the extensive alteration of rocks at shallow levels to kaolinite and alunite. For this, acid-sulfate waters are necessary.

Oxidation of Gases to Produce Acid-Sulfate Waters

Acid-sulfate waters of very low pH are produced by direct atmospheric oxidation of sulfidic gases above the water table (White, 1957) as indicated in Figures 11.1 and 11.2. In a following section, it is shown that oxidation of gases by dissolved oxygen in cold ground water cannot produce acid-sulfate waters because too much concurrent dilution precludes a low pH. The physical concept of the oxidation calculation is illustrated in Figure 11.3a, which shows the gas condensing in a fracture as it traverses a temperature gradient. The liquid condensate remains in contact with excess (non-condensible) gas which continues to flow through the fracture while atmospheric oxygen gains access from above. The oxygen reacts at $95°C$ with the two-phase mixture (liquid plus gas, Fig. 11.8b) remaining after condensation (i.e., we do not exclude the remaining non-condensible gases from reacting with oxygen along with the constituents dissolved in the liquid). The calculation is effected by numerically titrating O_2 gas into the condensate plus gas at $95°C$ and 1.01 bar with computation of overall heterogeneous equilibrium after each titration step.

From the standpoint of numerical computation, carrying out a continuous oxygen titration reaction such as this one (Fig. 11.10) and others described below entails overcoming a significant obstacle at the transition from reduced to oxidized conditions (e.g., .0145 moles O_2, Fig. 11.10). At this point, the concentrations of many reduced sulfide species (HS^-, H_2S, and complexes involving these) become so small that their molalities cannot be reliably computed using normal machine double precision (64-bit representation of variables). At the same point, concentrations of aqueous O_2 and H_2O_2 became large. Under such oxidizing conditions, we really do not care to know about sulfide molalities less than 10^{-50}, so it is convenient to recast all redox equilibria in terms of the H_2O-O_2 redox pair in place of the HS^--$SO_4^=$ pair used under reducing conditions (Reed, 1982). This amounts to substituting O_2 for HS^- as a component species, then providing a whole different set of equilibrium constants for the various redox equilibria. We are able to overlap the computations slightly at the transition point to verify that all computed molalities match through the transition.

Figure 11.10b shows that after reacting less than .02 moles of O_2 with 353 grams of condensate plus gas, the pH changes from 4.5 to 2.06 as sulfuric acid is produced from H_2S

$$H_2S + 2O_2 \rightarrow H_2SO_4 \quad (13)$$
(g and aq) (g) (aq)
(sulfuric acid)

The overall process occurs in two steps. The first is oxidation of methane and hydrogen (Fig. 11.10a), which are fully oxidized by nearly .01 mole of O_2. In nature, this oxidation is almost certainly insignificant because of kinetic barriers to oxidation of reduced carbon and hydrogen at low temperature (see review of this question by Drummond and Ohmoto, 1985). We allowed this oxidation because it has no quantitative effect on results except to delay the oxidation of H_2S. While methane and hydrogen oxidize (Fig. 11.10a), pH remains constant (Fig. 11.10b) and sulfates begin a gradual increase in molality. Once the hydrocarbon oxidation is complete, oxygen attacks H_2S, producing sulfuric acid and a catastrophic increase in acidity to a pH of 2.06 with consequent shifts in sulfide, carbonate, and sulfate equilibria as shown in Figure 11.10b. As H_2S oxidation proceeds, the oxidation products dissolve in the aqueous phase, depleting H_2S from the gas (Fig. 11.10a). (The calculation itself provides no information as to whether the actual oxidation of H_2S occurs in the gas or aqueous phase, or at their interface.) The final disappearance of H_2S gas (at .0145 moles O_2) allows O_2 gas to accumulate for the first time in the gas phase. A resultant water composition is given in Table 11.2. The particular pH attained depends on how much H_2S is oxidized. If only the H_2S originally dissolved in the aqueous phase were oxidized (10^{-4} m, Fig. 11.10b), the pH would not have dropped below 3.4. Since many acid-sulfate waters have pH's between 2 and 3 and some are less than 2 (White et al., 1971; Rowe et al., 1973), it appears that oxidation of H_2S gas that did not originally dissolve in the condensate contributes to the sulfuric acid in natural acid-sulfate springs. This conclusion applies in general, even though it is based here on a single, specific water plus gas composition, because our Broadlands-like composition is unusually rich in sulfide gas and represents an extreme case.

Clearly such an acidic water would quickly attack its host rock producing kaolinite and leaching base cations. Among such dissolved constituents, Al^{3+} is significant with respect to the gangue in gold ores. This point is discussed below in connection with mixing of acid-sulfate waters with the boiled liquid (process E, Figs. 11.1 and 11.2), but first we discuss gas condensation and oxidation in meteoric ground water.

Reaction of Gases with Meteoric Ground Water

In order to evaluate the potential of oxygenated ground water for producing acid waters to react with gold-bearing boiled waters, we titrated a New Zealand meteoric ground water (Table 11.2, col. 1, Timperley, 1983) containing 8 ppm dissolved O_2 into the gas phase left over from boiling at 101°C and 1 bar. This produces a neutral bicarbonate-sulfate water (Fig. 11.11) of the sort discussed by White et al. (1971) which they postulated formed by just such a reaction. As shown in Figure 11.11, the basic pattern of change in sulfide and sulfate is the same as in Figure 11.10, except that the process is "stretched out" because 60 kg of ground water containing 8 ppm O_2 (2.5×10^{-4} moles/kg H_2O, Table 11.2) are required to oxidize all CH_4, H_2, and H_2S in the 353 grams of gas from boiling. Figure 11.11 shows a gradual increase in aqueous $SO_4^=$ as H_2S oxidizes, finally resulting in a sulfate-carbonate water with a pH of 6.3 at 99°C (Table 11.2, Fig. 11.11). The gas phase changes composition in this reaction in exactly the same way as shown in Figure 11.10a, except that the positions of slope change are "stretched out" relative to Figure 11.10, just as in the aqueous phase (Fig. 11.11).

In the ground water titration reaction, pyrite forms continuously between 0 and 58 kg of titrated water and is replaced by hematite at 58 kg (Fig. 11.11) where O_2 becomes abundant in the aqueous phase. The pyrite forms as iron, supplied by the cold ground water, and reacts with H_2S supplied by the boiling hydrothermal water

$$4Fe^{2+} + 7H_2S + SO_4^= \rightarrow 4FeS_2 + 4H_2O + 6H^+ \quad (14)$$
(aq,   (gas   (aq)   (pyrite)   (aq)   (aq)
g.w.) or aq)

To the extent that inflowing ground waters carry significant concentrations of iron, this process is likely to be important in producing a "pyritic halo" on epithermal ore deposits as indicated in Figure 11.12. (Furthermore, it would operate equally well to produce the well-known pyritic halos on porphyry copper deposits.) Kraynov et al. (1982) show that iron concentrations in continental ground waters commonly range from a few ppm to several tens of ppm. Such concentrations, when reacted with H_2S, are sufficient to produce an excess of pyrite (mostly in veins) beyond that which is produced by sulfidation of iron originally in the host rock itself.

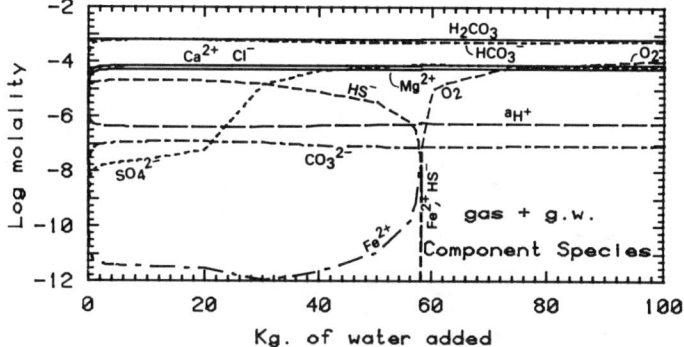

Figure 11.11. Calculated composition of the aqueous phase resulting from mixing and reaction of oxygenated ground water with the condensate plus gas from the boiled water, producing a neutral carbonate-sulfate water. Pyrite precipitates in the reduced interval (0-58 kg) and hematite precipitates in the oxidized interval (above 58 kg).

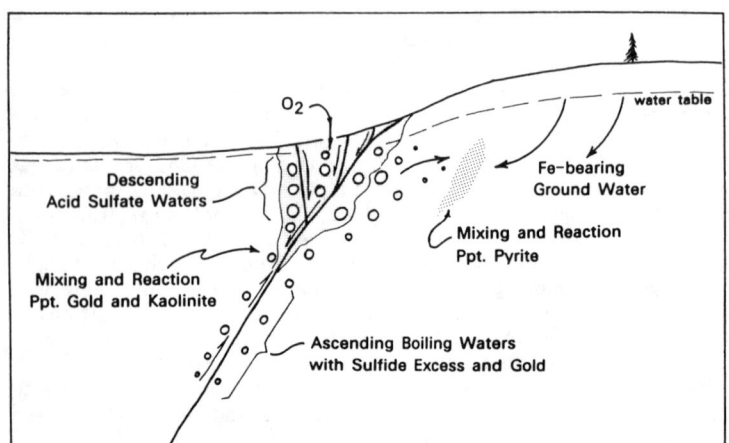

Figure 11.12. Schematic diagram of fluid-mixing zones in the upper part of a boiling hydrothermal system. In the center of the diagram, ascending, boiled, gold-bearing waters (as in Figs. 11.3 and 11.4 and Table 11.1) encounter descending acid-sulfate waters produced above the water table (Figs. 11.8 and 11.10). The waters mix and react precipitating gold (Fig. 11.13). The descending waters also alter the wall rock to kaolinite, particularly in the hanging wall of the vein. The right-hand side of the diagram shows mixing and reaction of boiled gases (with cold, recharge ground water), producing a neutral carbonate-sulfate water (Fig. 11.11). If the recharge water is iron rich, reaction of H_2S gas with iron could produce pyrite, which would appear as a "pyritic halo" on the ore deposit.

Gold Precipitation from Mixing of Acid-Sulfate Water with Boiled Aqueous Phase

In a preceding section on boiling and gold precipitation, we pointed out that boiling does not cause gold to precipitate from a sulfide-excess water with low gold concentration. Such waters carry their gold to the near-surface hot-spring environment where gold may precipitate when the $Au(HS)_2^-$ complex is destroyed by oxidation of the sulfide. Another reaction, possibly of greater significance than oxidation itself to near-surface gold precipitation, is acidification of gold-bearing boiled waters by descending acid-sulfate waters (process E, Figs. 11.1 and 11.2; Fig. 11.12). In this calculation, we numerically titrate at 95°C the acid-sulfate water produced by oxidation of the boiled gas (see above and Fig. 11.10) into the liquid remaining after boiling of our original water with mineral fractionation (Fig. 11.3). Results are shown in Figure 11.13. A steady increase in H^+ activity as acid water is added is apparent in Figure 11.13b. This dominates all chemical reactions. Acidification causes cinnabar to precipitate as reaction 10 (see above) is displaced to the left. After approximately 0.35 kg of acid water is added to the original 0.650 kg of boiled water, a gas phase separates (Fig. 11.13a) as CO_2 and H_2S are produced by reactions 1 and 2 (see above). Analogous reactions for HF and HCl are also displaced toward the associated form by acidification. Because the gas phase is necessarily a solution, other gases join H_2S, CO_2, HF, and HCl as shown in Figure 11.13a. Separation of H_2S gas combines with the acidification to break up the dominant complex of copper, $Cu(HS)_2^-$, forcing precipitation of covellite (Fig. 11.13c)

$$10H^+ + 8Cu(HS)_2^- + SO_4^{2-}$$
(aq)   (aq)   (aq)
$$\rightarrow 8CuS + 4H_2O + 9H_2S \quad (15)$$
(covellite) (aq) (gas and aq)

Further acidification similarly destroys the dominant gold complex, $Au(HS)_2^-$, forcing gold precipitation

$$6H^+ + 8Au(HS)_2^- + 4H_2O \rightarrow 8Au^0 + SO_4^{2-} + 15H_2S \quad (16)$$
(aq) (aq) (aq) (gold) (aq) (aq and gas)

The detailed numerical results show that the dominant effect in both of these reactions is displacement of bisulfide into aqueous H_2S by acidification (note declining HS^-, Fig. 11.13b), not the escape of sulfide to the gas phase, although the latter does remove 10 percent of the original aqueous bisulfide.

For this calculation, we used numerical acid-sulfate water (Table 11.2), which had not yet reacted with its numerical host rock. Consequently our acid-sulfate water lacked significant Al^{3+}, in contrast to the high concentrations measured in natural acid-sulfate waters (e.g., 4 to 30 ppm at Yellowstone, Rowe et al., 1973). If Al^{3+} were present in the acid water when it mixed with the hot, ascending boiled water, kaolinite would be expected to precipitate

$$2Al^{3+} + 2SiO_2 + 5H_2O$$
(aq, descending) (aq, ascending) (aq)
$$\rightarrow Al_2Si_2O_5(OH)_4 + 6H^+ \quad (17)$$
(kaolinite) (aq)

Such kaolinite would precipitate along with gold as

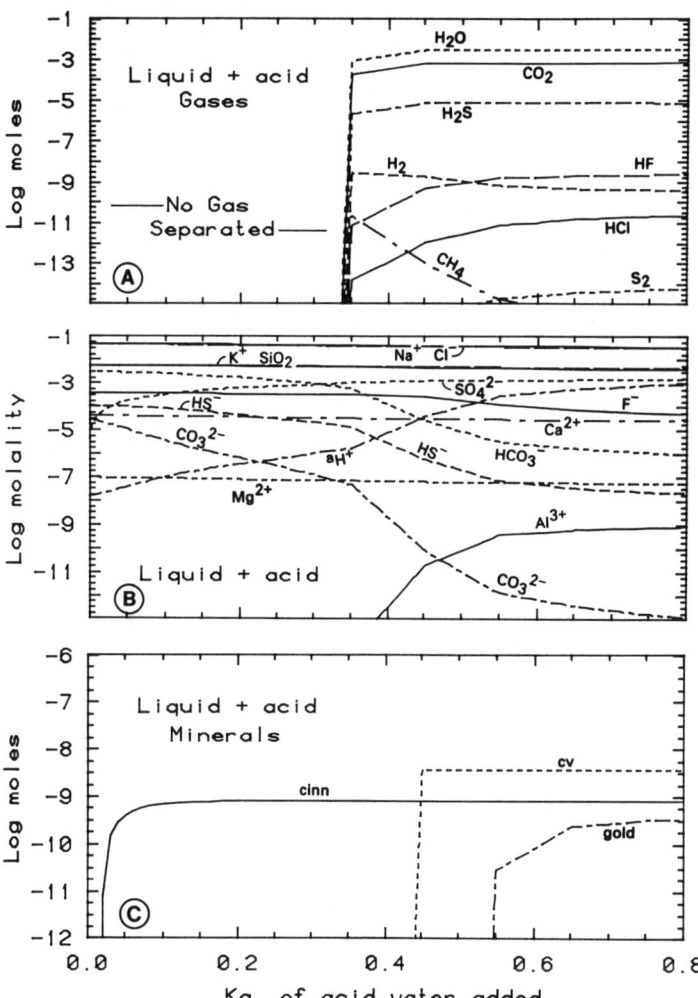

Figure 11.13. Calculated reaction and precipitation of gold upon mixing of acid-sulfate water with the liquid remaining at 100°C and 1 bar from boiling of the Broadlands-like water. a). Composition of the gas phase which separates as a consequence of acidification. b). Molalities of individual species in the aqueous phase. Activity of hydrogen ion is also shown. c). Moles of minerals to precipitate from 650 grams of boiled liquid.

vein gangue (Fig. 11.12) where the waters mix. An intimate mixture of gold with pure white kaolinite between bands of vein quartz has been reported by Kurt Katsura (U. Oregon, personal communication, 1985) in upper parts of veins in the Champion Mine, Bohemia District, Oregon. Kaolinite with gold in other such occurrences elsewhere might be mistaken for alteration clay "washed in" from altered wall rock.

Gold is commonly associated with arsenic and antimony sulfides around certain hot springs and in some epithermal ores (White, 1981; Henley, 1985, this volume.) Henley (1985, this volume) suggests that the As and Sb sulfides precipitate owing to acidification, and Ellis (1969) has suggested that the colloidal precipitating sulfides scavenge gold. Our calculations indicate that gold, too, precipitates by acidification (reaction 16); thus, the association of arsenic-antimony and gold (and cinnabar, Fig. 11.13c) may actually be a consequence of co-precipitation of all from sulfide complexes by acidification.

Gold Precipitation from Mixing of Oxygenated Ground Water with Boiled Aqueous Phase

When gold precipitates by acidification (as above, Fig. 11.13), atmospheric oxidation is indirectly responsible because oxidation of H_2S produces the requisite sulfuric acid. The critical gold-precipitating reaction (16), however, is driven by acid, not by oxygen. Another reaction that is capable of driving gold precipitation is oxidation of sulfide by oxygen dissolved in cold ground water. In this section, we discuss such a reaction coupled simultaneously to the cooling of boiled water by the cold ground water.

This calculation is accomplished by numerically titrating (Fig. 11.2) the 25°C New Zealand meteoric ground water containing 8 ppm dissolved O_2 (Table 11.2) into the 100°C gold-bearing water remaining from boiling (Fig. 11.3). In addition to mass accounting, the enthalpy and consequent temperature of the mixture is calculated from the masses and known enthalpies of the mixing fluids as discussed by Reed (1985). Thus, as cold water is titrated into hot, temperature decreases from 100°C toward 25°C. The results of this process are shown in Figure 11.14, using temperature on the abcissa. These graphs are best read from right to left, reflecting progressive addition of cold water to hot water.

There are two stages of reaction apparent in Figure 11.14, corresponding to reduced, then oxidized, conditions as also found in other oxidation calculations (Figs. 11.5, 11.10, and 11.11). The transition occurs at 65°C (Fig. 11.14), at which point a sufficient amount of dissolved oxygen had been added from the cold water to oxidize to sulfate all sulfides originally in the boiled water. At temperatures above 65°C, minuscule quantities of sphalerite, galena, acanthite, and chalcopyrite precipitate as the cold ground water chills the mixture, wringing all residual base metal from the boiled water. These sulfides precipitate despite dilution because cooling by 35°C (at constant pH, Fig. 11.14b) reduces their solubilities by one or more orders of magnitude while concurrent dilution only reduces metal concentrations to less than half. This same effect would cause sulfides to precipitate by cold water dilution at high temperatures, too, and may be an important cause of base-metal precipitation in epithermal veins without boiling. Pyrite precipitates abundantly (Fig. 11.14a) as iron from the ground water reacts with sulfide in the hot water. At just above 70°C, gold precipitates because aqueous sulfide concentration has been sufficiently decreased by oxidation to destroy the gold-bisulfide complex according to the following reaction

$$8Au(HS)_2^- + 30\ O_2 + 4H_2O$$
$$\text{(aq)} \qquad \text{(aq)} \qquad \text{(aq)}$$
$$\longrightarrow 8Au^0 + 16SO_4^= + 24H^+ \qquad (18)$$
$$\text{(gold)} \quad \text{(aq)} \qquad \text{(aq)}$$

Further oxidation of sulfide to sulfate results in dissolution of all sulfides at 65°C after the addition of .56 kg of cold water to an initial .65 kg of boiled liquid.

Figure 11.14a does not show the spectacular mineral replacement sequence that is compressed into a fraction of a degree of temperature change around 65°C where log f_{O_2} changes from -55 to -4. In this interval (where pyrite dissolves and hematite precipitates, Fig. 11.14a), the following reactions occur: (a) chalcopyrite is replaced by bornite, then bornite by chalcocite, chalcocite by native copper, and copper by aqueous Cu^{2+}; (b) acanthite is replaced by native silver, then silver dissolves as aqueous AgCl and $AgCl_2^-$; and (c) cinnabar is replaced by native mercury, then mercury dissolves in aqueous chloride and hydroxide complexes. Clearly these results are applicable to understanding the weathering of sulfide ore deposits, but this goes beyond the scope of the present study.

It is apparent in Figure 11.14b that the total molalities of copper, silver, and mercury decrease as pyrite, chalcopyrite, cinnabar, and acanthite precipitate owing to cooling and destruction of their sulfide complexes by oxidation. Near 65°C, just before complete oxidation of the sulfides, each of these molalities reaches a minimum as the metals are stored in sulfide minerals and native metals. Upon complete oxidation, these solids redissolve as sulfides

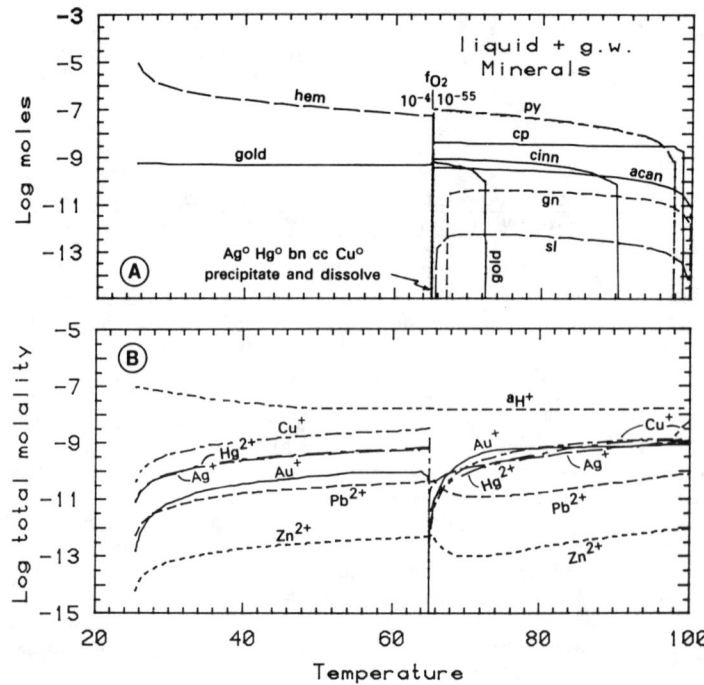

Figure 11.14. Calculated reaction and temperature change upon mixing of the 100°C boiled water with a 25°C oxygenated ground water. a). Moles of minerals to precipitate from 650 grams of initial boiled liquid. Oxygen fugacity changes abruptly from 10^{-55} to 10^{-4} at the point where all sulfide is oxidized to sulfate. In that minuscule interval, additional minerals precipitate as indicated and discussed in text. See Figure 11.3 for mineral abbreviations. b). Total molalities of ore metals in the aqueous phase. The sharp downturn at low temperature is a consequence of the extreme dilution by cold ground water.

are oxidized and the metals are complexed by chloride. The molalities of copper, silver, and mercury shift back to maximum values which, because of the dilution, are not as high as the starting values (before mixing). Further decreases in metal molalities occur as dilution proceeds. The lead and zinc molalities show a slightly different trend because galena and sphalerite redissolve before oxidation is complete. Although gold is still present after oxidation of the sulfides, the molality of gold follows the same trend as copper, silver, and mercury because gold is carried in a bisulfide complex. The gold molality also drops sharply at 65°C, before shifting back to a higher value. The higher gold molality results when, in the absence of sulfide, native gold is oxidized in the presence of Cl^- to a gold chloride complex according to the following reaction

$$2Au^0 + 1/2O_2 + 4Cl^- + 2H^+ \rightarrow 2AuCl_2^- + H_2O \quad (19)$$

This explains why the amount of native gold is slightly lower after oxidation (Fig. 11.14a). Eventually, further oxidation would redissolve all the gold into $AuCl_4^-$ and Au^{+++}; however, in our case, the concentrations of auric species were insignificant throughout the entire mixing process.

Upon complete oxidation, the only remaining minerals are hematite and gold. At the low temperatures involved here, an assemblage of goethite and gold would probably prevail in the natural environment. Thus, the final product of the mixing and oxidation is an assemblage of gold and iron oxide, with all other ore metals dissolved out. This may occur in an active hot-spring environment, as suggested in Figure 11.1, producing an assemblage that may be mistaken for one formed by weathering.

SUMMARY

Using calculations of simultaneous, multi-phase chemical equilibria, we have explored some of the constraints that local thermodynamic equilibrium places on the mineral content and metal zoning in epithermal ore deposits. We focused on a dilute, gas-rich Broadlands-type water, but many of our conclusions can be applied to other waters. A summary of the principal conclusions relevant to epithermal ore genesis follows.

Boiling, with consequent release of CO_2 gas, causes pH to increase by more than one unit. This drives precipitation of base-metal sulfides and carbonates. Most of the CO_2 originally in the hot water at high pressure separates into the gas phase as boiling traverses the first 30 degrees and 33 bars along the boiling curve. This concentrates the steepest pH increase in the same interval, resulting in a maximum rate of sulfide precipitation over a relatively narrow temperature-depth interval (Fig. 11.3b). Additional sulfide continues to precipitate throughout boiling to 100°C and 1 bar, but the quantities are quite small.

Concomitant escape of H_2S gas during boiling, combined with precipitation of sulfide minerals, removes aqueous sulfide, tending to destroy the gold-bisulfide complex. In a sulfide-deficient system, aqueous sulfide may be sufficiently depleted by H_2S escape and sulfide precipitation to cause gold to precipitate along with sulfides. The requisite sulfide deficiency is more likely to prevail in base-metal-rich waters, where precipitation of metal sulfides can remove a large proportion of aqueous sulfide. To be metal rich, the waters must be chloride rich. Thus, deposits of gold with vein sulfides are likely to have formed from saltier waters than deposits of gold in near-surface hot-spring environments.

The temperature decrease that accompanies boiling drives precipitation of feldspar, white mica, and chlorite gangue as the complexes of Al and Fe are destabilized. Throughout all but the highest temperature part of the boiling zone, the high pH favors formation of gangue feldspar instead of mica. Iron-rich chlorite forms instead of Mg-rich chlorite because cooling destabilizes iron complexes (particularly $Fe(OH)_4^-$) more sharply than Mg complexes. Also, at the high temperatures of the source fluids, the concentration of iron is similar in magnitude to those of magnesium (e.g., Arnorsson et al., 1983), in the presence of alteration chlorite.

The depth of boiling shifts dramatically depending on pressure. Thus, as pressure fluctuates owing to self-sealing and re-breaking, the depth interval of boiling changes, resulting in spatial overlapping of radically different chemical regimes. The chemical differences between high- and low-pressure boiling regimes are exaggerated by the dramatic differences between super-isoenthalpic and sub-isoenthalpic boiling (Fig. 11.7). These effects undoubtedly contribute significantly to the "telescoping" of zoning in epithermal deposits and to alternating ore and waste bands in veins. The same effects could also produce "mineralizing" and "barren" stages in ore formation, as indicated by crosscutting vein relationships.

Condensation of the boiled gases at 95°C produces a carbonic acid solution with a pH of 4.5. Oxidation of the condensate plus remaining H_2S-bearing gas by atmospheric O_2 produces acid-sulfate water of very low pH (pH 2 to 4, depending on how much H_2S oxidizes), capable of altering wall rock to kaolinite. Oxidation of the boiled gas by mixing and reaction with oxygenated ground water produces a sulfate-carbonate water of near-neutral pH. The latter calculation showed that acid-sulfate waters cannot form by oxidation of gases by dissolved O_2 because concurrent dilution precludes the high concentrations of sulfuric acid needed to achieve very low pH. Thus, acid-sulfate waters must form above the water table, as suggested by White (1957).

Mixing of acid-sulfate water (or acid-condensate water) with the gold-bearing liquid remaining from boiling causes gold to precipitate, as acidification drives off H_2S gas and shifts aqueous bisulfide to H_2S, destroying the $Au(HS)_2^-$ complex. Other metals that are complexed in bisulfides, particularly Hg^{2+}, As^{3+}, and Sb^{3+}, also precipitate upon acidification, accounting for their common association with gold in acid hot-spring environments. Gold also precipitates upon oxidation of $Au(HS)_2^-$ in the boiled water by dissolved O_2 in ground water.

Although we have investigated in some detail many of the processes that may form ore in volcanic epithermal environments, there are many processes to examine. Among the most pressing of these are: (a) boiling of salty, sulfide-deficient waters to investigate precipitation of assemblages of sulfide minerals with gold; (b) mixing of cold, meteoric ground water with hot (250°C), metal-bearing water without allowing boiling, to investigate this mechanism of vein-sulfide formation (results shown in Fig. 11.14 indicate that it may be quite effective); and (c) boiling of metal-bearing waters with provision for handling Hg, Sb, and As in the gas phase, to evaluate the relative importance of gas vs. liquid transport of these metals; and (d) reaction of the various liquids and condensates with volcanic host rocks, to evaluate more fully the relationships between wall-rock alteration and ore fluid chemistry.

ACKNOWLEDGMENTS

We would like to thank Bill Gemuts and the Anaconda Minerals Company for supporting much of the work presented here. The University of Oregon provided computing funds for the project for which we are most appreciative.

REFERENCES

Arnorsson, S., Gunnlaugsson, and Svavarsson, H., 1983, The chemistry of geothermal waters in Iceland. II. Mineral equilibria and independent variables controlling water compositions: Geochimica et Cosmochimica Acta, v. 47, p. 547-566.

Barnes, H. L., 1979, Solubilities of ore minerals; in Barnes, H. L. (ed.), Geochemistry of Hydrothermal Ore Deposits: Second Edition, John Wiley and Sons, New York, p. 404-460.

Bastin, E. S., 1922, Bonanza ores of the Comstock Lode, Virginia City, Nevada: U.S. Geological Survey, Bulletin 735, p. 41-63.

Berger, B. R., and Eimon, P., 1983, Conceptual models of epithermal precious-metal deposits; in Shanks, W. C., III (ed.), Cameron Volume on Unconventional Minerals Deposits: Society of Mining Engineers, American Institute of Mining and Metallurgy, p. 292-305.

Bourcier, W. L., and Barnes, H. L., 1981, Stabilities of hydrothermal zinc complexes (abs.): Geological Society of America, Abstracts with Programs, p. 414.

Browne, P. R. L., and Ellis, A. J., 1970, The Ohaki-Broadlands hydrothermal area, New Zealand: Mineralogy and related geochemistry: American Journal of Science, v. 269, p. 97-131.

Buchanan, L. J., 1981, Precious-metals deposits associated with volcanic environments in the southwest; in Dickinson, W. R., and Payne, W. D. (eds.), Relations of Tectonics to Ore Deposits in the South Cordillera: Arizona Geological Society Digest, v. 14, p. 237-262.

Drummond, S. E., 1981, Boiling and mixing of hydrothermal fluids: chemical effects on minerals precipitation: Unpublished Ph.D. thesis, The Pennsylvania State University, 380 p.

Drummond, S. E., and Ohmoto, H., 1985, Chemical evolution and mineral deposition in boiling hydrothermal systems: Economic Geology, v. 80, p. 126-147.

Elder, J. W., 1965, Physical processes in geothermal areas; in Lee, W. H. K (ed.), Terrestrial Heat Flow: American Geophysical Union, p. 211-239.

Ellis, A. J., 1969, Present-day hydrothermal systems and mineral deposition: Ninth Common Mining and Metallurgy Congress, American Institute of Mining and Metallurgy, p. 1-30.

Ellis, A. J., and Golding, R. M., 1963, The solubility of carbon dioxide above 100°C in water and in sodium chloride solutions: American Journal of Science, v. 261, p. 47-60.

Ewers, G. R., and Keays, R. R., 1977, Volatile and precious metal zoning in the Broadlands Geothermal Field, New Zealand: Economic Geology, v. 72, p. 1337-1354.

Facca, G., and Tonani, F., 1967, The self-sealing geothermal field: Bulletin of Volcanology, v. 30, p. 271-273.

Fournier, R. O., 1985, The behavior of silica in hydrothermal solutions; in Berger, B. R., and Bethke, P. M. (eds.), Geology and Geochemistry of Epithermal Systems: Society of Economic Geologists, Reviews in Economic Geology, v. 2.

Giggenbach, W. F., 1980, Geothermal gas equilibria: Geochimica et Cosmochimica Acta, v. 44, p. 2021-2032.

Goguel, J., 1982, The behavior of vapor-dominated reservoirs: Geothermics, v. 11, p. 3-13.

Grant, M. A., 1977, Broadlands--A gas-dominated geothermal field: Geothermics, v. 6., p. 9-29,

Grant, M. A., Truesdell, A. H., and Manon, M., 1984, Production induced boiling and cold water entry in the Cerro Prieto Geothermal Reservoir indicated by chemical and physical measurements: Geothermics, v. 13, p. 117-140.

Hedenquist, J. W., and Henley, R. W., 1985, Hydrothermal eruptions in the Waiotapu geothermal system, New Zealand: Their origin, associated breccias and relation to precious metal mineralization: Economic Geology, v. 80, p. 1640-1668.

Helgeson, H. C., 1969, Thermodynamics of hydrothermal systems at elevated temperatures and pressures: American Journal of Science, v. 267, p. 729-804.

Helgeson, H. C., Delaney, J. M., Nesbitt, H. W., and Bird, D. K., 1978, Summary and critique of the thermodynamic properties of rock-forming minerals: American Journal of Science, v. 278-A, p. 1-229.

Helgeson, H. C., and Kirkham, D. H., 1974a, Theoretical prediction of the thermodynamic behavior of aqueous electrolytes at high pressures and temperatures. I. Summary of thermodynamic/electrostatic properties of the solvent: American Journal of Science, v. 274, p. 1089-1198.

Helgeson, H. C., and Kirkham, D. H., 1974b, Theoretical prediction of the thermodynamic properties of aqueous electrolytes at high pressures and temperatures. II. Debye-Huckel parameters for activity coefficients and relative partial molal properties of the solute: American Journal of Science, v. 274, p. 1199-1261.

Helgeson, H. C., and Kirkham, D. H., 1976, Theoretical prediction of the thermodynamic behavior of aqueous electrolytes at high pressures and temperatures. III. Equation of state for aqueous species at infinite dilution: American Journal of Science, v. 276, p. 97-140.

Helgeson, H. C., Kirkham, D. H., and Flowers, G. C., 1981, Theoretical prediction of the thermodynamic behavior of aqueous electrolytes at high pressures and temperatures. IV. Calculation of activity coefficients, and apparent molal and standard and relative partial molal properties to 600°C and 5 kb: American Journal of Science, v. 281, p. 1249-1516.

Henley, R. W., 1985, The geothermal framework for epithermal deposits; in Berger, B. R., and Bethke, P. M. (eds.), Geology and Geochemistry of Epithermal Systems: Society of Economic Geologists, Reviews in Economic Geology, v. 2.

Henley, R. W., and Brown, K. L., 1985, A practical guide to the thermodynamics of geothermal fluids and hydrothermal ore deposits; in Berger, B. R., and Bethke, P. M. (eds.), Geology and Geochemistry of Epithermal Systems: Society of Economic Geologists, Reviews in Economic Geology, v. 2.

Henley, R. W., and Ellis, A. J., 1983, Geothermal systems, ancient and modern: Earth Science Reviews, v. 19, p. 1-50.

Henley, R. W., Truesdell, A. H., and Barton, P. B., Jr., 1984, Fluid-mineral Equilibria in Hydrothermal Systems: Socity of Economic Geologists, Reviews in Economic Geology, v. 1, 267 p.

Keith, T. C., White, D. E., and Beeson, M. H., 1978, Hydrothermal alteration and self-sealing in Y-7 and Y-8 drill holes in northern part of Upper Geyser Basin, Yellowstone National Park, Wyoming: U.S. Geological Survey, Professional Paper 1054-A.

Kerrick, D. M., and Darken, L., 1975, Statistical thermodynamic models for ideal oxide and silicate solid solutions, with application to plagioclase: Geochimica et Cosmochimica Acta, v. 39, p. 1431-1442.

Kraynov, S. R., Solomin, G. A., Vasil'kova, I. V., Kraynova, L. P., Ankudinov, Ye. V., Gudz, Z. G., Shpak, T. P., and Zakutin, V. P., 1982, Geochemical types of iron-bearing groundwaters of near-neutral reaction: Geochemistry International, v. 19, no. 2, p. 70-89.

Mahon, W. A. J., and Finlayson, J. B., 1972, The chemistry of the Broadlands geothermal area, New Zealand: American Journal of Science, v. 272, p. 48-68.

Mahon, W. A. J., McDowell, G. D., and Finlayson, J. B., 1980, Carbon dioxide: its role in geothermal systems: New Zealand Journal of Science, v. 23, p. 133-148.

Muffler, J. P., White, D. E., and Truesdell, A. H., 1971, Hydrothermal explosion craters in Yellowstone National Park: Geological Society of America, Bulletin 82, p. 723-740.

Nelson, C. E., and Giles, D. L., 1985, Hydrothermal eruption mechanisms and hot spring gold deposits: Economic Geology, v. 80, p. 1633-1639.

Reed, M. H., 1982, Calculation of multicomponent chemical equilibria and reaction processes in systems involving minerals, gases and an aqueous phase: Geochimica et Cosmochimica Acta, v. 46, p. 513-528.

Reed, M. H., 1985, Formation of massive sulfide, anhydrite and talc deposits by sea floor hot spring reaction with seawater. In review.

Reed, M. H., and Spycher, N., 1983, Calculated effects of boiling, cooling and oxidation on precipitation of ore and gangue minerals from a hydrothermal solution. Fourth International Symposium on Water-Rock Interaction, Extended Abstracts, p. 405-408.

Reed, M. H., and Spycher, N., 1984, Calculation of high temperature pH and mineral equilibria in hydrothermal waters, with application to geothermometry and studies of boiling and dilution: Geochimica et Cosmochimica Acta, v. 48, p. 1479-1492.

Rowe, J. J., Fournier, R. O., and Morey, G. W., 1973, Chemical analysis of thermal waters in Yellowstone National Park, Wyoming, 1960-65: U.S. Geological Survey, Bulletin 1303, 31 p.

Schwartzenbach, G., and Widmer, M., 1966, Die Loslichkeit von Metallsulfiden, II. Silbersulfid 1, Helvetica Chimica Acta, v. 49, p. 111-123.

Seward, T. M., 1973, Thio complexes of gold and the transport of gold in hydrothermal ore solutions: Geochimica et Cosmochimica Acta, v. 37, p 379-399.

Steven, T. A., and Eaton, G. P., 1975, Environment of ore deposition in the Creede mining district, San Juan Mountains, Colorado: I. Geologic, hydrologic and geophysical setting: Economic Geology, v. 70, p. 1023-1037.

Timperley, M. H., 1983, Phosphorous in spring waters of the Taupo Volcanic Zone, North Island, New Zealand: Chemical Geology, v. 38, p. 287-306.

Truesdell, A. H., 1979, Aquifer boiling may be normal in exploited high-temperature geothermal systems; in Remay, H. J. and Kruger P. (eds.), Proceedings of Fifth Workshop on Geothermal Reservoir Engineering, p. 299-303.

Walther, J. V., and Helgeson, H. C., 1977, Calculation of the thermodynamic properties of aqueous silica and the solubility of quartz and its polymorphs at high pressures and temperatures: American Journal of Science, v. 277, p. 1315-1351.

Weissberg, B. G., 1969, Gold-silver, ore-grade precipitates from New Zealand thermal waters: Economic Geology, v. 64, p. 95-108.

Weissberg, B. G., Browne, P. R. L., and Seward, T. M., 1979, Ore metals in active geothermal systems; in Barnes, H. L. (ed.), Geochemistry of Hydrothermal Ore Deposits: Second Edition, John Wiley and Sons, New York, p. 738-780.

White, D. E., 1957, Thermal waters of volcanic origin: Geological Society of America Bulletin, v. 68, p. 1637-1658.

White, D. E., 1981, Active geothermal systems and

epithermal ore deposits: Economic Geology, 75th Anniversary Volume, p. 392-423.

White, D. E., Muffler, L. J. P., and Truesdell, A. H., 1971, Vapor-dominated hydrothermal systems compared with hot-water systems: Economic Geology, v. 66, p. 75-97.

Wolery, T. J., 1979, Calculation of chemical equilibrium between aqueous solution and minerals; the EQ3/6 software package: U.C.R.L.-52658: Lawrence Livermore Laboratory.

Chapter 12
USING GEOLOGICAL INFORMATION TO DEVELOP EXPLORATION STRATEGIES FOR EPITHERMAL DEPOSITS

Samuel S. Adams

INTRODUCTION

This paper presents a rationale for (a) the integrated use of geologic information in the development of exploration strategies, (b) an increased awareness of the impact of human factors on the geologists' use of geologic information in exploration, and (c) the effective use of available geologic information through the development of more predictive and reliable models for exploration. This is accomplished, first, by illustrating how geologic information is used in the development of exploration strategies, and how other strategic factors influence the selection and use of geologic information in exploration. Second, human factors are identified and discussed that most significantly influence how geologists use geologic information. Finally, an approach for the organization and use (i.e., modeling) of geologic information in mineral exploration is presented. This modeling approach is referred to as a data-process-criteria model, and it is illustrated with examples for the hot-spring-type epithermal precious-metal deposit.

SOME CONSIDERATIONS IN THE USE OF GEOLOGIC INFORMATION IN EXPLORATION

Concepts presented in this paper are based upon certain assumptions or approaches to the use of geologic information in exploration. These approaches are not the only ones that might be used and are not shared by all exploration geologists. To aid the reader in understanding the writer's biases in later discussions, the most important of these approaches are briefly discussed below.

There is more geologic information available to most exploration projects than an exploration geologist has time to consider. In selecting what information to use, whether it is LANDSAT imagery or stable-isotope data, it is helpful to ask the following types of questions. Do I need it? What will it do for me? What questions do I have that it might answer? Are the data reliable? Do the data contribute to achieving my organization's objectives? Will they likely lead me to conclusions or just more questions? Am I satisfied with our exploration success, or do I need some better methods? Are we likely to need some new exploration criteria to stay competitive? How do the data fit into the information base or model I already have? Answers to these questions will help assure that the information assembled will contribute to the exploration program.

Geologists disagree on the appropriateness interpretation in the use of geologic exploration. Some argue that only empirical observations are sufficiently unambiguous and reliable for use in exploration, and that considerations of geologic processes and ore genesis are uncertain, risky, and have not contributed to exploration success (Ridge, 1983). Others contend that an understanding of empirical observations significantly enhances the use of geologic information in exploration. The writer's bias is that understanding and interpretation improve the use of geologic data in exploration, provided the interpretations are based on adequate and reliable data. It seems inevitable that the use of process interpretations and scientific concepts in exploration will expand as exploration becomes more difficult and costly. The gradual shift from predominantly empirical exploration toward greater use of geologic interpretations will favor those geologists who can deal with chemical, physical, and hydrologic processes, yet remain focused on the exploration objectives of the organization.

The application of geologic data and concepts to exploration is by analogy. Exploration geologists search for repetitions of geologic observations and relations known to be associated with examples or analogs of the deposit type sought. This is the most appropriate approach to exploration in most situations and is the basis for the data-process-criteria model presented in a later section of this paper.

Geologists do much of their work tacitly or intuitively in their minds based on experience; so that, to many observers, exploration geology seems as much an art as a science. This has been an appropriate and successful approach for years as data and understanding for mineral deposits and exploration regions have been evolving. In the writer's opinion, this is now changing dramatically as significant increases occur in both (a) geologic data and concepts for mineral deposits and exploration regions, and (b) the difficulty and cost of, as well as the demands on, exploration. Aspects of these changes are schematically illustrated in Figure 12.1. The decreases in exposed but undiscovered deposits, unexplored areas, and previously unrecognized deposit types makes exploration more costly and difficult. It seems that the greatest opportunity for offsetting these factors is through the use of geologic data and concepts in more predictive and reliable ore-deposit models. This will require a shift from tacit and intuitive approaches to more explicitly documented models.

As used in this paper, the term model refers to

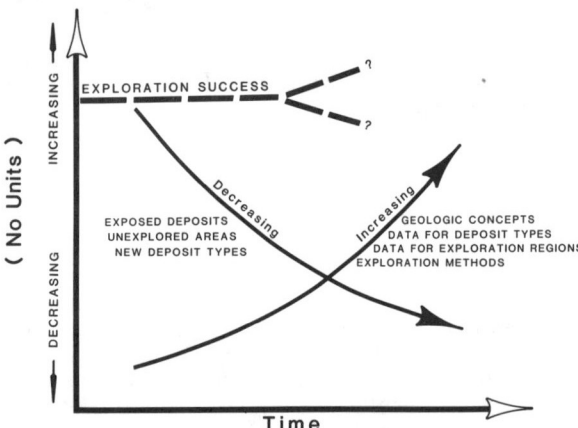

Figure 12.1. Schematic representation of changes in aspects of mineral exploration with time.

an assemblage of geologic data or geologic data and concepts. Few geologists share precisely the same definitions of model or the various types of models used in mineral exploration. Some geologists consider models too risky and unreliable for use in exploration. In the writer's opinion, problems associated with models arise from their imprecise definition, development, and use, not from the models themselves. The use of models in exploration, therefore, must emphasize explicit definitions and techniques that promote more reliable development and application of models. The data-process-criteria model presented in the final section of this paper addresses these requirements.

STRATEGIC FACTORS

Of the numerous factors that affect exploration, 11 factors have been selected as most significant in the development of an exploration strategy. Each of these 11 factors is shown in Figure 12.2 and briefly discussed below. Geologic information is one of the strategic factors, and like the others, it impacts and is impacted by every other strategic factor. Achievement of the goals and objectives of an organization requires the balanced consideration of each strategic factor.

Development of an exploration strategy considers the strategic factors in the approximate sequence presented in Figure 12.2. The objectives of the organization are first identified and potentially profitable commodities selected. The collection and interpretation of geologic information is then coordinated with the organization's financial resources, staff capabilities, considerations of regulations and land availability, competitor activity, and previous exploration. Once exploration methods have been identified for a particular deposit type in a particular region, the opportunities and risks of the proposed program can be evaluated. In practice, there are much iteration and reconsideration among the strategic factors. Many organizations develop and exploration strategy in parallel with other strategies; for example, a strategy to acquire identified deposits.

Organization Objectives

The goal of most organizations employing exploration methods is profit, which may be achieved through a variety of objectives. Examples of organization objectives that employ exploration are listed in Table 12.1. These objectives are used in various combinations that reflect the size, maturity, and history or culture of an organization. Non-profit organizations, such as governments, employ exploration methods for land-use planning and to identify sources for mineral supply.

An exploration strategy is most likely to achieve its objectives if the individual objectives are clearly defined and communicated to the exploration organization. If the objectives are unclear or unrealistic, it will be difficult for the best of exploration organizations to be successful. For example, during the last decade, the mineral exploration divisions of more than a dozen oil companies explored for epithermal precious-metal deposits in the United States. Even though some of the companies were "successful" in acquiring, discovering, and even developing gold and silver properties, over half are now out of the mineral business, and most of the remaining oil companies have severely cut back their exploration programs. The objectives that these organizations pursued proved to be unrealistic for their corporations in terms of scale of operation, return on investment, and discovery potential. The reader is referred to Snow and Mackenzie (1981), Bailly (1979), and Kyle (1984) for further discussions of exploration organization objectives. The exploration geologist is responsible for understanding the organization objectives, determining within his area of expertise if they can be achieved, and contributing to development of the best

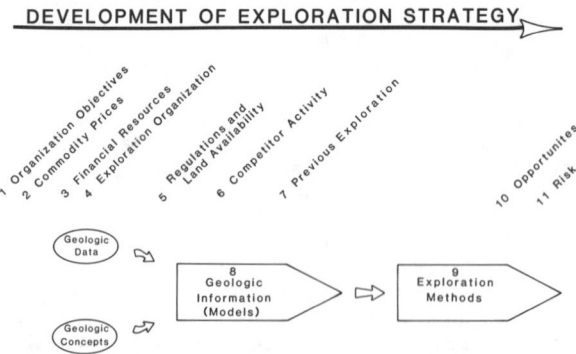

Figure 12.2. Approximate sequence for consideration of geologic information and the other ten strategic factors in developing an exploration strategy.

Table 12.1--Examples of organization objectives that employ exploration

For Profit Organizations

Maintain organization's business

    Develop feed for existing milling facility

    Develop new production facilities

Increase organization's business

    Develop additional production facilities

    Expansion into new countries and exploration areas

Integrate organization's refining, fabricating, and marketing business back into production

Identify and acquire lands and properties with high mineral potential for sale or joint venture

Exploit existing organization assets

    Forest lands, railroad right-of-ways, ranch lands, mineral holdings

    Technology--exploration methods of hardware, research facilities

    Information--data files for exploration regions, properties, and mineral-deposit types

    Personnel--knowledgeable geologists, engineers, metallurgists, landmen, managers, negotiators, and businessmen

Nonprofit Organizations

    Evaluate mineral potential of lands for land-use planning and decisions

    Identify source of mineral supply

---

exploration strategy. Only in this way can the correct geologic information for the correct deposit type(s) and the correct exploration area be selected.

Commodity Prices

Metals prices at the time of production will determine the profitability of most mining operations. Metal prices reflect both supply and demand, and although geologists and geologic information contribute little to predicting the latter, they do contribute to predicting the former. For example, data for world gold production (Table 12.2) and geologic information for the gold deposits of Russia and South Africa suggest that, barring political changes, deposits in these countries will continue to dominate world gold production and prices for the foreseeable future. New gold deposits, therefore, must be of geologic types that are profitable at current gold prices and capable of surviving periods of lower prices. This will include certain bonanza and low-cost, open-pit, disseminated epithermal deposits. The grade, geometry, tonnage, etc. of many epithermal deposits, however, preclude economic production at these gold prices. Gold-price forecasting would be vastly more speculative and risky, were it not for worldwide geologic and economic information for the various types of gold deposits. Geologists contribute to more accurate price forecasting by assuring that geologic information for gold deposits, deposit types, and the potential for new discoveries is as complete and accurate as possible.

The risks associated with commodity price forecasts are considered in the selection of

Table 12.2--World mine production of gold in 1984

Country	Production	
	Thousands of Troy Ounces	Percentage of Total
South Africa	21,905	47.9
USSR	8,650	18.9
Canada	2,614	5.7
United States	1,902	4.2
China	1,900	4.1
Brazil	1,611	3.5
Australia	1,199	2.6
Papua, New Guinea	835	1.8
Philippines	773	1.7
Colombia	735	1.6
Others	3,606	8.0
World Total	45,730	100.0

Source: American Bureau of Metal Statistics (1985)

organization objectives and strategies for their achievement. The decisions that result will determine the deposit types, geologic domains, and geographic areas appropriate for exploration and for which geologic information will be required.

Financial Resources

Geologic information and financing for exploration programs are closely interrelated. Geologic information for deposit types, proposed exploration areas and exploration methods are used to estimate budget requirements for exploration programs. The annual funding and duration of an exploration program will, in turn, determine the size of the exploration organization, the types of targets pursued, and the geologic information and exploration methods used. If an organization's resources are limited, it may require a particularly advantageous geologic opportunity (see below) to complete successfully. Finally, geologic information for the deposit type sought will indicate the probable size, complexity, and cost of development, which must be reconciled with the investment and risk the company is prepared to make in property development. For example, in remote areas such as portions of British Columbia, development costs may exceed a company's guidelines or capabilities. This has led one organization (Shillabeer, 1985) to devise a screening procedure to identify geologically favorable areas that meet development cost, environmental, and land availability criteria. Funding levels for exploration organizations may reflect a multitude of factors, including a percentage of corporate earnings, excess available funds, exploration management requests, current metal markets, and the increasing or decreasing confidence in and performance of the exploration organization.

Studies of exploration and discovery costs have recently been made for Canada (Cranstone, 1980, 1983; Mackenzie, 1984), Australia (Mackenzie and Woodall, 1983; Mackenzie and Bilodeau, 1984), the United States (Rose and Eggert, 1983; Eggert, 1983), and South Africa (Kyle, 1984). In general, these studies indicate the high average cost of discovering a mineral deposit and the uneven discovery success among exploration organizations. In Australia, for example, data for the period 1955-1978 showed that the average cost of finding an economic mineral deposit was $38 million, and that the average financial return on mineral exploration was most unattractive (Woodall, 1984a). In order for mineral exploration to be an attractive business venture, therefore, an organization must expect and must be able to achieve better than average results. Geologic information for mineral deposit types and exploration areas is critical for estimating the likelihood of acceptable exploration results and the financial requirements to achieve them.

Exploration Organization

The size, experience, and "culture" of the exploration organization will determine, in large part, how it responds to the organization objectives. Some organizations seem most successful at grass-roots exploration, while others excel at the acquisition and development of properties. Western Mining Corporation, for example, is a discovery company that was instrumental in delineating, among others, the Kambalda nickel deposit, the Yeelirrie uranium deposit, and the Olympic Dam copper-uranium-gold deposit (Woodall, 1984b). Ranchers Exploration and Development Company, now wholly owned by Hecla Mining Company, has been successful in exploration but has excelled in the acquisition, development, and mining of properties such as the Escalante epithermal silver deposit in southwestern Utah. Some organizations are better adapted to working on particular deposit types or, in particular, geologic terrains. For example, Lac Minerals has been remarkably successful in the discovery of Archean stratiform greenstone-belt gold deposits. According to Valliant (1985), a familiarity with the nature of gold occurrences in the Malartic area, and geologic concepts developed by Ridler (1970) and Hutchinson (1976) led successively to discoveries at Bousquet (6.6 million tons at 0.14 ounces per ton Au), Doyon (zone 2 contains 8.8 million tons at 0.149 ounces per ton Au), and Hemlo (47 million tons at 0.174 ounces per ton Au). Distinctive capabilities, styles, and objectives such as these companies display may override some of the other strategic factors and strongly influence the exploration strategy and the geologic information that different organizations use in exploration. The emphasis at Homestake Mining Company, for example, is on maintaining and strengthening its position as the premiere U.S. gold mining company (Anderson, 1982). Despite its presumed exploration advantage through ownership of the Homestake mine in South Dakota, the company is looking for various deposit types that meets its tonnage, grade, and earnings requirements. This program has led to the discovery of the McLaughlin deposit in California.

Regulations and Land Availability

State, local, and federal governments establish tax, environmental, and other regulations that affect the availability of land for exploration and mining and the profitability of mining operations. Geologic information, on the other hand, is essential for establishing the mineral potential of both public and private lands for purposes of land-use planning and the development of regulations. Geologic information for the Thunder Mountain epithermal district in central Idaho, for example, was instrumental in the exclusion of the district from the River of No Return Wilderness area. Geologic information for the proposed Cabinet Mountains Wilderness area in Montana has demonstrated high mineral potential for portions of the area (Banister et al., 1981). The mineral potential is now being weighed against other potential land uses to determine if the lands will be included in the national Wilderness system. Difficulty in bringing the significant Crandon massive sulfide deposit, discovered by the Exxon Corporation in northern Wisconsin, into production illustrates the importance of tax and environmental regulations in states without strong mining traditions. Prior to initiating exploration in such states, geologic information and the potential for discovery must be carefully weighed against the risk of prohibitive regulations.

Competitor Activity

Exploration is always competing with other organizations, and with the earth as well. What competitors do may affect the success of one's programs by pre-empting favorable lands, hiring the most competent geologists, and seizing promising joint-venture opportunities. Exploration strategy should include a realistic assessment of what the competition is doing with its geologic information; hence, what geologic information and exploration methods will be needed to meet and surpass competitors. Newmont Mining Company, for example, discovered the Carlin gold deposit in 1962, and subsequently discovered Maggie Creek, Gold Quarry, and Rain. Sixteen significant sediment-hosted epithermal gold deposits now have been discovered, and the Newmont deposits may contain as much as half the total gold resources in the sediment-hosted deposit type. Any company considering exploration for this deposit type will do well to consider the geologic information Newmont possesses and the exploration use to which it has been and is being put. To underestimate a successful competitor is a serious error, but any organization can be outdistanced through better use of geologic information, other conditions being equal.

Previous Exploration

A realistic assessment of the nature and extent of previous exploration on proposed exploration lands will indicate what geologic information and capabilities (i.e., budgets, exploration methods, time frames, etc.) will be required in a new program to improve on previous exploration and offer a reasonable chance of success. For example, in the early 1960's, Kennecott conducted a regional stream-sediment geochemical program within the area of the Belt Supergroup in northern Montana and Idaho. This program is credited with the discovery of the Troy deposit (Clark, 1971) and other stratiform copper-silver deposits. Recently, new concepts for the formation of these deposits have been published (Lange and Sherry, 1983). Any new exploration programs for stratiform copper-silver deposits in this region will have to improve upon these earlier programs and concepts and whatever exploration use has been made of them. Although it may be difficult to obtain reliable information on previous exploration programs, such information is worth the considerable effort it takes to obtain it.

Geologic Information

Geologic data and concepts make up the geologic information of exploration. This information includes data and concepts for the basic sciences, mineral-deposit types, geologic terranes in which the deposits occur, and regions considered for exploration. Information from previous exploration programs is equally important, as discussed in the previous paragraph. Geologic information includes all information relevant to exploration, such as geophysics, rock and stream-sediment geochemistry, topography, remote sensing, geomorphology, and the concepts of processes involved in mineral deposit formation (to mention only a few examples). The information ranges in scale and type from mountain ranges and concepts of plate tectonics to fluid inclusions and concepts of boiling and metal precipitation. All of it is potentially relevant to exploration; much of it ultimately will prove unnecessary, and much may in practice be missing. The exploration geologist uses available geologic information to most efficiently and effectively pursue the organization objectives and to identify additional information that is worth collecting.

The creative use of geologic information in exploration is illustrated by the discovery of the Olympic Dam deposit in South Australia by Western Mining Corporation (Haynes, 1979; Woodall, 1984b; and Lalor, 1984). In 1972, the company commenced a study of the Proterozoic and Lower Paleozoic rocks of South Australia in search for a sediment-hosted copper deposit, similar to those in the Zambian Copperbelt. During the preceding three years, a conceptual model for this deposit type had been developed. The model required assembling geologic information for particular geologic characteristics from the literature, regional reconnaissance, records of previous exploration programs filed with the State Survey, regional gravity and magnetic surveys, and lineament studies. By July, 1975, the first drill hole had intersected a blind mineralized interval (114 feet at 1.05% Cu), at an approxoimate depth of 900 feet. Subsequent drilling led to the delineation of reserves, reported in 1983 to be 500 million tons of ore grading 2.5% copper, 0.08% U_3O_8, 0.017 ounces per ton gold, and 0.17 ounces per ton silver. Olympic Dam, therefore, is a major discovery, and one of the world's great mineral deposits.

Many geologists have tended to attribute the discovery of Olympic Dam to luck. The discovery certainly included some unexpected, or lucky, aspects, including the presence of the deposit in significantly older sediments than had been expected, and geologic characteristics of the deposit that differ significantly from typical sediment-hosted copper deposits (i.e., the abundance of breccia and the high concentrations of iron, uranium, gold, and rare-earth elements). Also, despite the success of the first drill hole, the high-grade zone (2.2% Cu) was not encountered until drill hole 10 in November, 1976.

On the other hand, the first drill hole was sited after the careful development of the model and interpretation of much regional geologic data, and it did discover a major deposit. The target area met the principal requirements of the model including regionally altered basalts, numerous copper anomalies, magnetic and gravity anomalies and lineaments. Furthermore, within the Zambian Copperbelt are stratiform iron deposits, uranium deposits, and structurally controlled gold deposits. Although the currently known metal accumulations are not coextensive, Olympic Dam-like deposits may yet be discovered in the Copperbelt. In the writer's opinion, fine geologic work put Western Mining in the correct area, so that their perseverance and share of good luck were able to overcome the equivalent amount of bad luck that must attend every exploration program.

Exploration Methods

Exploration methods are procedures and techniques such as mapping, geochemical surveys, drilling and core logging, airborne geophysical surveys, etc., through which geologic information is gathered and a model is applied to an exploration area. Exploration methods are specifically selected to collect geologic data in order to test for the essential criteria of a particular mineral deposit model in a particular exploration area. Exploration strategy is implemented through the exploration methods. Cost effective, technically sound geologic, geochemical, and geophysical methods are essential for the success of any program, but they will not salvage a program for which the other strategic factors have not been properly assessed.

Opportunities

Every exploration program faces the challenge of recognizing an overlooked opportunity at a known prospect or discovering a completely new deposit. Exploration strategy should include, therefore, one or more specific and compelling reasons why the program will succeed where others have failed. These reasons are referred to as strategic opportunities, and they may include a new geologic concept, an area that has not previously been explored, or a new exploration method. During the last decade, the newly appreciated hot-spring and sediment-hosted (i.e., "Carlin"-type) type deposits offered strategic exploration opportunities for comparatively unexploited deposit types in the western United States. Without at least one such strategic opportunity an exploration program is apt to fail.

Unique opportunities arise from time to time in each of the strategic factors, causing exploration to be temporarily focused or accelerated by that factor. For example, if a company has excess profits, financial resources for exploration may suddenly increase, and the exploration program is temporarily accelerated. Alternatively, a new geologic concept, such as recognition of the geologic setting of the hot-spring epithermal deposits may be perceived to offer such opportunity for new discoveries that exploration priorities are rearranged. Yet other programs become redirected in response to breakthroughs in exploration hardware and methods. Incremental improvements in airborne EM, for example, will presumably lead to the discovery of several new massive sulfide deposits.

Risk

The final strategic factor is exploration risk, which is assessed for the other strategic factors, both individually and collectively. Risk is typically reviewed intermittently throughout the development of an exploration strategy. A critical assessment of risk is made when the geologic information has been compiled, a model developed, and strategic opportunities and exploration methods identified. At this point, the discovery potential and exploration costs for specific areas can be roughly estimated, and management must determine if the risks warrant pursuing the program. If not, further research is initiated, or the project deferred. In some cases, even in the face of a high risk/benefit ratio, the decision is made to proceed in some modest way, leading to progressively deeper involvement in a dubious program.

Strategic factors are rarely, if ever, in balance. First, we have yet to define that balance and how to recognize it. Second, the skills and biases of organizations and individuals cause inevitable imbalances. Finally, responsibilities for the various strategic factors usually reside in different parts of an organization, thereby challenging one of our weakest organizational attributes, communication. One of the exploration geologist's responsibilities is to be familiar with all of the strategic factors and do what can be done to achieve a successful balance.

HUMAN FACTORS

A geologist's use of geologic information in exploration is not an entirely objective scientific and technical exercise. As information is accumulated, interpreted, and used, most geologists become personally involved to some extent. Pragmatic, empirical, logical, and rational performance becomes inevitably entwined with personal experiences, intuitions, biases, judgments, motives, feelings, and reactions to risks. This reflects the less than totally rational and predictable influences of the human brain, mind, and emotions, on people's work, and it is not all bad. However, research on the physiology and functions of the brain (Restak, 1984) and in behavioral psychology (Kahneman et al., 1982) have shown that, under certain circumstances, human behavior is at variance with empirical and rational evidence and one's best intentions. Most geologists can cite examples of jumping to inadequately supported conclusions, defending illogical positions, pursuing irrational programs, taking inordinate risks or refusing to take reasonable ones, refusing to acknowledge and correct errors, and making professional decisions for largely personal reasons. Research suggests (Kahneman et al., 1982) people are neither entirely in control or aware of these tendencies and their consequences. Some of these behaviors may profoundly affect a geologist's use of geologic information and these behaviors are briefly mentioned below. Of particular importance to the use of geologic information in exploration are behaviors related to personal objectives, education and training, problem-solving methods, intuition and creativity, uncertainty, and aversion to loss.

Personal Objectives

How successfully geologic information is utilized depends in part on the geologist's personal objectives and why he or she is an exploration geologist. Some geologists seem to be in exploration for a variety of reasons that may not necessarily contribute much to the organization's objectives, such as love of the outdoors or science, difficulty with the pre-med curriculum, math, or physics, the romance of exploration, or simply the need for a job. Such personal objectives are certainly not tailored to many exploration positions, which require an interest in and working knowledge of the diverse strategic factors discussed above and a respect for the inevitable organizational and managerial functions and responsibilities. Sometime early in their careers, therefore, exploration geologists must reconcile where they thought they were going professionally with where they seem to be headed, and decide if they like that career direction. If they decide to commit themselves to the evolving opportunities in exploration, most geological scientists can expect to learn a great deal about the nongeological aspects of planning and directing (strategy) and executing (tactics) exploration programs.

To survive in exploration, geological research must be closely integrated within successful exploration programs. The defunct mineral exploration research functions at Shell Oil, Kennecott, Anaconda, and Exxon, to mention only a few, illustrate the liability of detached and inflated scientific functions. Geologists who prefer science must either embrace the exploration objectives or pursue careers in governmental agencies, teaching, or corporate research. Those who prefer more outdoor work have numerous options. Geologists who fail to make an overt decision and simply remain in exploration organizations risk becoming frustrated and burdensome to those exploration programs. The current educational and employment systems guarantee that a career in exploration geology will require some transition from the preparation and expectations of the academic environment to the reality of the exploration employment. How successfully geologists assess their personal objectives and capabilities and make this transition will influence their subsequent success in applying geologic information to the objectives of their organizations.

Education and Training

Geologists entering minerals exploration generally have been educated as scientists, not as explorationists, because that is what colleges and universities do. A rigorous scientific education does not prepare a geologist for the exploration environment in which success requires a working knowledge of the various strategic factors and applied exploration practices in addition to geological sciences. Geologists entering exploration must embrace a new set of values. Scientific training must accept a balanced, important, but nonetheless subordinate role. The extent to which geologists can accommodate to this environment depends upon their personal flexibility and objectives and how effectively the company approaches the training.

Once again, Western Mining Corporation is a valuable example. As described by Woodall (1984b), extraordinary success of the company has depended in part on confidence in science and scientists.

> Ore-deposit models which have both empirical and theoretical support instill the greatest confidence.
>
> ... Earth science research and ore-deposit models are only relevant if they give us a sounder basis for that confidence, make us bolder and more perceptive explorers, and help us to be more confident in the recognition of either the close proximity of ore or a new ore environment. But we can follow knowledge and reason just so far, then comes the act of faith, the leap beyond the sure path. Whether we are ultimately able to take that step is a test of our ultimate confidence in science and ore-deposit models.

Western Mining has created an exploration atmosphere which exudes confidence in and respect for science and scientists, supports their continued education and creative work, and capitalizes on their productivity. Most other mining companies avoid even the mention of the words science and scientists.

Problem Solving

Rational empiricism is our cultural method of choice for acquiring knowledge and solving problems. Throughout schooling, students are rewarded for returning factual (empirical) and logical (rational) answers. Geologists are trained to apply the scientific method, which employs both observations and experience (empirical) and hypotheses (reasoning). The accomplishments of science testify to the power of the method. But, as with any method, there are opportunities for misapplying rational empiricism.

Rational empiricism is most applicable when adequate data are available to measure, describe, and define a system. Most geologic systems are incompletely documented, requiring the use of considerable judgment, inference, and intuition with the geologic information. Geologic systems also tend to be complicated, requiring generalization and simplification. Even under these conditions, skilled geologists may be able to make interpretations and predictions that are useful in exploration. In other cases, it is preferable to collect additional information, if the cost can be justified.

The mere use of an approach that approximates the scientific method may promote overconfidence in the data and the conclusions, regardless of their quality. There is a risk in using data bases that are incomplete or inaccurate, as they may lead to inappropriate hypotheses and conclusions. Poorly thought out concepts and interpretations may become confused with scientifically derived hypotheses. Geologic reasoning in exploration is inductive, employing data from analog deposits to define general characteristics and hypotheses for a deposit type. Such reasoning is, at best, inferential, suggestive, and permissive. It requires continual testing as new data come available. Yet, as discussed below, there is a natural tendency to place unwarranted confidence in one's interpretations under uncertainty and then refuse to modify the interpretations in the face of new data. Rational empiricism and the scientific method are applicable to minerals exploration, but the hypotheses and predictions will tend to be more uncertain and risky than in the basic sciences for which the approaches were developed.

Most misapplications of rational empiricism are the fault of the user, not the approach. Overdependence on rational empiricism may lead to a loss of mental flexibility and creativity (see below). There is also the risk that the approach may be used as a protection against criticism, rather than to promote understanding and the organization's objectives. Rigidly insisting on the rational empirical approach just to satisfy a psychological need for security jeopardizes intuition (Goldberg, 1983).

Nothing in the foregoing is meant to impugn the use of rational empiricism in exploration. The predictive power of the approach is essential if exploration is to benefit more from existing data. The foregoing issues illustrate, however, that rational empiricism is no panacea, and that there are many opportunities for misapplications in mineral exploration.

Intuition and Creativity

Intuition and creativity are both potential attributes and threats for exploration programs. Guided by existing empirical data, they can significantly extend what previously has been accomplished in exploration. Based upon incomplete or inaccurate data, or unfounded concepts, they waste time and exploration funds. Both intuition and creativity have contributed to exploration along with empiricism, rationalism, and luck. Most geologists have experienced those "flashes of genius" or inspirations that have no obvious, rational origin or explanation, yet move us toward the solution of a problem. Most geologists consider their geologic work to be intuitive and creative, at least part of the time. The use of geologic information in exploration relies heavily on empirical geologic data, intermittently on interpretations we make about those data, and less frequently on intuitive or creative insights we have about those data. Although infrequent, intuitive breakthroughs that are consistent with empirical data may produce important exploration opportunities.

Intuition and creativity may become too independent, leading to resistance to regimentation, documentation, personal appraisal systems, models, and even other people's ideas. There may be a compulsion to do the job in our own special way and a feeling that other geologists can be empirical and rational, and anyone can be lucky, so our creativity and intuition are the best reflections of our special skills. A geologist's professional value and self-esteem are not measured by creativity and intuition, but by how effectively he or she balances the use of empiricism, rationalism, luck, intuition, and creativity.

Some geologists resist any rigorous rational empirical approach to exploration as a threat to their creativity and intuition. In fact, rational empiricism should promote rather than encumber creativity and intuition. The history of great discoveries illustrates that intuitive breakthroughs come to those knowledgeable in their fields, and that luck comes most often to those who have prepared themselves. Louis Pasteur put it this way: "In the fields of observation, chance favors only the prepared mind." Crick and Watson (Watson, 1969) discovered the double helix, not plate tectonics, because biology was their field of life-long study. Within that field, they were able to be creative and intuitive because of, and in spite of, their knowledge. Intuition and creativity seem first to absorb the best that empirical observations and rational interpretations have to offer and then to reorganize them in inspirational ways. In working geologic information, therefore, geologists must be willing to digest and evaluate existing facts and interpretations, allow their "processor" to make what it can of them, and then ruthlessly evaluate the validity of the results and their value to exploration. To indulge in the elevation of intuition over existing empirical observations, logical and rational thought, and good fortune, is to substitute personal objectives and emotions for the organization's objectives. To ignore or to fail to promote creativity and intuition is to condemn exploration programs to what has been known and what has been done without provision for the inevitable changes in exploration opportunities and circumstances.

Uncertainty

Exploration geology is uncertain due to incomplete data and more or less uncertain interpretations for complex, ill-defined geologic systems. Explorationists accept these uncertainties as risks in making exploration decisions. We like to think that we do the best we can with the information we have, but how good are our decisions? Are we really doing as well as can be expected, or are there biases and errors that make exploration less successful than it might be? Unfortunately, we lack adequate data from successful and unsuccessful exploration projects to answer this question. The best we can do at the present time is to review some evidence of how people behave under uncertainty and compare this with our own experience with exploration programs.

Behavioral studies indicate there is something deep within the human mind that abhors uncertainty. Kahneman illustrated the point this way (McKean, 1985):

> There's a strong overconfidence effect. Suppose I take you into a darkened room and show you a circle with no distance cues and ask you how big it is. You don't know whether it's a small circle very close or a large circle far away. If the mind worked like a computer, it would say it doesn't know the answer. But people always have a firm feeling about the circle size, even when they know they are probably wrong.

> What you see here is a classic example of how the human mind suppresses uncertainty. We are not only convinced that we know more about our politics, our businesses, and our spouses than we really do, but also that what we don't know must be unimportant.

Experiments conducted by Gazzaniga and reported by Restak (1984) demonstrate the aversion to uncertainty in another way. It has long been known that the left side of the brain controls functions of the right side of the body and vice versa. Therefore, the left hemisphere receives information from the right eye and operates the right hand. It is also known that severe epileptic fits can be arrested by severing the corpus callosum, a bundle of nerve fibers joining the two hemispheres of the brain. In one experiment, a patient whose corpus callosum had been severed,

> ... was shown two pictures, one projected to the left hemisphere (a chicken claw) and one to the right hemisphere (a snow scene). He was asked to indicate from a series of pictures what picture he had just seen. (He) selected a picture of a chicken with his right hand and a picture of a snow shovel with his left. When asked why he had made these particular selections, (he) remarked: "That's easy. The chicken claw goes with the chicken, and you need a shovel to clean out the chicken's shed."

In this instance, (his) left hemisphere employed its language superiority to construct a plausible, "logical" explanation for the choices he had selected. But the explanation was wrong, an error that suggests to Michael Gazzaniga that our speech and language systems routinely attempt to interpret rather than simply report our activities. If (his) explanation is at all typical of the rest of us, it suggests that the reasons we give for our own behavior may not be the salient ones at all.

Subsequent experiments by Gazzaniga demonstrated that the phenomena also occurs in normal people. The logic-language dominant left hemisphere of the brain, therefore, avoids being without an explanation by routinely concocting "logical" explanations that may be erroneous.

Based upon these examples, the impact of uncertainty on a geologist's use of geologic information may be considerable. Although geologists may revere the scientific method and the concept of multiple-working hypotheses, the discomfort of uncertainty promotes the acceptance of the first reasonable explanation for a set of data. This may result in greater emphasis on explanations than on accuracy and truth. The same tendency is described by Goldberg (1983):

> Knowing feels good. There is a certain tension created by ignorance, an incompleteness in an unresolved problem ... When the answer comes, there is a feeling of restoration.

Important insight into some of the ways geologists may use geologic information under uncertainty in exploration is provided by behavioral studies reported by Tversky and Kahneman (1982). They note that people strive to achieve a coherent interpretation of the events that surround them, and that cause-effect relations seem to achieve this goal. In other words, people build models in which observations are explained as either the cause or the effect of other observations. This is precisely what exploration geologists do when they interpret observations for a mineral deposit in terms of the processes that probably formed the characteristics of the deposit. In their studies, Tversky and Kahneman (1982) investigated how subjects intuitively build and use causal models, and how they respond when additional and conflicting information are supplied. The results are disquieting in that they indicate people may introduce errors in their use of models. Their results are also unconformably familiar, at least to the writer.

Tversky and Kahneman (1982) found that once a causal model has been constructed, there is a reluctance to revise it in the face of new data, regardless of the uncertainty in the original model, and regardless of the veracity of the new data. Instead, the causal model is used to explain the new data with little or no revision of the model, regardless of how inconsistent and condemning the new data may be. It seems that our highly developed explanatory skills and fluency of causal thinking inhibit the revision and correction of causal models in the face of new data. This suggests that exploration geologists will tend to protect causal models from new geologic observations, using the model to explain away the data rather than performing the indicated revisions on the causal model. Somehow this seems intuitive and probably explains why many geologists tend to resist the use of models. The work of Tversky and Kahneman (1982) also suggests correlaries to this phenomenon. First, people emphasize data that fit cause-effect relations, and place little or no weight on other data. Second, people commonly overpredict from highly uncertain models. Third, inferences from causes to consequences are made with greater confidence than inferences from consequences to causes. All of these observations have implications for geologists' use of geologic information in exploration and encourage the development and use of modeling approaches that suffer as little as possible from the foregoing shortcomings. This is addressed in the approach to modeling presented later in this paper.

Response to uncertainty may reflect one's rational empirical education, during which we were rewarded for having logical, factual answers to every question. The biggest risk to most people is looking foolish, being humiliated, appearing stupid; we have a need to be right. Techniques for using geologic information that minimize these types of problems will improve exploration. The modeling methodology discussed in a later section has been designed to minimize these problems.

Aversion to Loss

People attempt to avoid loss, even when it can be shown that alternatives may ultimately lead to greater losses. "There is something in the human mind that so abhors loss that giving up some quantity of money, commodity, or privilege is never fully offset by an equivalent gain," according to Kahneman and Tversky (McKean, 1985). In other words, loss looms larger than gain. This may explain, for example, why some exploration projects are kept alive by exploration managers or project geologists long after the accumulated geologic information should have ended them. Sponsors are unwilling to accept the termination and loss of a project. Legitimate remaining opportunities need to be pursued, but the human tendency to selectively disregard information and keep certain projects going is obviously a problem.

Aversion to loss may influence the use of geologic information in other ways. For example, geologists may act defensively or aggressively to criticism of their project out of fear of losing their budgets, their projects, or the respect of their colleagues. Colleagues, on the other hand, may overcriticize projects, out of fear that they will lose standing in the organization if the projects appear to successful. Despite our best intentions, geologic information may, from time to time, be used for personal purposes. Since these misuses are probably inevitable, geologic models should be designed to minimize the impact of personal motives and the other human factors.

DEVELOPMENT OF MINERAL-DEPOSIT MODELS

The responsibility of the exploration geologist is to predict where a mineral deposit is likely to occur, the likelihood that the deposit will be there, and the grade and tonnage the deposit is likely to contain. In conjunction with regional geologic information, the mineral-deposit model is the most powerful tool a geologist has for making these predictions. The predictiveness of a deposit model is derived from the geologic characteristics of analogs of the deposit type, the relations among these characteristics, and the level of confidence the geologist has in both. A carefully constructed model enables a geologist to make better predictions about mineral deposits and better estimates of the confidence he is entitled to have in those predictions. Modeling brings order out of chaos. Until geologic observations have been related one to another in a causal model, they are just so many bits of data. In this section, modeling terminology, the organization of geologic information for mineral-deposit types, and the development of mineral-deposit models are reviewed.

A <u>mineral-deposit type</u> is characterized by geologic and economic characteristics compiled from analogs of that deposit type. The characteristics are sufficiently different from those of other deposit types to justify the deposit type's distinction for exploration. A <u>mineral-deposit model</u> is the representation of data and interpretations for a mineral-deposit type. Geologists have constructed such models informally in their minds and more explicitly in writing for decades. Written models soon become outdated, but they are necessary records for communication, documentation, and updating. Exploration geologists update models in their minds as new information is accumulated. These tacit models are applied to exploration as they evolve, generally without written documentation; hence, without critical review by the geologist or geologic colleagues. As long as this approach achieves the organization objectives, there is no compelling reason to change it. However, not everyone builds and applies tacit models with equal success. Models benefit greatly from the critical review of explicit written versions by geologists and their colleagues.

The development of a mineral-deposit model is coordinated with the development of exploration strategy. Both benefit from the development and refinement of the other. The approximate coordination between model development steps and consideration of strategic factors is shown in Figure 12.3. Initial selection of a deposit type for exploration is determined by (a) the organization objectives, (b) commodity price expectations, (c) financial resources of the organization for exploration and development, (d) skills and preferences of the exploration organization, and (e) availability of land for that deposit type. The level to which the model must be developed to be successful in exploration will be determined principally by current competitor activities and the level of previous exploration on the available lands. The construction of the deposit model will be based upon existing geologic information and newly collected data, if the costs can be justified. Iteration between the evolving deposit model and exploration methods will attempt to assure that the methods significantly improve upon competitor's approaches and previous exploration. Finally, evaluation of the model's reliability for exploration will be coordinated with consideration of opportunities and risks the model and the exploration methods present for the exploration strategy. The model development steps are more fully discussed in the following section.

The mineral-deposit model is a critical link between the strategic factors, in particular geologic information, exploration methods, exploration organization skills, previous exploration, and competitor activity. New geologic information for a particular deposit type, for example, impacts the other strategic factors through their mutual relations to the deposit model.

No reasonably complete or consistent format has been developed for presenting a mineral-deposit model. Descriptions of a deposit type commonly reflect the author's particular expertise, or interest in, for example, fluid inclusions, wallrock alteration, or

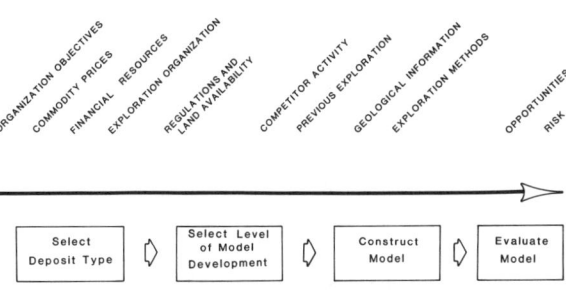

Figure 12.3. Approximate coordination between consideration of strategic factors and model development steps.

regional structure, rather than a comprehensive presentation of the geologic information required for exploration. Models published by the U.S. Geological Survey (e.g., Erickson, 1982; Cox, 1983), the Geological Survey of Canada (Eckstrand, 1984), and the Geological Association of Canada (e.g., Morganti, 1981) reflect significant progress and interest in model compilations, but they generally are not yet adequate for current exploration. The modeling format presented in the final section of this paper is intended to be a step toward fulfilling this need.

All models involve assumptions and interpretations. They are transient, ephemeral representations of what is known at the time, and are only as reliable as the data and science on which they are based. They are valuable springboards for exploration, new intuition, and improvement. However, no model is beyond continued improvement and challenge; and, as discussed in an earlier section, we are far better at building causal models than we are at correcting, validating, and improving them.

How models are envisioned determines, in large measure, how they are used and what they will contribute to exploration programs. To use an analogy, I prefer to think of a model not as a destination but as a mode of travel. It is not the written description of the model that is so valuable; it is a geologist's familiarity and experience with its data and concepts. The better one understands a model's data, interpretations, and uncertainties, the more predictable and reliable will be the model's application to exploration. The advantage of a geologist building his or her own model is that the model's contents presumably will be well known. The danger in using someone else's model is that time will never be taken to sufficiently scrutinize its components to justify the risk of its use in exploration. This, of course, need not be true. Perhaps it is the inconsistent format of models that discourages their critical review and use. Even one's own models require persistent modification and updating, as new data and interpretation become available.

Organization of Geologic Information

Geologic information for exploration is most conveniently organized by mineral-deposit type and exploration region (not considered here). Organization by deposit type is appropriate for exploration for three principal reasons. First, valuable occurrences of metals, minerals, and rocks are found in reasonably distinctive geologic settings, each with its unique assemblage of geologic characteristics and exploration guides. Exploration for a deposit type, therefore, is tailored to these characteristics. Second, the organization's objectives commonly specify particular commodities, ore grades, tonnages, and profitabilities that can be met only by certain deposit types. Finally, exploration for mineral deposits incorporates useful geologic information from all geologic disciplines, scales of observation, and methods of collection. Thus, the mineral deposit itself is the most appropriate framework for information organization. Within the deposit type, geologic information is commonly further organized by subdiscipline (i.e., mineralogy, structure, alteration, geochemistry, geophysics, etc.) and scale of observation (i.e., regional tectonic setting, volcanic-plutonic province, vein structure, fluid inclusion, etc.), which facilitates application in the various stages and at the various scales of exploration programs.

Model Terminology

Geologists differ in their use of the term model and in their definitions of the various model types used in mineral-deposit work. Definitions employed in this paper reflect, wherever possible, the most common prior and current usage. For a glossary of model terms, the reader is referred to Adams (1985). A model represents some combination of data and concepts, presented as text, formulae, graphics, and/or physical simulation. This definition recognizes two principal aspects of a model: its content and its method of presentation. A mineral-deposit model represents data and concepts of processes interpreted to have formed the data for a mineral deposit or mineral-deposit type. An empirical model (synonyms include factual model and occurrence model) represents only data and observations. It does not represent interpretations or concepts. A conceptual model (synonyms include genetic model, interpretive model, causal model, and process model) represents data and interpretations. A characteristic model represents data, with or without interpretations, for a particular characteristic of a mineral deposit (examples include structural model, alteration model, and depositional model). A methodological model represents data, with or without concepts, and a particular sequence of modeling steps to achieve an objective. Doctors use methodological models in clinical diagnosis. Mechanics use them to diagnose electrical and mechanical malfunctions in automobiles. Examples in geology include exploration and evaluation models, which customarily prescribe a sequence of steps to achieve an objective. The data-process-criteria model (DPC model) presented in the final section of this paper is a methodological model.

Level of Model Development

A mineral-deposit model is usually developed only to the level of completeness that is necessary for exploration success. In exploration, the requirements for a model are (1) that it improves on previous exploration on the proposed exploration lands, and (2) that it surpasses the exploration competitors. These requirements define a minimum model-development level. Sources for the development of a model are the (1) data, and (2) concepts for a mineral-deposit type, and (3) exploration methods applicable to their field detection. The maximum development level is the highest possible model-development level supported by current data, concepts, and exploration methods. The minimum and maximum levels, of course, are somewhat hypothetical and can be estimated only through a knowledge of the requirements (minimum level) and sources (maximum level) listed above. In the preparation of a model, the level to which the data, concepts, and exploration methods are developed depends upon (a) the level of competition and previous exploration (minimum development level), (b) the availability of data, concepts, and exploration methods for the deposit type (maximum development level), and (c) the purpose for which the model will be used. For example, a model prepared for mineral exploration will be developed to a higher level than a model prepared for resource assessment in support of land-use planning and commodity supply studies. The current model-development level is the actual or current level to which data, concepts, and exploration methods have been assembled and organized for a mineral-deposit type.

The level of development of a mineral-deposit model may range from a preliminary or low-level model to an advanced, or high-level model. A low-level model contains geologic data but few, if any, concepts, and generally is based upon only a few analog deposits. As more analogs are identified and described, the data base increases, permitting the identification of formation processes and new exploration guides in which confidence is high. This supports a higher level model. The exploration geologist must develop the model to a level that meets the exploration requirements, but no more. The collection and interpretation of unnecessary data, and the pursuit of time-consuming and unproductive scientific studies, are to be avoided. In practice, the appropriate levels of model development is pursued by trial and error. For example, exploration for sediment-hosted type deposits ("Carlin-type") has, at various times, employed geologic characteristics such as proximity to a thrust fault, lower plate rocks, and jasperoids. We now know that each of these geologic characteristics is informative, but that successful exploration requires a higher level model with more predictive criteria.

The minimum, maximum, and current levels of model development increase with time, but at different rates. New data, concepts, and exploration methods may produce abrupt increases in the maximum development level, as shown schematically in Figure 12.4. Increases in the minimum development

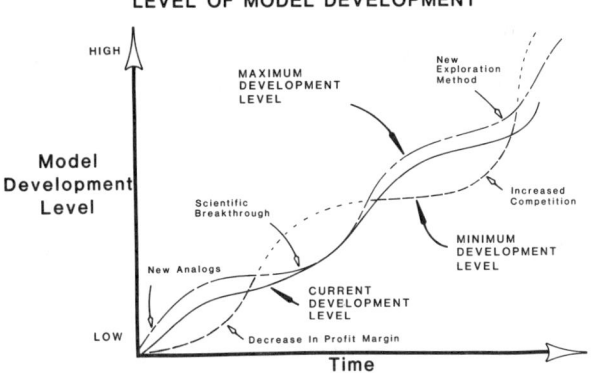

Figure 12.4. Schematic representation of minimum model development level required for exploration, maximum development level supported by geologic data and concepts and exploration methods, and current model development level (Adams, 1985).

level may be caused by, for example, decrease in profit margin and increased competition.

The maximum development level may not be sufficient to accommodate the minimum development level, if exploration has taken advantage of existing data, concepts, and exploration methods (note two time periods in Figure 12.4 where minimum level attempts to exceed maximum level). During these periods, exploration geologists will either create new geologic information and exploration methods through research or postpone exploration. When the maximum development level exceeds the minimum development level, the current development level is raised above the latter, after which exploration should capitalize on the current model without further investment in model development.

The development of a model is discontinued as soon as the minimum development level has been reached; i.e., when the exploration guides are a sufficient improvement upon previous exploration in the exploration area and upon the exploration guides being used by competitors. In practice, model development is generally not discontinued until readily available data and concepts have been incorporated. Model development may also be discontinued, or research expanded, when the minimum development level reaches the maximum development level; i.e., current geologic information and exploration methods have been exploited in exploration. Discontinuation of model development for these two reasons is shown schematically in Figure 12.5 for the steps of a data-process-criteria model (discussed in the next section). When a model ceases to be successful in exploration (i.e., the minimum development level rises above the current development level), model development is resumed.

The concept of level of model development has been discussed above. The concept of variability between models is now introduced. Each mineral deposit model constructed is a unique combination of (1) the purpose for which the model was developed, (2) the three sources for the maximum development level, and (3) the two requirements of the minimum development level. The five principal factors that contribute to variability between models are shown schematically in Figure 12.6 as spokes in the model development wheel. The reader will recognize data, concepts, and exploration methods as sources for the maximum development level, and competition and previous exploration as the requirements of the minimum development level. In Figure 12.7, model development wheels are shown for six hypothetical modeling situations. The spokes in the wheels are identical to those in Figure 12.6. The relative length of each spoke indicates the emphasis placed on that factor in the development of that model. In the top row are models that might have been developed by an academic research geologist, a U.S. Geological Survey resource assessment geologist, and a mineral

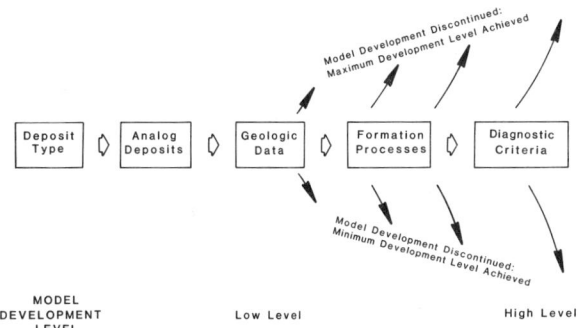

Figure 12.5. Schematic representation of steps in the preparation of a data-process-criteria (DPC) model and discontinuation of model development due to achievement of maximum or minimum model development level (Adams, 1985).

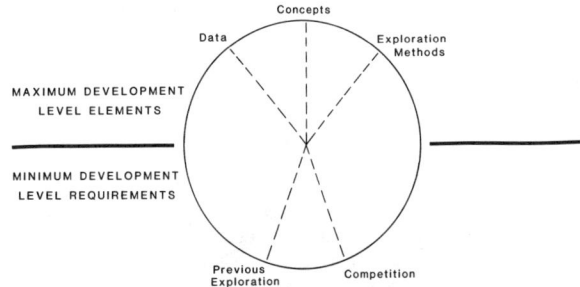

Figure 12.6. Schematic representation of the factors that influence variability between mineral deposit models.

VARIATIONS IN MODEL DEVELOPMENT WHEELS

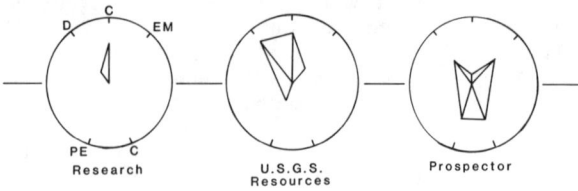

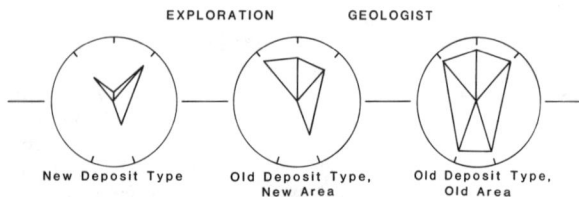

Figure 12.7. Schematic representation of variations between models developed for different purposes and exploration situations.

prospector. The research model emphasizes geologic concepts with only sufficient data to support the hypotheses and no consideration of exploration methods or factors reflecting a minimum model development level. A model developed by a U.S. Geological Survey ore deposit specialist is apt to consider a broad data base, a range of deposit formation concepts, and exploration methods to the extent that they (a) are necessary for development of the scientific model and (b) support resource assessment studies. The prospector's model, by contrast, places minimal emphasis on geologic concepts, but emphasizes all other factors, particularly previous exploration and competition.

The three hypothetical models at the bottom of Figure 12.7 might have been developed by an exploration geologist for three distinct situations. The first wheel reflects the limited data and virtual absence of concepts for a new deposit type. Since the deposit type has so recently been recognized, previous exploration is not a concern, but there apparently are other organizations that are exploring for the deposit type. The second wheel depicts a model for an old deposit type for which there is considerable available data and formation concepts. The model is to be applied in an unexplored area; hence, previous exploration is of no concern, but there is no exploration competition. In the final wheel is depicted the common exploration situation where the model for exploration for a well-known deposit type is to be applied in a previously explored area, with considerable current competition. In this latter case, strategic opportunities in geologic information and exploration methods are almost essential.

The development of a mineral-deposit model for science is a special situation, and a researcher presumably will push the model to the highest development level that time, talent, and resources permit. This will, from time to time, present exploration geologists with a scientific windfall that permits a significant increase in the current development level of a model for exploration.

DATA-PROCESS-CRITERIA MODEL

A data-process-criteria (DPC) model for a mineral-deposit type contains (1) the geologic characteristics of analogs of the deposit type (data), (2) the processes that are interpreted to have formed the geologic characteristics of the deposit (process), and (3) the most reliable and informative geologic characteristics (criteria) for exploration. Exploration geologists have, more or less informally, used this general approach in exploration for years. A DPC model differs from the historic modeling approach in being more explicit (i.e., a consistent, written format) in all the modeling steps, and in its more rigorous selection and assignment of importance to the criteria for exploration. The more explicit and rigorous DPC modeling approach is justified by (a) increasing complexity of geologic information for mineral deposits, (b) increasing cost and difficulty of exploration, (c) the need for more predictive and productive exploration methods, and (d) the tendency for human factors to erode the reliability and effectiveness of informally prepared models.

Most mineral deposits were formed by sequences of geologic processes that generally included some combination of chemical, physical, and hydrologic processes. Each geologic process that is essential for the formation of a mineral-deposit type and its geologic and economic characteristics is referred to, herein, as a formation process. Most formation processes produce geologic characteristics. The objective of a DPC model is to identify (a) all formation processes for a mineral-deposit type, and (b) geologic characteristics that can be used to determine whether or not each of the processes operated in an exploration area. The presence or absence of geologic evidence for each formation process indicates the favorability of the exploration area for the occurrence of the deposit type.

The steps in the preparation of a DPC model are illustrated below with simplified examples for the hot-spring-type epithermal precious-metal deposit. There is no attempt to present a complete or even state-of-the-art hot-spring model, but only to use this deposit type to illustrate the preparation of a DPC model. The simplified examples are based upon publications for the analog deposits (see Table 12.3) and articles that summarize the geology of this deposit type (Berger and Eimon, 1983; Giles and Nelson, 1982; Nelson and Giles, 1985). Only those portions of the model useful for illustration are included. The DPC model preparation steps (Fig. 12.8) include definition of a mineral-deposit type, compilation of analog deposits, selection of geologic data, data-process linking, identification of formation processes, evaluation of data-process links, selection of diagnostic criteria, and evaluation of the model.

Table 12.3--Some analog deposits for the hot-spring epithermal precious-metal deposit type*

Deposit Name	Location	Total Reserves			
		Tons (millions)	Au (oz/ton)	Ag (oz/ton)	
Borealis	Nevada	2.5	0.08	0.62	Strachan et al. (1982) Reid (1984)
Buckhorn	Nevada	5.1	0.045	0.6	Monroe (1984)
Hasbrouck	Nevada	5	0.10		Graney (1984)
Round Mtn.	Nevada	195	0.043	0.08	Tingley and Berger (1985)
Sleeper	Nevada	3.7	0.13	0.80	Engineering and Mining Journal (1985)
McLaughlin	California	20	0.16		
Wau	Papua, New Guinea	District Production:	0.58 million ounces Au		Sillitoe et al. (1984)
Iwato	Japan		0.13	0.20	Saito and Sato (1978)

*Analog deposits display evidence of having formed within few hundred meters of the paleosurface, according to criteria presented in the text.

Definition of a Mineral-Deposit Type

A mineral-deposit type is characterized by geologic and economic characteristics and formation processes that have been compiled from analogs of the deposit type and that are sufficiently different from those of other deposit types to justify the distinction of the deposit type for exploration. This pragmatic definition emphasizes that if the object of exploration is a particular assemblage of geologic and economic characteristics, and not the characteristics of even a closely related deposit type, a distinct mineral-deposit model is required. For example, if the exploration objective is simply a precious-metal-bearing deposit, one deposit type that addresses all epithermal precious-metal deposits will qualify. If, on the other hand, the exploration objective is a very large, highly profitable underground mining operation, only models for certain bonanza-vein type deposits will qualify.

The general geologic and economic characteristics of a mineral-deposit type should be roughly defined before model development begins. The definition is accompanied by a brief summary of the geologic characteristics that distinguish it from, and that it shares with, related or similar deposit types. For example, hot-spring-type deposits are differentiated from other epithermal deposits by a variable assemblage of geologic characteristics that suggest deposit formation occurred within a few hundred meters of a paleosurface. These characteristics include (1) siliceous sinter, (2) fumarolic mineral precipitates, (3) hydrothermal eruption breccia that includes vent breccia and some combination of ejecta blankets, fall-back breccia, and hydrofracture breccia, (4) a fracture stockwork, commonly peripheral to the vent breccia, and (5) wallrock alteration characteristic of acid leaching above a ground-water table. None of these characteristics is unique to the hot-spring-type deposits; each occurs in at least some examples of other epithermal deposit types. Also, most of these characteristics are absent from some hot-spring-type deposits; only hydrothermal vent breccia and stockworks have been persistently identified in hot-spring-type deposits. Their absence from some deposits may be due to erosion. Hot-spring-type deposits also differ from most other epithermal deposits in their generally higher concentrations of Hg, Tl, Sb, and Ba, in and above the gold-bearing zone. Hot-spring-type deposits intermittently or persistently share additional geologic characteristics with other epithermal deposit types, including (1) fracture control of mineralization, (2) banded open-space vein filling that includes abundant fine-grained quartz, pyrite or marcasite, and adularia, and (3) wallrock alteration that includes propylitization, adularia alteration, and silicification.

Compilation of Analog Deposits

Analog deposits are the example deposits of a mineral-deposit type that provide geologic and

STEPS IN DPC MODEL

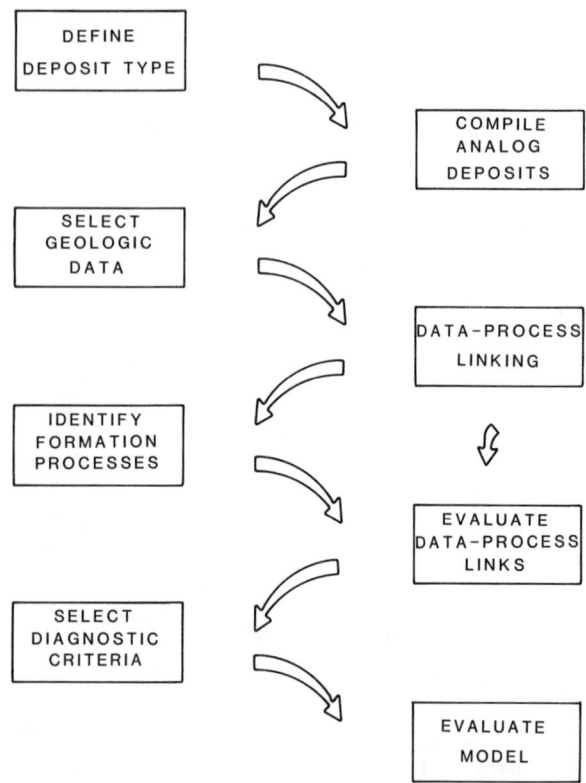

Figure 12.8. Steps in the preparation of a data-process-criteria (DPC) model.

analog deposits. If only the higher grade, larger tonnage deposits are of interest, the unique geologic characteristics of those deposits must be identified so that only they are discovered with the model.

Analog deposits for the hot-spring-type deposit are compiled in Table 12.3. Only deposits for which information was readily available are included. Preparation of the simplified hot-spring model was based on publications for these analog deposits, but particularly on the summary article by Nelson and Giles (1985).

Selection of Geologic Data

Geologic data for construction of a mineral-deposit model are derived from analog deposits and geologic environments in which the deposits occur. Factual data are developed through mapping, thin section study, chemical analyses, etc., and is initially incorporated into the model without interpretation. Sources of data include the literature, professional colleagues, and visits to the more accessible, exposed, and controversial deposits to test the assembled data and concepts and collect missing data.

The collection of geologic data from the analogs involves both judgment and bias. A conscious effort is made to collect relevant data of all types (i.e., structure, geomorphology, petrography, etc.) and scales (i.e., regional tectonic setting to detailed deposit characteristics) and to avoid the tendency to overemphasize the collection of data types that are familiar to the observer.

Data collection is guided by hypotheses for the formation of the geologic characteristics of the deposit type. This approach permits more efficient testing of hypotheses during the data collection and minimizes the collection of unnecessary data, even where the initial hypotheses prove to be wrong. Iteration between data collection, hypotheses, and the testing of hypotheses proceeds until the data are exhausted, confidence in the formation processes becomes sufficiently high, or the model is discontinued because of loss of interest.

The geologic data may be organized in a variety of ways. Tables or matrices in which both the data and the analog deposits are shown have been useful. The data base should reflect both the variability among geologic characteristics and the most typical characteristics.

Some structural geologic data for selected analogs of the hot-spring-type deposit are compiled in Table 12.4. The data are incomplete and simplified and are presented for illustrative purposes only. In a complete model, all geologic characteristics of the analog deposits, including regional geologic setting, host rocks, intrusions, alteration, etc. would be presented in a similar matrix-like table.

During data accumulation, geologic characteristics may be identified that appear to offer a significant advantage in exploration. Such geologic characteristics can be immediately applied to exploration, if confidence is high that the potential value of their use exceeds the risks of their use prior to completion of the model.

economic characteristics for construction of a mineral-deposit model. Analog deposits should reflect both the variability of the geologic and economic characteristics and the most likely or typical characteristics of the deposit type. The more analog deposits that are included in a model, particularly from widely separated areas, the greater will be the model's reliability for exploration. As few as a dozen analogs may suffice, if the characteristics display little variability. If, on the other hand, the analogs show considerable variability in grade, tonnage, structural control, alteration, etc., a few dozen analogs may be required to develop a reliable model. High variability also may indicate that the selected deposit type should be divided into two or more separate deposit types. All readily available analogs should be included in the model.

Both the geologic and economic characteristics must be compiled from, and representative of, the same group of analog deposits. Deposits discovered in exploration can be expected to have about the same grade and tonnage distribution as occurs among the

Table 12.4--Selected structural geologic data for some analogs of the hot-spring type deposit

Data	General Description	Borealis	Hasbrouck	Round Mountain	Wau	McLaughlin
References	Berger and Eimon (1983); Giles and Nelson (1982); Nelson and Giles (1985)	Reid (1984); Strachan (1985); Tooker (1985)	Graney (1984); Tooker (1985)	Tingley and Berger (1985); Tooker (1985)	Sillitoe et al. (1984)	Nelson and Giles (1985); Tooker (1985).
Regional/ District Structure	Fracture zones, faults, caldera ring and graben fractures, joints, and areas of doming and intense fracturing.	N-trending normal faults and NE- and E-trending shear zones	N-trending steeply dipping faults; caldera ring fractures (?)	NW- and NE-trending, steeply dipping faults, fractures and low-angle fractures; joints	Low-angle normal fault and Maar ring fault	Not described.
Hydrothermal Eruption Breccia						
Ejecta Blanket and fallback breccia	Ejecta breccia blanket up to several meters thick, generally restricted to immediate vent area but covering up to a square mile; fragments may show episodic silicification prior to brecciation; may be eroded in fossil systems.	Possibly deposited contemporaneously with talus deposits	Not described	Not described	Thickness up to 60 m	Not described
Vent breccia and hydro-fracture breccia	Pipelike, sheetlike, tabular, dikelike and podiform masses of breccia; millimeters to tens of feet in thickness and diameter; pebble dikes and brecciated vein walls; isolated to multiple overlapping or neighboring breccia bodies; variably matrix and clast supported; very fine-grained to boulders.	Multiple stages	Irregular bodies in silicified rock	Linear and irregular pipelike bodies related to fractures and fracture intersections; multiple stages	Irregular veins and pods	Present
Stockwork	Vein stockworks, generally in previously silicified and/or adularia-altered wallrock.	In silicified breccia	In silicified rock and silicified breccias	Closely related to fractures and breccias	Some within hydrothermal breccias	Present

Data-Process Linking

Once geologic data for the analogs have been selected, the data are used to identify probable processes that formed the deposit type. The processes are identified by linking each geologic datum to every possible process by which it may have been formed. Although this data-process linking step precede the final selection of formation processes, it is more conveniently discussed in the following section.

Identification of Formation Processes

The formation of most mineral deposits requires that a particular sequence of chemical, physical, and hydrologic processes operated and in a particular sequence. Each process that is <u>essential</u> for some significant geologic or economic characteristic of the deposit type is referred to as a formation process. The absence of even one formation process precludes formation of the deposit type as it is presently defined. The objective in modeling is to identify the formation processes; the objective in exploration is to identify areas where all of the formation processes operated. Since the formation processes generally have long since ceased to operate, they must be identified by the geologic characteristics they produced. In the DPC modeling process, the formation processes are identified from the geologic data. Then the most reliable, informative, and easily measurable geologic characteristics (diagnostic criteria) are then selected for the identification of where, in an exploration area, the formation processes operated.

The identification of a formation process may need to be augmented by information on the intensity, duration, and repetition with which the process operated. In a hot-spring-type deposit, for example, ore grade may be determined, in part, by the intensity of fracturing and the volume of hydrothermal fluid flow. Evidence for (a) the number of episodes of brecciation and (b) the extent of alteration of breccia fragments associated with each episode could provide information on these two processes.

The identification and evaluation of formation processes probably will require, in addition to skills at developing and assembling field and laboratory data, a working knowledge of chemical, physical, and hydrologic processes. Few geologists enjoy such a broad, formal education; hence, there is a continual need to learn during employment and to collaborate with knowledgeable colleagues. Geologists inexperienced with these processes may find it helpful to build models in teams with colleagues whose skills augment their own or they may restrict their modeling to empirical models. The misinterpretation of formation processes through the misapplication of chemical, physical, and hydrologic processes not only wastes valuable time, but risks substituting erroneous conceptual programs for more reliable, empirical ones.

Formation processes are identified for all the geologic data collected from the analog deposits. Each geologic observation is linked to one or more processes, from which it might have formed. Linking is recorded in a diagram that becomes an integral part of the model. Selection of formation processes can be illustrated with a simplified example for the hot-spring-type deposit. Figure 12.9 presents geologic data for analogs of the deposit type that includes a breccia and geologic characteristics A, B, C, D, and E. Possible processes that may have formed the breccia have been identified. These possible processes include intrusion, hydrothermal eruption, sedimentation, and tectonism, and each is linked to breccia. Characteristics of these particular breccias, including (a) their generally circular shape in map view, (b) sharp vertical contacts with enclosing rocks, (c) absence of an igneous component, and (d) multiple altered and brecciated clasts suggest the breccias were formed by episodic hydrothermal eruption. Hydrothermal eruption has been selected, therefore, as the preferred formation process for the breccia. Tectonism is the most likely alternate process, because all analog deposits are associated with faults and fractures of various orientations. It may subsequently be shown that tectonism is also a preferred process, perhaps an antecedent to hydrothermal explosion. Note that in this simplified example, hydrothermal explosion was also supported by geologic data A and C. Processes selected as preferred processes are commonly, but not always, supported by multiple geologic data. Conversely, possible processes supported by multiple data from the analog deposits are not always judged to be essential to the formation of significant geologic and/or economic characteristics of the deposit type. The possible processes intrusion, sedimentation, and tectonism were not selected as preferred processes for the formation of the breccia, but in a complete model they may have been selected as preferred processes for other geologic characteristics of hot-spring deposits. The processes metal leaching, alteration, and metal precipitation would have been supported by numerous additional data in a complete hot-spring model.

In Table 12.5, three preferred processes and their alternate processes for the simplified hot-spring model are compiled. Such a table presents the rationale for selection of the preferred processes and is an important component of a data-process-criteria model.

Once the preferred processes have been selected, they are summarized in a diagram of data-process links. Structural geologic data, previously compiled in Table 12.4, are linked to three probable formation processes for the simplified hot-spring model in Figure 12.10. Each line from a geologic datum to a process

SELECTION OF FORMATION PROCESSES

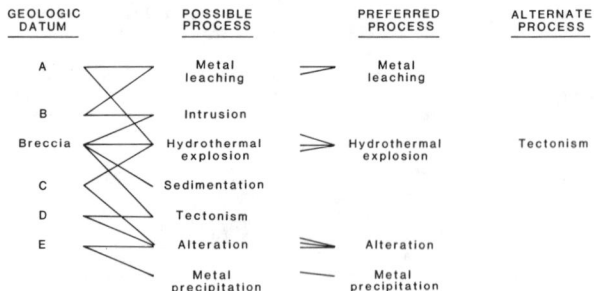

Figure 12.9. Simplified example of the selection of formation processes for the hot-spring-type deposit.

SOME DATA-PROCESS LINKS FOR THE HOT-SPRING MODEL

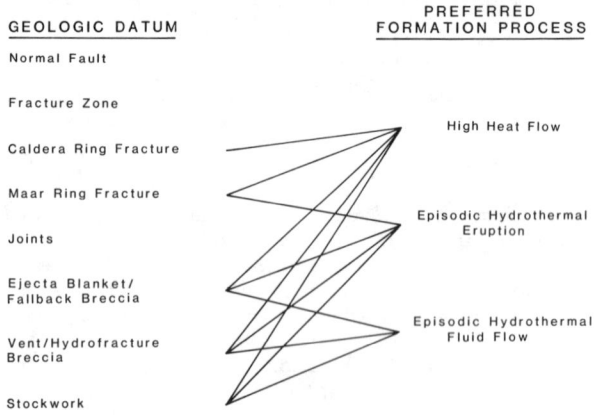

Figure 12.10. Data-process links for selected structural data and three possible formation processes for the hot-spring deposit model.

Table 12.5—Three preferred and alternate processes for the simplified hot-spring model

Preferred Process	Alternate Process(es)	Comments
High heat flow		Indicated by the effects of hydrothermal fluids including brecciation, alteration, mineralization, and metallization; may be due to intrusion or, less likely, high regional heat flow.
	None	Given the geologic data, no alternate to high heat flow can be imagined.
Episodic hydrothermal eruption		Indicated by episodic brecciation and alteration.
	Intrusion	A probable process elsewhere in model as source of high heat flow; unlikely as alternate process for hydrothermal eruption, because of absence of igneous components in breccias and numerous effects of hydrothermal fluids.
	Tectonism	A probable process elsewhere in model as cause of faulting and fracturing; unlikely as alternate process for hydrothermal eruption because of shapes of breccia bodies and lack of relations to significant faults and fractures; possibly an antecedent process for hydrothermal eruption.
	Sedimentary brecciation	Unlikely because of breccia shapes and crosscutting relations with wallrocks and associated hydrothermal effects.
Episodic convection of hydrothermal fluids		Indicated by episodic brecciation, alteration, mineralization, and metallization.
	Ground-water circulation	Unlikely that nonhydrothermal fluids produced brecciation, alteration, mineralization and metallization because of metal solubility considerations and fluid-inclusion data.

reflects the interpretation that the datum formed from the process, and that the datum is a field indication for the process. In a complete hot-spring model, the structural data shown would also link to some processes not shown, and data not shown would link to some of the processes shown. During the identification of the formation processes, all geologic data assembled for the analogs should be linked to processes. To help avoid premature and prejudicial selection of formation processes, no data should be eliminated at this stage becuase it is thought to be related to nonessential processes. Processes that are later found to be nonessential to deposit formation can be eliminated from the model.

The identification of formation processes may be aided by some additional techniques. For example, thinking through the sequence metal source-transport-precipitation-enrichment-preservation may help identify additional and less obvious processes. Another technique is to think of processes as having resulted from or having been made possible by antecedent processes and as having produced or permitted subsequent processes, and then attempt to identify these additional processes. For example, consider an enargite-gold epithermal ore shoot. Processes antecedent to the hydrothermal process that precipitated the metals probably included some combination of regional tectonic processes, intrusion of a pluton, development of a ground-water source, development of hydrothermal circulation cells, fracturing of the host rock, and so forth. Processes subsequent to metal precipitation may have included hydrothermal leaching and redistribution of mineralization, enrichment of the vein below a weathering or leaching zone, displacement of the vein along faults, and the formation of topographically elevated erosional remnants of silicified wall rock. Only antecedent and subsequent processes that are essential to deposit formation are included in the final model.

Finally, a glossary of geologic processes, preferably arranged by categories, may help the search for essential formation processes. For example, the identification of processes that may have precipitated precious metals in an epithermal deposit might be aided by reference to a glossary of processes that included (a) changes within the hydrothermal fluid, such as boiling, mixing, and cooling, and (b) a variety of wallrock reactions.

Data-process linking promotes identification of all possible and then probable and alternate processes that could reasonably account for each geologic characteristic of the analog deposits. When the linking has been completed, confidence will be highest in the possible processes that are linked to several geologic characteristics. Confidence also will be highest in processes that are chemically, physically, and hydrologically reasonable in the envisioned deposit-forming environment. These processes are selected as preferred processes. It is advisable to retain at least one alternate process for each probable process, and more alternate processes for more uncertain preferred processes. Maintaining the best possible alternate processes sharpens conceptualization of all processes, resulting in the selection of preferred processes that are most likely and most reliable. Without continual challenge, ill-defined or inaccurate preferred processes will persist in a model.

Geologic and economic characteristics produced by formation processes may be either relatively numerous and obvious or few and obscure. At the same time, processes responsible for geologic and economic characteristics of analog deposits may be identified with relative ease or with great difficulty. For example, processes that control barren versus mineralized veins, high-grade zones and large- versus small-tonnage deposits may be difficult to identify. It is important, nonetheless, to attempt to identify all essential formation processes and criteria by which they can be recognized, but particularly those that produced the significant economic characteristics of the deposit type.

The identification of formation processes entails greater uncertainty than any other step in the construction of a DPC model. Because the formation processes so significantly affect the reliability of a model for exploration, particular effort is made to evaluate the reliability of the formation processes. This is customarily done after the formation processes have been selected and again during the final model evaluation step. The evaluation of the formation processes attempts to determine the extent to which the following conditions have been met: (1) at least one process has been identified for every geologic characteristic compiled from the analog deposits; (2) each formation process is essential to deposit formation, as evidenced by its data-process links; (3) each process is confirmed, where possible, by multiple data-process links; (4) techniques such as a glossary of process terms, sequential process linking, and antecedent-subsequent processes were used to develop the list of preferred and alternate processes; (5) processes were iteratively tested against data collection; and (6) each process is sound scientifically and in the geologic environment in which is it interpreted to have operated.

Evaluation of Data-Process Links

Application of a mineral-deposit model in exploration requires geologic criteria that reliably indicate the likelihood for the occurrence of the deposit type in an exploration area. In the case of a low-level, empirical model, the criteria are empirical geologic characteristics that occur with the analog deposits. In a higher level or more conceptual model, the criteria are also geologic characteristics of the analog deposits, but they have been shown to have formed from particular deposit formation processes. These geologic characteristics are referred to as diagnostic criteria, and each one provides information on the presence or absence of one or more formation processes in an exploration area. The evaluation of geologic data/characteristics for selection of diagnostic criteria is discussed below. Exploration use of empirical observations for lower level models has been common in exploration and is not discussed in this paper.

Once formation processes have been identified, as described in the previous section, each geologic characteristic linked to them is evaluated for the strength of the evidence it provides for the process and, therefore, suitability of the characteristics for a diagnostic criterion. To qualify as a diagnostic criterion, a geologic characteristic must provide strong evidence that a formation process did or did not operate in an exploration area. Evaluation of the data-process links is, therefore, the next step in model development.

Some geologic characteristics are formed only by one geologic process and are always found with that process. Such characteristics are particularly valuable evidence for that process because, if present, they indicate that the process operated and, if absent, they indicate that the process did not operate. One may say, therefore, that they are necessary evidence for the process to have operated and sufficient evidence that it did. Other geologic characteristics are always formed by a particular process, but they also are formed by other processes. These characteristics are necessary, but less sufficient, evidence for the process. Characteristics that are not always formed by a particular process and are not formed by other processes are sufficient but less necessary evidence for that process. Finally, characteristics that are not always formed by a particular process and are formed by other processes are both less necessary and less sufficient evidence for that particular process. The task, therefore, is to identify the most necessary and most sufficient geologic characteristics for each formation process. These characteristics are selected as diagnostic criteria because they are the most informative and reliable criteria for determining if a formation process operated in a particular area. Information on all the formation processes will, in turn, indicate the likelihood that a deposit will occur in an area.

Estimation of the necessity and sufficiency of a criterion for a formation process employs the rules shown in Table 12.6. Necessity and sufficiency are evaluated semiquantitatively (i.e., high, intermediate, and low) instead of numerically, because the estimates are inexact and subjective. In Figure 12.10, structural data are linked to three preferred formation processes for the hot-spring deposit-type model. On the left side of Figure 12.11, the necessity and sufficiency of each of these links has been estimated and placed next to the link. Each geologic datum with high necessity and/or high sufficiency for one or more process has been selected as a diagnostic criterion and listed on the right side of Figure 12.11. Let us now review how necessity and sufficiency were estimated for some of these links.

Caldera ring fractures are important indicators for some processes not shown in Figure 12.10; for example, intrusion, volcanism, and fracturing, but they are also linked to high heat flow. We will evaluate the

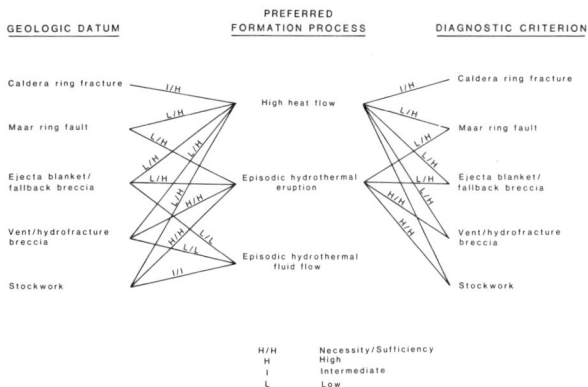

Figure 12.11. Necessity and sufficiency estimates for selected data-process links and selected diagnostic criteria for the hot-spring deposit model.

necessity of this link, i.e., the necessity of a caldera ring fracture as evidence that high heat flow occurred in an area. Using the rule for necessity in Table 12.6, is a caldera ring fracture formed by or associated with high heat flow consistently, in some/many cases or rarely? It was decided that caldera ring fractures occur in areas of high heat flow in some/many cases; hence, its necessity is intermediate. In other words, in only some/many cases do areas that show evidence of high heat flow (defined for our purposes as evidence of

Table 12.6--Rules for the estimation of necessity and sufficiency of a geologic datum or characteristic for the occurrence of a formation process to which the datum is linked in a data-process-criteria model

If the criterion is formed by or associated with the process:	The Necessity is:
Consistently	High
In some/many cases	Intermediate
Rarely	Low
If the criterion (a) is produced only by the process (i.e., not produced by plausible alternative processes) and (b) reflects the significant aspects of the process:	The Sufficiency is:
Common/consistently	High
In some/many cases	Intermediate
Rarely	Low

shallow intrusion, hydrothermal alteration, hydrothermal mineralization and metallization, etc.) occur with caldera ring fractures. Therefore, a caldera ring fracture is not highly necessary for high heat flow to have occurred. Now let us estimate the sufficiency of the link, using the sufficiency rule in Table 12.6. It was concluded that a caldera ring fracture is consistently produced only in areas of high heat flow; hence, its sufficiency for high heat flow is high. In other words, caldera ring fractures are found in areas that consistently show evidence of high heat flow; hence, the presence of a caldera ring fracture is compelling evidence (i.e., highly sufficient evidence) that high heat flow occurred in the area. The absence of a caldera ring fracture is negative but ambiguous evidence for the absence of high heat flow, whereas the presence of a caldera ring fracture is highly sufficient evidence for the occurrence of high heat flow in an area.

Consider next the link from stockwork veining to high heat flow. It was concluded that stockwork veining rarely occurs in areas of high heat flow; hence, its necessity for the process is low. The sufficiency of stockwork veining for high heat flow is high, however, because stockwork veining is rarely found in areas that have not experienced high heat flow. The link between stockwork veining and episodic hydrothermal fluid flow has intermediate necessity, because stockwork veining is associated with this process in only some/many cases. The sufficiency of stockwork veining is also intermediate, because stockwork veining is associated with areas that show no evidence of episodic hydrothermal fluid flow in some/many cases. By contrast, vent/hydrofracture breccia is consistently associated with the hydrothermal eruption process and does not form by other processes; hence, its necessity and sufficiency are high. The other links have been evaluated in a similar manner.

Selection of Diagnostic Criteria

Geologic data in Figure 12.11 have been selected as diagnostic criteria that have high necessity and/or sufficiency for one or more processes. These geologic characteristics are most informative for determining if a formation process operated in an exploration area. Note that the ejecta blanket/fallback and vent/hydrofracture breccias were not selected as criteria for episodic hydrothermal fluid flow. Although these geologic characteristics are highly sufficient for hydrothermal eruption, they do not provide direct information on hydrothermal fluid flow, for which the evidence must indicate large volumes of fluid flow. In a complete model for the hot-spring deposit type, diagnostic criteria for hydrothermal fluid flow would include (a) extent of wallrock alteration, (b) stages of wallrock alteration, (c) open-space mineralization, and related criteria. The episodic nature of hydrothermal eruption and hydrothermal fluid flow would also have to be confirmed by specific evidence, such as multiple stages of crosscutting fractures, banded mineralization, episodically altered wallrocks, and so forth.

The most useful diagnostic criteria for exploration are those that (1) have high sufficiency and/or necessity, and (2) are easily and inexpensively measured by exploration methods. Criteria that have high necessity and sufficiency for a process may be all that is required to determine if the process operated. In practice, it is best to confirm the presence or absence of a process with at least two criteria. Some criteria have high necessity and/or sufficiency for more than one process; hence, they are especially useful.

Evaluation of Data-Process-Criteria Model

The systematic evaluation of a DPC model is the last and most important step in model development. As discussed earlier under Human Factors, there is a natural human tendency to build causal models without adequate regard for their validity. To minimize this risk, no model should be applied to exploration without an explicit assessment and documentation of its reliability, hence the risks associated with its use. The writer has developed a series of tests that evaluate each step in model development. Since all modeling steps, including the selection of geologic data for analogs, include subjective judgment, evaluation of a model and its suitability for exploration is also subjective. The evaluation tests are briefly summarized below.

Each of the seven previous modeling steps (Fig. 12.8) are evaluated for accuracy, completeness, and the confidence that seems warranted by the results of the step. Tests for accuracy emphasize the veracity of (a) the uniqueness of the deposit type, (b) the appropriateness of the analog deposits, (c) the geologic and economic data, (d) the scientific principles used in selecting the formation processes, (e) the preferred and alternate processes, (f) estimates of necessity and sufficiency, and (g) the diagnostic criteria selected.

Tests for completeness determine if (a) geologic and economic similarities among and differences between the deposit type and related deposit types are as completely documented as current data permits and exploration warrants; (b) all available analog deposits have been compiled; (c) all types (mineralogical, tectonic, structural, stratigraphic, etc.) and relevant scales (regional to microscopic) of geologic data for the analogs are represented; (d) analogs accurately reflect the range of geologic and economic characteristics of the deposit type; (e) preferred and alternate processes have been identified for all geologic characteristics of the analog deposits; (f) necessity and sufficiency have been estimated for all data-process links; and (g) if the most informative (highest necessity and/or sufficiency) and reliable geologic data have been selected as diagnostic criteria.

Confidence in the results of each modeling step is based upon the tests for completeness and accuracy and certain additional factors for some steps. For example, the reliability of a formation process depends, in large part, on the number of data to which it is linked, and general scientific and geologic confirmation of the process, including the conditions under which it operates, how it operates, and the geologic characteristics it produces under various geologic conditions. Confidence in the diagnostic criteria depends upon the reliability of their necessity and their sufficiency estimates and upon the reliability of the processes themselves.

Every mineral-deposit model entails uncertainties due to incompleteness and the possibility of errors in data and interpretations. A carefully prepared and evaluated data-process-criteria model permits evaluation of the uncertainties and more responsible exploration decisions. The alternate approach, namely the use of selected empirical observations, also entails uncertainties, but without the benefit of knowing where in the model they are, their probable significance, and how they might be minimized. Existing data and concepts will not permit the development of high-level models for all deposit types, necessitating that exploration either proceeds with empirical observations or awaits the development of new data and concepts. Even when only empirical or low-level models have been prepared, they are available to guide continued research and capitalize on advancements as they occur.

Application of the Data-Process-Criteria Model to Exploration

Once a model has been evaluated and judged sufficiently reliable for use in exploration, specific exploration methods are selected for applying the model to exploration areas (see earlier discussion under Strategic Factors). Exploration methods are selected to provide the most rapid and inexpensive field and laboratory data for each diagnostic criterion. Methods applicable to exploration for hot-spring deposits might include, for example, (a) use of existing regional mapping to identify volcanic-plutonic provinces, structures (calderas, faults, fracture zones, etc.) and altered zones; (b) ground reconnaissance to identify individual hydrothermal systems (alteration and mineralization) and evidence of near-surface hydrothermal processes (sinter, ejecta blankets, fallback breccias, vent breccias, hydrothermal breccias, stockworks, etc.), and (c) geochemical sampling for evidence of anomalous concentrations of Hg, Tl, Sb, Ba, Au, and Ag. Rapid and inexpensive methods may not be available for all diagnostic criteria, and the risk of proceeding without them (the other diagnostic criteria may suffice) will have to be weighed against more expensive and time-consuming methods. Once a suite of exploration methods has been matched to the model's diagnostic criteria, strategic exploration opportunities are identified and, finally, the risk of applying the model to the proposed exploration areas is assessed (see Strategic Factors). If there are specific reasons why the program might be successful where others have failed or failed to try, exploration should proceed.

Application of a DPC model to exploration commences at whatever scale is dictated by previous exploration and current competition (see Strategic Factors). In regions relatively unexplored for a deposit type, data for regional and generally inexpensive diagnostic criteria are collected first, as a screening tool. If processes for these criteria can be confirmed, for example, volcanism and plutonism in the case of hot-spring deposits, exploration proceeds to more detailed observations. Since volcanic-plutonic regions and the hydrothermal systems they contain are reasonably well known in the United States, strategic exploration opportunities for hot-spring deposits will most likely be on the more detailed field reconnaissance scale. Reconnaissance and detailed mapping for evidence of sinter, the various hydrothermal eruption breccias, stockwork, and multiple stages of each will determine the likelihood that the episodic hydrothermal eruption process operated in rocks close to the surface (a requirement for an open-pit deposit). The mapping method will also determine the likelihood of episodic hydrothermal fluid flow by collecting data for diagnostic criteria that indicate the extent and number of stages of alteration and mineralization.

Exploration attempts to identify exploration areas in which data for diagnostic criteria indicate that all formation processes of a deposit type operated. Such areas are highly prospective for the occurrence of a deposit and warrant detailed exploration by more expensive methods, such as trenching, drilling, geophysical surveys, and so forth. In the case of the hot-spring deposit type, an area with evidence for some combination of sinter or ejecta blankets, episodic brecciation, episodic hydrothermal fluid flow, and metallization (anomalous concentrations of some combination of Au, Hg, Tl, Sb, and Ag) is highly prospective. Conversely, data that reliably document the absence of a formation process condemn the area for occurrences of the deposit type as it is presently defined. Deposits are formed by chains of geologic processes, each of which is essential to deposit formation. Regardless of how extensively several formation processes may have operated, for example, volcanism, plutonism, regional fracturing and faulting, hydrothermal alteration, mineralization, and metallization, the absence of one formation process makes the occurrence of a deposit unlikely. For example, the absence of near-surface episodic hydrothermal brecciation from the preceeding process sequence makes the occurrence of a hot-spring-type deposit unlikely. Obviously other deposit types may be present in areas that have experienced these processes, and one may wish to explore for them, but the potential is low for the original target type. It should be noted that changing deposit types in the midst of an exploration program generally entails significantly different economic characteristics that may or may not meet the organization's objectives. Changing deposit types is also a human technique for avoiding loss (see Human Factors).

Experience with various mineral-deposit models indicates that it is generally easier to identify high-necessity criteria for a formation process than high-sufficiency criteria. This is because many geologic characteristics are always formed by a particular process (therefore they are highly necessary), but the characteristics are either also formed by other processes or they do not document all critical aspects of the process. For example, an outcrop of quartz-adularia mineralization documents hydrothermal fluid flow, but it does not necessarily document the episodic fluid flow associated with episodic hydrothermal explosion that is apparently necessary for hot-spring-type deposits. The outcrop, therefore, may represent a deeper bonanza vein. Therefore, quartz-adularia is of only intermediate sufficiency for the episodic hydrothermal fluid-flow process and must be combined with other diagnostic criteria, such as multiple banded mineralization and multiple crosscutting mineralized fractures, to build a case for episodic hydrothermal fluid flow. The greater abundance of high-necessity criteria tends to establish the permissiveness of an area for formation processes and deposit occurrence, but only high-sufficiency criteria or assemblages of intermediate-sufficiency criteria can establish the likelihood or favorability for formation processes and occurrence of a deposit.

It is preferable, although not always possible, to confirm the presence or absence of a process in an area with multiple diagnostic criteria. This significantly increases the reliability of decisions to continue or abandon exploration programs. Where data for key diagnostic criteria are lacking, it may be possible to justify the cost of collecting the data by drilling, trenching, or other methods that have a high likelihood of providing the necessary data. The availability of such data for diagnostic criteria should significantly increase or decrease the favorability of the area. The collection of data that do not document a high-necessity or high-sufficiency criterion generally increases confusion without reducing uncertainty. If the cost and risk of additional data collection become too great prior to discovery or condemnation of an area, the project is deferred until methods can be found to continue exploration at lower risk and cost.

Summary of Data-Process-Criteria Model

The eight steps of the data-process-criteria model begin with the definition of a deposit type that presumably meets the strategic factors of (1) the organization objectives, (2) a potentially profitable commodity, (3) the organization's financial resources, and (4) the skills and style of the exploration organization. Completion of the model identifies the most informative and reliable diagnostic criteria for exploration and evaluates the reliability of the entire model for exploration. Many aspects of the model have been used routinely by exploration geologists. However, completion of each step in the model as described above is believed to have important advantages for exploration, the most significant of which are summarized below.

1. The DPC model is based upon sound geologic observations for analogs of the deposit type. This data base constrains and promotes the most reliable geologic interpretations.
2. The explicit written model promotes communication with colleagues, peer review, and, thereby, the development of concensus state-of-the-art models.
3. The model format promotes multiple working hypotheses and consideration of all relevant data in an attempt to correct the human tendency to build causal models with negligible regard for their validity. In this way, the DPC model promotes the search for accuracy and truth, rather than simply the search for explanations.
4. The usefulness of geologic characteristics for exploration is determined through the estimation of necessity and sufficiency of diagnostic criteria for identifying where in exploration areas deposit formation processes have operated.
5. The selection of diagnostic criteria for exploration is justified and documented by the linking of criteria to one or more deposit formation process(es) from which the criteria are interpreted to have formed.
6. The model identifies the minimum diagnostic criteria (i.e., one or more for each formation process) for which exploration data will be required, in order to assess the favorability of an area for a deposit. Exploration must include the highest necessity and highest sufficiency criteria possible for each formation process.
7. The model approach identifies deposit types for which only empirical or low-level models can be developed with existing data and concepts.
8. The evaluation of multiple data-process links and the selection of multiple diagnostic criteria permit selection of the most informative, reliable, and least costly criteria for exploration.
9. Preparation of a DPC model identifies where research leading to new data and concepts is most likely to yield new diagnostic criteria for exploration.
10. The model approach helps avoid collection of geologic data that do not contribute significantly to evaluation of favorability for a deposit in exploration areas.
11. The approach promotes the critical evaluation of the model itself and estimation of the reliability with which it can be used in exploration.

CONCLUSIONS

Geologic information, in the form of data and concepts, plays an integral and more complex role in strategies for mineral exploration than is generally appreciated. In addition to its obvious use in selecting and evaluating exploration areas, geologic information significantly influences (1) selection of organization objectives, (2) selection of commodities for exploration, (3) estimation of financial resources required for successful exploration programs, (4) organization skills required for particular types of exploration and the deposit types best suited to particular exploration organizations, (5) regulatory burdens that particular exploration programs and mining operations can tolerate, (6) evaluation of competitor activities, (7) previous exploration on proposed exploration lands, (8) selection of exploration methods, (9) development of exploration opportunities, and (10) assessment of exploration risks. The contribution of geologic information to the evaluation of each of these strategic factors is as critical to the development of successful exploration strategy as it is to the more obvious geologic exploration functions.

Consideration of human factors indicates that exploration will benefit from modeling techniques that help geologists control and guide certain human behaviors and tendencies in their use of geologic observations and interpretations. It has been demonstrated that most humans instinctively (1) build causal models to explain relations among observations, (2) tend to disregard the validity of these models, and (3) resist correcting the models even when new data clearly indicate the models are in error. These behaviors appear to be too fundamental to human nature to expect much change in people's basic behavioral patterns. However, if geologists are aware of these behaviors and use modeling techniques that minimize their damaging influences, the use of geologic information in exploration should benefit. The data-process-criteria model is intended to be a step in this direction.

Finally, the data-process-criteria model promotes the more rigorous and reliable use of geologic information by making explicit (1) the deposit type, (2) its analogs, (3) geologic characteristics of the analogs, (4) processes responsible for the geologic characteristics and the deposit type, (5) the most reliable and informative criteria for exploration, and (6) the relative importance of each criterion for evaluating exploration favorability. The explicitness of DPC models promotes the critique and improvement of the model by the model author and colleagues. Multiple working hypotheses are promoted by a series of techniques at every step in model development in an attempt to develop models that are accurate rather than simply explanatory. It is hoped that the model format not only promotes the natural human tendency to develop causal (i.e., process) models, but to discontinue their development in favor of empirical models when limited data and concepts warrant. Emphasis on model evaluation should help define the level of model development that can be supported by existing data and concepts and demonstrate that inadequacies in the model are due to limitations in data and concepts, not in the geologist's skill. This, in turn, may make it easier for geologists to accept the limitations of their models and accept corrections when new geologic information warrants.

REFERENCES

Adams, S. S., 1985, Mineral-deposit modeling in exploration: International Workshop on Gold Deposit Modeling in Exploration, September 23-25, 1985, Colorado School of Mines, Golden, Colorado, 39 p.

American Bureau of Metal Statistics, Inc., 1985, Non-ferrous metal data 1984: Secaucus, New Jersey, p. 108-109.

Anderson, J. A., 1982, Gold--its history and role in the U.S. economy and the U.S. exploration program of Homestake Mining Company: Mining Congress Journal, p. 51-58.

Bailly, P. A., 1979, Managing for ore discoveries: Mining Engineering, v. 31, no. 6, p. 663-671.

Bailly, P. A., 1982, Risk and the economic geologist: Economic Geology, v. 77, no. 3, p. 728-734.

Banister, D. P., Weldin, R. D., Zilka, N. T., and Schmauch, S. W., 1981, Economic appraisal of the Cabinet Mountains Wilderness, Lincoln and Sanders Counties, Montana, in Mineral resources of the Cabinet Mountains Wilderness, Lincoln and Sanders Counties, Montana: U.S. Geological Survey, Bulletin 1501, Chapter D, p. 53-77.

Berger, B. R., and Eimon, P. I., 1983, Conceptual models of epithermal precious-metals deposits, in Shanks, W. C., III (ed.), Cameron Volume on Unconventional Mineral Deposits: Society of Mining Engineers, p. 191-205.

Clark, A. L., 1971, Strata-bound copper sulfides in the Precambrian Belt Supergroup, northern Idaho and northwestern Montana: Society of Mining Geologists Japan, Special Issue 3, p. 261-267.

Cox, D. P., 1983, Mineral resource assessment of Colombia--Ore deposit models: U.S. Geological Survey, Open-File Report 83-423.

Cranstone, D. A., 1980, Canadian ore discoveries 1946-1978; a continuing record of success: Internal Report, Department of Energy, Mines and Resources, Ottawa, 22 p.

Cranstone, D. A., 1983, Canadian mineral discovery experience since World War II, in The Economics of Mineral Exploration: International Institute of Applied Systems Analysis, Laxenburg, Austria.

Eckstrand, O. R., Editor, 1984, Canadian mineral-deposit types--A geologic synopsis: Geological Survey of Canada, Economic Geology Report 36, 86 p.

Eggert, R. G., 1983, Base and precious-metal exploration by major corporations, in The Economics of Mineral Exploration: International Institute of Applied Systems Analysis, Laxenburg, Austria.

Engineering and Mining Journal, 1985, p. 152.

Erickson, R. L., compiler, 1982, Characteristics of mineral deposit occurrences: U.S. Geological Survey Open-File Report 82-795, 248 p.

Giles, D. L., and Nelson, C. E., 1982, Epithermal lode gold deposits of the Circum-Pacific Rim: American Association of Petroleum Geologists, Circum-Pacific Energy and Mineral Resource Conference, Honolulu, August, 1982.

Goldberg, P., 1983, The Intuitive Edge: Los Angeles, Jeremy P. Tarcher, Inc., 241 p.

Graney, J. R., 1984, Controls of alteration and precious-metal mineralization in a fossil hydrothermal system, Hasbrouck Mountain, Nevada (abs.): Geological Society of America, Abstracts with Programs, v. 16, no. 6, p. 523.

Haynes, D. W., 1979, Geological technology in mineral resource exploration, in Keisall, D. F., and Woodcook, J. T. (eds.), Mineral Resources of Australia: Third Invitation Symposium, Australian Academy of Technological Sciences, p. 73-95.

Hutchinson, R. W., 1976, Lode gold deposits--The case for volcanogenic derivation: Pacific Northwest Metals and Mining Conference, State of Oregon Department of Geology and Mining Industries, Fifth Gold and Money Session and Gold Technical Session, p. 64-105.

Kahneman, D., Slovic, P., and Tversky, A., Editors, 1982, Judgment Under Uncertainty--Heuristics and Biases: New York, Cambridge University Press, 555 p.

Kyle, D. L., 1984, Successful mineral discovery--the statistics, philosophy and strategy: Transactions

Geological Society of South Africa, v. 87, no. 2, p. 181-197.

Lalor, J. H., 1984, The discovery of the Olympic Dam copper-uranium deposit: Western Mining Corporation Limited, Parkside, South Australia, Australia, 5 p.

Lange, I. M., and Sherry, R. A., 1983, Genesis of the sandstone (Revett) type of copper-silver occurrences in the Belt Supergroup of northwestern Montana and northeastern Idaho: Geology, v. 11, p. 643-646.

Mackenzie, B. W., 1984, Economic mineral exploration targets: Centre for Resource Studies, Working Paper No. 28, Queen's University, Ontario, 63 p.

Mackenzie, B. W., and Bilodeau, M. L., 1984, Economics of mineral exploration in Australia: Australian Mineral Foundation, Glenside, v. XXVI, 171 p.

Mackenzie, B. W., and Woodall, R., 1983, Economic productivity of base-metal exploration in Australia and Canada, in Economics of Mineral Exploration: International Institute of Applied Systems Analysis, Laxenburg, Austria, 64 p.

McKean, K., 1985, Decisions, decisions: Discover, June, p. 22-31.

Monroe, S. C., 1984, Geology and geochemistry of the Buckhorn mine, Eureka County, Nevada (abs.): Geological Society of America, Abstracts with Programs, v. 16, no. 6, p. 599.

Morganti, J. M., 1981, Ore deposit models--4. sedimentary-type stratiform ore deposits--Some models and a new classification: Geoscience Canada, v. 8, no. 2, p. 65-75.

Nelson, C. E., and Giles, D. L., 1985, Hydrothermal eruption mechanisms and hot-spring gold deposits: Economic Geology, v. 80, no. 6, p. 1633-1639.

Reid, R. F., 1984, The geology of the Borealis deposit (abs.): Geological Society of America, Abstracts with Programs, v. 16, no. 6, p. 632.

Restak, R. M., 1984, The Brain: New York, Bantam Books, 371 p.

Ridge, J. D., 1983, Genetic concepts versus observational data in governing ore exploration: CIMM Bulletin, v. 76, no. 852, p. 47-85.

Ridler, R. H., 1970, Relationship of mineralization to volcanic stratigraphy in the Kirkland-Larder Lakes area, Ontario: Proceedings Geological Association of Canada, v. 21, p. 33-42.

Rose, A. W., and Eggert, R. G., 1983, Planning and success of mineral exploration in the United States, in Economics of Mineral Exploration: International Institute of Applied Systems Analysis, Laxenburg, Austria, 42 p.

Saito, M., and Sato, E., 1978, On the recent exploration at the Iwato Gold Mine: Mining Geology, v. 29, p. 191-202.

Shillabeer, J. H., 1985, Two procedures for development of exploration tactics in British Columbia: CIMM Bulletin, v. 78, no. 878, p. 63-67.

Sillitoe, R. H., Baker, E. M., and Brook, W. A., 1984, Gold deposits and hydrothermal eruption breccias associated with a maar volcano at Wau, Papua, New Guinea: Economic Geology, v. 79, p. 638-655.

Snow, G. G., and Mackenzie, B. W., 1981, The environment of exploration--economic, organizational, and social constraints, in Skinner, B. J. (ed.), Seventy-Fifth Anniversary Volume: Society of Economic Geologists, El Paso, The Economic Geology Publishing Company, p. 871-896.

Strachan, D. G., 1985, Geologic discussion of the Borealis gold deposit, Mineral County, Nevada, in Tooker, E. W. (ed.), Geologic Characteristics of Sediment- and Volcanic-Hosted Disseminated Gold Deposits--Search for an Occurrence Model: U.S. Geological Survey, Bulletin 1646, p. 89-94.

Strachan, D. G., Pettit, P. M., and Reid, R. F., 1982, The geology of the Borealis gold deposit, Mineral County, Nevada (abs.): Geological Society of America, Abstracts with Programs, v. 14, no. 7, p. 627.

Tingley, J. V., and Berger, B. R., 1985, Lode gold deposits of Round Mountain, Nevada: Nevada Bureau of Mines and Geology, Bulletin 100, 62 p.

Tooker, E. W., 1985, Discussion of the disseminated gold-occurrence model, in Tooker, E. W. (ed.), Geologic Characteristics of Sediment- and Volcanic-Hosted Disseminated Gold Deposits--Search for an Occurrence Model: U.S. Geological Survey, Bulletin 1646, p. 107-149.

Tversky, A., and Kahneman, D., 1982, Causal schemes in judgments under uncertainty, in Kahneman, D., Slovic, P., and Tversky, A., Judgment Under Uncertainty--Heuristics and Biases: New York, Cambridge University Press, p. 117-128.

Valliant, R., 1985, The Lac discoveries: Canadian Mining Journal, May, p. 39-47.

Watson, J. D., 1969, The Double Helix: New York, New American Library.

Woodall, R., 1984a, Success in mineral exploration--A matter of confidence: Geoscience Canada, v. 11, no. 1, p. 41-46.

Woodall, R., 1984b, Success in mineral exploration--Confidence in science and ore-deposit models: Geoscience Canada, v. 11, no. 3, p. 127-133.

Selected conversion factors*

TO CONVERT	MULTIPLY BY	TO OBTAIN	TO CONVERT	MULTIPLY BY	TO OBTAIN
Length			**Pressure, stress**		
inches, in	2.540	centimeters, cm	lb in^{-2} (= lb/in^2), psi	7.03×10^{-2}	kg cm^{-2} (= kg/cm^2)
feet, ft	3.048×10^{-1}	meters, m	lb in^{-2}	6.804×10^{-2}	atmospheres, atm
yards, yds	9.144×10^{-1}	m	lb in^{-2}	6.895×10^3	newtons (N)/m^2, N m^{-2}
statute miles, mi	1.609	kilometers, km	atm	1.0333	kg cm^{-2}
fathoms	1.829	m	atm	7.6×10^2	mm of Hg (at 0° C)
angstroms, Å	1.0×10^{-8}	cm	inches of Hg (at 0° C)	3.453×10^{-2}	kg cm^{-2}
Å	1.0×10^{-4}	micrometers, μm	bars, b	1.020	kg cm^{-2}
Area			b	1.0×10^6	dynes cm^{-2}
in^2	6.452	cm^2	b	9.869×10^{-1}	atm
ft^2	9.29×10^{-2}	m^2	b	1.0×10^{-1}	megapascals, MPa
yds^2	8.361×10^{-1}	m^2	**Density**		
mi^2	2.590	km^2	lb in^{-3} (= lb/in^3)	2.768×10^1	gr cm^{-3} (= gr/cm^3)
acres	4.047×10^3	m^2	**Viscosity**		
acres	4.047×10^{-1}	hectares, ha	poises	1.0	gr cm^{-1} sec^{-1} or dynes cm^{-2}
Volume (wet and dry)			**Discharge**		
in^3	1.639×10^1	cm^3	U.S. gal min^{-1}, gpm	6.308×10^{-2}	l sec^{-1}
ft^3	2.832×10^{-2}	m^3	gpm	6.308×10^{-5}	m^3 sec^{-1}
yds^3	7.646×10^{-1}	m^3	ft^3 sec^{-1}	2.832×10^{-2}	m^3 sec^{-1}
fluid ounces	2.957×10^{-2}	liters, l or L	**Hydraulic conductivity**		
quarts	9.463×10^{-1}	l	U.S. gal day^{-1} ft^{-2}	4.720×10^{-7}	m sec^{-1}
U.S. gallons, gal	3.785	l	**Permeability**		
U.S. gal	3.785×10^{-3}	m^3	darcies	9.870×10^{-13}	m^2
acre-ft	1.234×10^3	m^3	**Transmissivity**		
barrels (oil), bbl	1.589×10^{-1}	m^3	U.S. gal day^{-1} ft^{-1}	1.438×10^{-7}	m^2 sec^{-1}
Weight, mass			U.S. gal min^{-1} ft^{-1}	2.072×10^{-1}	l sec^{-1} m^{-1}
ounces avoirdupois, avdp	2.8349×10^1	grams, gr	**Magnetic field intensity**		
troy ounces, oz	3.1103×10^1	gr	gausses	1.0×10^5	gammas
pounds, lb	4.536×10^{-1}	kilograms, kg	**Energy, heat**		
long tons	1.016	metric tons, mt	British thermal units, BTU	2.52×10^{-1}	calories, cal
short tons	9.078×10^{-1}	mt	BTU	1.0758×10^2	kilogram-meters, kgm
oz mt^{-1}	3.43×10^1	parts per million, ppm	BTU lb^{-1}	5.56×10^{-1}	cal kg^{-1}
Velocity			**Temperature**		
ft sec^{-1} (= ft/sec)	3.048×10^{-1}	m sec^{-1} (= m/sec)	°C + 273	1.0	°K (Kelvin)
mi hr^{-1}	1.6093	km hr^{-1}	°C + 17.78	1.8	°F (Fahrenheit)
mi hr^{-1}	4.470×10^{-1}	m sec^{-1}	°F − 32	5/9	°C (Celsius)

Divide by the factor number to reverse conversions.
Exponents: for example 4.047×10^3 (see acres) = 4,047; 9.29×10^{-2} (see ft^2) = 0.0929.

ISBN 0-9613074-1-2